PHASE EQUILIBRIA IN BINARY HALIDES

PHASE EQUILIBRIA IN BINARY HALIDES

V. I. Posypaiko and
E. A. Alekseeva

Compiled and Edited by

H. B. Bell
University of Strathclyde
Glasgow, Scotland

Translated from Russian by

B. Indyk

IFI/PLENUM • NEW YORK—WASHINGTON—LONDON

Library of Congress Cataloging in Publication Data

Posypaiko, V. I. (Viktor Iosifovich)
Phase equilibria in binary halides / V. I. Posypaiko and E. A. Alekseeva; compiled and edited by H. B. Bell; translated from Russian by B. Indyk.
p. cm.
Includes bibliographical references.
ISBN 978-1-4684-9026-8 ISBN 978-1-4684-9024-4 (eBook)
DOI 10.1007/978-1-4684-9024-4
1. Fused salts—Tables. 2. Halides—Tables. I. Alekseeva, E. A. II. Bell, H. B. (Henry Bradley) III. Title.
QD198.P674 1987
546'.73'0212—dc19 87-29267
CIP

Softcover reprint of the hardcover 1st edition 1987

A Division of Plenum Publishing Corporation
233 Spring Street, New York, N.Y. 10013

FOREWORD

Molten salts have been used for many years in a large number of industrial applications, and interest in them has increased markedly in the present century. There is a vast amount of experimental data published on molten salt systems, and much of this is due to Russian workers.

In 1961 the Russian Academy of Sciences published a manual dealing with work in the period 1886-1955. These were updated in 1979 with the publication of a 3-volume collection of binary phase diagrams.

The present volume is a translation of the Russian books but is restricted to the systems involving halides. In order to keep the length of the book within bounds the data have been recast. The Russian work treats each system in detail, giving data, methods of study, references, and figures, where available, all together. In this translation all the information is given, but to keep the length down the methods have been classified, the references collected at the end of the volume, and the data given in tabular form. All the figures for halide systems in the Russian volume are included. In the tables, concentrations are given in mole percent; in some cases these have been converted from weight percent in the original Russian. Where the Russian work was in tabular form, these have been translated as given. Where the figures exist, these are referred to in the text. No attempt has been made to edit the data and these are as in the Russian work.

For the more complex systems, tables of liquidus and solidus temperatures are given, and where transformations in the solid state occur, these are quoted. For simple systems a list of characteristic invariant points is given. Data for physical parameters, i.e., density, conductivity, refractive index, and structure, are listed where available.

It is hoped that this book will make a useful reference for data on the halide systems and that it can be easily referred to.

H. B. Bell
University of Strathclyde
Glasgow, Switzerland

GUIDE TO USING THIS BOOK

The different methods of studying phase equilibria are given on pages viii and ix under the heading Classification of Methods. There are eight methods, numbered 1 to 8. Each method, where applicable, is further subdivided, with each subdivision indicated by a letter appended to the number. The entries in the Methods column of the tables refer to this classification. A list of abbreviations used in the table follows on page x. Finally, beginning on page xi, a table of contents lists the systems studied in this work and the pages on which they appear.

CLASSIFICATION OF METHODS

1. VISUAL

   1a. Refractories
   1b. Refractive index
   1c. Polarized light
   1d. High temperature microscope
   1e. Optical crystallography
   1f. Microstructure
   1g. Photometric
   1h. Diffractometric
   1i. Dispersion of light

2. THERMAL

   2a. Thermographic
   2b. Heating curves
   2c. Cooling curves
   2d. Differential thermal analysis
   2e. Thermogravimetric
   2f. Cryoscopic
   2g. Thermal conductivity
   2h. Calorimetric
   2i. Vapor pressure

3. CRYSTALLOGRAPHIC

   3a. X-ray analysis
   3b. X-ray phase observation
   3c. Directional crystallization
   3d. Radiation crystallography
   3e. X-ray-graphical
   3f. Change in crystallographic cell dimension

4. ELECTRICAL

   4a. Resistivity
   4b. Polarography
   4c. Potentiometric

5. CHEMICAL

   5a. Electrophotometric mass analysis
   5b. Mass-spectrometric analysis
   5c. Chemical analysis

6. PHYSICAL

6a. Pycnometric
6b. Dilatometric
6c. Tensometric
6d. Hardness
6e. Filtration
6f. Viscosity
6g. Vapor pressure
6h. Impact compression
6i. Surface tension
6j. Dilatometric
6k. Zone melting
6ℓ. Molar volume
6m. Solubility

7. RADIOACTIVE

8. DISTRIBUTION COEFFICIENT

ABBREVIATIONS USED IN THE TABLES

Eut.	-	Eutectic
Per.	-	Peritectic
Comp.	-	Compound
M.P.	-	Melting point
Segn.	-	Segregation
Incongr.	-	Incongruent
Trans.	-	Transition
Congr.	-	Congruent
Cub.	-	Cubic
Orthorh.	-	Orthorhombic
Min.	-	Minimum
P.T.	-	Phase transformation

TABLE OF CONTENTS

AgBr

2nd Component	T°C	Conc. mol %	Phases involved	Def.	Method	Ref.
AgI			Fig. 1			1
	630		AgI.4AgBr(1)	Comp.		1
	340	32	B23 + B1 + 1	Eut.		
CsBr	207	33	AgBr, $CsAgBr_2$	Eut.	2d,3b	2
	265	50	$CsAgBr_2$	Comp.		
	256	54	$CsAgBr_2$,Cs_2AgBr_3	Eut.		
	275	58	Cs_2AgBr_3, CsBr	Per.		
	419	0	AgBr	M.P.	1	3
	191	32.5	AgBr,AgBr.2CsBr	Eut.		
	270	51.5	AgBr.2CsBr, CsBr	Per.		
CuBr			Fig. 2			4
			Fig. 3			5
	350	47	AgBr, CuBr	Eut.	1c,3b	
	341	38	AgBr, CuBr	Eut.	1c,3b	
$InBr_3$	233	40	AgBr, $InBr_3$	Eut.	1,2a	6
	>408	2-30	AgBr, $InBr_3$	Segn.		
	436	100	$InBr_3$	M.P.		
	430	0	AgBr	M.P.		

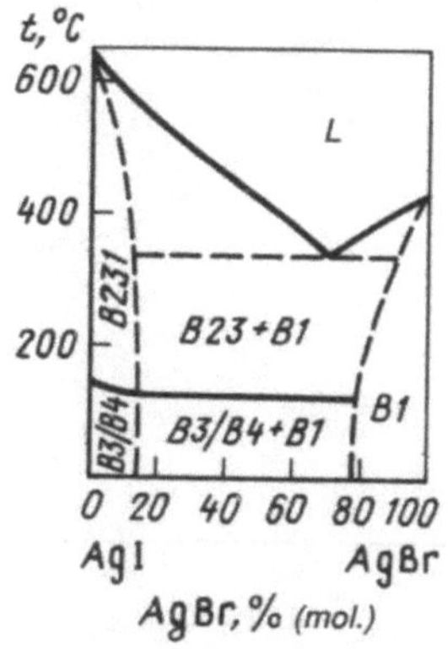

Fig. 1. AgI-AgBr system.

AgBr

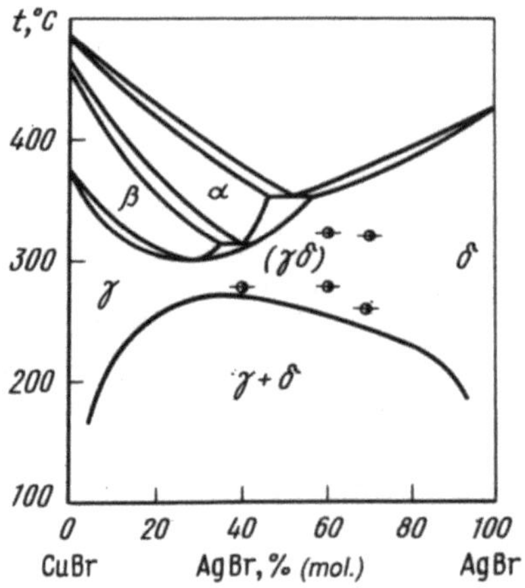

Fig. 2. CuBr-AgBr system.

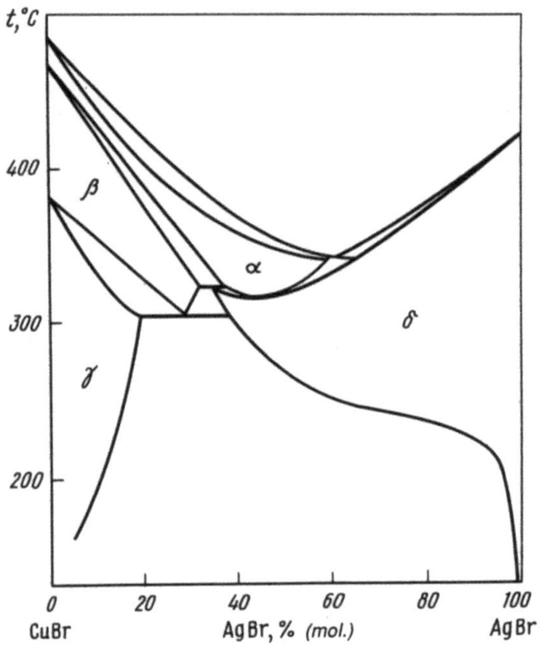

Fig. 3. CuBr-AgBr system.

AgBr

2nd Component	T°C	Conc. mol %	Phases involved	Def.	Method	Ref.
KBr			AgBr, KBr	Eut.	2d,3b	2
LiBr	430	24	Sol. Sln. complete range	Eut.	1	7
	<350		Sol. Sln. separates		2,3b	8
NaBr			Complete sol.sln. no minimum		1,2 2d,3	2,9 10
RbBr	307		Rb_2AgBr_3, AgBr	Per.	2,2a,3b	2
	228	30	Rb_2AgBr_3, RbBr	Eut.		
$TeBr_4$	419	0	AgBr	M.P.	2a,2e,3b	11
		80	AgBr, $TeBr_4$	Eut.		
	388	100	$TeBr_4$	M.P.		
	408	5-55	AgBr, $TeBr_4$	Segn.		

AgCl

2nd Component	T°C	Conc. mol %	Phases involved	Def.	Method	Ref.
$BiCl_3$	184	55	AgCl, $BiCl_3$	Eut.	2d	12
	468	0	AgCl	M.P.		
	236	100	$BiCl_3$	M.P.		
CsCl	310	54.5	CsCl, CsCl.AgCl	Eut.	1a,2	13,14
	312		CsCl.AgCl	Comp.		
	260	31.5	CsCl.AgCl,AgCl	Eut.		
CuCl	260	46	AgCl, CuCl	Eut.		16
$CuCl_2$	338	35	AgCl, $AgCl.CuCl_2$		1	15
	405	50	$AgCl.CuCl_2$	Comp.		
	392	55	$AgCl.CuCl_2$, $CuCl_2$			

AgCl

2nd Component	T°C	Conc. mol %	Phases involved	Def.	Method	Ref.
$GaCl_3$	76	93.9	$GaCl_3$, $AgCl.GaCl_3$	Eut	1,2b	17
	213	60.6	$AgCl.GaCl_3$,AgCl	Per.		
		0-55	Segregation			
$InCl_3$	369	33	AgCl,$InCl_3$	Eut.		16
	372	39.9	$InCl_3$, Ag_3InCl_6	Eut.	1,2b,2c	18
	388	25	Ag_3InCl_6	Comp.		
	380	20	Ag_3InCl_6, AgCl	Eut.		
	335		Sol. state trnsfr		2b	
	295		$\alpha \rightleftharpoons \beta$-$Ag_3InCl_6$		2c	
KCl	315	30	KCl, AgCl	Eut	2	19
NaCl			Fig. 4			20
			Unbroken sol. sln.- no min.		1	20
		45	Max.curvature of solidus			21
	180	10-80	Segn.of sol soln.			
	<198		Segregation		2d,3b	10
$NbCl_5$	252		Ag[$NbCl_6$]	M.P. (in-congr.)	1,2d	22

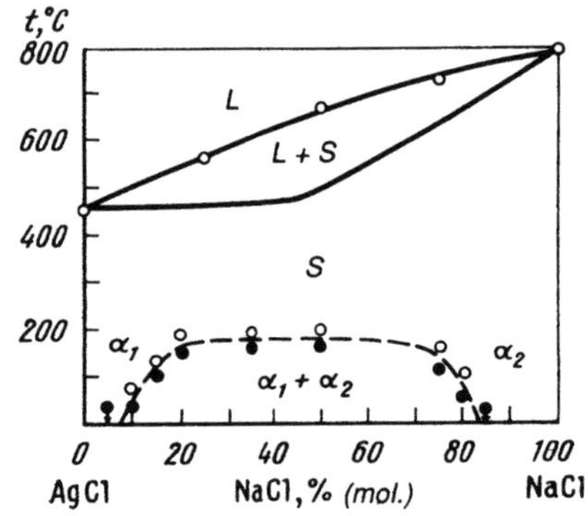

Fig. 4. AgCl-NaCl system.

AgCl

2nd Component	T°C	Conc. mol %	Phases involved	Def.	Method	Ref.
$PbCl_2$	314	39.98	$AgCl, PbCl_2$	Eut.	2c	23
	497	100	$PbCl_2$	M.P.		
	455	0	AgCl	M.P.		
	310	42	$AgCl, PbCl_2$	Eut.	2	24
	311.5	38.5	$AgCl, PbCl_2$	Eut.	2	25
	498	100	$PbCl_2$	M.P.	1	26
	459	0	AgCl	M.P.		
	314	60	$AgCl, PbCl_2$	Eut.		
RbCl	238	37	AgCl, RbCl	Eut.	1	27
$SbCl_5$	455	0	AgCl	M.P.	2d	28
	265	7	$AgCl, AgSbCl_6$	Per.		
	2		$AgSbCl_6, SbCl_5$	Eut.		
	3	100	$SbCl_5$	M.P.		
$TaCl_5$	374		$Ag[TaCl_6]$	M.P. (in-congr.)	1,2d	29
	266		$\alpha \rightleftharpoons \beta$	Trans.		
	304		$\beta \rightleftharpoons \gamma$	Trans.		
$TeCl_4$	455	0	AgCl	M.P.		30
	224	100	$TeCl_4$	M.P.		
	219	97.4	$AgCl, TeCl_4$	Eut.		
	455	0-80	Decomposition			
TlCl	495		AgCl.TlCl(liq)	Comp.	2e	31
	430	100	TlCl	M.P.	1	32
	252	60	TlCl, AgCl.2TlCl	Per.		
	210	45	AgCl.2TlCl, AgCl	Eut.		
	459	0	AgCl	M.P.		

AgI

2nd Component	T°C	Conc. mol %	Phases involved	Def.	Method	Ref.
CdI_2	335			Eut.	2a,3b	33
		95	Sol. sln.			
		54	α-AgI sol.sln.			
	170		S.sln. α-AgI + CdI_2 ⇌ S.sln. β-AgI			
	145		γ-Ag_2CdI_4			
	95		γ ⇌ β-Ag_2CdI_4			
CsI			Existence of AgI.CsI		2a	34
	195	26	AgI,$CsAg_2I_3$	Eut.	2d,3b	1
	210	~35	$CsAg_2I_3$	Comp.		
	195	39	$CsAg_2I_3$, Cs_2AgI_3	Eut.		
	253	48	Cs_2AgI_3,CsI	Per.		
	123	0	AgI (Polymorph. trans.)			
	247	66.7	CsI,2CsI.AgI	Per.	2a,3b	35
	180	40	2CsI.AgI,α-AgI	Eut.		
	146	0	AgI (Polymorph. trans.)			
GaI_3			Fig. 5			36
	250	60	β-AgI,$AgGaI_4$	Per.	2,3b,le	36

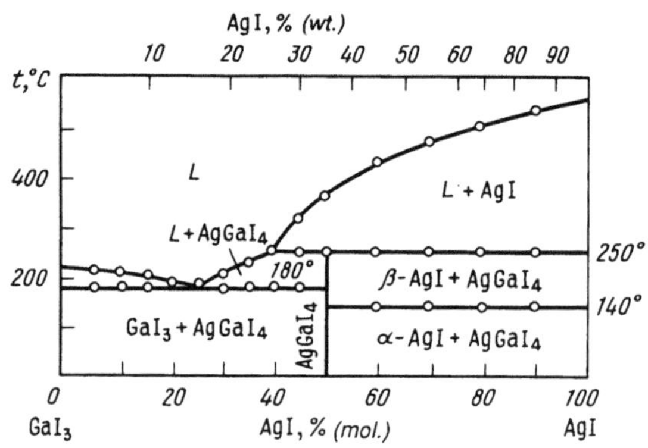

Fig. 5. GaI_3-AgI system.

AgI

2nd Component	T°C	Conc. mol %	Phases involved	Def.	Method	Ref.
GaI_3	180	76	$AgGaI_4$,GaI_3	Eut.		36
	140	0	AgI (Polymorph. trans.)			
HgI_2	120-252		$AgHgI_3$ Ag_2HgI_4 (trans.)	Comp.	2d,3b	33
InI_3			Fig. 6			37
	153	90	AgI,InI_3	Eut.	2	37
		25-67	Separation			
KI	253	25	AgI,KAg_4I_5	Per.	1,2,2d	1
	238	29.5±1	KAg_4I_5,KI	Eut.		
	147	0	$\alpha \rightleftharpoons \beta$-AgI s.s. transf.			
	130		K_2AgI_3			
	38		KAg_4I_5	M.P.		
	260	25	AgI, conditional comp.			38
			5AgI.KI(1)	Per.	1d,2d	
	240.5	30	KI,compound 1	Eut.		

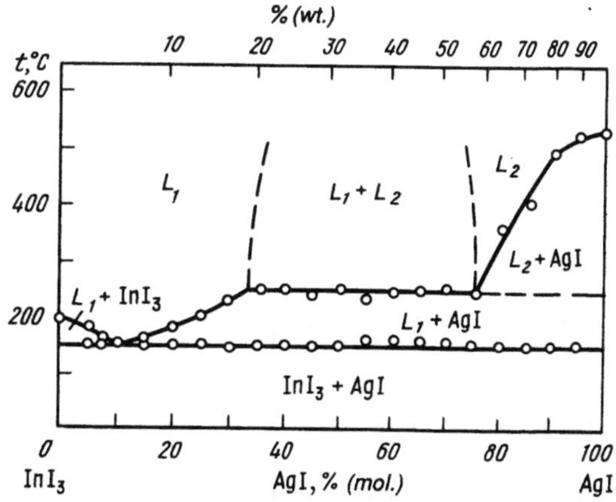

Fig. 6. InI_3-AgI system.

AgI

2nd Component	T°C	Conc. mol %	Phases involved	Def.	Method	Ref.
NH_4I	234	27	$AgI, NH_4Ag_4I_5$	Per.	2d,3b	1
	198	34	$NH_4Ag_4I_5$, $(NH_4)_2AgI_3$	Eut.		
	270	48	$(NH_4)_2AgI_3$, NH_4I	Per.		
NaI	390	48	NaI,AgI AgI - transforms	Eut.	2d,3b	1
RbI	228	25	$AgI, RbAg_4I_5$	Per.	2d,3b	1
	197	30	$RbAg_4I_5$, Rb_2AgI_3	Eut.		
	298	42	$RbAgI_3$, RbI	Per.		
TeI_4	557	0	α-AgI	M.P.	2a,2e, 6c,3b	39
	250	84	α-AgI, TeI_4	Eut.		
	280	100	TeI_4	M.P.		
	147	0	AgI	Phase trnsf.		
TlI	227		$TlAgI_2$, AgI	Per.	2d,3b	1
	202	31	$TlAgI_2$, Tl_2AgI_3	Eut.		
	331		Tl_2AgI_3	Comp.		
	316	70	Tl_2AgI_3, TlI	Eut.		

$AlBr_3$

$AlCl_3$	67	36-40	$AlBr_3, AlCl_3$	Eut.	1,6c	40
AlI_3	186	100	AlI_3	M.P.	2a	41
	97	0	$AlBr_3$	M.P.		
	25	70	Complete range sol. sln. with min.			

$AlBr_3$

2nd Component	T°C	Conc. mol %	Phases involved	Def.	Method	Ref.
CsBr	95	1	$AlBr_3$, $Cs_2Al_7Br_{23}$	Eut.	2	42
	148	21.0	$Cs_2Al_7Br_{23}$, $CsAl_2Br_7$	Eut.		
	154	23.4	$CsAl_2Br_7$, $CsAlBr_4$	Per.		
	357	50.0	$CsAlBr_4$	Comp.		
	330	54.5	$CsAlBr_4$, Phase of unknown compn.	Eut.		
		1-21	Separation range			
	88	~1.0-1.2	$AlBr_3$, $2AlBr_3CsBr$	Eut.	2b,2c,3b	43
	149	30	$2AlBr_3.CsBr$, $AlBr_3.CsBr$	Per.		
	340	50	$AlBr_3.CsBr$	Comp.		
	330	52.5	$AlBr_3.CsBr$,CsBr	Eut.		
	130	1.2-25	Separation range			
$GeBr_4$	20	10	$(AlBr_3)_4$, $GeBr_4$,$AlBr_3$	Eut.	2d	44
KBr	91	26	$K_2Al_7Br_{23}$, KAl_2Br_2	Eut.	2	42
	96	32	KAl_2Br_7,$KAlBr_4$	Per.		
	192.5	50	$KAlBr_4$	Comp.		
	185.0	53.2	$KAlBr_4$, Phase of indeter. comp.	Eut.		
		<20	Liquid range			
NaBr			Existence of $AlBr_3.NaBr$ shown		4b	45 46
RbBr			Fig. 7			42
	74	1-2	$AlBr_3$, $Rb_2Al_7Br_{23}$	Eut.	2	42
	107	22	$Rb_2Al_7Br_{23}$, $RbAl_2Br_7$	Eut.		
	110	25.8	$RbAl_2Br_7$,$RbAlBr_4$	Comp.		
	278	50	$RbAlBr_4$	Per.		
	261	52.7	$RbAlBr_4$, Phase of indeter. compn.	Eut.		
		2-21	Melting range	Eut.		

$AlBr_3$

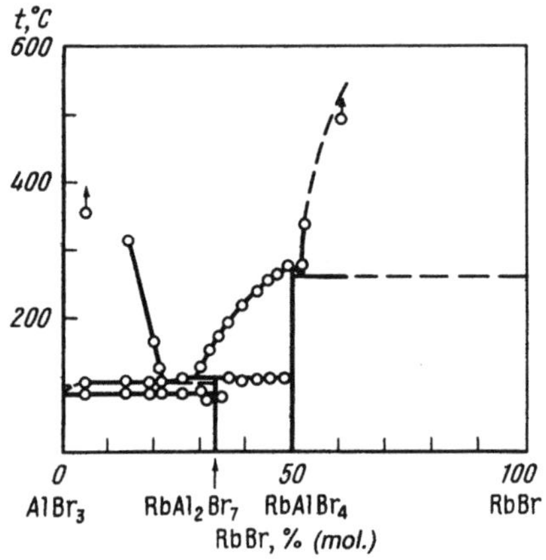

Fig. 7. $AlBr_3$-RbBr system.

2nd Component	T°C	Conc. mol %	Phases involved	Def.	Method	Ref.
RbBr	83	25-26	$AlBr_3$, $2AlBr_3.2RbBr$	Eut.	2b,2c,3b	43
	98	30	$2AlBr_3.RbBr$, $AlBr_3.RbBr$	Per.		
	285	50	$AlBr_3.RbBr$	Comp.		
	267	52	$AlBr_3.RbBr$,RbBr	Eut.		
	95	0.8-25	Melting range			
			Monotectic reaction with crystallization of $AlBr_3$			
		25	$3AlBr_3.RbBr$	Comp.		

$AlCl_3$

2nd Component	T°C	Conc. mol %	Phases involved	Def.	Method	Ref.
AlI_3	186	100	AlI_3	M.P.	2a	49
	194.5	0	$AlCl_3$	M.P.		
	110	60	$AlCl_3$, AlI_3	Eut.		
$BaCl_2$	180	18.3	$(BaCl_2)_3$, $BaCl_2$,$AlCl_3$	Eut.	1	46
	193	0	$AlCl_3$	M.P.		
	962	100	$BaCl_2$	M.P.		

$AlCl_3$

2nd Component	T°C	Conc. mol %	Phases involved	Def.	Method	Ref.
$BeCl_2$	90-100	52	Sol. sln. $AlCl_3$ + 10% $BeCl_2$ Sol. sln. $BeCl_2$ + 23% $AlCl_3$	Eut.	2a,3b	47
$BiCl_3$			Fig. 8			48
	173	32	$AlCl_3$, $BiCl_3.AlCl_3$	Eut.	1,2,6c, 3b,4a	48
	184	50	$BiCl_3.AlCl_3$	Comp.		
	160	65	$BiCl_3.AlCl_3$, $BiCl_3$	Eut.		
	194	25	Fast cooling - decomposition			
	152		$BiCl_2.AlCl_3$ - Polymorph. trans.			

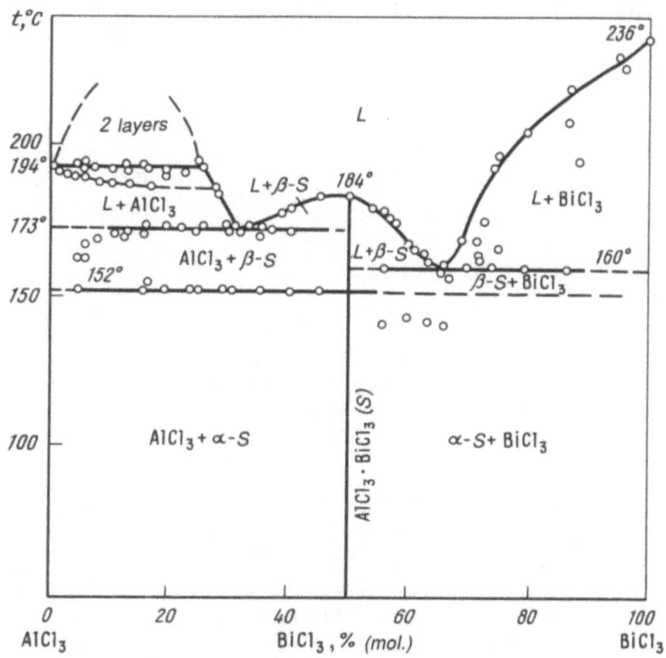

Fig. 8. $AlCl_3$-$BiCl_3$ system.

$AlCl_3$

2nd Component	T°C	Conc. mol %	Phases involved	Def.	Method	Ref.
CsCl	334	60	CsCl, $CsAlCl_4$	Eut.	2,6c	50
	377	55.8	$CsAlCl_4$	Comp.		
	148	30	$CsAlCl_4$, $AlCl_3$	Eut.		
	451	100	$CsCl_3$ (Polymorph. trans.)			
	150	-	$AlCl_3$, $CsAlCl_4$	Eut.	2	51
	364	50	$CsAlCl_4$	Comp.		
	337	57.4	CsCl, $CsAlCl_4$	Eut.		
	150	28.7	$AlCl_3$, $CsAlCl_4$	Eut.	2	52
	378	50	$CsAlCl_4$	Comp.		
	350	55.7	$CsAlCl_4$, CsCl	Eut.		
	0.18-0.4		Segregation			
	193	0	$AlCl_3$	M.P.	2d,1	53
	158	27	$AlCl_3$, $CsAlCl_4$	Eut.		
	377	50	$CsAlCl_4$	Comp.		
	344	55	$CsAlCl_4$, α-CsCl	Eut.		
	482	-	α-CsCl, β-CsCl	Per.		
	646	100	β-CsCl	M.P.		
	193	0-20	Segregation			
$FeCl_2$	178	13	$AlCl_3$, $2AlCl_3.FeCl_2$	Eut.	2a,3b,6c	54
	218	33.3	$2AlCl_3.FeCl_2$	Comp.		
	208	42	$2AlCl_3.FeCl_2$, $FeCl_2$	Eut.		
$FeCl_3$			Fig. 9			55
		0-100	Complete range of sol. sln.		2,6c	55
			Fig. 10			56
	203	7	$AlCl_3$, $FeCl_3$	Per.		56
		12	Sol.sln.($AlCl_3$)A			
		42	Sol.sln.($FeCl_3$)B			
			Two phase: A - ($AlCl_3$) & B - ($FeCl_3$)			

$AlCl_3$

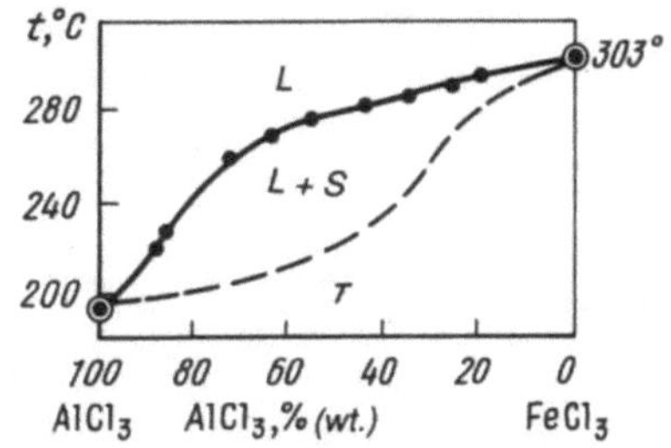

Fig. 9. $AlCl_3$-$FeCl_3$ system.

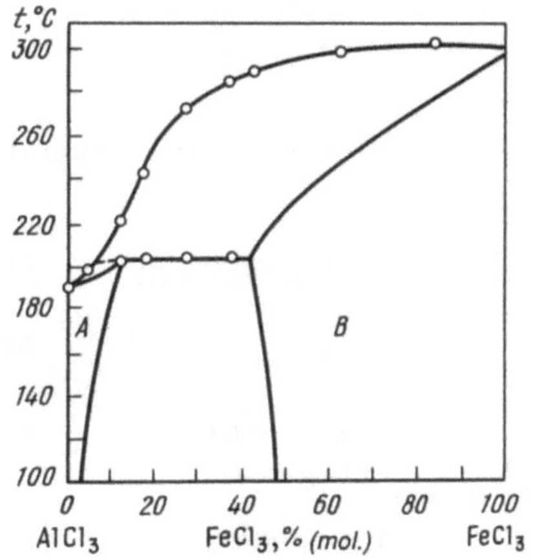

Fig. 10. $AlCl_3$-$FeCl_3$.

2nd Component	T°C	Conc. mol %	Phases involved	Def.	Method	Ref.
$GaCl_2$	84	70	$AlCl_3$, $GaCl_2$	Eut.	2a	57
	194	<40	Segregation			
	59	100	$GaCl_2$ - Polymorph. trans.			
$GaCl_3$	74	94	$AlCl_3$, $GaCl_3$	Eut.	2d, 1	58
	197	0	$AlCl_3$	M.P.		
	78	100	$GaCl_3$	M.P.		
			Fig. 11			59
	74.5	94.3	$AlCl_3$, $GaCl_3$	Eut.	2d,3b	59
		<96.5	Sol. sln.			

$AlCl_3$

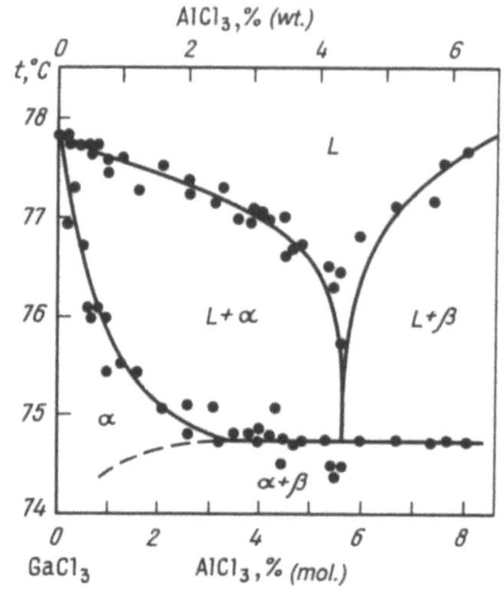

Fig. 11. Part of $GaCl_3$-$AlCl_3$ system.

2nd Component	T°C	Conc. mol %	Phases involved	Def.	Method	Ref.
$GeCl_4$		~100	$GeCl_4$	Eut.	2d,3c	60
$HfCl_4$	122	2	$AlCl_3$, $HfCl_4$	Eut.	2	61
	150	4-70	Segregation			
	189	13.4	$AlCl_3$, $HfCl_4$	Eut	1	62
	183	6	$AlCl_3$, $HfCl_4$	Eut.	2	63
	416	~100	$HfCl_4$ - Probably phase trans.			
InCl			Fig. 12			64
	164	~30	$AlCl_3$, $InAlCl_4$	Eut.	2	64
	174	50	$InAlCl_4$	Comp.		
	196	23	Segregation			
	190	0	$AlCl_3$	M.P.	2,1d	65
	188	20	$AlCl_3$,$3AlCl_3$.InCl	Eut.		
	250	25	$3AlCl_3$.InCl	Comp.		
	142	30	$3AlCl_3$.InCl, $AlCl_3$.InCl	Eut.		
	394	50	$AlCl_3$.InCl	Comp.		
	204	70	$AlCl_3$.InCl - phase x	Eut.		

$AlCl_3$

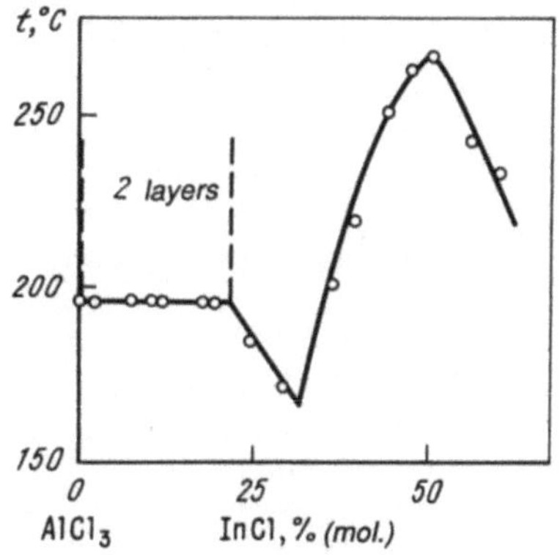

Fig. 12. Part of $AlCl_3$-InCl system.

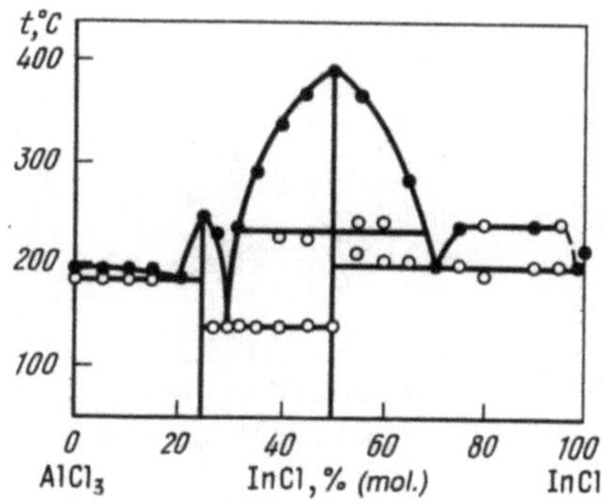

Fig. 13. $AlCl_3$-InCl system.

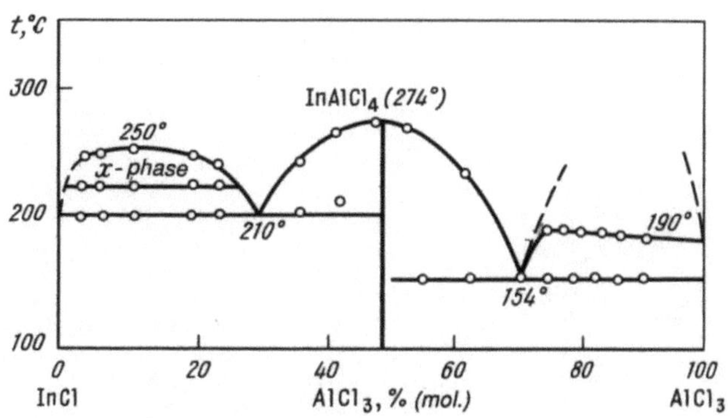

Fig. 14. InCl-$AlCl_3$ system.

$AlCl_3$

2nd Component	T°C	Conc. mol %	Phases involved	Def.	Method	Ref.
InCl	216	100	InCl	M.P.		65
	190	<15	Segregation			
	238		$AlCl_3$.InCl → Polym.phase trans.			
			Fig. 13			66
	188	20	$AlCl_3$,$3AlCl_3$.InCl	Eut.	2a,5	
	250	25	$3AlCl_3$.InCl	Comp.		
	138	30	$3AlCl_3$.InCl, $AlCl_3$.InCl	Eut.		
	394	50	$AlCl_3$.InCl	Comp.		
	200	70	$AlCl_3$.InCl, InCl	Eut.		
	238		$AlCl_3$.InCl → Polym.phase trans.			
			Fig. 14			67
	192	0	$AlCl_3$	M.P.	2d	67
	154	30	$AlCl_3$, $InAlCl_4$	Eut.		
	274	50	$InAlCl_4$	Comp.		
	210	70	$InAlCl_4$ - phase x*	Eut.		
		Close to 100	Phase x, InCl	Eut.		
	190	0-25	Segregation No InCl.$3AlCl_3$			
	156	31.5	$InAlCl_4$, $AlCl_3$	Eut.	2,3b	68
	255	50	$InAlCl_4$	Comp.		
	203	68	$AlCl_3$, $InAlCl_4$	Eut.		
	223	76-99	Segregation			
	195	0-25.5	Segregation			

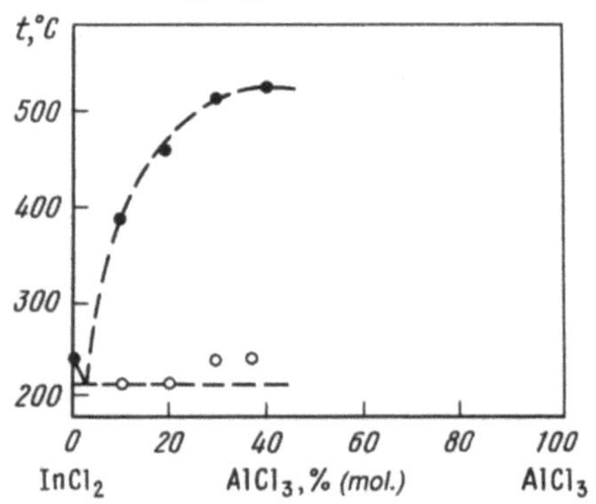

Fig. 15. $InCl_2$-$AlCl_3$ system.

$AlCl_3$

2nd Component	T°C	Conc. mol %	Phases involved	Def.	Method	Ref.
$InCl_2$			Fig. 15			66
	210	97		Eut.	2a,5	66
		<60	Glass			
$InCl_3$		85-100	Components insoluble		2a	66
			Complete insolubility		2a	69
KCl	128	33	$KAlCl_4$, $AlCl_3$	Eut.	2	70
	256	50	$KAlCl_4$	Comp.		
	250	51	KCl, $KAlCl_4$	Eut.		
$LaCl_3$	160	20	$AlCl_3$, $AlCl_3.LaCl_3$	Eut.	2b,2c	71
	180	30	$LaCl_3$, $LaCl_3.AlCl_3$	Per.		
LiCl	80	35	$AlCl_3$, $LiAlCl_4$	Eut.	2a,3b	72
	132	50	$LiAlCl_4$	Comp.		
	128	~50	$LiAlCl_4$, LiCl	Eut.		
	612	100	LiCl	M.P.		
$MoCl_5$	121	38	$AlCl_3$, $MoCl_5$	Eut.	2c	73

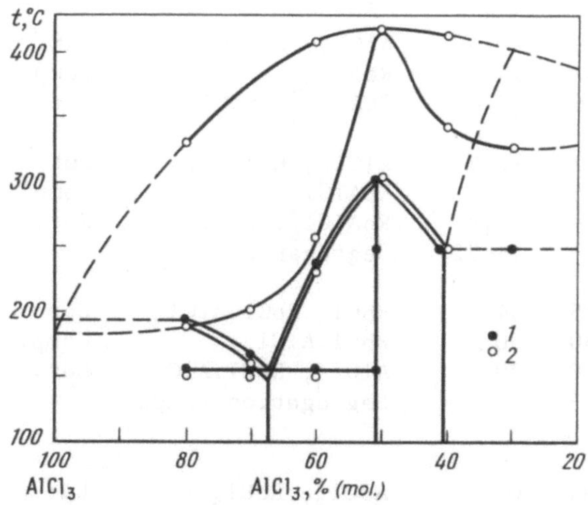

Fig. 16. Part of $AlCl_3$-NH_4Cl system.

$AlCl_3$

2nd Component	T°C	Conc. mol %	Phases involved	Def.	Method	Ref.
NH_4Cl			Fig. 16			74
	168	33	$AlCl_3$(?), $NH_4Cl.AlCl_3$	Eut.	2,6c	
	303	50	$NH_4Cl.AlCl_3$	Comp.		
	248	59	$NH_4Cl.AlCl_3$, NH_4Cl	Eut.		
	420-2		$NH_4Cl.AlCl_3$	B.P.		
NaCl	108	39	$NaAlCl_4$,$AlCl_3$	Eut.	2	70
	151	50	$NaAlCl_4$	Comp.		
	150	50-51.2	NaCl, $NaAlCl_4$	Eut.		
	193	0	$AlCl_3$	M.P.	1	75
	108	18	$NaCl.AlCl_3$,$AlCl_3$	Eut.		
	152	-	$NaCl.AlCl_3$	Comp.		
	152	30	NaCl,$NaCl.AlCl_3$	Eut.		
	800	100	NaCl	M.P.		
$NbCl_5$	125	46	Sol.sln.of $NbCl_5$ (8%) in $AlCl_3$ with sol.sln. of $AlCl_3$(10%) in $NbCl_5$	Eut.	2,6c	76
RbCl	198	-	$AlCl_3$, $RbAlCl_4$	Eut.	2	77
	332	50	$RbAlCl_4$	Comp.		
	316	53.8	$RbAlCl_4$, RbCl	Eut.		
	147	27.9	$AlCl_3$, $RbAlCl_4$	Eut.	2	78
	336	50	$RbAlCl_4$	Comp.		
	319	52.8	$RbAlCl_4$, RbCl	Eut.		
		18-22	Segregation range			
	315	54	RbCl, $RbCl.AlCl_3$	Eut.	2d	79
	830	50	$RbCl.AlCl_3$	Comp.		
	148	30	$AlCl_3$, $RbCl.AlCl_3$	Eut.		
	193	0-20	Segregation range			
$ReCl_5$	194	~0	$AlCl_3$, $ReCl_5$	Eut.	2a	80

$AlCl_3$

2nd Component	T°C	Conc. mol %	Phases involved	Def.	Method	Ref.
$SbCl_3$	69.9	91.0	$AlCl_3$,$SbCl_3$	Eut.	1	81
$SbCl_5$	190	0	$AlCl_3$	M.P.	2d	82
	3	100	$SbCl_5$	M.P.		
	~3	~100	$AlCl_3$, $SbCl_5$	Eut.		
$SeCl_4$	163.8	50	$AlCl_3$, $SeCl_4$	Comp.	1	83
	100	34.8	$AlCl_3$,$SeCl_4$.$AlCl_3$	Eut.		
$SmCl_3$	190	10	$AlCl_3$, $SmCl_3$.$AlCl_3$	Eut.	2b,2c,6c	84
	242	15	$SmCl_3$.$AlCl_3$,$SmCl_3$	Per.		
$SnCl_4$	-33	~100	$AlCl_3$, $SnCl_4$	Eut.	2c	85
		~100	$AlCl_3$, $SnCl_4$	Eut.	1	86
$TaCl_5$	117	38.8	Sol.sln.of $AlCl_3$ (17%) in $TaCl_5$ with sol.sln.of $TaCl_5$ (15%) in $AlCl_3$	Eut.	1,6c	76
$TeCl_4$	9.0	37.5	$AlCl_3$, $TeCl_4$ ACl_3	Eut.	1	87
	9.3	50	$TeCl_4$.$AlCl_3$	Comp.		
	4.0	60.5	$TeCl_4$.$AlCl_3$,$TeCl_4$	Eut.		
$ThCl_4$	250	>0.3	Complete insolubility		2,6c	88
	194	~0.0	$AlCl_3$,$ThCl_4$	Eut.		
$TiCl_3$			Sol.sln. $AlCl_3$-$TiCl_3$		3b	89
			Sol.sln. $AlCl_3$-$TiCl_3$		3b	90

$AlCl_3$

2nd Component	T°C	Conc. mol %	Phases involved	Def.	Method	Ref.
$TiCl_4$	-26	98.1	$TiCl_4$,$AlCl_3$	Eut.	1	91
	-23.1	100	$TiCl_4$	M.P.		
	-24	>99.63	$AlCl_3$,$TiCl_4$	Eut.	2	92
	18	99.63	Solubility of $AlCl_3$ in $TiCl_4$	Liquid		
	80	94.7	" " "	"		
	55	98.5	" " "	"		
	24	>99.5	Al_2Cl_6,$TiCl_4$	Eut.	2	93
TlCl	160	29.5	$AlCl_3$, $2AlCl_3$.TlCl	Eut.	2d, 1d	94
	170	30.0	$2AlCl_3$.TlCl, Tl[$AlCl_4$]	Per.		
	308	50.0	Tl[$AlCl_4$]	Comp.		
	248	59.4	Tl[$AlCl_4$], TlCl	Eut.		
	190	<25	Segregation			
WCl_5	194	0.0	$AlCl_3$	M.P.	2d,3b	95
	158	30.0	α-WCl_5,$AlCl_3$	Eut.		
	240	100.0	β-WCl_5	M.P.		
	180	100.0	Phase transf.	-		
WCl_6	168±2	28	$AlCl_3$,α-WCl_6	Eut.	2c	96
	222	100	γ → β Phase transf.			
	174	100	β → α " "			
$ZrCl_4$			Fig. 17			97
	116	10	$AlCl_3$,$ZrCl_4$	Eut.	2,6c	97
	127	14	Segregation			
	165	25	$(AlCl_3)_2$,$ZrCl_4$	Eut.	2,3b	98
	189	10	$AlCl_3$, $ZrCl_4$	Eut.	1	99
	176	14	$(AlCl_3)_2$, $ZrCl_4$ Eut.	2a		100
	135	0	$AlCl_3$ - Phase transf.			

AlF_3

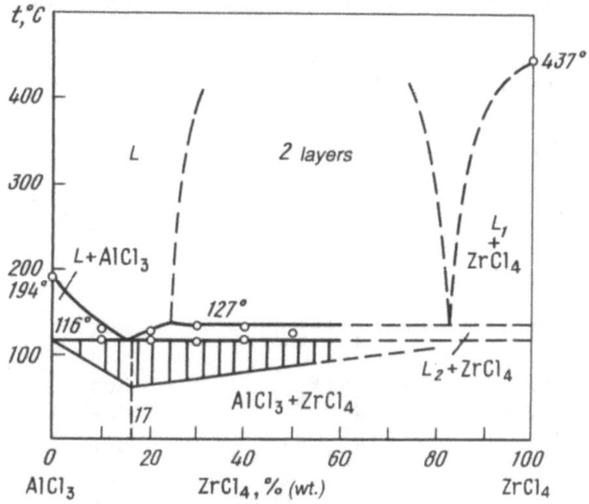

Fig. 17 $AlCl_3$-$ZrCl_4$ system.

2nd Component	T°C	Conc. mol %	Phases involved	Def.	Method	Ref.
BaF_2	855±12		$BaAl_2F_8$	M.P.	3,3b	101
	895 + 8		$BaAlF_5$	M.P.		
	760 ± 8		$BaAlF_5$ → Phase transf.			
	970±10		$BaAl_2F_{12}$	M.P. (cong)		
	855±10		$Ba_3Al_2F_{12}$ → Phase transf.			
	905±10		$Ba_5Al_2F_{16}$	M.P.		
	900±10		$Ba_9Al_2F_{24}$	M.P.		
	910	35	AlF_3, $3BaF_2.2AlF_3$	Eut.	2d,3b	102
	994	50	$3BaF_2.2AlF_3$	Comp.		
	946	65	$3BaF_2.2AlF_3$,BaF_2	Eut.		
	1353	100	BaF_2	M.P.		
CaF_2	828	62.5	CaF_2,β-$CaAlF_5$	Eut.	2d,3b,1f	
	881	55	$CaAlF_5$,AlF_3	Per.		
	670-690		$CaAlF_5$ → Phase transf.			

AlF_3

2nd Component	T°C	Conc. mol %	Phases involved	Def.	Method	Ref.
CaF_2	829	62.5	CaF_2,$CaAlF_5$	Eut.	2,3b	103
	893	52.5	$CaAlF_5$,AlF_3	Per.		
	820	64	AlF_3,CaF_2	Eut.	1	104
EuF_2	845±15		$EuAlF_5$	M.P.	3b,3,2b	105
KF	810	16.6	KF,K_3AlF_6	Eut.	1	106
						107
	656	95	KF,K_3AlF_6	Eut.	2d,3b	
	985	66.6	K_3AlF_6	Comp.		
	559±2	55	K_3AlF_6,$KAlF_4$	Eut.		
	574±1	50	$KAlF_4$,AlF_3	Eut.		108
	-15		K_3AlF_6 → Phase transf.	Cub.-orthorh.		
LiF	711	19	AlF_3,Li_3AlF_6	Eut.	2	109
	711	85	LiF,Li_3AlF_6	Eut.	2,3b	110
	782	75	Li_3AlF_6	Comp.		
	708	65.0	Li_3AlF_6,AlF_3	Eut.		
	710	73.3	LiF,Li_3AlF_6	Eut.	2d	111
	783	50	Li_3AlF_6	Comp.		
	702	42.0	Li_3AlF_6,AlF_3	Eut.		
			$3LiF.AlF_3$ – existence proved Refr.index 1.3714 ± 0.0001	2d,1b		112
MgF_2	1000	57.5	AlF_3,MgF_2	Eut.	2,1f	113
	1265	100	MgF_2	M.P.		
MnF_2	830	80	MnF_2,$AlMnF_5$	Eut.	2d,3b	114
	875	75	$AlMnF_5$,α-AlF_3	Per.		
	440	0	AlF_3 – Phase change	Cub.-orthorh.		

AlF_3

2nd Component	T°C	Conc. mol %	Phases involved	Def.	Method	Ref.
NaF	885	86	NaF,3NaF.AlF_3	Eut.	2c	115
	685	54	5NaF.3AlF_3,AlF_3	Eut.	2a,1	116
	725	60	3NaF.AlF_3, 5NaF.3AlF_3	Per.		
	1100	75	3NaF.AlF_3	Comp.		
	885	86	NaF,3NaF.AlF_3	Eut.		
	996	100	NaF	M.P.		
	565	71-100	NaF → Phase transf.			
	705	47	$NaAlF_4$,AlF_3	Eut.	2d,3b	117
	735	50	$NaAlF_4$	Comp.		
	680	55	$Na_5Al_3F_{14}$, $NaAlF_4$	Eut.		
	730	60	Na_3AlF_6, $Na_5Al_3F_{14}$	Per.		
	1000	75	Na_3AlF_6	Comp.		
	880	88	NaF,Na_3AlF_6	Eut.		
	729	60	Na_3AlF_6,Niolint	Per.	2	118
	684	55	Niolint, AlF_3	Eut.		
	585	12.7	NaF,Na_3AlF_6	Eut.	1	119 120
PbF_2	618±20		$PbAlF_6$	M.P.	3b,3,2d	105
	614±20		$PbAl_2F_8$	M.P.		
	560±12		$Pb_2Al_2F_{24}$ Sol.sln.(octahedral)	M.P.		

AlI_3

2nd Component	T°C	Conc. mol %	Phases involved	Def.	Method	Ref.
CsI	186	0	AlI_3	M.P.	2	121
	166		AlI_3,$CsAlI_4$	Eut.		
	260	50	$CsAlI_4$	Comp.		
	240	55	$CsAlI_4$,CsI	Eut.		
	521	100	CsI	M.P.		
	164	15	AlI_3,CsI.2AlI_3	Eut.	2e	122
	205	33	CsI.2AlI_3	Comp.		
	185	36.0	CsI.2AlI_3,CsI.AlI_3	Eut.		
	250	50.0	CsI.AlI_3	Comp.		
	240	53.5	CsI.AlI_3,CsI	Eut.		
	232		CsI.AlI_3 → Phase transf.			

AlI_3

2nd Component	T°C	Conc. mol %	Phases involved	Def.	Method	Ref.
GaI_3			Complete range of sol. sln. Liquidus - str.line passing through M.P.s of components.		2,3b	123
InI	190	0	AlI_3	M.P.	2,3b	124
	142	38	$InI.AlI_3,AlI_3$	Eut.		
	194	50	$InI,InI.AlI_3$	Per.		
	365	100	InI	M.P.		
		0-15) 85-100)	Segregn. possible			
KI	98	30.0	$AlI_3,KAlI_4$	Eut.	2	121
	202	50.0	$KAlI_4,KI$	Per.		
	105	32.5	$AlI_3,AlI_3.KI$	Eut.	2a	122
	210	50.0	$AlI_3.KI$	Comp.		
	207	52.5	$AlI_3.KI,KI$	Eut.		
LiI	136	26.5	$AlI_3,AlI_3.LiI$	Eut.	2	121
	237	50	AlI_3,LiI	Comp.		
	227	51.8	$AlI_3.LiI,LiI$	Eut.		
NaI	160	25	$AlI_3,NaAl_3I_{10}$	Per.	2	121
	139	28.3	$NaAl_3I_{10}$, $NaAl_2I_7$	Eut.		
	164	34	$NaAl_2I_7$, $NaAlI_4$	Per.		
	220	48	$NaAlI_4,NaI$	Per.		
	123	30	$AlI_3,NaI.AlI_3$	Eut.	2a	122
	220	50	$NaI.AlI_3$	Comp.		
	215	52	$NaI.AlI_3,NaI$	Eut.		
Nb_3I_8		<50	Complete insolubility in liquid as well as in solid state.		2d	125

AlI_3

2nd Component	T°C	Conc. mol %	Phases involved	Def.	Method	Ref.
RbI	114		AlI_3,$RbAlI_4$	Eut.	2	121
	190.5	50	$RbAlI_4$,RbI	Per.		
	135	29	AlI_3,RbI.AlI_3	Eut.	2a	122
	205	50	RbI.AlI_3	Comp.		
	198	~52	RbI.AlI_3,RbI	Eut.		
TeI_4	184	0	Al_2I_6	M.P.	2,3b	126
	116	~40	Al_2I_6, Al_2I_6.2TeI_4	Eut.		
	172	57.4	Al_2I_6.2TeI_4, TeI_4	Per.		
	280	100	TeI_4	M.P.		

$AsCl_3$

2nd Component	T°C	Conc. mol %	Phases involved	Def.	Method	Ref.
$SbCl_3$	-19	0	$AsCl_3$	M.P.	1	127
	73.5	100	$SbCl_3$	M.P.		
	~-19	~0	$AsCl_3$,$SbCl_3$	Eut.		
$TeCl_4$	224	100	$TeCl_4$	M.P.	1,2d	128
	16	0	$AsCl_3$,$TeCl_4$	Eut.		
	100	11 →	Solubility			
	150	15	Solubility			
$TiCl_4$	-57	48.1	$AsCl_3$,$TiCl_4$	Eut.	1,4a	129
	-16	0	$AsCl_3$	M.P.		
	-23.1	100	$TiCl_4$	M.P.		

AsI_3

2nd Component	T°C	Conc. mol %	Phases involved	Def.	Method	Ref.
FeI_2	145	0	AsI_3	M.P.	2d	130
	570	100	FeI_2	M.P.		
	138	5	AsI_3,FeI_2	Eut.		
	570	30-100	Segregation			

AsI_3

2nd Component	T°C	Conc. mol %	Phases involved	Def.	Method	Ref.
GaI_3	130	94	AsI_3,GaI_3	Eut.	2,3b	123
InI_3	142	0	AsI_3	M.P.	2d	131
	210	100	InI_3	M.P.		
	130	35	AsI_3, Sol.sln.of 11% AsI_3	Eut.		
TeI_4	145	0	AsI_3	M.P.	2d	130
	227	100	TeI_4	M.P.		
	130	~5	AsI_3,TeI_4	Eut.		

BCl_3

$GeCl_4$	-49.7	100	$GeCl_4$	M.P.	2d	133
	-108.5	0	BCl_3	M.P.		
	-116	24	BCl_3,$GeCl_4$	Eut.		
		100–98.6	Solid soln. range			
$TeCl_4$	<178.8		Crit.temp. BCl_3 → the components insoluble		1,2d	132

BF_3

BaF_2			Existence of $Ba(BF_4)_2$,$BaBF_5$	3,3b,5b		134

$BaBr_2$

2nd Component	T°C	Conc. mol %	Phases involved	Def.	Method	Ref.
$BaCl_2$	841	10	Complete range of sol. sln.	Min.	1	137
CsBr	560		$2CsBr.BaBr_2$	M.P.	1	135
	642		$BaBr_2$, $2BaBr_2.CsBr$	Per.	1,2d,3b	136
	575	62	$2BaBr_2.CsBr$, $BaBr_2.2CsBr$	Eut.		
	588	67	$BaBr_2.2CsBr$	Comp.		
	558	80	$BaBr_2.2CsBr$, CsBr	Eut.		
KBr	633	59.5	KBr, $2KBr.BaBr_2$	Eut.	1	137
	612	30.5	$2KBr.BaBr_2$, $BaBr_2$	Eut.		
	609	48	$BaBr_2$, $BaBr_2.2KBr$	Eut.	2d,1,3b	136
	632		$BaBr_2.2KBr$	Comp.		
LiBr	470	79	LiBr, $BaBr_2$	Eut.	1	137
	469	77	LiBr, $BaBr_2$	Eut.	2d,1,3b	136
NaBr	592	70	NaBr, $BaBr_2$	Eut.	1	137
	596	70	NaBr, $BaBr_2$	Eut.	1	138
	594	58	NaBr, $BaBr_2$	Eut.	2d,1,3b	136
RbBr	580	48.5	$BaBr_2$, $2RbBr.BaBr_2$	Eut.		135
	610		$2RbBr.BaBr_2$	Comp.		
	590	78.0	$2RbBr.BaBr_2$, $RbBr_2$	Eut.		
	612	81	RbBr, $2RbBr.BaBr_2$	Eut.	1,,2d,3b	136
	628	69	$2RbBr.BaBr_2$	Comp.		
	612	55	$2RbBr.BaBr_2$, $RbBr.2BaBr_2$	Eut.		
	629		$2BaBr_2.RbBr$, $BaBr_2$	Per.		

$BaCl_2$

2nd Component	T°C	Conc. mol %	Phases involved	Def.	Method	Ref.
BaF_2	840	15	$BaCl_2$,BaF_2.$BaCl_2$	Eut.	2d	169
	1008	50	BaF_2.$BaCl_2$	Comp.		
	940	73	BaF_2.$BaCl_2$, BaF_2	Eut.		
$BiCl_3$	236	100	Complete insolubility	M.P.	2d	139
$CaCl_2$	616	57	$(CaCl_2)_x$.$(BaCl_2)_y$,$BaCl_2$	Per.	1	140
	592	64	$CaCl_2$,$(CaCl_2)_x$.$(BaCl_2)_y$	Eut.		
	770	100	$CaCl_2$	M.P.		
	962	0	$BaCl_2$	M.P.	1	141
	628	53	$CaCl_2$.$BaCl_2$, $BaCl_2$	Per.		
	597	63	$CaCl_2$,$CaCl_2$.$BaCl_2$	Eut.		
	774	100	$CaCl_2$	M.P.		
	616	57	$BaCl_2$.$CaCl_2$, $BaCl_2$	Per.	1	142
	590	64	$CaCl_2$,$BaCl_2$.$CaCl_2$	Eut.		
$CeCl_3$	962	0	β-$BaCl_2$	M.P.	2a,3b	143
	930		α-$BaCl_2$, β-$BaCl_2$	Per.		
	800	50	Ba_3CeCl_9, α-$BaCl_2$	Per.		
	700	68	$CeCl_3$, Ba_3CeCl_9	Eut.		
	820	100	$CeCl_3$	M.P.		
			Fig. 18			144
	962	0	α-$BaCl_2$	M.P.	2a,3b	144
	795	46	Sol.sln. in α-$BaCl_2$ and β-$BaCl_2$			

$BaBr_2$

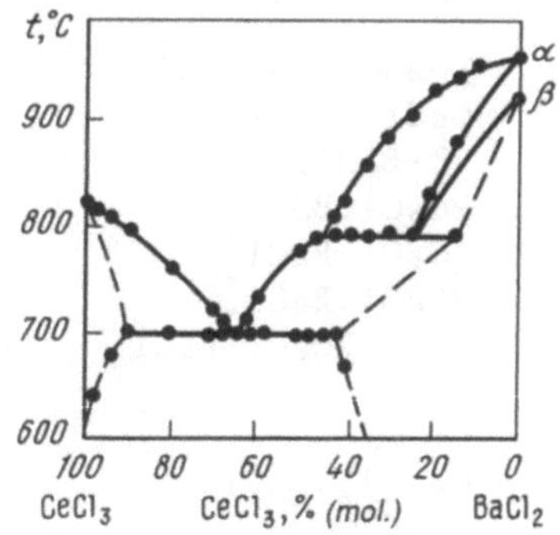

Fig. 18. $CeCl_3$-$BaCl_2$

2nd Component	T°C	Conc. mol %	Phases involved	Def.	Method	Ref.
$CeCl_3$	700	65	Sol.sln. of $BaCl_2$ in $CeCl_3$. 18% $BaCl_2$, sol.sln. in β-$BaCl_2$	Eut.		144
	818-820	100	$CeCl_3$	M.P.		
CsCl	600	60.0	CsCl.$BaCl_2$, $BaCl_2$	Per.	2a	145
	582	63.3	2CsCl.$BaCl_2$, CsCl.$BaCl_2$	Eut.		
	588	66.7	2CsCl.$BaCl_2$	Comp.		
	543	78.7	CsCl,2CsCl.$BaCl_2$	Eut.		
	438-442	100	Phase transf.			
	546	62.5	α-Cs_2BaCl_4, $BaCl_2$	Eut.	1,2a	146
	570	66.6	α-Cs_2BaCl_4	Comp.		
	540	82.5	α-CsCl, α-Cs_2BaCl_4	Eut.		
	642	100	α-CsCl	M.P.		
	480	100	Phase transf. α ⇌ β CsCl			
	506	66.6	Phase transf. α ⇌ β Cs_2BaCl_4			

$BaCl_2$

2nd Component	T°C	Conc. mol %	Phases involved	Def.	Method	Ref.
CsCl	918		β-$BaCl_2$,α-$BaCl_2$	Per.	1,2d	147
	598	62	$CsCl_2.BaCl_2$, β-$BaCl_2$	Per.		
	582	64	$2CsCl.BaCl_2$, $CsCl.BaCl_2$	Eut.		
	588	66.7	$2CsCl.BaCl_2$	Comp.		
	556	78.7	CsCl,$2CsCl.BaCl_2$	Eut.		
$GaCl_2$	350	~96	$Ga_2Cl_4.BaCl_2$, $BaCl_2$	Per.	2a,3b	148
	170	~100	Ga_2Cl_4, $Ga_2Cl_4.BaCl_2$	Eut.		
	58	100	Ga_2Cl_4 → Phase transf.			
			Fig. 19			149
$GaCl_3$	300	67	$BaGaCl_5$, Phase x	Per.	2e	149
	135	82	$Ba(GaCl_4)_2$, $BaGaCl_5$	Per.		
	76	99.25	Sol. sln.			
	98	91-97	Segregation			
			Fig. 20			150
$InCl_3$	435	54	$Ba(InCl_6)_2$, Sol. sln. of $BaCl_2$.7% $InCl_3$	Per.	2a	150
	412	59	$InCl_3$, $Ba(InCl_6)_2$	Eut.		
	288		Decomposition of $Ba_6InCl_{15}(6BaCl_2.InCl_3)$			
KCl	650	57.3	$K_2Cl_2.BaCl_2$, $BaCl_2$	Eut.	1	151
	660	66.7	$K_2Cl_2.BaCl_2$	Comp.		
	658	73.3	KCl,$K_2Cl_2.BaCl_2$	Eut.		

$BaCl_2$

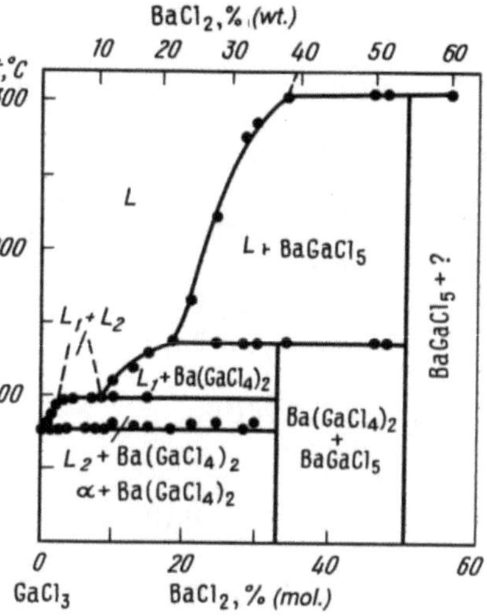

Fig. 19. Part of $GaCl_3$-$BaCl_2$ system.

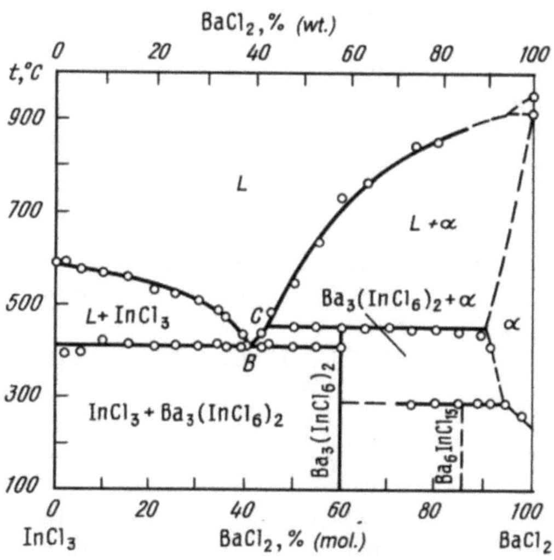

Fig. 20. $InCl_3$-$BaCl_2$ system.

$BaCl_2$

2nd Component	T°C	Conc. mol %	Phases involved	Def.	Method	Ref.
KCl	770	100	KCl	M.P.	1	152
	962	0	$BaCl_2$	M.P.		153
	658	88.5	$BaCl_2.2KCl$, KCl	Per.		137 142
	660-662	66.7	$BaCl_2.2KCl$	Comp.		
	650	57.7	β-$BaCl_2$, $BaCl_2.2KCl$	Eut.		
	915-920	6.0	α-$BaCl_2$,β-$BaCl_2$	Per.		
	648	60.3	$BaCl_2.2KCl$, $BaCl_2$	Eut.	1	154
	658	68.7	$BaCl_2.2KCl$	Comp.		
	643	73.0	KCl,$BaCl_2.2KCl$	Eut.		
	924	8	α-$BaCl_2$,β-$BaCl_2$	Per.	2	155
	648	57.1	K_2BaCl_4,β-$BaCl_2$	Eut.		
	661	66.7	K_2BaCl_4	Comp.		
	658	74.1	KCl,K_2BaCl_4	Eut.		
LiCl	516	79.7	LiCl,$BaCl_2$	Eut.	1	137
	962	0	$BaCl_2$	M.P.	1	156
	605	100	LiCl	M.P.		
	514	80.0	$BaCl_2$,LiCl	Eut.		
	572	93	Phase transf.			
$MgCl_2$	600	41.5	$MgCl_2.2BaCl_2$, $BaCl_2$	Per.	2,3,3b	157 158
	558	55.7	$MgCl_2$, $MgCl_2.2BaCl_2$	Eut.		
NaCl	654	61.7	NaCl,$BaCl_2$	Eut.	1	137
	654	62	NaCl,$BaCl_2$	Eut.	1	159
	800	100	NaCl	M.P.		160
	926	0	$BaCl_2$	M.P.		
	654	70	NaCl,$BaCl_2$	Eut.	1,2a	161
	962	0	$BaCl_2$	M.P.		
	800	100	NaCl	M.P.		

$BaCl_2$

2nd Component	T°C	Conc. mol %	Phases involved	Def.	Method	Ref.
$NdCl_3$	962	0	β-$BaCl_2$	M.P.	2a,3b	162
	930	-	α-$BaCl_2$,β-$BaCl_2$	Per.		143
	736	45	Ba_3NdCl_9, α-$BaCl_2$	Per.		
	624	72	$NdCl_3$, Ba_3NdCl_9	Eut.		
		100	$NdCl_3$	M.P.		
$PrCl_3$	962	0	β-$BaCl_2$	M.P.	2a,3b	
	930		α-$BaCl_2$,β-$BaCl_2$	Per.		
	766	42	Ba_3PrCl_9, α-$BaCl_2$	Per.		
	664	65	$PrCl_3$,Ba_3PrCl_9	Eut.		
	795	100	$PrCl_3$	M.P.		
$PuCl_3$	648	64	$PuCl_3$, Ba_3PuCl_9	Eut.	2,1c	163
	758	38	Ba_3PuCl_9, sol. sln. $BaCl_2$.10% $PuCl_3$	Per.		
RbCl	962		$BaCl_2$	M.P.	1,2a	161
	638		$BaCl_2$,$BaCl_2$.2RbCl	Per.		
	634		$BaCl_2$.2RbCl,RbCl	Eut.		
	716		RbCl	M.P.		
			Existence of $BaCl_2$.2RbCl confirmed by X-ray phase observ.			
$TaCl_5$	216	100	$TaCl_5$.$BaTaCl_7$	Eut.	2e	164
	362	95	$BaTaCl_7$,$BaCl_2$	Per.		
	292		$BaTaCl_7$ → Phase transf.			
$ThCl_4$	362	0	$BaCl_2$	M.P.	2d	165
	650	41	$BaCl_2$, $3BaCl_2.ThCl_4$	Per.		
	585	57	$3BaCl_2.ThCl_4$, $3BaCl_2.2ThCl_4$	Per.		
	535	65	$3BaCl_2.2ThCl_4$,$ThCl_4$	Eut.		
	768	100	$ThCl_4$	M.P.		

$BaCl_2$

2nd Component	T°C	Conc. mol %	Phases involved	Def.	Method	Ref.
UCl_4	583	42	$BaCl_2$,$BaUCl_6$	Per.	2b	166
	434	58	Ba_2UCl_6,UCl_4	Eut.		
	600±2		$BaCl_2$, $2BaCl_2.UCl_4$	Per.	2d,3b	167
	415±2		$2BaCl_2.UCl_4$, UCl_4	Eut.		
YCl_3	532	70	YCl_3, $3BaCl_2.2YCl_3$	Eut.	2	168
	680	40	$3BaCl_2.2YCl_3$	Comp.		
	669	30	$3BaCl_2.2YCl_3$, $BaCl_2$	Eut.		
	962	0	β-$BaCl_2$	M.P.	2a,3b	143
	930		α-$BaCl_2$,β-$BaCl_2$			
	664	36	Ba_3YCl_9, α-$BaCl_2$	Per.		
	532	70	YCl_3,Ba_3YCl_9	Eut.		
	700	100	YCl_3	M.P.		
$ZnCl_2$	962	0	$BaCl_2$	M.P.	1	159
	470	55	$ZnCl_2.BaCl_2$, $BaCl_2$	Per.		
	318	∿100	$ZnCl_2$, $ZnCl_2.BaCl_{21}$	Eut.		
	320	100	$ZnCl_2$	M.P.		

BaF_2

2nd Component	T°C	Conc. mol %	Phases involved	Def.	Method	Ref.
BeF_2			Table I			170
	920	28	BaF_2,γ-$BaBeF_2$	Eut.	2b,2c	170
	1080	50	γ-$BaBeF_4$	Comp.	3b	
	870		γ-$BaBeF_4$, β-$BaBeF_4$	Per.		
	600	98-99	β-$BaBeF_4$,BeF_2	Eut.		
	400	80)				
		93)	Phase transf. BeF_2			
		95)				
		97)				

BaF_2

TABLE I. BaF_2-BeF_2 system.

BeF_2 %	T°C								Solid Phases
	Liquidus		Eutectic		Monotectic		α = β $BaBeF_4$		
	Heat	Cool	Heat	Cool	Heat	Cool	Heat	Cool	
5	-	-	920	915	-	-	-	-	BaF_2
10	-	-	920	910	-	-	340	-	
15	1160	1160	915	912	-	-	345	330	
18	1085	1100	918	915	-	-	360	-	
20	1060	1075	920	918	-	-	330	340	
23	1010	1000	918	915	-	-	400	-	
25	960	-	915	915	-	-	-	-	
27	-	-	920	916	-	-	-	-	
30	-	-	915	913	-	-	350	-	γ-$BaBeF_4$
33	970	970	920	918	-	-	-	-	
35	1000	1010	918	915	-	-	340	335	
38	1045	1050	920	915	-	-	-	-	
40	1050	1065	915	912	-	-	335	-	
43	1070	1075	920	915	-	-	-	-	
45	1080	1080	920	918	-	-	-	345	
48	1080	1080	-	-	-	-	-	-	
50	1080	1080	-	-	-	-	360	-	
52	1080	1080	-	-	-	-	-	-	
55	1075	1070	-	-	780	-	-	-	
57	1050	1050	-	-	780	775	-	-	
60	1030	1020	615	-	777	775	-	-	
63	990	990	-	-	780	780	-	-	
65	955	955	600	-	775	775	-	-	
70	900	900	610	-	775	765	-	-	
72	860	857	600	-	780	770	-	-	β-$BaBeF_4$
75	825	815	610	-	780	765	-	-	
77	800	795	-	-	775	775	-	-	
80	-	-	614	-	775	775	-	-	
82	-	-	-	-	775	770	-	-	
84	-	-	-	-	775	775	-	-	
85	-	-	605	0	780	770	-	-	
87	-	-	-	-	777	770	-	-	
90	-	-	600	-	775	770	-	-	
93	-	-	600	0	775	765	-	-	
95	765	-	600	0	0	0	0	0	

BaF_2

2nd Component	T°C	Conc. mol %	Phases involved	Def.	Method	Ref.
CaF_2	1022	50	Complete range of solid sln.	Min.	1	171
			Fig. 21			172
	1037±5	50	Complete range of solid sln.	Min.	2d,3b	172
	915		Sol. sln. segregates			
	945		CaF_2 $\alpha \rightleftharpoons \beta$ transf.			
	915		Sol. sln. $BaF_2 \cdot (\alpha\text{-}CaF_2) \rightleftharpoons$ sol.sln. $BaF_2 \cdot$ 1-2% CaF_2 + sol.sln.$\alpha\text{-}CaF_2 \cdot$ 0.5% BaF_2			
	1075-945		Sol.sln: $BaF_2\text{-}(\beta\text{-}CaF_2) \rightleftharpoons BaF_2\text{-}(\alpha\text{-}CaF_2)$			
CeF_3			Sol.slns. of fluoride and "tisonite" structure		3b	173

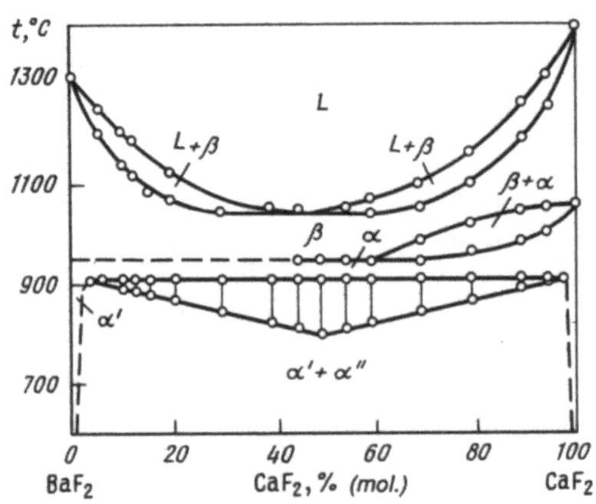

Fig. 21. BaF_2-CaF_2 system.

BaF_2

2nd Component	T°C	Conc. mol %	Phases involved	Def.	Method	Ref.
CoF_2	845	35.5	$CoBa_2F_6$, BaF_2	Per.	2d,3b	174
	825	38.5	$CoBaF_4$, $CoBa_2F_6$	Eut.		
	861	50.0	$CoBaF_4$	Comp.		
	835	58.5	$Co_3Ba_2F_{10}$, $CoBaF_4$	Eut.		
	841	60.0	CoF_2, $Co_3Ba_2F_{10}$	Per.		
CsF	628	84	BaF_2, CsF	Eut.	1,2a	175
CuF_2	1050	42.7	β-CuF_2, BaF_2	Eut.	2	
	1070	50.0	α-CuF_2, β-CuF_2	Per.		
ErF_3			Incomplete sol.sln.of fluorite type at BaF_2 end with "tisonite" type at ErF_3 and		3b	173
FeF_3	800± 10		$BaFeF_5$	M.P.	3b,3	177
	870± 12		$Ba_3(FeF_6)_2$	M.P.		
	1355	0	BaF_2	M.P.	2d,3b,6e	178
	900	27.5	BaF_2, $Ba(FeF_6)_2$	Eut.	1b	
	975±5	40	$Ba_3(FeF_6)_2$	Comp.		
	815	45	$Ba_3(FeF_6)_2$, $BaFeF_5$	Eut.		
	825±5	50	$BaFeF_5$	Comp.		
	775	57.5	$BaFeF_5$, FeF_3	Eut.		
	1355± 15	100	FeF_3	M.P.		
			S.G. of $BaFeF_5$ at 25°C = 4.31			
			μ at λ = 589nm = 1.51-1.52			

BaF_2

2nd Component	T°C	Conc. mol %	Phases involved	Def.	Method	Ref.
GdF_3			Fig.22			179
	1350	11.5	Sol.sln.(BaF_2)	Max.	2d,3b	179
	1025	36	" " (BaF_2. 30% GdF_3 with BaF_2.GdF_3)	Eut.		
	1072	50	BaF_2.GdF_3 (cubic modific.)	Comp.		
	1050	66	BaF_2.GdF_3, sol. sln. of GdF_3 23% BaF_3	Eut.		
	1060±5	95-100	$\alpha \rightleftharpoons \beta$-$GdF_3$ - Phase transf.			
	995±5		BaF_2.GdF_3 $\rightleftharpoons$ sol.sln.(BaF_2) + sol.sln.(GdF_3)			
			The components exhibit sol.slns. of hexagonal modification		3b	173
HoF_3			Limited sol.slns. of fluorite type (BaF_2) with "tisonite" and mixture of the two		3b	173
InF_3			$BaInF_5$ (monoclinic structure)		3b	180
	630±12		$\alpha \rightleftharpoons \beta$-$Ba_3(InF_6)_2$ - Phase transf. $Ba_3(InF_6)_2$ - tetragonal struct.			
LaF_3			Fluoride and "tisonite" (hexagonal) structure		3b	173
			Fig. 23			181
	1482±7	37	BaF_2 fluoride str. sol. sln.	Max.	2d	181
	1335±7	67	Sol.sln. (BaF_2. 55% LaF_3), sol.sln. (LaF_3. 15% BaF_2 - "tisonite" structure)			

BaF_2

2nd Component	T°C	Conc. mol %	Phases involved	Def.	Method	Ref.
LiF	920	62.2	β-BaF_2,α-BaF_2	Per.	1	182
	827	64	$LiF.BaF_2$, β-BaF_2	Per.		
	760	83.8	β-LiF,$LiF.BaF_2$	Eut.		
	812	93	α-LiF, β-LiF	Per.		
	848	100	α-LiF	M.P.		
	765	83.5	$BaF_2.LiF$,LiF	Eut	2a	171
	850		BaF_2, $BaF_2.LiF$	Per.		

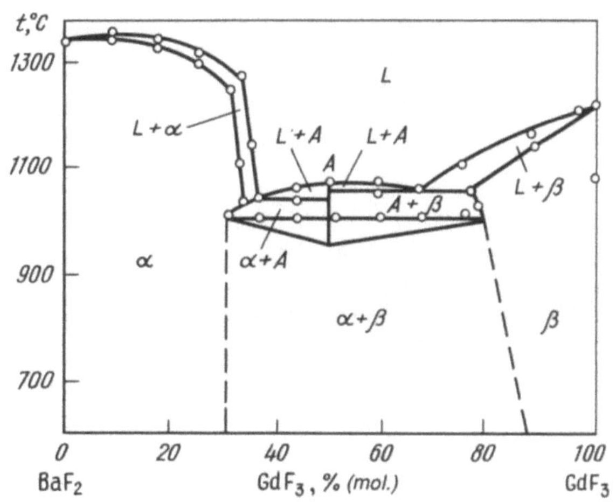

Fig. 22. BaF_2-GdF_3 system.

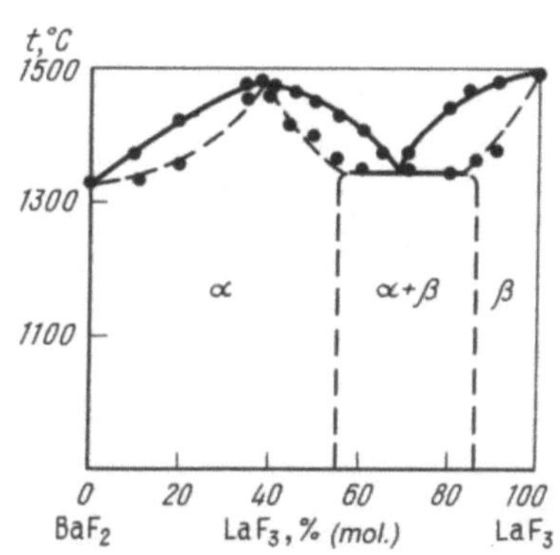

Fig. 23. BaF_2-LaF_3 system.

BaF_2

2nd Component	T°C	Conc. mol %	Phases involved	Def.	Method	Ref.
LuF_3			Limited sol. sln. (hexagonal)		3b	173
MgF_2	942	53.5	MgF_2, $2MgF_2.BaF_2$	Per.	1	183
	912	40	$2MgF_2.BaF_2$, BaF_2	Eut.		
	885	38	BaF_2, $BaF_2.2MgF_2$	Eut.	2	176
	920	52.3	$BaF_2.2MgF_2$	Comp.		
	915	62	$BaF_2.2MgF_2$, MgF_2	Eut.		
MnF_2	750		$MnBaF_4$	M.P.	2d,3b	184
			Fig. 24			185
	1355± 10	0	BaF_2	M.P.	2d,3b	185
	735	~45	Sol.sln.(BaF_2.15% MnF_2) + sol.sln. ($BaMnF_4$.49%MnF_2)	Eut.		

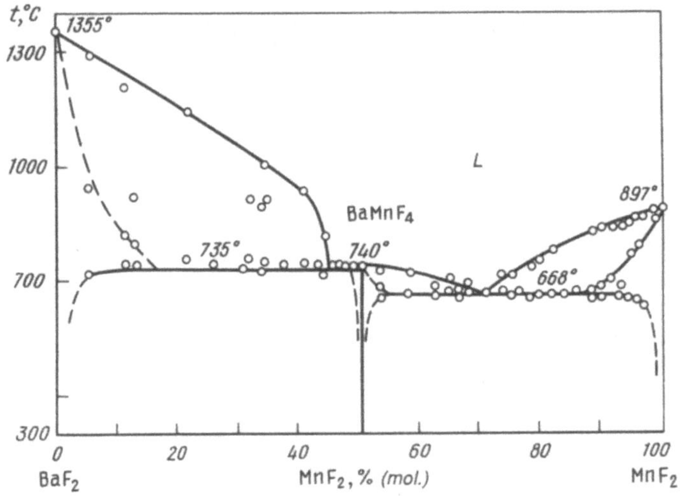

Fig. 24. BaF_2-MnF_2 system.

BaF_2

2nd Component	T°C	Conc. mol %	Phases involved	Def.	Method	Ref.
MnF_2	740	50	$BaMnF_4$	Comp.		185
	668	71.0	Sol.sln.($BaMnF_4$.53%MnF_2)+ sol.sln.(MnF_2.12%BaF_2)	Eut.		
	897±10	100	MnF_2	M.P.		
NdF_3			Sol.sln.of fluorite and "tisonite" structure		3b	173
NiF_2	890		$NiBa_2F_6$,BaF_2	Per.	2d,3b	184
	882	34	$NiBaF_4$, $NiBa_2F_6$	Eut.		
	968	56	$Ni_3Ba_2F_{10}$, $NiBaF_4$	Eut.		
	977	58	Ni_3BaF_8, $Ni_3Ba_2F_{10}$	Per.		
	1000	60.5	NiF_2,Ni_3BaF_8	Per.		
ScF_3			Fig. 25			186
	1340 ± 5	0	BaF_2	M.P.	2d,3b	186
	1550±15	100	ScF_3	M.P.		
	965	39		Comp.		
	920±5	25	BaF_2 + Comp.	Eut.		
	865±5	59	Comp. + ScF_3	Eut.		
	920	10	Solubility of comp. in BaF_2			
	865	80	Solubility of comp. in ScF_3			
			Compound: monoclinic str. a = 15.14 ± 0.05Å) b = 7.27 ± 0.03Å) β = 102° 30' c = 7.00 ± 0.03Å) density: theo. 5.05 exp. 5.00			
	855		Phase transform. of compound			
	840		Decomposition of sol.sln. on the BeF_2 base			
	800		Decomposition of sol. sln. on the ScF_3 base ~100° supercooling over complete range			

BaF_2

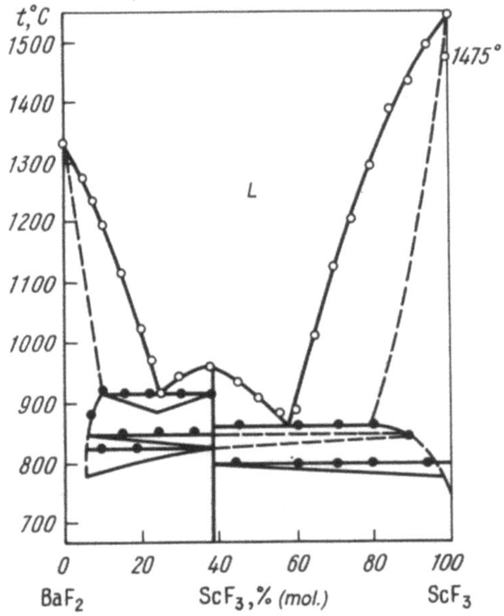

Fig. 25. BaF_2-ScF_3 system.

2nd Component	T°C	Conc. mol %	Phases involved	Def.	Method	Ref.
SmF_3			Limited sol. sln. hexagonal str.		3b	173
SrF_2		~33	Complete range of sln. maximum hardness		6d	187
TbF_3			Limited sol. sln. hexagonal str.		3b	173
TiF_3	890±10		$BaTiF_5$	Comp. M.P.	2d,3b	188
	844±12		$Ba_3(TiF_6)_2$	Comp. M.P.	3	

BaF_2

2nd Component	T°C	Conc. mol %	Phases involved	Def.	Method	Ref.
TuF_3			Limited sol.sln. BaF_2 (fluoride type)		3b	175
			Limited sol.sln. TuF_3 ("tisonite" type)			
UF_3	~1400		Sol.slns. with Peritectic		3b	189
		80	Solubility of BaF_2 in UF_3			
		50	Solubility of UF_3 in BaF_2			
YF_3		<40	Sol.slns. of fluoride str.		3b	190
			BaF_2 sol.sln. - fluoride str.		3b	173
			YF_3 sol. sln. - "tisonite"			
			Between the two - mixture			
	900		$Ba_3(YF_6)_2$, BaF_2	Eut.	2d,3b	191
	1020	40	$Ba_3(YF_6)_2$	Comp.		
	920		$BaYF_5$, $Ba_3(YF_6)_2$	Eut.		
	940	50	$BaYF_5$	Comp.		
	920		YF_3, $BaYF_5$	Eut.		
			Fig. 26			
	1340±5	0	BaF_2	M.P.	2d,3b	192
	1135±5	100	YF_3	M.P.		
	1015	50	BaF_2.40%YF_3 sol.sln., ε sol.sln.	Per.		
	970	60	ε sol.sln., BaY_2F_8	Eut.		
	985	66.6	BaY_2F_8	Comp.		
	975	75.0	BaY_2F_8, β-YF_3	Eut.		
	1065	100	Phase transf.(I⇌II)			

BaF_2

2nd Component	T°C	Conc. mol %	Phases involved	Def.	Method	Ref.
YbF_3			Fig. 27 BaF_2 sol. sln., fluoride str. YbF_3 sol. sln. "tisonite" Mixture of the two between		3b	173

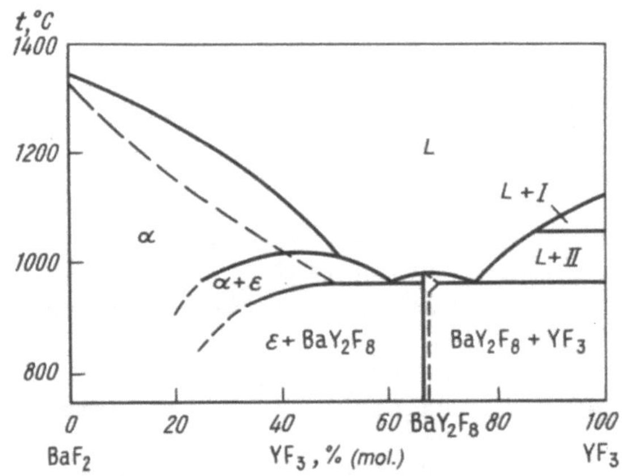

Fig. 26. BaF_2-YF_3 system.

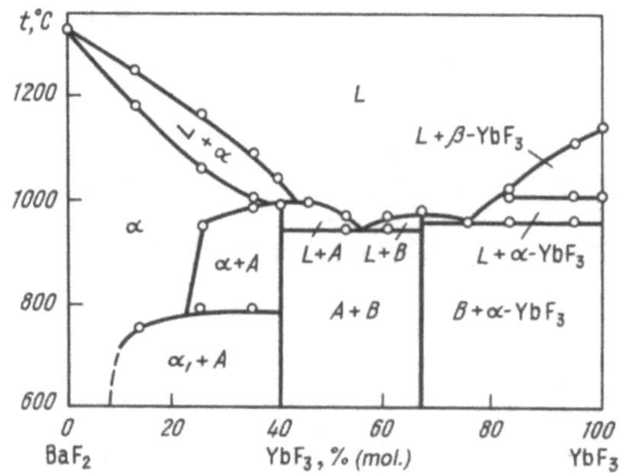

Fig. 27. BaF_2-YbF_3 system.

BaF_2

2nd Component	T°C	Conc. mol %	Phases involved	Def.	Method	Ref.
YbF_3	995	45	BaF_2.37%YbF_3 sol. sln., $3BaF_2.2YbF_3$(A)	Per.	2d,3b	193
	950	54	$3BaF_2.2YbF_3$, $BaF_2.2YbF_3$	Eut.		
	985	66.6	$BaF_2.2YbF_3$(B)	Comp.		
	965	75	$BaF_2.2YbF_3$, α-YbF_3	Eut.		
	1000	82	α-YbF_3,β-YbF_3	Per.		
ZnF_2	795	40.5	Ba_2,ZnF_6,BaF_2	Per.	2d,3b	194
	762	46.5	$BaZnF_4$,Ba_2ZnF_6	Eut.		
	785	50.0	$BaZnF_4$	Comp.		
	775	55.0	$Ba_5Zn_6F_{22}$, $BaZnF_4$	Per.		
	752	64.0	$Ba_2Zn_3F_{10}$, $Ba_5Zn_6F_{22}$	Per.		
	740	67.5	$Ba_2Zn_7F_{18}$, $Ba_2Zn_3F_{10}$	Eut.		
	780	75.0	ZnF_2, $Ba_2Zn_7F_{18}$	Per.		
			Ba_2ZnF_6 - "square" structure			
			$BaZnF_4$ - orthorhombic			
			$Ba_2Zn_3F_{10}$ - monoclinic			
			$Ba_2Zn_7F_{18}$ - triclinic			

BaI_2

2nd Component	T°C	Conc. mol %	Phases involved	Def.	Method	Ref.
SrI_2	446	77.5	SrI_2.5%BaI_2 sol.sln., BaI_2. 53.4% SrI sol.sln.	Eut.	2b,2c	195

$BeCl_2$

2nd Component	T°C	Conc. mol %	Phases involved	Def.	Method	Ref.
BeF_2	306	72.5	BeF_2,$BeCl_2$	Eut.	2b	205
			Fig. 28			205
			Table II			
$GaCl_3$			Fig. 29			196
	64.5	95.5	Complete sol. sln. range	Min.	2a	196
KCl	363	24.5	$KBeCl_5$,$BeCl_2$	Eut.	2d,4a	197
	398	33.4	KBe_2Cl_5	Comp.		
	300	52.1	K_2BeCl_4, KBe_2Cl_5	Eut.		
	536	66.7	K_2BeCl_4	Comp.		
	528	72.5	KCl,K_2BeCl_4	Eut.		
LiCl			Li_2BeCl_4 → incongruent melting			198
	320	64	$BeCl_2$,Li_2BeCl_4	Eut.		

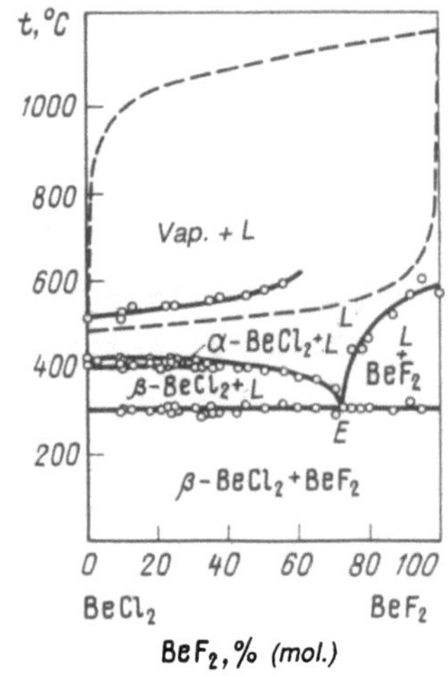

Fig. 28. $BeCl_2$-BeF_2 system.

$BeCl_2$

TABLE II. $BeCl_2$-BeF_2 system.

BeF_2* (mol %)	T°C						
				Transformation			
					$BeCl_2$		
	Boil.	Liq.	Eut.	BeF_2**	$\alpha' \to \beta'$	$\beta' \to \beta$	$\beta \to \alpha$
0	508	421	-	-	250	334	412
2.5	-	414	-	-	250	320	-
9.27	510	-	300	-	250	345	408
9.63	524	416	304	109	254	358	404
17.65	-	420	300	109	254	358	412
22.54	540	418	301	116	234	-	406
25	-	415	306	108	250	348	406
35.45	-	-	300	-	246	-	400
37.5	560	-	296	104	245	345	404
45	562	400	314	106	-	340	-
50.23	575	396	308	-	256	-	-
60	-	380	308	106	-	-	-
65	-	370	309	-	254	-	-
70.28	-	354	292	104	-	-	-
72.5(E)	-	306	306	-	-	-	-
75	-	445	305	-	250	-	-
80	-	468	306	-	-	-	-
86.61	-	520	305	109	-	-	-
91.66	-	568	320	104	-	-	-
95	-	604	304	-	-	-	-
100	-	572	-	110	-	-	-

* Primary phases α-$BeCl_2$ from 0-70.28%. BeF_2 from 75-100%

** Transformation of BeF_2 from low to high temp. form.

$BeCl_2$

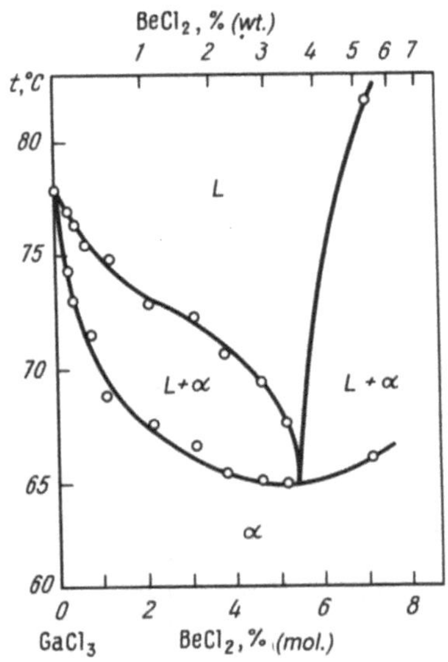

Fig. 29. Part of $GaCl_3$-$BeCl_2$ system.

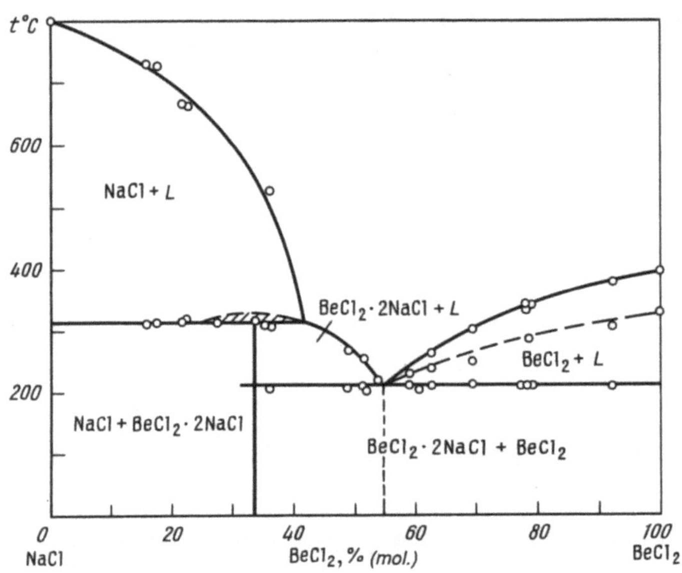

Fig. 30. NaCl-$BeCl_2$ system.

$BeCl_2$

2nd Component	T°C	Conc. mol %	Phases involved	Def.	Method	Ref.
$MoCl_5$	194	100	$MoCl_5$	M.P.	2d,1	199
		100	$BeCl_2$,$MoCl_5$	Eut.		
NaCl			Fig.30			200
	210	45	$BeCl_2$.2NaCl, $BeCl_2$	Eut.	2b,2c	
	315	58	NaCl,$BeCl_2$.2NaCl	Per.		
			The cross-hatched area in the fig. lies below $BeCl_2$.2NaCl curve. From 50-100% $BeCl_2$ transf. α ⇌ β $BeCl_2$ resulting in apparent increase in NaCl content.			
	215	50		Eut.		201
$NbCl_5$	180	56	$BeCl_2$,$NbCl_5$ sol.sln.		2d,1	202
$TaCl_5$			Fig.31			202
	174	52	$BeCl_2$,$TaCl_5$ sol.sln.	Eut.	2d,1	

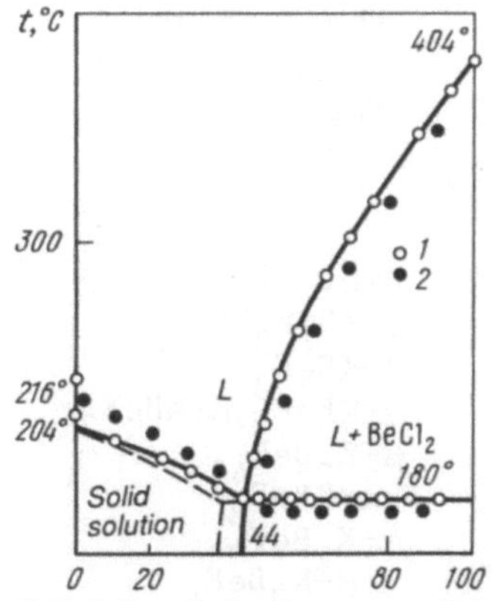

Fig. 31. $NbCl_5$,$TaCl_5$-$BeCl_2$ system.

$BeCl_2$

2nd Component	T°C	Conc. mol %	Phases involved	Def.	Method	Ref.
UCl_4	300±2	25	$BeCl_2$, UCl_4	Eut.	2d,3b	203
WCl_6	404	0	$BeCl_2$	M.P.	2d	204
	256	100	WCl_6	M.P.		
	~256	~100	$WCl_6.BeCl_2$	Eut.		
	224	100	$\alpha \rightleftharpoons \beta$ Phase transf.			
	174	100	$\beta \rightleftharpoons \gamma$ "			

BeF_2

2nd Component	T°C	Conc. mol %	Phases involved	Def.	Method	Ref.
CsF	367	22.5	β-$CsBe_2F_5$, BeF_2	Eut.	2d,2b, 2c,3b	206
	480	33	$\alpha CsBe_2F_5$	Comp.		
	450		α-$CsBe_2F_5$, β-$CsBe_2F_5$	Per.		
	393	41.6	α-$CsBeF_3$, $CsBe_2F_5$	Eut.		
	475	50	α-$CsBeF_3$	Comp.		
	449	52	α-Cs_2BeF_4, α-$CsBeF_3$	Eut.		
	793	67	α-Cs_2BeF_4	Comp.		
	659	78	α-Cs_2BeF_4, β-Cs_3BeF_5	Per.		
	617	84	α-Cs_3BeF_5, β-Cs_3BeF_5	Per.		
	598	86	CsF,α-Cs_3BeF_5	Eut.		
	404		$\alpha \rightleftharpoons \beta$ Cs_2BeF_4 → Phase transf.			
	360		$\alpha \rightleftharpoons \beta$ $CsBeF_3$ → Phase transf.			
	140		$\beta \rightleftharpoons \gamma$ $CsBeF_3$			
KF			Fig. 32			207
	340±5	40	β-$KBeF_3$,BeF_2	Eut.	2b,2c,3b	208
	406±5	50	β-$KBeF_3$	Comp.		
	400		α-$KBeF_3$,β-$KBeF_3$	Per.		
	400	∿52	β-K_2BeF_4, α-$KBeF_3$	Eut.		
	700		α-K_2BeF_4, β-K_2BeF_4	Per.		
	791±5	∿67	α-K_2BeF_4	Comp.		
	740		K_3BeF_5, α-K_2BeF_4	Per.		

BeF_2

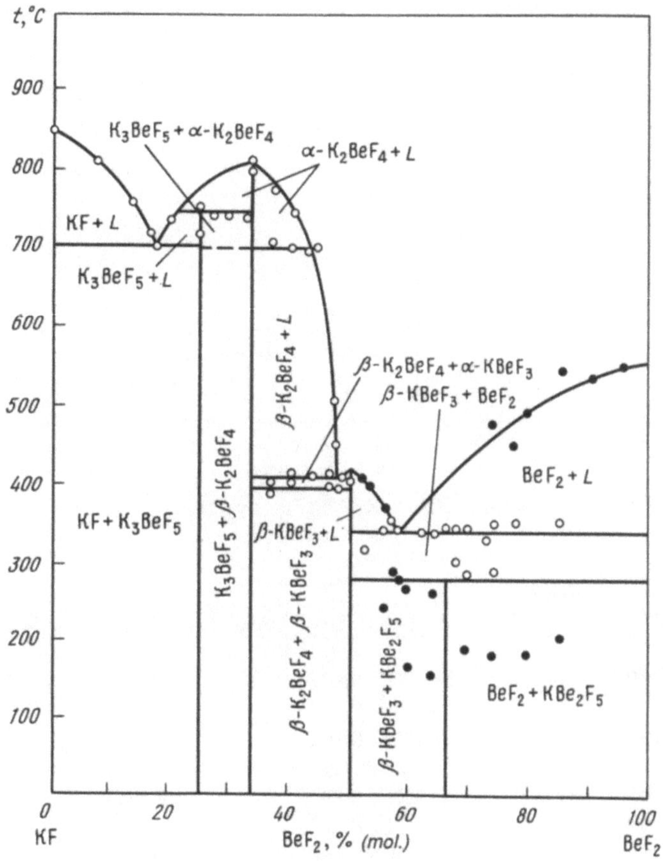

Fig. 32. KF-BeF_2 system.

2nd Component	T°C	Conc. mol %	Phases involved	Def.	Method	Ref.
KF	700	82	KF,K_3BeF_5	Eut.		207
	275		KBe_2F_5 $KBeF_3$ + BeF_2 → Solid transf.			208
			Fig. 33			
	330	27	β-KF.2BeF_2,BeF_2	Eut.		
	350	33.4	α-KF.2BeF_2	Comp.		
	335	30;40	β-KF.2BeF_2, α-KF.2BeF_2	Per.		
	330	42	KF.BeF_2, β-KF.2BeF_2	Eut.		

BeF_2

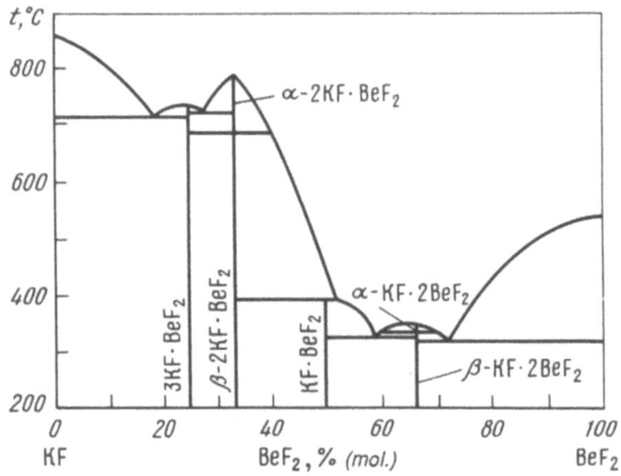

Fig. 33. KF-BeF_2 system.

2nd Component	T°C	Conc. mol %	Phases involved	Def.	Method	Ref.
KF	399	48	β-2KF.BeF_2, KF.BeF_2	Per.		207, 208
	698	60	α-2KF.BeF_2, β-2KF.BeF_2	Per.		
	795	67	α-2KF.BeF_2	Comp.		
	708	73	3KF.BeF_2, α-2KF.BeF_2	Eut.		
	720	75	3KF.BeF_2	Comp.		
	710	82	KF, 3KF.BeF_2	Eut.		
			Fig. 34			209
	705±5	81	KF,K_3BeF_5	Eut.	2a	209
	735±5	76	K_3BeF_5, α-K_2BeF_4	Per.		
	800±5	66.7	α-K_2BeF_4	Comp.		
	346	28	KBe_2F_5,BeF_2	Eut.	2	210
	353	33	KBe_2F_5	Comp.		
	327	42	$KBeF_3$,KBe_2F_5	Eut.		
	220		Q-BeF_2 (Quartz-like) → Phase transf.			

BeF_2

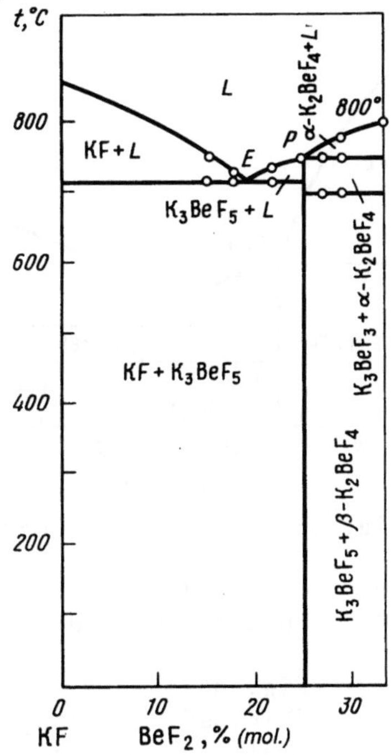

Fig. 34. Part of $KF-BeF_2$ system.

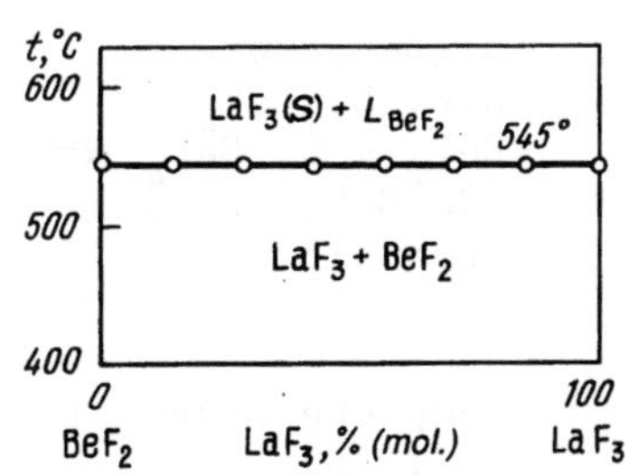

Fig. 35. BeF_2-LaF_3 system.

2nd Component	T°C	Conc. mol %	Phases involved	Def.	Method	Ref.
LaF_3			Fig. 35			209
	>1000	0-100	LaF_3 insoluble in liquid BeF_2		2a	

BeF_2

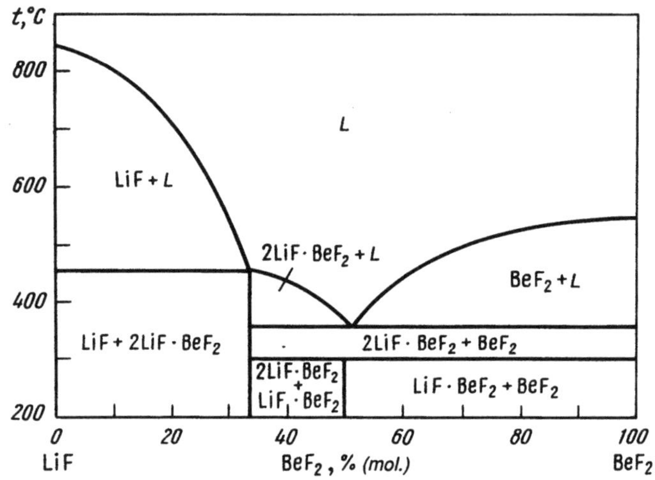

Fig. 36. LiF–BeF_2 system.

2nd Component	T°C	Conc. mol %	Phases involved	Def.	Method	Ref.
LiF			Fig. 36 $2LiF.BeF_2 \rightarrow$ incongruent melting compound $2LiF.BeF_2 + BeF_2 \rightleftharpoons 2LiF.BeF_3 \rightarrow$ sol. state transform.		2,3b	
	360	48	Li_2BeF_4, BeF_2	Eut.	2a,3a	211
	458	66	LiF, Li_2BeF_4	Per.		
	280	50	$BeF_2 + Li_2BeF_4 \rightleftharpoons 2LiBeF_3 \rightleftharpoons$ sol. state transform.			
MgF_2			Complete range of sol.sln.		2d,3b 1f,5	212
NaF			Fig. 37			213
	650	33	β-BeF_2,α-BeF_2	Per.	2a,3b	213
	420	34	Q_b-BeF_2,β-BeF_2 (Q_b – high temp. structure, quartz-like)	Per.		
	340	36	$NaBeF_3$,Q_b-BeF_2	Eut		

BeF_2

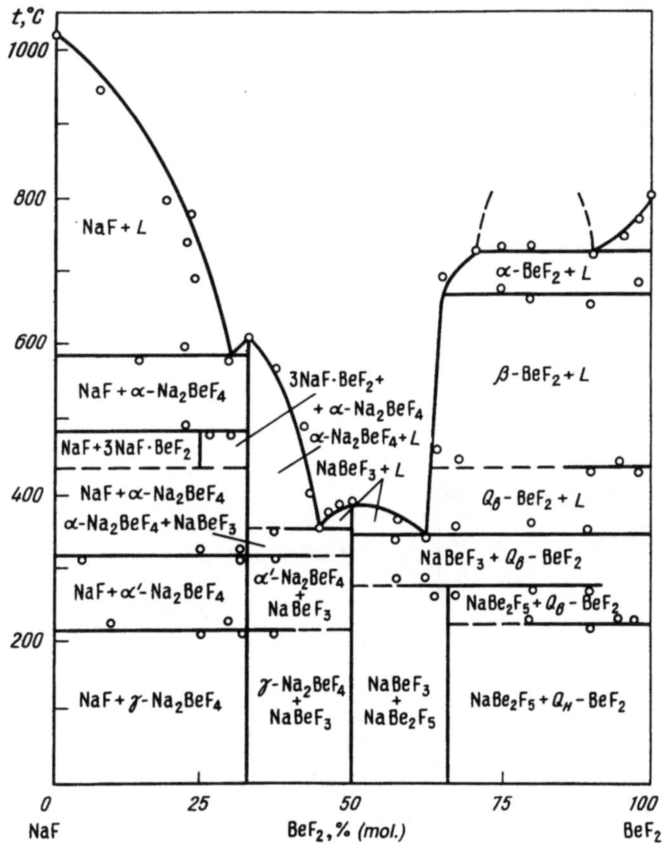

Fig. 37. $NaF-BeF_2$ system.

2nd Component	T°C	Conc. mol %	Phases involved	Def.	Method	Ref.
NaF	377	50	$NaBeF_3$	Comp.		213
	343	56	Na_2BeF_4, $NaBeF_3$	Eut.		
	610±5	66.7	Na_2BeF_4	Comp.		
	580	77	NaF, Na_2BeF_4	Eut.		
	710	10-35	Segregation			

BeF_2

2nd Component	T°C	Conc. mol %	Phases involved	Def.	Method	Ref.
NaF	220		Q_b-BeF_2 ⇌ Q_h-BeF_2		Transformation	213
	270		$NaBeF_3$ + Q_b-BeF_2 ⇌ $NaBe_2F_5$			
	220		α'-Na_2BeF_4 ⇌ γ-Ne_2BeF_4			
	326		α-Na_2BeF_4 ⇌ α'-Na_2BeF_4			
	480		NaF + α-Na_2BeF_4 ⇌ 3NaF.BeF_2			
	480-430		Compound 3NaF.BeF_2			
			Q_h → low temp. quartz type			
RbF			Fig. 38			214
	425	22.5	Sol.sln.of BeF_2 in $RbBe_2F_5$	Eut.	2d,3b	214

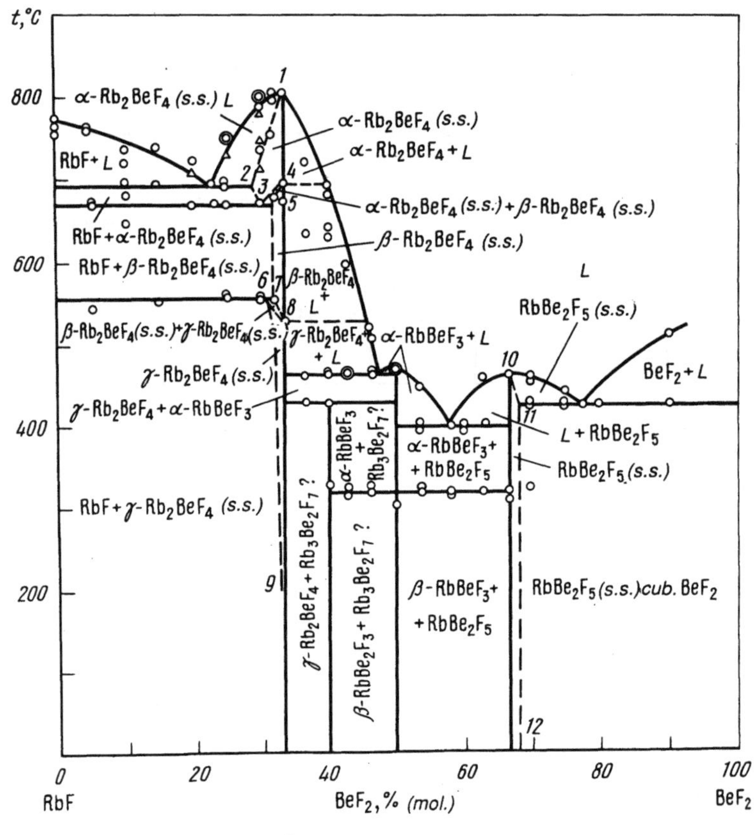

Fig. 38. RbF-BeF_2 system.

BeF_2

2nd Component	T°C	Conc. mol %	Phases involved	Def.	Method	Ref.
RbF	462	33.3	$RbBe_2F_5$	Comp.	le	214
	396	42	$RbBeF_3$, $RbBe_2F_5$	Eut.		
	465	50	$RbBeF_3$	Comp.		
	462	52.5	γ'-Rb_2BeF_4, $RbBeF_3$	Eut.		
	807	66.7	α-Rb_2BeF_4	Comp.		
	693	77	RbF,α-Rb_2BeF_4 sol.sln.	Eut.		
	692		Rb_2BeF_4 $\alpha \rightleftharpoons \beta$ sol.state transf. $\alpha \rightleftharpoons \beta$ " " "			
	672	70	α-Rb_2BeF_4 sol.sln. $\rightleftharpoons$ β-Rb_2BeF_4 sol.sln. + RbF			
	552	69	RbF + β-Rb_2BeF_4 sol.sln. $\rightleftharpoons$ γ-Rb_2BeF_4 sol.sln.			
	427		γ-$RbBeF_4$ + $RbBeF_3$ $\rightleftharpoons$ $Rb_3Be_2F_7$			
	322		α-$RbBeF_3$ $\rightleftharpoons$ β-$RbBeF_3$ μ - pure $RbBe_2F_5$ = 1.332 sol.sln. = 1.329			
			Fig. 39			208
	400	20	Rb.2BeF_2,BeF_2	Eut.	Not given	208
	450	33	RbF.2BeF_2	Comp.		
	390	40	RbF.BeF_2, RbF.2BeF_2	Eut.		
	425	50	2RbF_2.BeF_2, RbF.BeF_2	Per.		
	800	67	2RbF_2.BeF_2	Comp.		
	735	72	3RbF.BeF_2, 2RbF_2.BeF_2	Eut.		
	740	75	3RbF.BeF_2	Comp.		
	680	83	RbF,3RbF.BeF_2	Eut.		
SrF_2			Fig.40			215
	582±27	0.1	$SrBeF_4$,BeF_2	Eut.	2,3b	215
	883±18	60	SrF_2,$SrBeF_4$	Eut.		
	774±11	7.5-25	Segrn.			

BeF_2

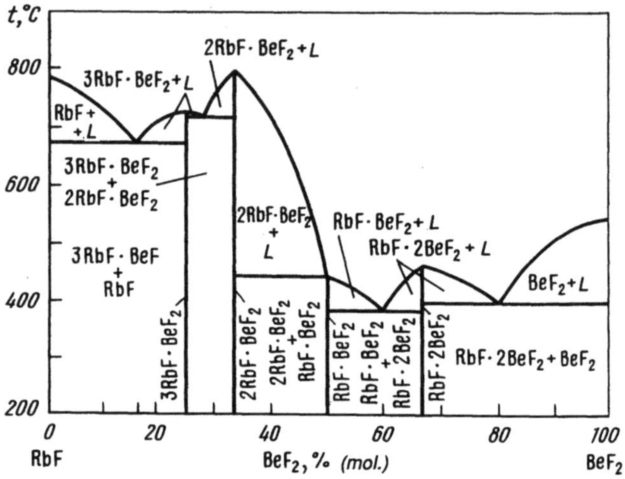

Fig. 39. $RbF-BeF_2$ system.

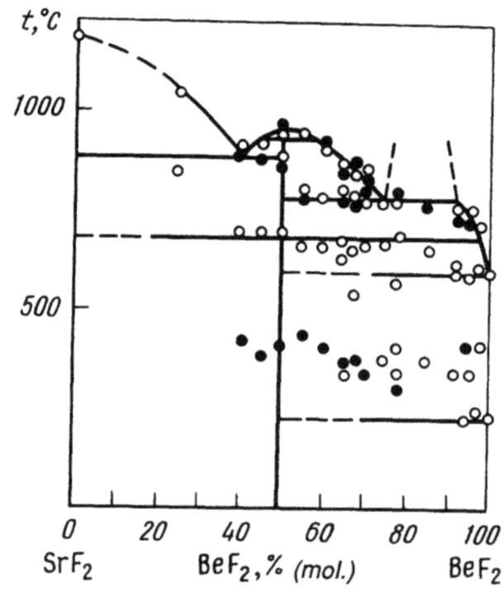

Fig. 40. SrF_2-BeF_2 system.

BeF_2

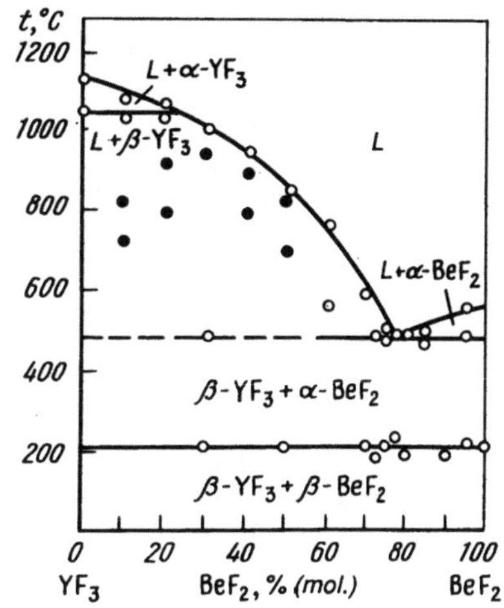

Fig. 41. YF_3–BeF_2 system.

2nd Component	T°C	Conc. mol %	Phases involved	Def.	Method	Ref.
			Phases of $SrBeF_4$			215
SrF_2	954		Melt ⇌ α (Stoichiometric)			
	923		α ⇌ β_B (Stoichiometric)			
	664		β_B ⇌ γ (Stoichiometric)			
	954		Melt ⇌ α (Nonstoichiometric)			
	923		α ⇌ β_B (Nonstoichiometric)			
	-120		β_B ⇌ β_H (Nonstoichiometric)			
	664		β_B ⇌ γ (Nonstoichiometric)			
	378		γ ⇌ α (Nonstoichiometric)			
	440	35	Undefined thermal effects on the heating curves			
	466	22.3	Undefined thermal effects on the heating curves			
	896	7.6	Undefined thermal effects on the heating curves			
	799	2.5	Undefined thermal effects on the heating curves			
ThF_4	527	2	BeF_2,ThF_4	Eut.	2,3b 1f	216
UF_4	535	~0	BeF_2,UF_4	Eut.	2d,1f 3b,6e	217
YF_3			Fig. 41			218
	220	0	BeF_2→Phase transf.		1,2a	219
	700-950	0-60	Incongruent melting		3a	

BeF_2

2nd Component	T°C	Conc. mol %	Phases involved	Def.	Method	Ref.
ZrF_4	535	7.5	ZrF_4, BeF_2	Eut.	2a,3b	220
	645		Segregation			
	~745	20	Critical point of segregation			

$BiBr_3$

2nd Component	T°C	Conc. mol %	Phases involved	Def.	Method	Ref.
$BiCl_3$	220	0	α-$BiBr_3$	M.P.	1,6f	223
	194	15.1	α-$BiBr_3$,β-$BiBr_3$	Per.	6g	
	173	50.5	β-$BiBr_3$, $BiCl_3$	Eut.		
	234	100	$BiCl_3$	M.P.		
BiI_3			Fig. 42			224
	204	0	$BiBr_3$	M.P.	2d,1	224
	408	100	BiI_3	M.P.		
			Complete range of sol. sln.			
CsBr	220	18	$BiBr_3$, CsBr.$BiBr_3$	Eut.	2,1,1e	221
	275	28	CsBr.$BiBr_3$, 3CsBr.2$BiBr_3$	Per.		
	640	60	3CsBr.2$BiBr_3$	Comp.		
	605	67	3CsBr.2$BiBr_3$, α-3CsBr.$BiBr_3$	Eut.		
	620	73	3CsBr.$BiBr_3$	Comp.		
	565	86	α-3CsBr·$BiBr_3$,csBr	Eut.		
	450		3CsBr.$BiBr_3$ → Solid phase transf.			

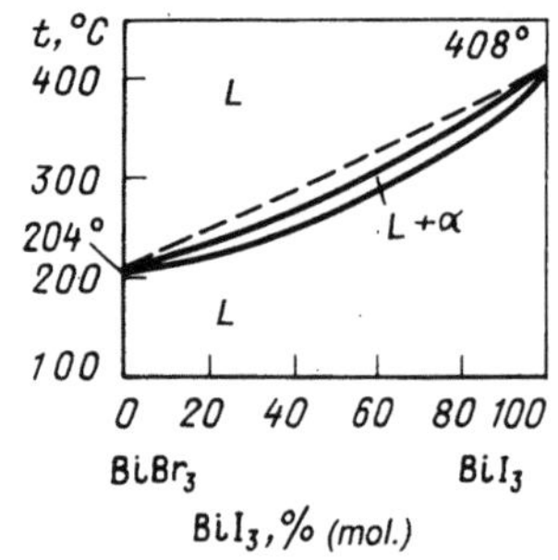

Fig. 42. $BiBr_3$-BiI_3 system.

$BiBr_3$

2nd Component	T°C	Conc. mol %	Phases involved	Def.	Method	Ref.
KBr	170	32	$BiBr_3$, $3KBr.2BiBr_3$	Eut.	2,1,3b	221
	390	55	$3KBr.2BiBr_3$, $2KBr.BiBr_3$	Per.		
	450	62	$2KBr.BiBr_3$, α-$4KBr.BiBr_3$	Per.		
	490	70	α-$4KBr.BiBr_3$,KBr	Per.		
	415		$4KBr.BiBr_3$ →Solid phase transf.			
RbBr	200	23	$BiBr_3$, $3RbBr.2BiBr_3$	Eut.	2,1	221
	520	58	$3RbBr.2BiBr_3$, α-$4RbBr.BiBr_3$	Per.		
	560	76	α-$4RbBr.BiBr_3$,RbBr,	Per.		
	440		$4RbBr.BiBr_3$ → Solid phase transf.			
$TeBr_4$	219	0	$BiBr_3$	M.P.	2d	222
	200	20	$TeBr_4$,$BiBr_3$	Eut.		
	386	100	$TeBr_4$	M.P.		
	180	0	$BiBr_3$ → Solid phase transf.			

$BiCl_3$

2nd Component	T°C	Conc. mol %	Phases involved	Def.	Method	Ref.
$CaCl_2$	236	0.0	$BiCl_3$	M.P.	2d	225
	<420	0-100	Complete immiscibility in solid state			
$CdCl_2$	236	0.0	$BiCl_3$	M.P.	2d	225
	<420	0-100	Complete immiscibility in solid state			
CsCl	215	26	$BiCl_3$, $3CsCl.2BiCl_3$	Eut.	2,1,3b	221
	580	57	$3CsCl.2BiCl_3$, α-$3CsCl.BiCl_3$	Per.	3a	
	670	74	α-$3CsCl.BiCl_3$	Comp.		
	560	89	α-$3CsCl.BiCl_3$, α-CsCl	Eut.		
	460	100	CsCl → Solid phase transf.			
	400		$3CsCl.BiCl_3$→ Solid phase transf.			

$BiCl_3$

2nd Component	T°C	Conc. mol %	Phases involved	Def.	Method	Ref.
CsCl	184	15	$BiCl_3$, $CsCl.3BiCl_3$	Eut.	1,2c	226
	198	25	$CsCl.3BiCl_3$	Comp.		
	194	28	$CsCl.3BiCl_3$,CsCl	Eut.		
$FeCl_3$	210-300	0-70	Eutectic present	Eut.	4a	227
$GaCl_3$	144	38	$GaCl_3.BiCl_3$, $BiCl_3$	Eut.	2,3b	228
	159	50	$GaCl_3.BiCl_3$	Comp.	4a,6f	
	70	77	$GaCl_3$, $GaCl_3.BiCl_3$	Eut.		
			Existence of Ga_2BiCl_9 indicated by elec. conduct. meas. but not by viscosity meas.			
	236	0.0	$BiCl_3$,	M.P.	2d	225
	147	35	$GaCl_3.BiCl_3$, $BiCl_3$	Eut.		
	162	50	$GaCl_3.BiCl_3$	Comp.		
	70	91	$GaCl_3$, $GaCl_3.BiCl_3$	Eut.		
	78	100	$GaCl_3$	M.P.		
			$BiCl_3$ melts tend to supercool			
	150	34.4	$BiCl_3$, $BiCl_3.GaCl_3$	Eut.	1	229
	161	50	$BiCl_3.GaCl_3$	Comp.		
	71	90.8	$BiCl_3.GaCl_3$, $GaCl_3$	Eut.		
$GeCl_4$	236	0.0	$BiCl_3$	M.P.	2d	225
			Complete solubility in liquid and solid state			
$HfCl_4$	235	0.0	$BiCl_3$	M.P.		230
	206	18.5	$BiCl_3$,$HfCl_4$	Eut.	1	
	432	100	$HfCl_4$	M.P.		

$BiCl_3$

2nd Component	T°C	Conc. mol %	Phases involved	Def.	Method	Ref.
HgCl	236	0.0	$BiCl_3$	M.P.	2d	225
	224	4.0	$BiCl_3$,HgCl	Eut.		
$HgCl_2$	236	0.0	$BiCl_3$	M.P.	2d	225
	190	38	$BiCl_3$,$HgCl_2$	Eut.		
	278	100	$HgCl_2$	M.P.		
$InCl_3$	230	0.0	$BiCl_3$,$InCl_3$	Eut.	2a	231

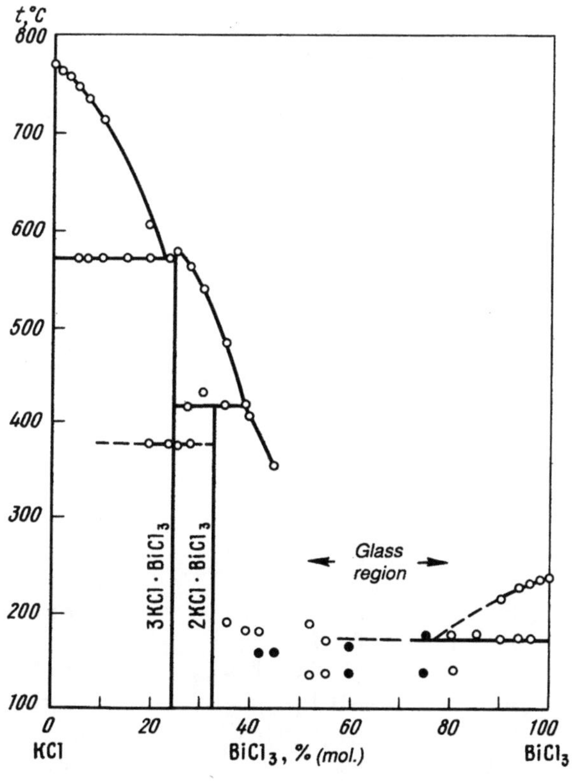

Fig. 43. KCl-$BiCl_3$ system.

$BiCl_3$

2nd Component	T°C	Conc. mol %	Phases involved	Def.	Method	Ref.
KCl			Fig. 43			232
	430	60	$3KCl.BiCl_3$, $2KCl.BiCl_3$	Per.	3b,2c	
	~580	75	$3KCl.BiCl_3$	Comp.		
	575	78	$KCl,3KCl.BiCl_3$	Eut.		
	135 168		} Thermal effects indicating eutectic			
		20-60	Glass Possibility of $KBiCl_4$			
	158	25	$BiCl_3$, $KCl.2BiCl_3$	Eut.	1,2c	226
	174.5	33	$KCl.2BiCl_3$	Comp.		
	158	38	$KCl.2BiCl_3$, $KCl.BiCl_3$	Eut.		
	206	50	$KCl.BiCl_3$	Comp.		
	204	51	$KCl.BiCl_3$, $2KCl.BiCl_3$	Eut.		
	410	60	$2KCl.BiCl_3$, $3KCl.BiCl_3$	Per.		
	580	77	$3KCl.BiCl_3$	Comp.		
	560	79	$3KCl.BiCl_3$,KCl	Eut.		
	180	30	$BiCl_3$, β-$3KCl.BiCl_3$	Eut.	2d,1	221
	380	53	β-$3KCl.BiCl_3$, α-$3KCl.BiCl_3$	Per.	3a	
	580	74	α-$3KCl.BiCl_3$	Comp.		
	570	78	α-$3KCl.BiCl_3$,KCl	Eut.		
LiCl	208	16.2	$LiCl,BiCl_3$	Eut.	1,2c	226
	236	0	$BiCl_3$	M.P.	1,2d,3b	223
	192	5	α-$LiCl.4BiCl_3$, $BiCl_3$	Eut.		
	224	20	α-$LiCl.4BiCl_3$	Comp.		
	202	35	$LiBiCl_4$, α-$LiCl.4BiCl_3$	Eut.		
	210	50	$LiBiCl_4$	Comp.		
	206	55	$LiCl,LiBiCl_4$	Eut.		
	613	100	LiCl	M.P.		
	114		$LiCl.4BiCl_3$ → Phase Transf.			

$BiCl_3$

2nd Component	T°C	Conc. mol %	Phases involved	Def.	Method	Ref.
LiCl			$LiBiCl_4$ - very hygroscopic, soluble in oxides, insoluble in benzene and toluene			233
			$LiCl.4BiCl_3$ unstable in air, soluble in conc. & dilute acids, slightly soluble in organic solvents			
$MgCl_2$	236	0.0	$BiCl_3$ Complete insolubility in solid	M.P.	2d	225
$MoCl_5$	184	40	$BiCl_3,MoCl_5$	Eut.	2d,1	234
	194	100	$MoCl_5$	M.P.		
NaCl	196	19	$BiCl_3$, $NaCl.2BiCl_3$	Eut.	1,2c	226
	226	33	$NaCl.2BiCl_3$	Comp.		
	229	42	$NaCl.2BiCl_3$, $NaCl.BiCl_3$	Per.		
	236	45	$NaCl.BiCl_3$, $2NaCl.BiCl_3$	Per.		
			Existence of $NaBiCl_4$		2a,6c	227
	196	20	$BiCl_3,NaBiCl_4$	Eut.	3b,4a	235
	243		$NaBiCl_4$	Comp.		236
	242	52	$NaBiCl_4$,NaCl	Eut.		
	132		$NaBiCl_4$ → Phase transf.			
$NbCl_5$			Fig. 44			237
	190	46	$NbCl_5,BiCl_3$ $BiCl_3$ rich melts supercool	Eut.	1,2e	237

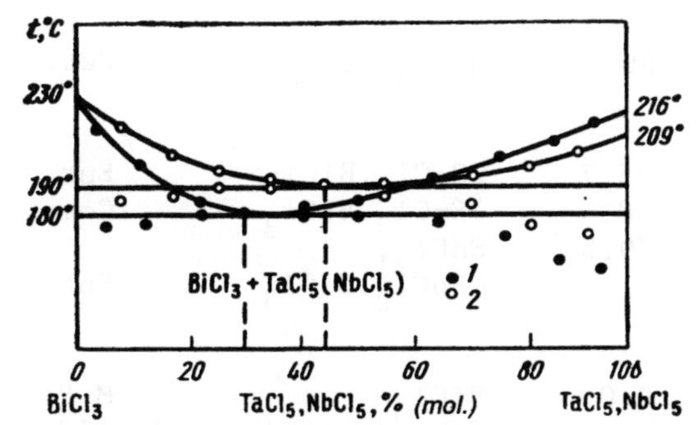

Fig. 44. $BiCl_3$-$TaCl_5$, $NbCl_5$ systems. 1 - $TaCl_5$, 2 - $NbCl_5$.

$BiCl_3$

2nd Component	T°C	Conc. mol %	Phases involved	Def.	Method	Ref.
$NdCl_3$	236	0.0	$BiCl_3$	M.P.	2d	225
			Complete insolubility in solid			
$PbCl_2$	232	0.0	$BiCl_3$	M.P.	2	238
	211.7	13.4	$BiCl_3$, phase x	Eut.		
	323.5	32.7	$PbCl_2$, phase x	Per.		
	501	100	$PbCl_2$	M.P.		
RbCl	172	20	$BiCl_3$, RbCl.$2BiCl_3$	Eut.	1,2c	226
	202	32	RbCl.$2BiCl_3$	Comp.		
	194	40	RbCl.$2BiCl_3$, RbCl.$BiCl_3$	Eut.		
	525	50	Rb1.$BiCl_3$	Comp.		
	402	58	RbCl.$BiCl_3$,RbCl	Eut.		
	200	29	$BiCl_3$, 3RbCl.$2BiCl_3$	Eut.	2d,1	221
	470	56	3RbCl.$2BiCl_3$, α-3RbCl.$BiCl_3$	Per.	3a	
	610	75	α-3RbCl.$BiCl_3$	Comp.		
	570	83	α-3RbCl.$BiCl_3$,RbCl	Eut.		
	395		3RbCl.$BiCl_3$ → Solid phase transf.			
$SbCl_5$	230	0.0	$BiCl_3$	M.P.	2d	239
	3	100	$BiCl_3$,$SbCl_5$	Eut.		
	3	100	$SbCl_5$	M.P.		
		20-70	Segregation			
$SeCl_4$	155	50	$BiCl_3$,$SeCl_4$	Eut.	1,3b,1e	240
$SnCl_2$	192	35.7	$2SnCl_2$.$BiCl_3$	Eut.	1	230
	230	68.4	$2SnCl_2$.$BiCl_3$	Comp.		
	220	79.5	$SnCl_2$, $SnCl_2$.$BiCl_3$	Eut.		
$SnCl_4$	236	0.0	$BiCl_3$	M.P.	2d	225
			Complete solid & liquid insolubility.			

$BiCl_3$

2nd Component	T°C	Conc. mol %	Phases involved	Def.	Method	Ref.
$SrCl_2$	236	0.0	$BiCl_3$	M..P.	2d	225
	420		Complete solid insolubility			
$TaCl_5$			Fig. 44			
	180	27	$BiCl_5$,$TaCl_5$	Eut.	1,2a	237
			$BiCl_5$ rich melts supercool			
		>60	Temp. of eutectic crystallization rises			
$TeCl_4$	171	54	$BiCl_3$,$TeCl_4$	Eut.	2a	241
$TiCl_4$	236	0.0	$BiCl_3$	M.P.	2d	225
			Limited solid & liquid solubility			
WCl_6	217	8.1	$BiCl_3$,WCl_6	Eut.	2	242
	228	100	WCl_6 $\gamma \rightleftharpoons \beta$	transformation		
	184	100	WCl_6 $\beta \rightleftharpoons \alpha$	"		
$ZnCl_2$			$BiCl_3$,$ZnCl_2$ (Temp. & comp. close to pure $BiCl_3$)	Eut.	1	230
$ZrCl_4$	212	16	$BiCl_3$,$ZrCl_4$	Eut.	1	230

BiI_3

2nd Component	T°C	Conc. mol %	Phases involved	Def.	Method	Ref.
CsI	439	0	BiI_3	M.P.	1,2a,3b	243
	365	14	BiI_3,$3CsI.2BiI_3$	Eut.		
	634	60	$3CsI.2BiI_3$	Comp.		
	541	78	$3CsI.2BiI_3$,CsI	Eut.		
	621	100	CsI	M.P.		
GaI_3	180	69	BiI_3,GaI_3	Eut.	2,3b	244
InI_3	165	92	InI_3,BiI_3	Eut.	2a	245

BiI_3

2nd Component	T°C	Conc. mol %	Phases involved	Def.	Method	Ref.
KI	439	0	BiI_3	M.P.	1,2d,3b	243
	327	32	BiI_3,3KI.2BiI_3	Eut.		
	409	58	3KI.2BI_3,KI	Per.		
	682	100	KI	M.P.		
RbI	439	0	BiI_3	M.P.	1,2d,3b	243
	330	17	BiI_3,3RbI.2BiI_3	Eut.		
	508	60	3RbI.2BiI_3	Comp.		
	451	73	2RbI.2BiI_3,RbI	Eut.		
	642	100	RbI	M.P.		
SbI_3			Complete range of solid solution		2d,3b	246
	160	90	Sol.slns. BiI_3 and SbI_3		2d,3b, 1f,5a	247
TeI_4	412	0	BiI_3	M.P.	2d,3b	248
	262	75	BiI_3,TeI_4	Eut.		
	280	100	TeI_4	M.P.		

$CaCl_2$

2nd Component	T°C	Conc. mol %	Phases involved	Def.	Method	Ref.
CaF_2			Fig. 45			284
	772	0	$CaCl_2$	M.P.	1,2d	284
	650(E)	20	$CaCl_2$, $CaCl_2$.CaF_2	Eut.		
	736(P)	40	$CaCl_2$.CaF_2,CaF_2	Per.		
	656	19	$CaCl_2$, $CaCl_2$.CaF_2	Eut.	1,2d	285
	740	42.5	$CaCl_2$.CaF_2,CaF_2	Per.		
	782±2	0.0	$CaCl_2$	M.P.	2c,2d	286
	644±2	24		Eut.		

$CaCl_2$

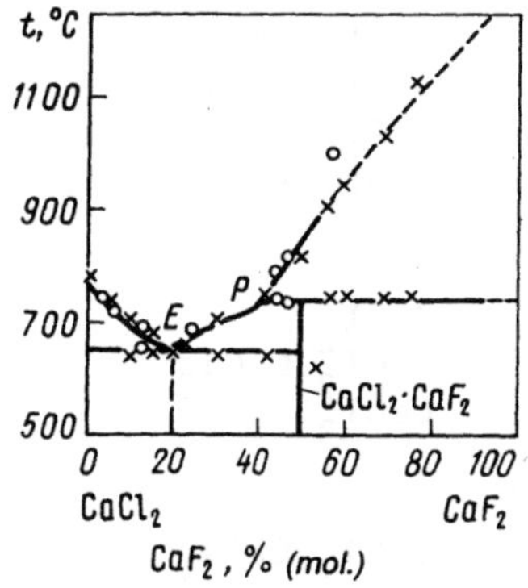

Fig. 45. $CaCl_2$-CaF_2 system.

TABLE III

CaI_2% (mol)	Temp. Liq.	Temp. Eut.	CaI_2% (mol)	Temp. Liq.	Temp. Eut.
100	788	-	37	558	556
90	754	554	32	588	557
80	718	554	25.1	628	556
70	680	556	18.1	662	555
60	644	557	16.1	680	555
52.5	614	556	13.6	696	554
45	587	555	7.5	732	554
44.5	583	555	0	778	-

Crystallizing phases: CaI_2: 100-44.5%, CaI_2, $CaCl_2$ at 37%, $CaCl_2$, 32-0%

$CaCl_2$

2nd Component	T°C	Conc. mol %	Phases involved	Def.	Method	Ref.
CaI_2			Table III			287
	556	37	$CaCl_2$.4%CaI_2 sol. sln.	Eut.	2a,3b	287
			CaI_2.10% $CaCl_2$ sol. sln.			
$CeCl_3$	772	0.0	$CaCl_2$	M.P.	2	249
	618	55	$CaCl_2$,$CeCl_3$	Eut.		
	802	100	$CeCl_3$	M.P.		
$CrCl_2$	619	42	$CrCl_2$,$CaCl_2$	Eut.	2a,6c	250
	619	42	$CrCl_2$,$CaCl_2$		2b,2c	251
					3b,3a	252
CsCl	772	0.0	$CaCl_2$	M.P.	2c	253
	705	11.0	CsCl.$CaCl_2$, $CaCl_2$	Eut.		
	910	50	CsCl.$CaCl_2$	Comp.		
	612	89	CsCl,CsCl.$CaCl_2$	Eut.		
	646	100	CsCl	M.P.		
	703		$CsCaCl_3$,$CaCl_2$	Eut.	1,2a	254
	908		$CsCaCl_3$	Comp.		
	604		CsCl,$CsCaCl_3$	Eut.		
CuCl	385	87	$CaCl_2$,CuCl	Eut.	1d,2	255
$GaCl_3$	253		Decomposition $CaGaCl_5$	Per.	2a	256
	150		$Ca(GaCl_4)_2$, $CaGaCl_5$	Per.		
	76.1	99.3	Sol.sln. $GaCl_3$.6%$CaCl_2$, $Ca(GaCl_4)_2$	Eut.		
$InCl_3$	478	65	Sol.sln.$CaCl_2$.23% $InCl_3$,$InCl_3$	Eut.	2b,2c	257

$CaCl_2$

2nd Component	T°C	Conc. mol %	Phases involved	Def.	Method	Ref.
KCl	635	14	$CaCl_2$.KCl,$CaCl_2$	Eut.	1	258
	745	33.3	$CaCl_2$.KCl	Comp.		
	598	60	KCl,$CaCl_2$KCl	Eut.		
	600	76.5	KCl,KCl.$CaCl_2$	Eut.	1	259
	752	50	KCl.$CaCl_2$	Comp.		
	640	26.5	KCl.$CaCl_2$,$CaCl_2$	Eut.		
	624	23.8	$CaCl_2$,$CaCl_2$.KCl	Eut.	1	260
		50.0	$CaCl_2$.KCl	Per.		
	596	73.4	$CaCl_2$.KCl,KCl	Eut.		
			Data of Menge,O.(Z.Anorg. Chem. 1911,72, p.162) reinforced		2	261
	630	15	$CaCl_2$, 2KCl.2$CaCl_2$	Eut.	1	262
	740	50	2KCl.2$CaCl_2$	Comp.		
	590	60	2KCl.2$CaCl_2$, KCl	Eut.		
$LaCl_3$	772	0.0	$CaCl_2$	M.P.	1	263
	630	28		Eut.		
	850	100	$LaCl_3$	M.P.		
LiCl	475	63	- Minimum of Liquidus -		1	259
	447	64	$CaCl_2$,β-LiCl	Eut.	1	264
	565	89	β-LiCl, α-LiCl	Per.		
	583	43	A kink on the $CaCl_2$ crys. curve undeciphered by authors			
	480	58	$CaCl_2$,LiCl	Eut.	2	265
	438		$CaCl_2$.LiCl → Solid state transf.			
$MgCl_2$	614	51	$MgCl_2$,$CaCl_2$ sol. sln.	Eut.	2,2d,3b	266
	620±2	52.8	$MgCl_2$, $CaCl_2$.17% $MgCl_2$ sol.sln.	Eut.	2d,3b,1e	267

$CaCl_2$

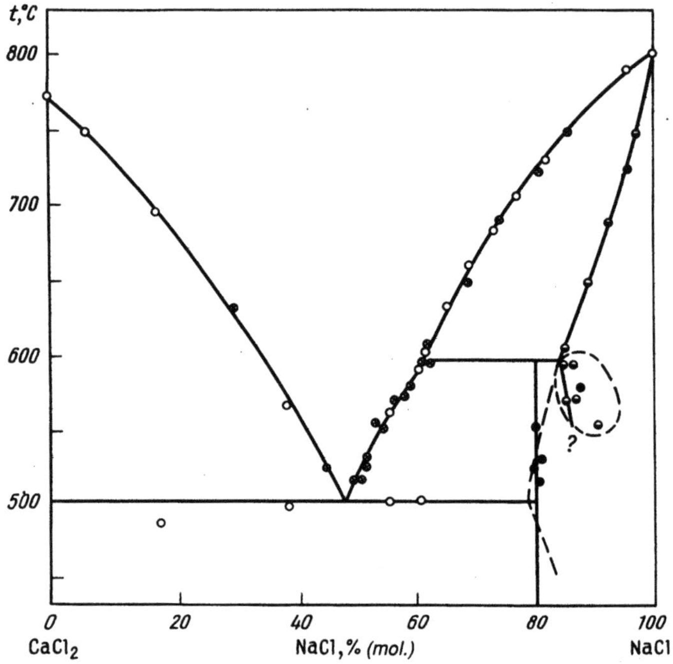

Fig. 46. $CaCl_2$-NaCl system.

2nd Component	T°C	Conc. mol %	Phases involved	Def.	Method	Ref.
NaCl			Fig. 46			268
	500±2	48	$CaCl_2$, $CaCl_2$.4NaCl	Eut.	2,	268
	594	62	$CaCl_2$.4NaCl, NaCl 20% $CaCl_2$ sol. sln.	Per.		
	600	15	$CaCl_2$ sol. in NaCl max.			
	596	11	Phase transf.		1	269
	494	48	NaCl, $CaCl_2$	Eut	2,2d,3b	266
	490	48	NaCl, $CaCl_2$	Eut.	1	264
$NbCl_5$	204	~100	$CaNbCl_7$,$CaCl_2$	Per.	2a	270
	204	~100	$NbCl_5$, $CaNbCl_7$	Eut.		

$CaCl_2$

2nd Component	T°C	Conc. mol %	Phases involved	Def.	Method	Ref.
$NdCl_3$	765	0.0	$CaCl_2$	M.P.	2	271
	585	43	$NdCl_3$,$CaCl_2$	Eut.		
	775	100	$NdCl_3$	M.P.		
$NiCl_2$	719	54	$CaCl_2$,$NiCl_2$	Eut.	2a,3b	272
$PbCl_2$	425	82	$CaCl_2$,$PbCl_2$	Eut.	1d,2	255
	772	0.0	$CaCl_2$	M.P.	2a,3b	273
	457±3	79±2	$CaCl_2$,$PbCl_2$	Eut.		
	497	100	$PbCl_2$	M.P.		
$PuCl_3$	774	0.0	$CaCl_2$	M.P.		274
	610	43	$CaCl_2$,$PuCl_3$	Eut.	2, 1c	
	767	100	$PuCl_3$	M.P.		
RbCl	772	0	$CaCl_2$	M.P.	2c	275
	678	17.5	RbCl.$CaCl_2$,$CaCl_2$	Eut.		
	855	50.0	RbCl.$CaCl_2$	Comp.		
	580	82.5	RbCl,RbCl.$CaCl_2$	Eut.		
	717	100	RbCl	M.P.		
$SrCl_2$	656	35	Sol.sln. range - min.		1	262
$TaCl_5$	218	100	$CaTaCl_7$,$CaCl_2$	Per.	2a	270
	216	100	$TaCl_5$,$CaTaCl_7$	Eut.		
$ThCl_4$	560	46	$CaCl_2$,$ThCl_4$	Eut.	2,6c	276
	782	0.0	$CaCl_2$	M.P.	2d	277
	560	52	$CaCl_2$,$ThCl_4$	Eut.		
	768	100	$ThCl_4$	M.P.		

$CaCl_2$

2nd Component	T°C	Conc. mol %	Phases involved	Def.	Method	Ref.
UCl_3	775±2	0.0	$CaCl_2$	M.P.	2d	278
	630±2	41	$CaCl_2$ 10% UCl_3 sol. sln, UCl_3 50% $CaCl_2$ sol. sln.	Eut.		279
	836±2	100	UCl_3	M.P.		
UCl_4	492	60	$CaCl_2$,UCl_4	Eut.	2c	280
	490±2	60	$CaCl_2$,UCl_4	Eut.	2d,3b	281
VCl_3	527	50	$CaCl_2$,VCl_3	Eut.	2	282
	575		$CaCl_2.6VCl_6$ → Solid phase trans.			
	514		Undefined thermal effects			
	577					
	656	5		Eut.	2a	283
	691	>9	Segregation			

CaF_2

2nd Component	T°C	Conc. mol %	Phases involved	Def.	Method	Ref.
CeF_3			Fig. 47			288
	1280	5	CaF_2,$9CaF_2.CeF_3$	Eut.	288	
	1370	10	$9CaF_2.CeF_3$	Comp.	2a	

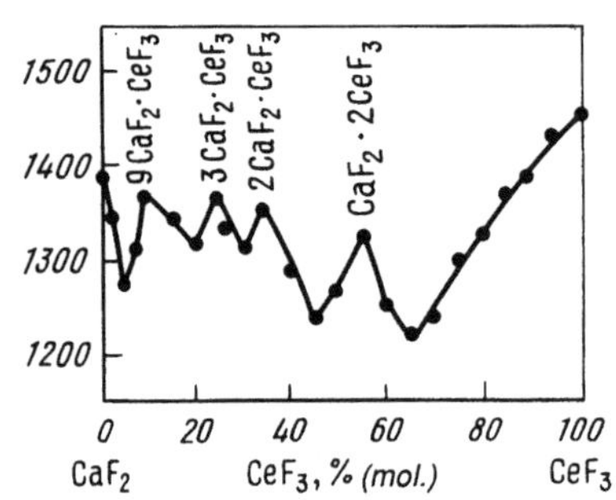

Fig. 47. CaF_2-CeF_3 system.

CaF_2

2nd Component	T°C	Conc. mol %	Phases involved	Def.	Method	Ref.
CeF_3	1330	20	$9CaF_2.CeF_3$, $3CaF_2.CeF_3$	Eut.	2a	288
	1365	25	$3CaF_2.CeF_3$	Comp.		
	1310	30	$3CaF_2.CeF_3$, $2CaF_2.CeF_3$	Eut.		
	1355	33.3	$2CaF_2CeF_3$	Comp.		
	1240	45	$2CaF_2.CeF_3$, $3CaF_2.4CeF_3$	Eut.		
	1325	57.0	$3CaF_2.4CeF_3$	Comp.		
	1220	65	$3CaF_2.4CeF_3$, CeF_3	Eut.		
			Components form sol. slns. of fluoride to hexagonal structure		3b	289
CsF	1062	17.0	$Cs_2F_2.2CaF_2$, CaF_2	Eut.	2c	290
	1079	20.0	$Cs_2F_2.2CaF_2$	Comp.		
	690	90.5	CsF, $Cs_2F_2.2CaF_2$	Eut.		
	1062	17.0	$Cs_2F_2.2CaF_2$, CaF_2	Eut.	1,2a	291
	1079	20.0	$Cs_2F_2.2CaF_2$	Comp.		
	690	90.5	CsF, $Cs_2F_2.2CaF_2$	Eut.		
	697	100	CsF	M.P.		
CuF_2			Existence of $CaCuF_4$ confirmed		3b	292

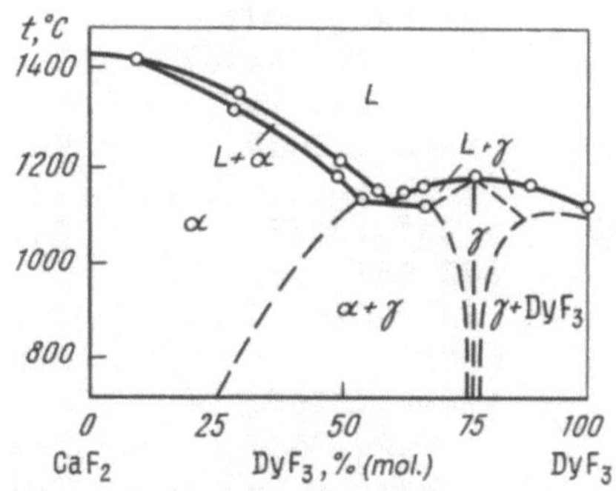

Fig. 48. CaF_2-DyF_3 system.

CaF_2

2nd Component	T°C	Conc. mol %	Phases involved	Def.	Method	Ref.
DyF_3			Fig. 48			293
	1425	0.0	CaF_2	M.P.	2d,3b,1e	293
	1125	58.0	CaF_2 sol.sln., γ-phase sol.sln.	Eut.		
	1175	75.0	γ-phase	Comp.		
	~1108	~100	γ-phase sol.sln., DyF_3	Eut.		
	1108	100	DyF_3	M.P.		
ErF_3			Sol.sln.(fluorite type) and limited sol.sln. (tisonite type)		3b	289
	1118±3	57	CaF_2 sol.sln., cubic sol.sln.	Eut.	2d,3b	294
	1151	75	Cubic sol.sln.	Max.		
	1100	90	Hexagonal (tisonite type) sol.sln. β-ErF_3	Eut.		
	1145	100	β-ErF_3	M.P.		
	1096	100	ErF_3 - Phase transf.			
FeF_3	790±10		$CaFeF_5$	M.P.	3,3b	295
GdF_3			Limited sol.sln.,hexagonal		3b	289
			Fig. 49			296
	1425	0.0	CaF_2	M.P.	2d,3b,3	296
	1210	58.0	46% GdF_3 sol.sln. in CaF_2, 67% GdF_3 (γ phase) sol.sln. in γ-CaF_2.3GdF_3	Eut.		
			(γ phase - cubic sol.sln. in CaF_2.3GdF_3)			
	1300	75-80	γ-CaF_2.3GdF_3	Comp.		
	1205	~100	87% GdF_3 sol.sln. in γ-CaF_2.3GdF_3	Eut.		
	1205	100	β-CdF_3	M.P.		
	1020		γ-CaF_2.3GdF_3 ⇌ δ-CaF_2.3GdF_3	Transf.		
	1020-1000		γ-phase ⇌ δ-phase	Transf.		
	1040		α ⇌ β GdF_3	Transf.		

CaF_2

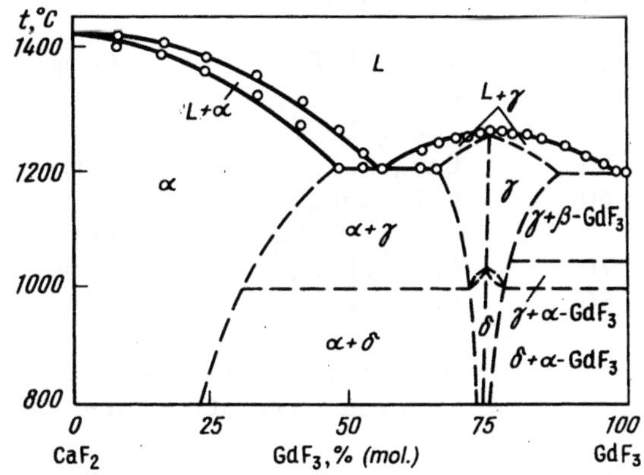

Fig. 49. CaF_2-GdF_3 system.

2nd Component	T°C	Conc. mol %	Phases involved	Def.	Method	Ref.
HoF_3			Sol.sln.- fluorite type			289
			Limited sol.sln. - tisonite type		3b	
InF_3	<650	$0 \leq x < 0.07$	Phase $(Ca_{1-x}In_x)F_{2+x}$		3b	
LaF_3			Fig. 50			288
	1450	4.8	$20CaF_2.LaF_3$	Comp.	2a	288
	1340	15	$20CaF_2.LaF_3$, $4CaF_2.LaF_3$	Eut.		

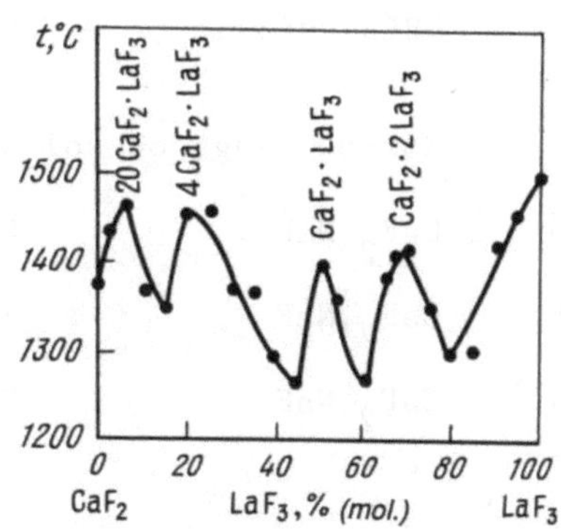

Fig. 50. CaF_2-LaF_3 system.

CaF_2

2nd Component	T°C	Conc. mol %	Phases involved	Def.	Method	Ref.
LaF_3	1460	20	$4CaF_2.LaF_3$	Comp.		288
	1260	40	$4CaF_2.LaF_3$, $CaF_2.LaF_3$	Eut.		
	1400	50	$CaF_2.LaF_3$	Comp.		
	1260	60	$CaF_2.LaF_3$, $CaF_2.2LaF_3$	Eut.		
	1420	70	$CaF_2.2LaF_3$	Comp.		
	1300	80	$CaF_2.2LaF_3$,LaF_3	Eut.		
			Sol.sln.fluorite and tisonite (hexagonal) structure		3b	289
	1300	60		Eut.	2d,3b	297
		70	CaF_2 sol.sln. in LaF_3			
		50	LaF_3 sol.sln. in CaF_2. Fluorite sol.sln. decomposes			
LiF	769	80.5	CaF_2,LiF	Eut.	2	298
	766	79.5	CaF_2,LiF	Eut.	1	299
	760±5	76	CaF_2,LiF	Eut.	1,2b,3b	300
LuF_3			Limited sol.sln. - hexagonal		3b	289
MgF_2	980	50	MgF_2,CaF_2	Eut.	2	301
NaF			Narrow range of sol.sln.		2a	302
	818	67.6	CaF_2,NaF	Eut.	1	269
	818	67.6	CaF_2,NaF	Eut.	1	303
	818	68.4	CaF_2,NaF	Eut.	1	304

CaF_2

2nd Component	T°C	Conc. mol %	Phases involved	Def.	Method	Ref.
NdF_3			Sol.slns.of fluorite & tisonite (hexagonal) structure		3b	289
	1275	58	Sol.sln.of 34% CaF_2 in NdF_3	Eut.	2d,3b	297
			Sol.sln.of 54% NdF_3 in CaF_2			
			Hexagonal sol.sln. transforms to fluorite with 5-8% NdF_3			
RbF	1100	42	$RbF.CaF_2$,CaF_2	Eut.	1,2	305
	1102	50	$RbF.CaF_2$	Comp.		
	760	90	RbF,$RbF.CaF_2$	Eut.		
SmF_3			Limited hexagonal sol.sln.		3b	289
	1200	56	Sol.sln.of 45.8% SmF_3 in CaF_2	Eut.	1e,2d, 3b	293
			Sol.sln.of 25% CaF_2 in SmF_3			
	1310	100	SmF_3	M.P.		
SrF_2			Complete range of sol.sln.		6d	306
TbF_3			Limited hexagonal sol.sln.		3b	289
TiF_3	890±20	50	$CaTiF_5$ (monoclinic)	M.P.	2d,3,3b	307
TuF_3			Fluorite sol.sln., limited tisonite (hexagonal) sol.sln.		3b	289

CaF_2

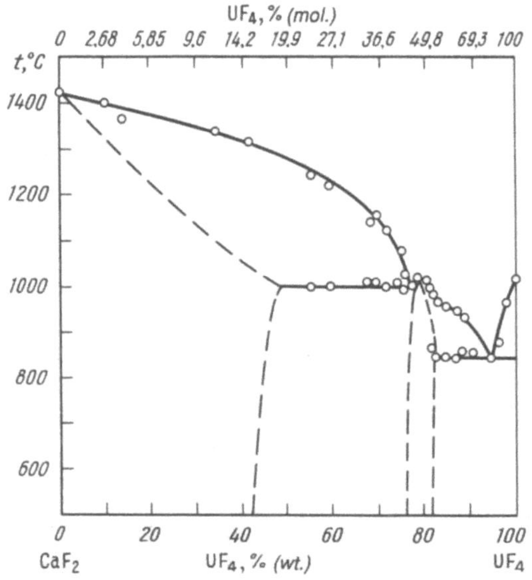

Fig. 51. CaF_2-UF_4 system.

2nd Component	T°C	Conc. mol %	Phases involved	Def.	Method	Ref.
UF_4			Fig. 51			308
	1420	0	CaF_2	M.P.	2a,3b,5	308
	997	44.96	Sol.sln.of 17% UF_4 in CaF_2, Sol.sln. on the base of $CaUF_6$	Eut.		
	1008	49.51	$CaUF_6$	Comp.		
	838	80.7	$CaUF_6$ sol.sln.,UF_4	Eut.		
	1005	100		M.P.		
			Up to 17% UF_4 - b.c.c. CaF_2			
		43.5-51.5	$CaUF_6$ based sol.sln.			
	910	100	Phase transf.			

CaF_2

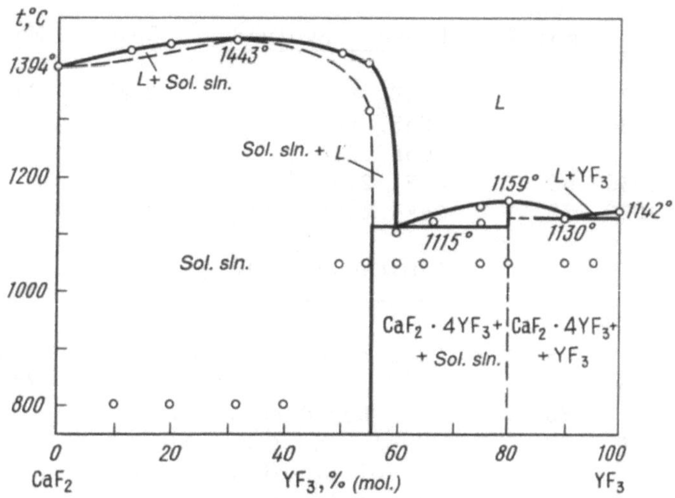

Fig. 52. CaF_2-YF_3 system.

2nd Component	T°C	Conc. mol %	Phases involved	Def.	Method	Ref.
YF_3			Fig. 52			309
	1443	33	CaF_2 sol.sln.	Max.	1,3b	309
	1115	60	55% YF_3 sol.sln. CaF_2, $CaF_2.4YF_3$			
	1159	80	$CaF_2.4YF_3$	Comp.		
	1130	90	$CaF_2.4YF_3$,YF_3	Eut.		
			$CaF_2.2YF_3$ $2CaF_2.5YF_3$ Fluorite and hexagonal sol.slns.		3b	310
	926		α-$CaYF_5$,CaF_2	Eut.	2d,3b	311
	980		YF_3,α-$CaYF_5$	Per.		
	905		$CaYF_5$	Transf.		
	1450	10	$9CaF_2.YF_3$	Comp.	2a	288
	1080	55	$9CaF_2.YF_3$, $2CaF_2.5YF_3$	Eut.		
	1150	·71.3	$2CaF_2.5YF_3$	Comp.		
	1010	85	$2CaF_2.5YF_3$, YF_3	Eut.		
			Fluorite sol.sln. and limited tisonite (hexagonal) sol.sln.		3a	289

CaF_2

2nd Component	T°C	Conc. mol %	Phases involved	Def.	Method	Ref.
YbF_3	1420	12	$9CaF_2.YbF_3$	Comp.	2a	288
	1020	60	$9CaF_2.YbF_3$, $CaF_2.2YbF_3$	Eut.		
	1050	66	$CaF_2.2YbF_3$	Comp.		
	1020	72	$CaF_2.2YbF_3$,YbF_3	Eut.		
			Limited sol.sln. - hexagonal		3b	289
		45	CaF_2 based sol.sln.		293	293
	1055	54		Eut.	2d,3b	
		80	YbF_3 based sol.sln.			
	960	93	sol.sln. YbF_3 ⇌ α sol.sln. CaF_2 + α sol.sln. YbF_3	Eut.		
	1000		α ⇌ β YbF_3			
	1145	100	YbF_3	M.P.		

CaI_2

2nd Component	T°C	Conc. mol %	Phases involved	Def.	Method	Ref.
CaF_2		49.3-70	CaF_2 crystallizes		2a,3b	287
	673	74	CaF_2 crystallizes	Eut.		
		78-100	CaI_2 crystallizes			
			Table IV			

TABLE IV

CaI_2 (mol)	Temp. Liq.	Temp. Eut.	CaI_2 (mol)	Temp. Liq.	Temp. Eut.
100	788	-	70	-	673
89.9	741	674	67.5	741	674
85	720	673	64.8	765	673
78	690	674	57.5	841	675
74	673	673	49.3	917	674

$CdBr_2$

2nd Component	T°C	Conc. mol %	Phases involved	Def.	Method	Ref.
CsBr	568	0	$CdBr_2$	M.P.	1	312
	373	38	$CdBr_2$, $CdBr_2.CsBr$	Eut.		
	425	51	$CdBr_2.CsBr$	Comp.		
	421	57	$CdBr_2.CsBr$, $CdBr_2.2CsBr$	Eut.		
	457	67	$CdBr_2.2CsBr$	Comp.		
	444	75	$CdBr_2.2CsBr$, $CsBr$	Eut.		
	384	36	$CdBr_2$, $CdBr_2.CsBr$	Eut.	2d	313
	421	50	$CdBr_2.CsBr$	Comp.		
	407	57	$CdBr_2.CsBr$, $CdBr_2.2CsBr$	Eut.		
	469	67	$CdBr_2.2CsBr$	Comp.		
	439	75	$CdBr_2.2CsBr$,$CsBr$	Eut.		
			$CdBr_2.CsBr$	Comp.	3b	314
			$CdBr_2.2CsBr$	Comp.		
	373		$CdBr_2$, $CdBr_2.CsBr$	Eut.	1,2a,3b	315
	425		$CdBr_2.CsBr$	Comp.		
	421		$CdBr_2.CsBr$, $CdBr_2.2CsBr$	Eut.		
	457		$CdBr_2.2CsBr$	Comp.		
	442		$CdBr_2.2CsBr$,$CsBr$	Eut.		
KBr	567	0	$CdBr_2$	M.P.	1	316
	354	44	$CdBr_2$, $KBr.2CdBr_2$	Per.		
	297	64	$KBr.2CdBr_2$,KBr	Eut.		
	360	43	$CdBr_2$, $KBr.2CdBr_2$	Per.	1	317
	300	63	$KBr.2CdBr_2$,KBr	Eut.		
	338	46	$CdBr_2$, $CdBr_2.KBr$	Eut.	2d	313
	346	50	$CdBr_2.KBr$	Comp.		
	299	63	$CdBr_2.KBr$,KBr	Eut.		

$CdBr_2$

2nd Component	T°C	Conc. mol %	Phases involved	Def.	Method	Ref.
LiBr			Continuous sol. sln.		2a	
NaBr	368	53.5	$CdBr_2$,NaBr	Eut.	1	319
	370	53	$CdBr_2$,NaBr	Eut.	2d	313
$PbBr_2$	340	92	$CdBr_2$,$PbBr_2$	Eut.	1	320
	376	100	$PbBr_2$	M.P.		
	340	92	$CdBr_2$,$PbBr_2$	Eut.	1	321
RbBr	322	46	$CdBr_2$, $CdBr_2$.RbBr	Eut.	2d	313
	336	50	$CdBr_2$.RbBr	Comp.		
	320	60	$CdBr_2$.RbBr, $CdBr_2$.2RbBr	Eut.		
	359	67	$CdBr_2$.2RbBr	Comp.		
	356		$CdBr_2$.2RbBr,RbBr	Eut.		
$TeBr_4$	566	0	$CdBr_2$	M.P.	2d	322
	344	85	$CdBr_2$,$TeBr_4$	Eut.		
	386	100	$TeBr_4$	M.P.		
TlBr	568	0	$CdBr_2$	M.P.	1	323
	392	41	$CdBr_2$,TlBr.$CdBr_2$	Eut.		
	398	49	$CdBr_2$.TlBr	Comp.		
	310	75	$CdBr_2$.TlBr,TlBr	Eut.		
	460	100	TlBr	M.P.		
	580 620 660		$CdBr_2$.TlBr: α ⇌ β; $CdBr_2$.3TlBr not confirmed		2a	318 324
			$CdBr_2$.TlBr: α ⇌ β		6f	
	310		$CdBr_2$, $CdBr_2$.TlBr	Eut.	1,2a	315
	398		$CdBr_2$.TlBr	Comp.	3b	
	392		$CdBr_2$.TlBr,TlBr	Eut.		

$CdBr_2$

2nd Component	T°C	Conc. mol %	Phases involved	Def.	Method	Ref.
$CdCl_2$			Unbroken sol.slns.		2a	324
CdI_2	385	95	Unbroken sol.sln.	Min.	1	323

$CdCl_2$

2nd Component	T°C	Conc. mol %	Phases involved	Def.	Method	Ref.
CsCl	458	29	$CdCl_2$.CsCl, $CdCl_2$	Eut.	1	325
	488	64	$CdCl_2$.2CsCl, $CdCl_2$.CsCl	Eut.		
	470	78	CsCl,$CdCl_2$.2CsCl	Eut.		
	473	21.5	$CsCd_5Cl_{11}$, $CdCl_2$	Per.	2d	326
	448	27.5	$CsCdCl_3$, $CsCd_5Cl_{11}$	Eut.		
	546	50	$CsCdCl_3$	Comp.		
	469	64.5	Cs_2CdCl_4, $CsCdCl_3$	Eut.		
	473	66.7	Cs_2CdCl_4	Comp.		
	453	73	Cs_3CdCl_5, Cs_2CdCl_4	Eut.		
	456	75	Cs_3CdCl_5	Comp.		
	454	77	CsCl,Cs_3CdCl_5	Eut.		
	465	100	$\alpha \rightleftharpoons \beta$ sol.state transf.			
	444	66.6	$\alpha \rightleftharpoons \beta$ sol.state transf.			
	360	66.6	Cs_2CdCl_4 decomposes			
	473		$Cs_3Cd_2Cl_7$ decomposes			
	460	21	α-$CdCl_2$,β-$CdCl_2$	Per.	1	327
	442	27	β-$CdCl_2$, $CdCl_2$.CsCl	Eut.		
	469	63	$CdCl_2$.CsCl, $CdCl_2$.2CsCl	Eut.		
	460	71	$CdCl_2$.2CsCl, $CdCl_2$.4CsCl	Eut.		
	460	76	$CdCl_2$.4CsCl,CsCl	Per.		

$CdCl_2$

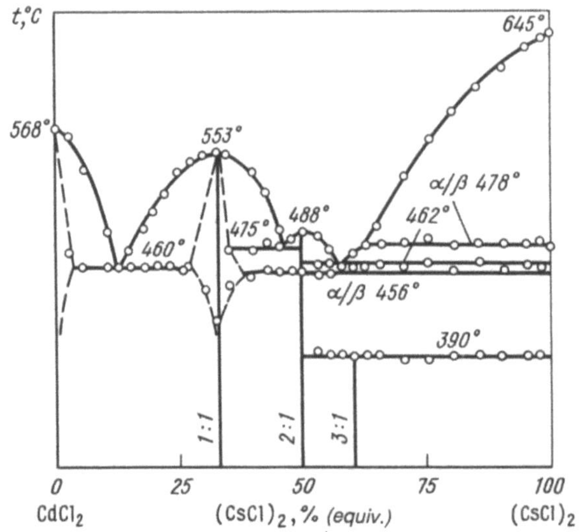

Fig. 53. $CdCl_2$-$(CsCl)_2$ system.

2nd Component	T°C	Conc. mol %	Phases involved	Def.	Method	Ref.
CsCl			Fig. 53			328
	568	0	$CdCl_2$	M.P.	1,2a	328
	460	23	5% CsCl sol.sln. in $CdCl_2$, 40% CsCl sol.sln. in $CsCl.CdCl_2$			
	553	51	$CsCl.CdCl_2$	Comp.		
	475	63	51% CsCl sol.sln. in $CsCl.CdCl_2$ α-$2CsCl.CdCl_2$	Eut.		
	488	67	α-$2CsCl.CdCl_2$	Comp.		
	462	73	α-$2CsCl.CdCl_2$, β-CsCl	Eut.		
	476	77	β-CsCl,α-CsCl	Per.		
	645	100	CsCl	M.P.		
	478	100	CsCl - Phase transf.			
	456	67	$2CsCl.CdCl_2$ - Phase transf.			
	390		$3CsCl.CdCl_2$ - formation of Comp.			
	458	29	$CdCl_2$.2CsCl, $CdCl_2$	Eut.	1	329
	565	50	$CdCl_2$.2CsCl	Comp.		
	488	64	$CdCl_2$.CsCl, $CdCl_2$.2CsCl	Eut.		
	495	33	$CdCl_2$.CsCl	Comp.		
	470	78	CsCl,$CdCl_2$.CsCl	Eut.		

$CdCl_2$

2nd Component	T°C	Conc. mol %	Phases involved	Def.	Method	Ref.
$GaCl_2$	170	~100	Ga_2Cl_4,$CdCl_2$	Eut.	1,3b,5	330
	500-540	0.5-0.6	$GaCl_2$ decomposition: $3GaCl_2 \rightarrow 2GaCl_3 + Ga$			
	58	100	Ga_2Cl_4 - Phase transf.			
$GaCl_3$	192	~70	$CdCl_2.2GaCl_3$, $CdCl_2$	Per.	2b	331
	52	~95	$GaCl_3$, $CdCl_2.2GaCl_3$	Eut.		
InCl	568	0	$CdCl_2$	M.P.	1,2a,3b	332
	360	47	β-$InCdCl_3$,$CdCl_2$	Eut.		
	369	50	β-$InCdCl_3$	Comp.		
	225		α-$InCdCl_3$, β-$InCdCl_3$	Per.		
	200	90	InCl,α-$InCdCl_3$	Eut.		
$InCl_3$			Fig. 54			333
	490	45	2% $CdCl_2$ sol. sln.in $InCl_3$ 2% $InCl_3$ sol. sln. in $CdCl_2$	Eut.	2a	333

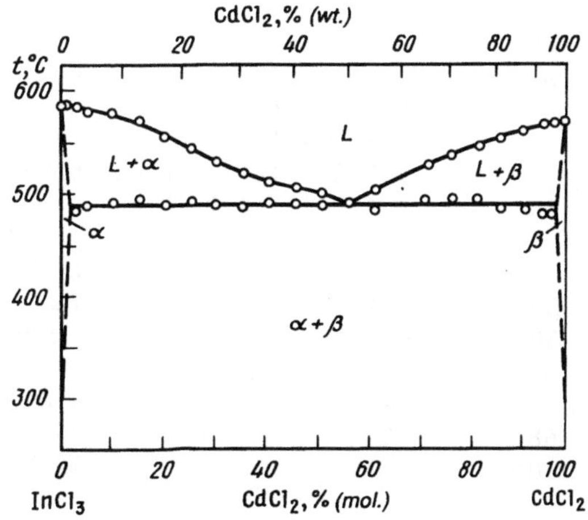

Fig. 54. $InCl_3$-$CdCl_2$ system.

$CdCl_2$

2nd Component	T°C	Conc. mol %	Phases involved	Def.	Method	Ref.
KCl	383	33	$CdCl_2$,$KCl.CdCl_2$	Eut.	1	334
	430	50	$KCl.CdCl_2$	Comp.		
	388	62	$KCl.CdCl_2$, β-$4KCl.CdCl_2$	Eut.		
	392	63.5	β-$4KCl.CdCl_2$, α-$4KCl.CdCl_2$	Per.		
	462	68.4	α-$4KCl.CdCl_2$,KCl	Per.		
	568	0	$CdCl_2$	M.P.	1	335
	396	33	$CdCl_2$,$KCl.CdCl_3$	Eut.		
	383	62	$KCl.CdCl_2$, $4KCl.CdCl_2$	Eut.		
	462	68	$4KCl.CdCl_2$,KCl	Per.		
	389	63	$4KCl.CdCl_2$ α ⇌ β transformation			
	375		$KCl.CdCl_2$ α ⇌ β transformation			
	460	22	α-$CdCl_2$,β-$CdCl_2$	Per.	1	336
	396	33	β-$CdCl_2$,$KCl.CdCl_2$	Eut.		
	428	50	$KCl.CdCl_2$	Comp.		
	383	62	$KCl.CdCl_2$, β-$4KCl.CdCl_2$	Eut.		
	389	63	β-$4KCl.CdCl_2$, α-$4KCl.CdCl_2$	Per.		
	382	12		Eut.		305
			Fig.55			337
	378	33	$CdCl_2$,$KCdCl_3$	Eut.	1,2a	337
	427	50	$KCdCl_3$	Comp.		
	383	61	$KCdCl_3$,K_4CdCl_6	Eut.		
	453	68	K_4CdCl_6,KCl	Per.		
	~383		α-K_4CdCl_6, β-K_4CdCl_6	Per.		
			Solid solutions in the range of $CdCl_2$ & $KCdCl_3$			
	568	0	$CdCl_2$	M.P.	1,2a	338
	397	35	$KCl.CdCl_2$,$CdCl_2$	Eut.	3b	
	428	50	$KCl.CdCl_2$	Comp.		
	385	64	$4KCl.CdCl_2$, $KCl.CdCl_2$	Eut.		
	476	68	KCl,$4KCl.CdCl_2$	Per.		
	770	100	KCl	M.P.		

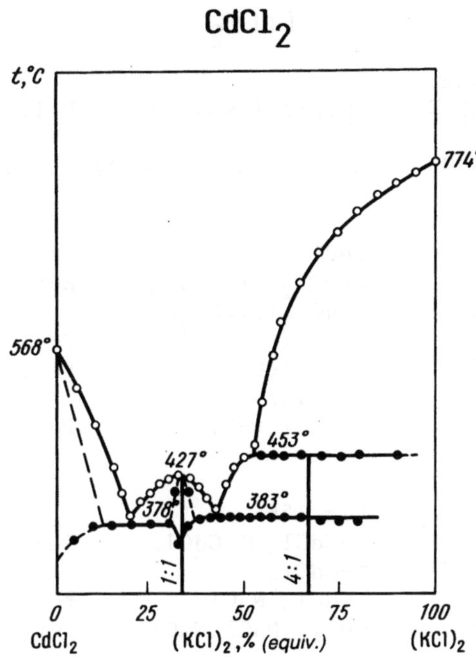

Fig. 55. $CdCl_2$-$(KCl)_2$ system.

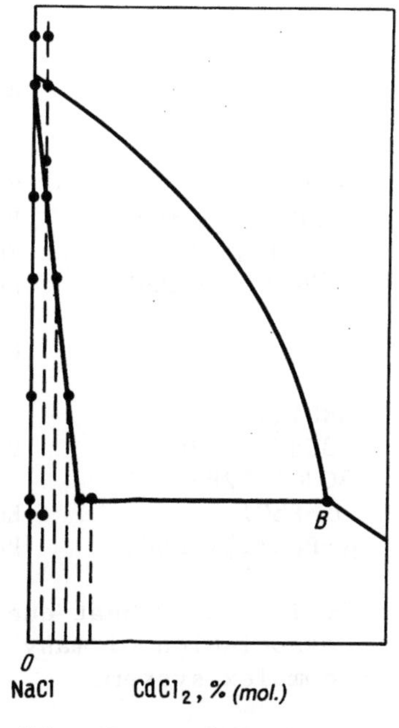

Fig. 56. Part of $NaCl$-$CdCl_2$ system.

$CdCl_2$

2nd Component	T°C	Conc. mol %	Phases involved	Def.	Method	Ref.
LiCl	498	36.7	Minimum of a complete range of solid sln.		1	319
	500	40	$CdCl_2$, $(CdCl_2).(LiCl_2)_3$	Eut.	1	339 340
	528	66.7	$(CdCl_2)_2.(Li_2Cl_2)_3$, $CdCl_2.(LiCl)_2$	Per.	1	
	547	75	$CdCl_2.(LiCl)_2$ β-LiCl	Per.		
	567	80	β-LiCl,α-LiCl	Per.		
NaCl			Fig. 56			341
	460	30	α-$CdCl_2$,β-$CdCl_2$	Per.	1	342
	387	45	β-$CdCl_2$, $CdCl_2$.NaCl	Eut.		343
	433	55	$CdCl_2$.NaCl,NaCl	Per.		
	380	44	$CdCl_2$,2NaCl.$CdCl_2$	Eut.	1	319
	420	54	2NaCl.$CdCl_2$,NaCl	Per.		
	433	95.5		Per.	2,4a	344
$NbCl_5$		~100		Eut.	1,2d	345
$PbCl_2$	568	0	α-$CdCl_2$	M.P.	1	335
	459	60	α-$CdCl_2$,β-$CdCl_2$	Per.		343
	385	78	β-$CdCl_2$,β-$PbCl_2$	Eut.		346
	410	82	β-$PbCl_2$, α-$PbCl_2$	Per.		
	368	62		Eut.		305
	440	64	$CdCl_2$, $3CdCl_2.2PbCl_2$	Per.	1,3b	340
	370	78	$3CdCl_2.2PbCl_2$, β-$PbCl_2$	Eut.		
	420	87	β-$PbCl_2$,α-$PbCl_2$	Per.		
	600		Evolution of heat due to association of many complex systems		2a	318

$CdCl_2$

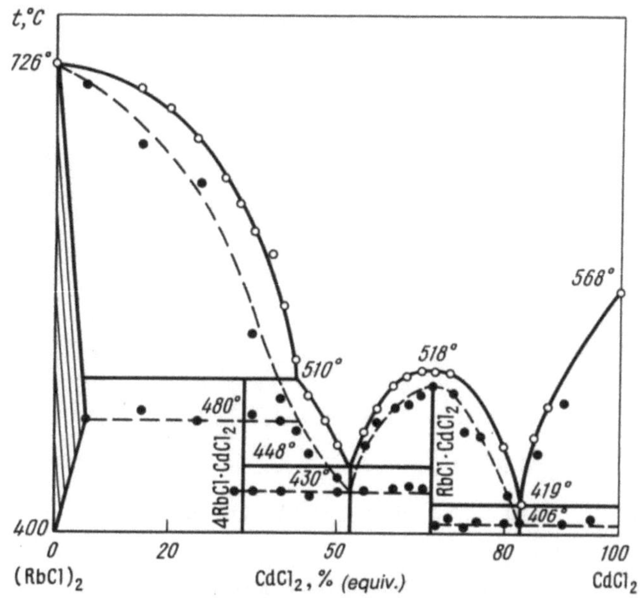

Fig. 57. $(RbCl)_2$-$CdCl_2$ system.

2nd Component	T°C	Conc. mol %	Phases involved	Def.	Method	Ref.
RbCl			Fig. 57			305
	451	60		Eut.		305
	425	27.5	$RbCd_5Cl_{11}$, $CdCl_2$	Per.	2d,3b	347
	414	30	$RbCdCl_3$, $RbCd_5Cl_{11}$	Eut.	2d,3b	
	499	50	$RbCdCl_3$	Comp.		
	456	61.5	$Rb_3Cd_2Cl_7$, $RbCdCl_3$	Per.		
	437	67	Rb_2CdCl_4, $Rb_3Cd_2Cl_7$	Per.		
	435	68	Rb_4CdCl_6, Rb_2CdCl_4	Eut.		
	471	72.5	RbCl, Rb_4CdCl_6	Per.		
	456		$Rb_4Cd_3Cl_{10}$	Comp.		
	568	0	$CdCl_2$	M.P.	1,2a,3b	338
	419	30	$RbCl.CdCl_2$, $CdCl_2$	Eut.		
	518	50	$RbCl.CdCl_2$	Comp.		
	448	64.5	$4RbCl.CdCl_2$, $RbCl.CdCl_2$	Eut.		
	510	73	5% $CdCl_2$ sol.sln.in RbCl, $4RbCl.CdCl_2$	Per.		
	726		RbCl	M.P.		

$CdCl_2$

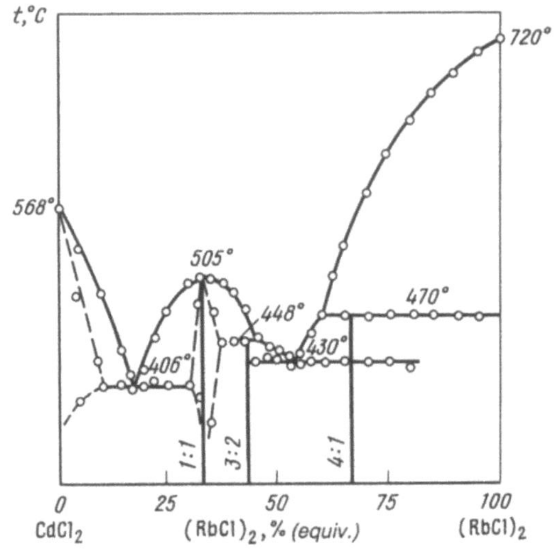

Fig. 58. $CdCl_2$-$(RbCl)_2$ system.

2nd Component	T°C	Conc. mol %	Phases involved	Def.	Method	Ref.
RbCl			Fig. 58			328
	568	0	$CdCl_2$	M.P.	1,2d	328
	406	31	18% RbCl sol.sln.in $CdCl_2$	Eut.		
			46% RbCl sol.sln.in $RbCl.CdCl_2$	Eut.		
	505	50.8	$RbCl.CdCl_2$	Comp.		
	448	63	37% RbCl in $RbCl.RbCl_2$, $3RbCl.2CdCl_2$	Per.		
	430	70	$3RbCl.2CdCl_2$, $4RbCl.CdCl_2$	Eut.		
	470	75	$4RbCl.CdCl_2$,RbCl	Per.		
	720	100	RbCl	M.P.		
	430		$4RbCl.CdCl_2$ (Transformation temp. falls in the eutectic)	Phase Trans.		
	568	0	α-$CdCl_2$	M.P.		348
	460	21	α-$CdCl_2$,β-$CdCl_2$	Per.		
	425	30	β-$CdCl_2$,$CdCl_2$.RbCl	Eut.		
	514	50	$CdCl_2$.RbCl	Comp.		
	441	68	$CdCl_2$.RbCl, $CdCl_2$.3RbCl	Eut.		
	514	75	$CdCl_2$.3RbCl,RbCl	Per.		
	726	100	RbCl	M.P.		

$CdCl_2$

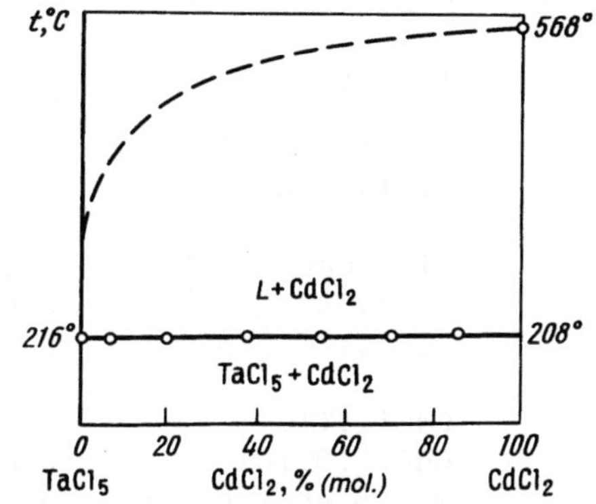

Fig. 59. $TaCl_5$-$CdCl_2$ system.

2nd Component	T°C	Conc. mol %	Phases involved	Def.	Method	Ref.
$SbCl_5$	3	~100	$CdCl_2$,$SbCl_5$	Eut.	2d	339
	3	100	$SbCl_5$	M.P.		
$SrCl_2$	480	38	Sol.sln.with 15 and 90% $SrCl_2$	Eut.	2d	349
$TaCl_5$			Fig. 59			345
			Eutectic almost coincident with M.P. of pure $TaCl_5$		1,2d	345
$TeCl_4$	570	0	$CdCl_2$	M.P.	1,2d	350
	214	96	$TeCl_4$,$CdCl_2$	Eut.		
	224	100	$TeCl_4$	M.P.		
TlCl	567	0	$CdCl_2$	M.P.	1	351
	410	20	TlCl.$CdCl_2$,$CdCl_2$	Eut.		
	326	62.5	TlCl,TlCl.$CdCl_2$	Eut.		
	568	0	$CdCl_2$	M.P.	1	352
	403	35	TlCl.$CdCl_2$,$CdCl_2$	Eut.		
	434	50	TlCl.$CdCl_2$	Comp.		
	306	78	TlCl,TlCl.$CdCl_2$	Eut.		
	430	100	TlCl	M.P.		

$CdCl_2$

2nd Component	T°C	Conc. mol %	Phases involved	Def.	Method	Ref.
TlCl	396	38	$Tl_2Cl_2.CdCl_2$, $CdCl_2$	Eut.	1	353
	430		$(TlCl)_2.CdCl_2$	Comp.		
	297	77	$TlCl,Tl_2Cl_2.CdCl_2$	Eut.		
	400	34.5	$TlCl.CdCl_2,CdCl_2$	Eut.	2a	354
	426	50	$TlCl,CdCl_2$	Comp.		
	298	70.5	$TlCl,TlCl.CdCl_2$	Eut.		
			$CdCl_2.TlCl$	Comp.	2a	318
			$CdCl_2.3TlCl$	Comp.		324
			No $\alpha \rightleftharpoons \beta$ transf. in $CdCl_2.TlCl$			
	568	0	$CdCl_2$	M.P.	1	329
	396	22	$CdCl_2.TlCl,CdCl_2$	Eut.		
		50	$CdCl_2.TlCl$	Comp.		
	300	63	$TlCl,CdCl_2.TlCl$	Eut.		
	429	100	TlCl	M.P.		
	372	52	Homomorphic transformation			
	396		$TlCl.CdCl_2$, $CdCl_2$	Eut.	1,2a,3b	329
	430		$TlCl.CdCl_2$	Comp.		
	290		$TlCl,TlCl.CdCl_2$	Eut.		
	372	52	Homomorphic transformation of TlCl			
$ZnCl_2$	300	92	$CdCl_2,ZnCl_2$		1	355
CdF_2	478	46	$CdCl_2,CdF_2$	Eut.	1	319
CdI_2	568	0	$CdCl_2$	M.P.	1	341
	360	33	$CdCl_2,CdI_2$	Eut.		
	388	100	CdI_2	M.P.		
	460		Homomorphic transformation of $CdCl_2$			

CdF_2

2nd Component	T°C	Conc. mol %	Phases involved	Def.	Method	Ref.
CrF_3	794	28	$CdCrF_5, CdF_2$	Eut.	2d,3b	356
	838	32.5	$CrF_3, CdCrF_5$	Per.		
KF	855	25.5	$CdF_2, KCdF_3$	Eut.	2d	357
	949	50	$KCdF_3$	Comp.	3b	
	728	60	$KCdF_3, K_3Cd_2F_7$	Per.		
	710	66.6	$K_3Cd_2F_7, K_2CdF_4$	Per.		
	694	82	K_2CdF_4, KF	Eut.		
	870	25	$KCdF_3, CdF_2$	Eut.	1	358
	970	50	$KCdF_3$	Comp.		
	706	80	$KCdF_3, KF$	Eut.		
LiF	650	55.6	CdF_2, LiF	Eut.	1,2a	358
NaF	592	55	NaF, CdF_2	Eut.	1	319 358
RbF	902	23.0	$CdF_2, RbCdF_3$	Eut.	2d,3b	359
	1018	50.0	$RbCdF_3$	Comp.		
	800	77.5	$RbCdF_3, Rb_3Cd_2F_7$	Per.		
	792	78.0	$Rb_3Cd_2F_7, Rb_2CdF_4$	Per.		
	713	88.0	Rb_2CdF_4, RbF	Eut.		

CdI_2

2nd Component	T°C	Conc. mol %	Phases involved	Def.	Method	Ref.
CsI	388	0	CdI_2	M.P.	1	360
	200	44	$2CsI.CdI_2, CdI_2$	Eut.		
	425	66.7	$2CsI.CdI_2$	Comp.		
	413	76	$CsI, 2CsI.CdI_2$	Eut.		
	640	100	CsI	M.P.		
	387	0	CdI_2	M.P.	2a	361
	210	45	$CdI_2, 2CsI.CdI_2$	Eut.		
	429	66.6	$2CsI.CdI_2$	Comp.		
	410	71	$2CsI.CdI_2, CsI$	Eut.		
	641	100	CsI	M.P.		

CdI_2

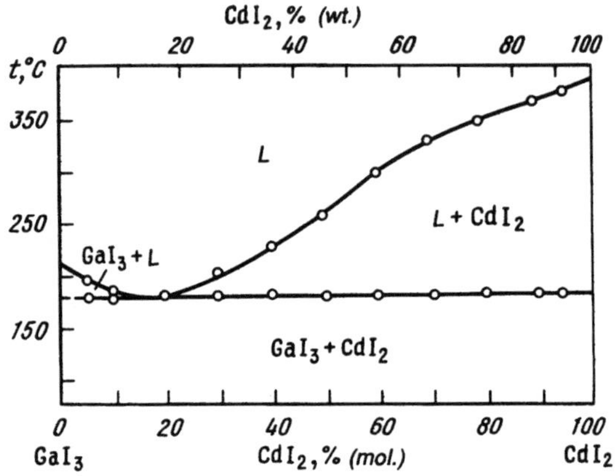

Fig. 60. GaI_3-CdI_2 system.

2nd Component	T°C	Conc. mol %	Phases involved	Def.	Method	Ref.
GaI_3			Fig. 60			362
	180	80	GaI_3,CdI_2	Eut.	2,3b	362
InI	250	55	In_4CdI_6,CdI_2	Eut.	2d,3b,4a	363
	320	80	In_4CdI_6	Comp.		
	310	85	InI,In_4CdI_6	Eut.		
	365	100	InI	M.P.		
	208		In_4CdI_6			
InI_2			Fig. 61			364
	387	0	CdI_2	M.P.	2a	364
	218	96.5	InI_2,CdI_2	Eut.	2a	
	226	100	InI_2	M.P.		
	161	100	InI_2	P.T.		
InI_3			Fig. 62			365
	386	0	CdI_2	M.P.	2c,2d	365
	190	37.5	$CdI_2.6InI_3$,CdI_2	Eut.		
	420	87.0	$CdI_2.6InI_3$	Comp.		
	196	98.5	InI_3,$CdI_2.6InI_3$	Eut.		
	198	100	InI_3	M.P.		
	338		CdI_2	P.T.		
	298		$CdI_2.6InI_3$			

CdI_2

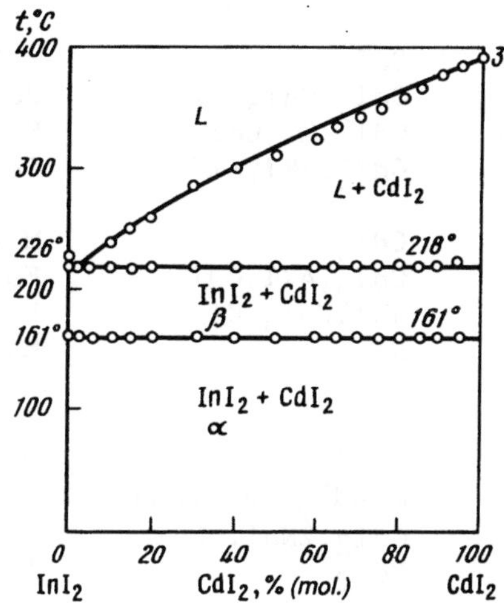

Fig. 61. InI_2-CdI_2 system.

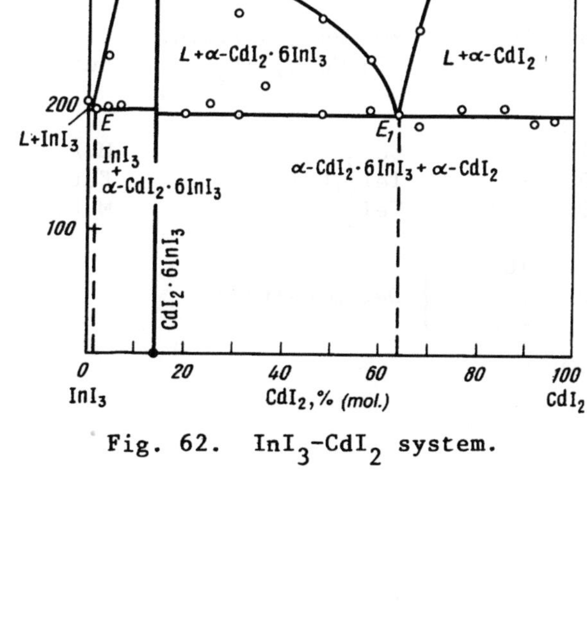

Fig. 62. InI_3-CdI_2 system.

CdI_2

2nd Component	T°C	Conc. mol %	Phases involved	Def.	Method	Ref.
KI	388	0	CdI_2	M.P.	1	366
	223	49	CdI_2,CdI_2.KI	Per.		
	213	53.5	CdI_2.KI,CdI_2.2KI	Eut.		
	272	62	CdI_2.2KI,KI	Per.		
	388	0	CdI_2	M.P.	2a	367
	210	52	CdI_2,4KI.CdI_2	Eut.		
	286	63	4KI.CdI_2.KI	Per.		
	680	100	KI	M.P.		
NaI	388	0	α-CdI_2	M.P.	1	341
	287	44	β-CdI_2,CdI_2.2NaI	Eut.		368
	371	54	CdI_2.2NaI, NaI	Per.		
	310		CdI_2	P.T.		
PbI_2	388	0	CdI_2	M.P.	1	369
	360	30	Continuous sol.sln.	Min.		346
	410	0	CdI_2	M.P.	1	370
	360	25	Continuous sol.sln.	Min.		
	386	100	PbI_2	M.P.		
RbI	388	0	CdI_2	M.P.	2a	361
	216	52	CdI_2,2RbI.CdI_2	Eut.		
	315	64	2RbI.CdI_2,RbI	Per.		
	643	100	RbI	M.P.		
SnI_2	305	70	20% SnI_2 sol.sln. in CdI_2,SnI_2	Eut.	2,4a	371
TeI_4	384	0	CdI_2	M.P.	2a	372
	260	92.5	TeI_4,CdI_4	Eut.		
	280	100	TeI_4	M.P.		
	350	10	Decomposition			
	350	70	Decomposition			
	350	92	Decomposition			

CdI_2

2nd Component	T°C	Conc. mol %	Phases involved	Def.	Method	Ref.
TlI	250	55	CdI_2.4TlI,CdI_2	Eut.	1	373
	354	79	TlI.CdI_2.4TlI	Per.		
	426	100	TlI	M.P.		
	388	0	CdI_2	M.P.	1	323
	255	44	TlI.CdI_2,CdI_2	Eut.		
	263	49	TlI.CdI_2	Comp.		
	256	54	$2Tl_2I_2$.CdI_2, TlI.CdI_2	Eut.		
	357	79	TlI,$2Tl_2I_2$.CdI_2	Per.		
	426	100	TlI	M.P.		
			Existence of TlI.CdI_2 confirmed		3b	374
ZnI_2			Continuous sol.sln.		4c,2a,6c	375

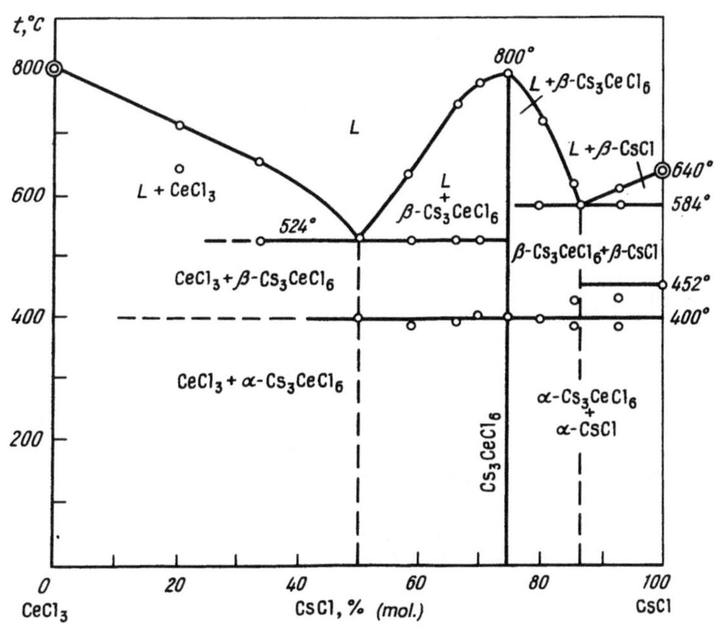

Fig. 63. $CeCl_3$-CsCl system.

$CeCl_3$

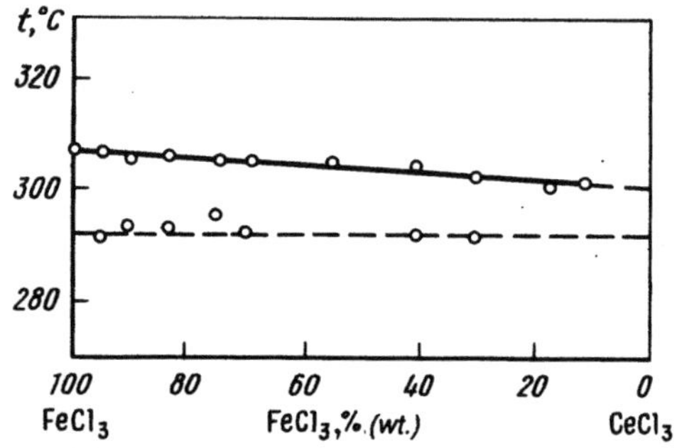

Fig. 64. $FeCl_3$-$CeCl_3$ system.

2nd Component	T°C	Conc. mol %	Phases involved	Def.	Method	Ref.
CsCl			Fig. 63			376
	524	50	$CeCl_3$, β-Cs_3CeCl_6	Eut.	2c	376
	800	74	β-Cs_3CeCl_6	Comp.		
	584	86	β-Cs_3CeCl_6, β-CsCl	Eut.		
	424-452		(452 for pure CsCl)	P.T.		
	390-400		Cs_3CeCl_6	P.T.		
$FeCl_2$	607	65	$CeCl_3$,$FeCl_2$	Eut.	2c	377
	631	58	$FeCl_2$,$CeCl_3$		2a,6c	378
$FeCl_3$			Fig. 64			378
			Insoluble in liquid and solid state.		2a,6c	378
	290-295		Secondary thermal effect due to dissociation of $FeCl_3$ to $FeCl_2$ and Cl_2			

$CeCl_3$

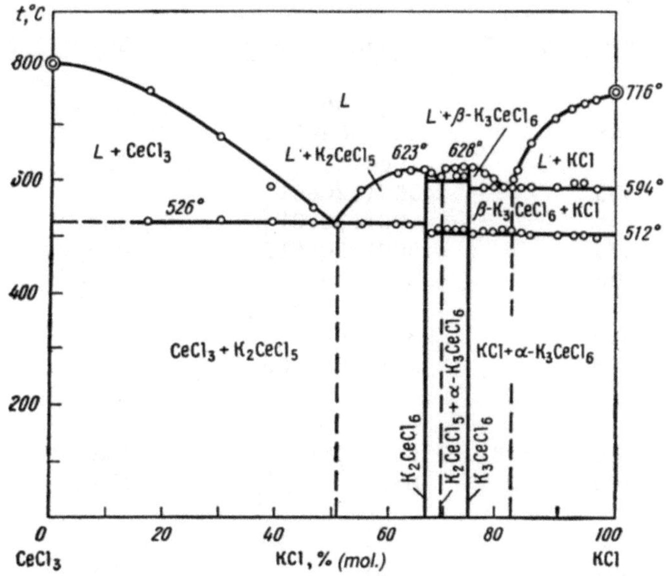

Fig. 65. $CeCl_3$-KCl system.

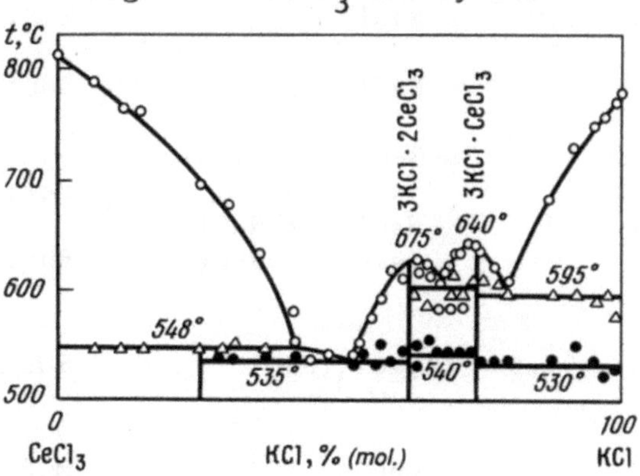

Fig. 66. $CeCl_3$-KCl system.

2nd Component	T°C	Conc. mol %	Phases involved	Def.	Method	Ref.
KCl			Fig. 65			376
	802	0	$CeCl_3$	M.P.	2c	376
	526	50.7	$CeCl_3$,K_2CeCl_5	Eut.		
	623	66.7	K_2CeCl_5	Comp.		
	618	69.8	K_2CeCl_5, β-K_3CeCl_6	Eut.		
	628	74.9	β-K_3CeCl_6	Comp.		
	596	80.8	β-K_3CeCl_6,KCl	Eut.		
	516		K_3CeCl_6	P.T.		

$CeCl_3$

2nd Component	T°C	Conc. mol %	Phases involved	Def.	Method	Ref.
KCl			Fig. 66			379
	548		$KCl.3CeCl_3$, $CeCl_3$	Per.	2a	379
	535		$3KCl.2CeCl_3$, $KCl.3CeCl_3$	Eut.		
	675		$3KCl.2CeCl_3$	Comp.		
	600		$3KCl.CeCl_3$, $3KCl.2CeCl_3$	Eut.		
	640		$3KCl.CeCl_3$	Comp.		
	595		$KCl,3KCl.CeCl_3$	Eut.		
$MgCl_2$	660	68.8	$CeCl_3,MgCl_2$	Eut.	2c	376
	708	100	$MgCl_2$	M.P.		
NaCl	802	0	$CeCl_3$	M.P.	2	380
	488	67.5	$CeCl_3$,NaCl	Eut.		
	800	100	NaCl	M.P.		
	510	60	NaCl,$CeCl_3$	Eut.	2a	381
	630-637	45	β-$CeCl_3$ sol.sln., α-$CeCl_3$	Per.	2a,3b	382
	503	57	NaCl,15% NaCl sol. sln. in β-$CeCl_3$	Eut.		
$SnCl_2$	240	97.5	$CeCl_3,SnCl_2$	Eut.	2c	377
$SrCl_2$	820	0	$CeCl_3$	M.P.	2a,3b	383
	640	43	$CeCl_3,SrCl_2$	Eut.		
	848	100	$SrCl_2$	M.P.		
	580		$xCeCl_3.ySrCl_2$	P.T.		
	525		$\alpha \rightleftharpoons \beta$ $xCeCl_3.ySrCl_2$	P.T.		
$ThCl_4$	640	6.06	$CeCl_3,ThCl_4$	Eut.	2	384

CeF_3

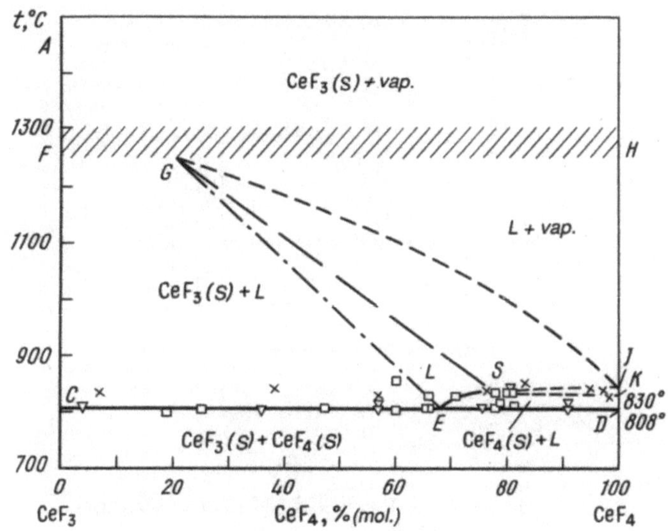

Fig. 67. CeF_3-CeF_4 system.

2nd Component	T°C	Conc. mol %	Phases involved	Def.	Method	Ref.
CeF_4			Fig. 67			385
	808	~70	CeF_3,CeF_4	Eut.	2,3b,le	385
			In the figure: >70% CeF_4 the line inconsistent with content of CeF_4,			
			The region CEA: complete decomposition not confirmed.			
CsF	650	7.5	CsF,Cs_3CeF_6	Eut.	2a	386
	938	25.0	Cs_3CeF_6	Comp.		
	754	33.0	Cs_3CeF_6,CeF_3	Eut.		
KF			Fig. 68			386
	788	74	$KCeF_4$,CeF_3	Per.	2a	386
	710	80	KF,$KCeF_4$	Eut.		
	628	75	3KF + CeF_3 $\rightleftharpoons$ K_3CeF_6	P.T.		
LiF	750	80	LiF,CeF_3	Eut.	2	387
	735	84.4	LiF,CeF_3	Eut.	2a	388

CeF_3

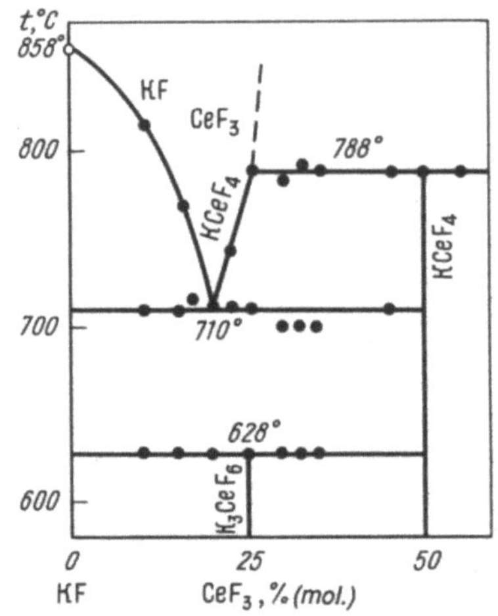

Fig. 68. Part of KF-CeF_3 system.

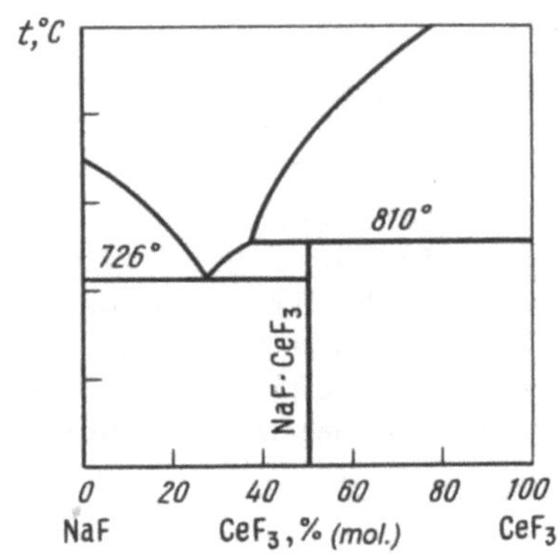

Fig. 69. NaF-CeF_3 system.

2nd Component	T°C	Conc. mol %	Phases involved	Def.	Method	Ref.
NaF			Fig. 69			389
	810	62	NaF.CeF_3,CeF_3	Per.	2	389
	726±3	72	NaF,NaF.CeF_3	Eut.	1d	390
	810	62	CeF_3.NaF,CeF_3	Per.	1f,2,3b	391
	726	72	NaF,NaF.CeF_3	Eut.		
SrF_2			Sol.slns. of fluorite and tisonite hex. struct.		3b	289

$CoBr_2$

2nd Component	T°C	Conc. mol %	Phases involved	Def.	Method	Ref.
$TeBr_4$	340	85	$CoBr_2$,$TeBr_4$	Eut.	2d	392
	386	100	$TeBr_4$	M.P.		

$CoCl_2$

2nd Component	T°C	Conc. mol %	Phases involved	Def.	Method	Ref.
CsCl	501	35	$CsCoCl_3$, $CoCl_2$	Eut.	2d	393
	547	50	$CsCoCl_3$	Comp.		
	521	57.25	Cs_2CoCl_4, $CsCoCl_3$	Eut.		
	597	67	Cs_2CoCl_4	Comp.		
	547	74	Cs_3CoCl_5, Cs_2CoCl_4	Eut.		
	549	75	Cs_3CoCl_5	Comp.		
	528	82	CsCl,Cs_3CoCl_5	Eut.		
	475-490		Cs_3CoCl_5 + Cs_2CoCl_4 $\rightleftharpoons$ $Cs_5Co_2Cl_9$	P.T.		
	444±5		Cs_2CoCl_4 + $CsCoCl_3$ $Cs_3Co_2Cl_7$	P.T.		
$CuCl_2$	362	~75	$CoCl_2$,α-$CuCl_2$	Eut.	2	394
	488	80	α-$CuCl_2$,β-$CuCl_2$	Per.		
$GaCl_3$			Fig. 70			395
	165	95.25	$CoCl_2.GaCl_3$, $CoCl_2$	Per.	2,3b	395
	76	>99.7	$GaCl_3$, $CoCl_2.GaCl_3$	Eut.		
$InCl_3$	525	49	45% sol.sln.of $CoCl_2$ in $InCl_3$, 4% sol.sln. of $InCl_3$ in $CoCl_2$	Eut.	2a	333
						333
KCl	362	52	$KCoCl_3$,$CoCl_3$	Per.	2d	393
	451	56.5	K_2CoCl_4,$KCoCl_3$	Eut.		
	436	67	K_2CoCl_4	Comp.		
	427	69.5	KCl,K_2CoCl_4	Eut.		
	343		$KCoCl_3$	P.T.		

$CoCl_2$

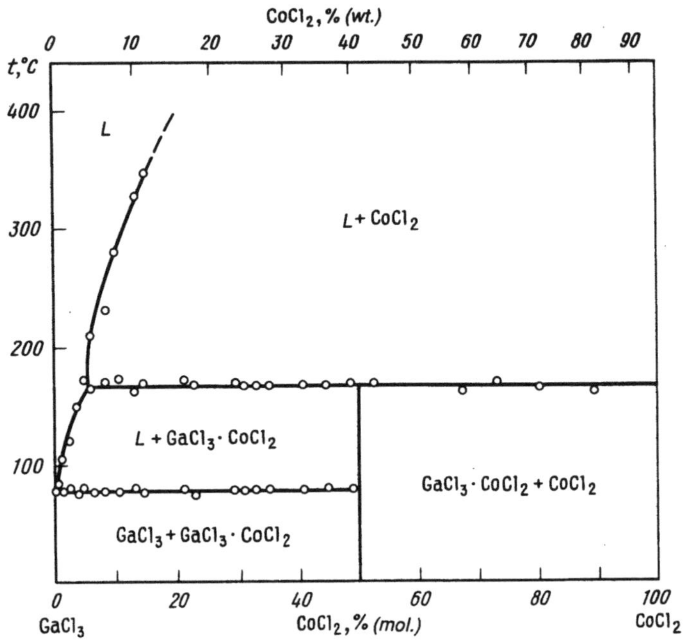

Fig. 70. $GaCl_3$-$CoCl_2$ system.

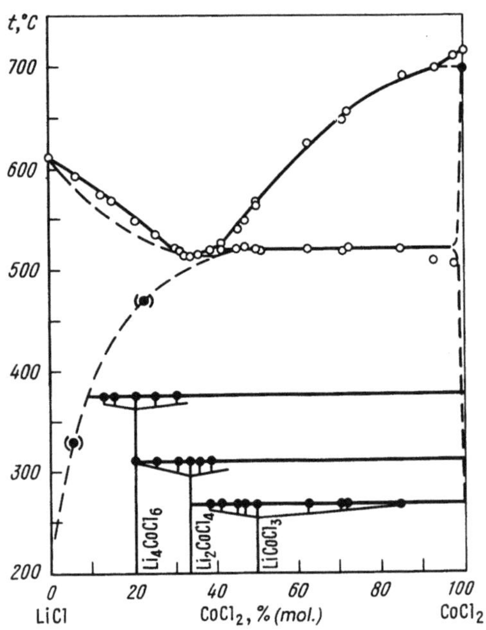

Fig. 71. LiCl-$CoCl_2$ system.

$CoCl_2$

2nd Component	T°C	Conc. mol %	Phases involved	Def.	Method	Ref.
LiCl			Fig. 71			393
	522	59	Sol.sln: LiCl-$CoCl_2$, 2% LiCl sol.sln. in $CoCl_2$	Per.	2d,3b	393
	512	67	Sol.sln: LiCl-$CoCl_2$	Min.		
	375		CoCl sol.sln.+ LiCl sol.sln. ⇌ Li_4CoCl_8	P.T.		
	31		Li_4CoCl_6 + $CoCl_2$ ⇌ Li_2CoCl_4	P.T.		
	268		Li_2CoCl_4+ $CoCl_2$ ⇌ $LiCoCl_3$	P.T.		
NaCl	370	63.5	NaCl,$CoCl_2$	Eut.	2,1f	396
	<350		2NaCl.$CoCl_2$	Comp.		
	368	60.5	$CoCl_2$,NaCl	Eut.	2d	393
	<350		$NaCoCl_4$	Comp.		
$NbCl_5$		100	~$NbCl_5$	Eut.	2d,1	345
$NiCl_2$	710	0	$CoCl_2$ Continuous sol.sln. with no extremes	M.P.	2	397
	989	100	$NiCl_2$	M.P.		
			Fig. 72			398
	680	7	Continuous sol.sln. (up to 30% $NiCl_2$)			398

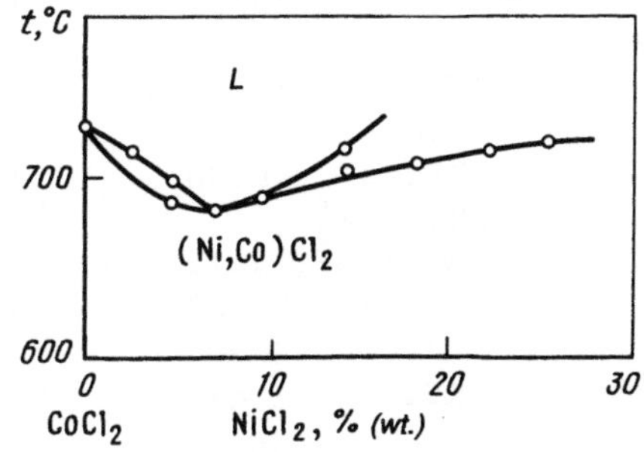

Fig. 72. Part of $CoCl_2$-$NiCl_2$ system.

$CoCl_2$

2nd Component	T°C	Conc. mol %	Phases involved	Def.	Method	Ref.
$NiCl_2$	721	1.4	$CoCl_2$,$NiCl_2$	Eut.	2d	399
			Confirmed eutectic	Eut.	3b	400
$RbCl_2$	466	41.7	$RbCoCl_3$,$CoCl_2$	Eut.	2d	393
	485	50	$RbCoCl_3$	Comp.		
	464	57.3	Rb_2CoCl_4,$RbCoCl_3$	Eut.		
	538	66.7	Rb_2CoCl_4	Comp.		
	503	72	Rb_3CoCl_5,Rb_2CoCl_4	Eut.		
	506	75	RbCl,Rb_3.$CoCl_5$	Per.		
	495-520		Rb_2CoCl_4	P.T.		
$TaCl_5$		100	Eut.close to pure $TaCl_5$	Eut.	2d,1	345
TlCl	422	48	$TlCoCl_3$,$CoCl_2$	Eut.	1,2a,3b	401
	430	50	$TlCoCl_3$	Comp.		
	346	68.5	Tl_2CoCl_4,$TlCoCl_3$	Per.		
	328	78.5	TlCl,Tl_2CoCl_4	Eut.		
	430	100	TlCl	M.P.		
			$TlCoCl_3$ - Lower symmetry then hexagonal or cubic			
	278		$TlCoCl_3$	P.T.		

CoF_2

2nd Component	T°C	Conc. mol %	Phases involved	Def.	Method	Ref.
KF	945	24	KF.$2CoF_2$,CoF_2	Eut.		402
	1042	34	KF.$2CoF_2$	Comp.		
	1009	43	KF.CoF_2,KF.$2CoF_2$	Eut.		
	1032	50	KF.CoF_2	Comp.		
	848	62.5	2KF.CoF_2,KF.CoF_2	Eut.		
	858	66.7	2KF.CoF_2	Comp.		
	750	81.0	13.7%CoF_2 sol.sln. in KF,2KF.CoF_2	Eut.		
	823		2KF.CoF_2 $\alpha \rightleftharpoons \beta$	P.T.		
	583		2KF.CoF_2 $\beta \rightleftharpoons \gamma$	P.T.		

CoF_2

2nd Component	T°C	Conc. mol %	Phases involved	Def.	Method	Ref.
SnF_2			Congruent melting at SnF_2:CoF_2 ratios: 14:1, 8:1, 3:2, 4:1, 2:1 Anions present: $Sn_2F_5^-$ & SnF_3^-		2c	403

CoI_2

2nd Component	T°C	Conc. mol %	Phases involved	Def.	Method	Ref.
InI_2			Fig. 73			364
	520	0	CoI_2	M.P.	2d	364
	200	65	$InI_2.2CoI_2$,CoI_2	Per.		
	163	85	InI_2,$InI_2.2CoI_2$	Eut.		
	226	100	InI_2	M.P.		
	161	100	InI_2	P.T.		
InI_3	520	0	CoI_2	M.P.	2d	404
	180	~94	CoI_2,InI_3	Eut.		
	208	100	InI_3	M.P.		
	140		$2InI_3.CoI_2$ (solid)	Comp.		

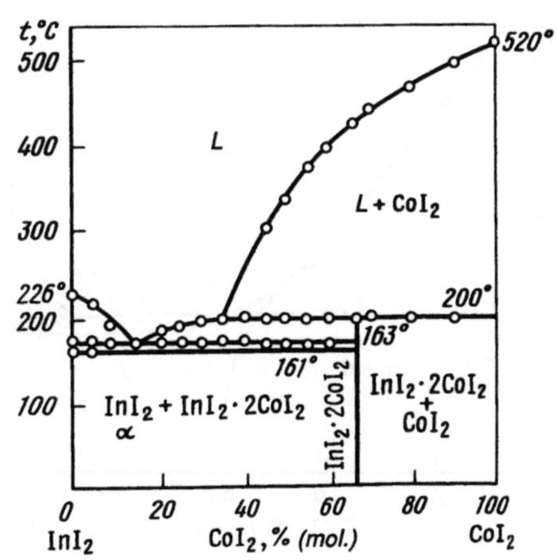

Fig. 73. InI_2-CoI_2 system.

$CrCl_2$

2nd Component	T°C	Conc. mol %	Phases involved	Def.	Method	Ref.
CsCl	582	28	$CsCrCl_3$,$CrCl_3$	Eut.	2d	405
	718	50	$CsCrCl_3$	Comp.		
	565	~67	Cs_2CrCl_4,$CsCrCl_3$	Per.		
	518	80.5	CsCl,Cs_2CrCl_4	Eut.		
	468	100	CsCl	P.T.		
	577	28.5	$CsCrCl_3$,$CrCl_2$	Eut.	2d,1f	406
	709	50	$CsCrCl_3$	Comp.		
	563	69.5	Cs_2CrCl_4,$CsCrCl_3$	Per.		
	515	80.5	α-CsCl,Cs_2CrCl_4	Eut.		
	454	100	CsCl	P.T.		
	577	28.5	$CsCrCl_3$,$CrCl_2$	Eut.	2d,3b	406a
	709	50	$CsCrCl_3$	Comp.		
	563	69.5	Cs_2CrCl_4,$CsCrCl_3$	Per.		
	515	80.5	α-CsCl,Cs_2CrCl_4	Eut.		
	454	100	CsCl	P.T.		
$FeCl_2$			Fig. 74			407
	680±1	61	10-12% $FeCl_2$ sol.sln. in $CrCl_2$; 44-47% $CrCl_2$ sol.sln. in $FeCl_2$		2a,3b,1e	407a
			Continuous sol.slns. Max. crystlln.interval 80°C		2d	408

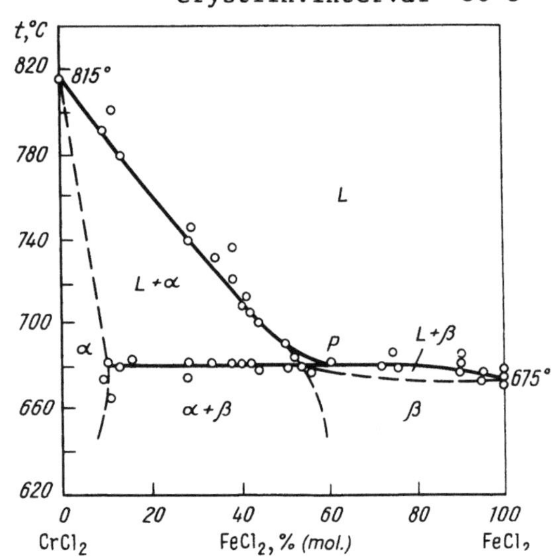

Fig. 74. $CrCl_2$-$FeCl_2$ system.

$CrCl_2$

2nd Component	T°C	Conc. mol %	Phases involved	Def.	Method	Ref.
KCl	472	36	$KCrCl_3$,$CrCl_2$	Eut.	2d	405
	495	50	$KCrCl_3$	Comp.		
	467	61	K_2CrCl_4,$KCrCl_3$	Eut.		
	481	66.7	K_2CrCl_4	Comp.		
	475	70	KCl,K_2CrCl_4	Eut.		
	466	39	KCl.$CrCl_2$,$CrCl_2$	Eut.	2d	409
	501	50	KCl.$CrCl_2$	Comp.		
	464	60.5	2KCl.$CrCl_2$, KCl.$CrCl_2$	Eut.		
	490	67	2KCl.$CrCl_2$	Comp.		
	473	69.7	KCl,2KCl.$CrCl_2$	Eut.		
	474	40	$KCrCl_3$,$CrCl_2$	Eut.	2d,3b	405
	494	50	$KCrCl_3$,$CrCl_2$	Comp.		
	462	58.5	K_2CrCl_4,$KCrCl_3$	Eut.		
	478	66.7	K_2CrCl_4	Comp.		
	470	66.7*	KCl,K_2CrCl_4 *According to drawing 31%	Eut.		
LiCl	551	60.7	Sol.sln.LiCl-$CrCl_2$ 4% LiCl sol.sln. in $CrCl_2$	Eut.	2d,3b	405
	290		$CrCl_2$ + LiCl sol.sln. ⇌ Li_4CrCl_6 or Li_3CrCl_5			
$MgCl_2$	703-712.5	52.63	Fig. 75 Solid solutions	Eut.	2a,3b,1e	406 407
$MnCl_2$			Fig. 76			407
	660±1	78	41-43% $CrCl_2$ sol. sln. in $MnCl_2$, 11-13% $MnCl_2$ sol.sln.in $CrCl_2$	Per.	2a,1e,3b 6c	407
NaCl	458		Na_3CrCl_5	Comp.	2	410
	437	53.7	$CrCl_2$.Na_3CrCl_5	Eut.		
	437	53.7	Na_3CrCl_5, $CrCl_2$	Eut.	2d,3b	405
	458	59.0	NaCl,Na_3CrCl_5	Per.		
	448	54	4NaCl.$CrCl_2$,$CrCl_2$	Eut.	2	411
	468	62	NaCl,4NaCl.$CrCl_2$	Per.		

$CrCl_2$

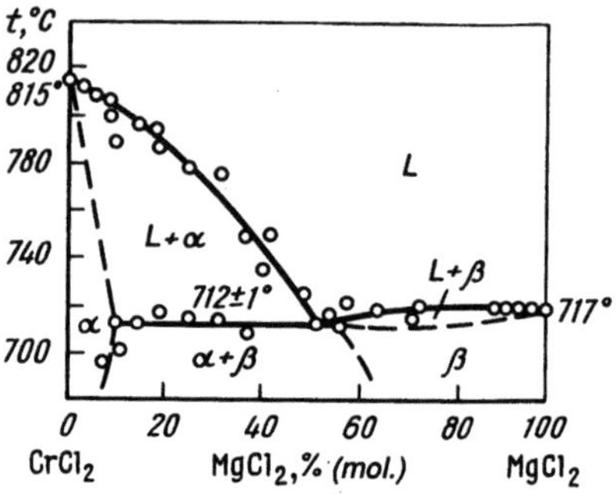

Fig. 75. $CrCl_2$-$MgCl_2$ system.

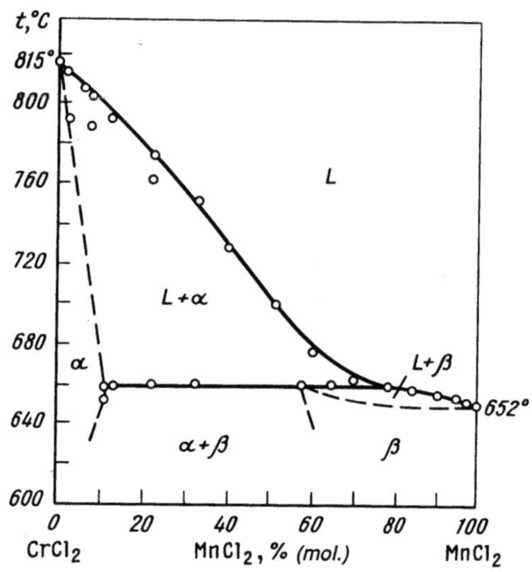

Fig. 76. $CrCl_2$-$MnCl_2$ system.

2nd Component	T°C	Conc. mol %	Phases involved	Def.	Method	Ref.
NaCl	440	57	Na_2CrCl_4,$CrCl_2$	Eut.	2a,3b	412
	450	59	NaCl,Na_2CrCl_4	Per.		
	435	53.5	Na_3CrCl_5,$CrCl_2$	Eut.	2a	413
	460	60	NaCl,Na_3CrCl_5	Per.		
	435	46.5	Na_3CrCl_5,$CrCl_2$	Eut.	2c,3b,	406
	460	59	NaCl,Na_3CrCl_5	Per.	le	407
			2NaCl.$CrCl_2$ (Confirmed)	Comp.	2d,le	414

$CrCl_2$

2nd Component	T°C	Conc. mol %	Phases involved	Def.	Method	Ref.
RbCl	525	34.2	$RbCrCl_3, CrCl_2$	Eut.	2d,3b	405
	606	50	$RbCrCl_3$	Comp.		
	545	63.2	$Rb_2CrCl_4, RbCrCl_3$	Eut.		
	551	66.7	Rb_2CrCl_4	Comp.		
	523	75.5	$RbCl, Rb_2CrCl_4$	Eut.		

$CrCl_3$

2nd Component	T°C	Conc. mol %	Phases involved	Def.	Method	Ref.
CsCl	805	47	$Cs_3Cr_2Cl_9, CrCl_3$	Eut.	2a	415
	894	60	$Cs_3Cr_2Cl_9$	Comp.		
	816	71	Cs_3CrCl_6, $Cs_3Cr_2Cl_9$	Eut.		
	838	75	Cs_3CrCl_6	Comp.		
	623	94.5	α-CsCl, Cs_3CrCl_6	Eut.		
	642	100	α-CsCl	M.P.		
	442	100	CsCl	P.T.		
$GaCl_3$	77.5		Solubility of $CrCl_3$ < 0.2%	Eut.	2	395
KCl			Fig. 77			416
	768	66.4	$3KCl.CrCl, CrCl_3$	Eut.	2,3b	416
	835	75	$3KCl.CrCl_3$	Comp.		
	692	88.8	$KCl, 3KCl.CrCl_3$	Eut.		

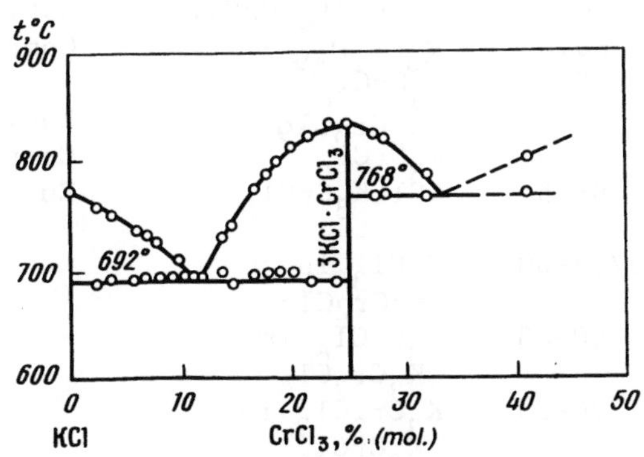

Fig. 77. Part of $KCl-CrCl_3$ system.

$CrCl_3$

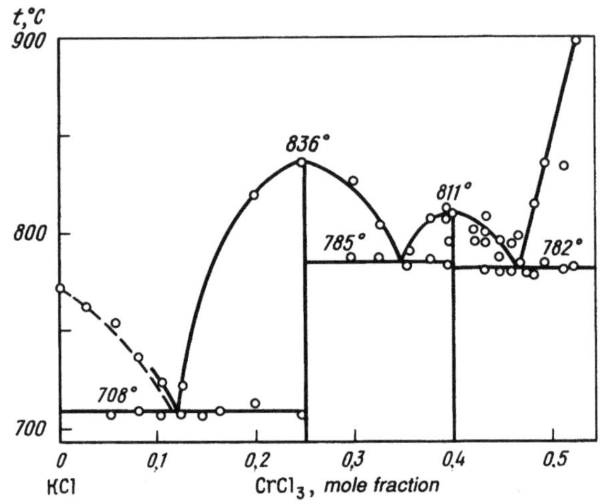

Fig. 78. Part of KCl-$CrCl_3$ system.

2nd Component	T°C	Conc. mol %	Phases involved	Def.	Method	Ref.
KCl	795	53.6	K_3CrCl_6,$CrCl_3$	Eut.	2c	417
	840	75	K_3CrCl_6	Comp.		
	700	89.3	KCl,K_3CrCl_6	Eut.		
			Fig. 78			418
	782		$K_2Cr_3Cl_9$,$CrCl_3$	Eut.	2b,2c,3a,	
	811		$K_3Cr_2Cl_9$	Comp.	3f	
	785		K_3CrCl_6,$K_3Cr_2Cl_9$	Eut.		
	836		K_3CrCl_6	Comp.		
	708		KCl,K_3CrCl_6	Eut.		
	780	54.0	$K_3Cr_2Cl_9$,$CrCl_3$	Eut.	2a	419
	813	60.0	$K_3Cr_2Cl_9$	Comp.		
	783	65.7	K_3CrCl_6, $K_2Cr_3Cl_9$	Eut.		420
	850	75.0	K_3CrCl_6	Comp.		
	700	88.5	KCl,K_3CrCl_6	Eut.		
		57.2-60	$CrCl_3$ in $K_3Cr_2Cl_9$		3b	421
		60-63	K_3CrCl_6 in $K_3Cr_2Cl_9$			
		70-75	$K_3Cr_2Cl_9$ in K_3CrCl_6			
		75-75.3	KCl in K_3CrCl_6			

$CrCl_3$

2nd Component	T°C	Conc. mol %	Phases involved	Def.	Method	Ref.
KCl	550	25.0	$CrCl_3.KCl, CrCl_3$	Eut.	2d,3b	422
	750	45.0	$CrCl_3.3KCl$, $CrCl_3.KCl$	Per.		
	805	75	$CrCl_3.3KCl$	Comp.		
	725	97.5	$KCl, CrCl_3.3KCl$	Eut.		
LiCl	574	~86-86.5	$Li_3CrCl_6, CrCl_3$	Per.	2a	419
	538	87	LiCl sol.sln., Li_3CrCl_6	Eut.		
		>87	LiCl sol.sln.			
	595	81-87	Decomposition			
$MgCl_2$	714	100	MgCl	M.P.	2c	417
	714		Components virtually immiscible	Eut.		
NaCl	570	68.6	$Na_3CrCl_6, CrCl_3$	Eut.	2c	417
	603	75	Na_3CrCl_6	Comp.		
	539	78.8	$NaCl, Na_3CrCl_6$	Eut.		
			In the vicinity of Na_3CrCl_6 viscosity increases			
	440		Na_3CrCl_6	P.T.		
			Fig. 79			418
	580		Na_3CrCl_6, $CrCl_3$	Eut.	2b,2c,3a, 2f	
	590		α-Na_3CrCl_6, β-Na_3CrCl_6	Per.		
	620		Na_3CrCl_6	Comp.		
	597		$NaCl, Na_3CrCl_6$	Eut.		
	582	69.0	$Na_3CrCl_6, CrCl_3$	Eut.	2a	419
	610	75.0	Na_3CrCl_6	Comp.		
	593	78.5	$NaCl, Na_3CrCl_6$	Eut.		
	562		Na_3CrCl_6	P.T.		
		70.6-75	$CrCl_3$ sol.slns. in Na_3CrCl_6		423 3b	
		75-75.7	NaCl sol.slns. in Na_3CrCl_6			

$CrCl_3$

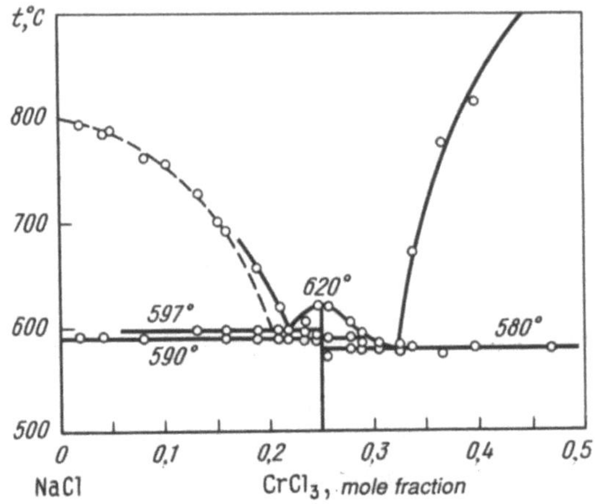

Fig. 79. Part of $NaCl-CrCl_3$ system.

2nd Component	T°C	Conc. mol %	Phases involved	Def.	Method	Ref.
NaCl	~450	37.5	$CrCl_3.NaCl$, $CrCl_3$	Eut.	2d,3b	422
	475	42.5	$CrCl_3.3NaCl$, $CrCl_3.NaCl$	Per.		
	575	42.5	$NaCl, CrCl_3.3NaCl$	Per.		
$NbCl_5$			$CrCl_3$ and $NbCl_5$ do not interact		2a	424
RbCl	821	50.2	$Rb_3Cr_2Cl_9, CrCl_3$	Eut.	2a	415
	874	60	$Rb_3Cr_2Cl_9$	Comp.		
	832	69.3	Rb_3CrCl_6, $Rb_3Cr_2Cl_9$	Eut.		
	868	75	Rb_3CrCl_6	Comp.		
	681	92	$RbCl, Rb_3CrCl_6$	Eut.		
	717	100	RbCl	M.P.		
	803	50.1	$CrCl_3$, $Rb_3Cr_2Cl_9$	Eut.	2a	425
	847	60	$Rb_3Cr_2Cl_9$	Comp.		
	808	69.3	$Rb_3Cr_2Cl_9$, Rb_3CrCl_6	Eut.		
	834	75	Rb_3CrCl_6	Comp.		
	680	92	$Rb_3CrCl_6, RbCl$	Eut.		

$CrCl_3$

2nd Component	T°C	Conc mol %	Phases involved	Def.	Method	Ref.
$SbCl_5$	3	100	$SbCl_5$	M.P.	2d	426
	~3	~100	$CrCl_3$,$SbCl_5$	Eut.		
$TaCl_5$			$TaCl_5$ & $CrCl_3$ do not interact		2a	427

CrF_2

2nd Component	T°C	Conc mol %	Phases involved	Def.	Method	Ref.
CrF_3	831±5	14	CrF_2 based sol.sln. + a "compound"	Eut.	2,1f,3b	424
	997±5	29	a "compound" + 10% CrF_2 sol.sln. in CrF_3			

CrF_3

2nd Component	T°C	Conc mol %	Phases involved	Def.	Method	Ref.
CsF	1080	46	CrF_3,$Cs_2Cr_5F_{17}$	Per.	2d,3b	428
	936	50	$Cs_2Cr_5F_{17}$,$CsCrF_4$	Per.		
	808	61	$CsCrF_4$,Cs_3CrF_6	Eut.		
	1162		Cs_3CrF_6	Comp.		
	684	98	Cs_3CrF_6,CsF	Eut.		
KF	1115		CrF_3,$K_2Cr_5F_{17}$	Per.	2d,3b	429
	925	50	$K_2Cr_5F_{17}$,α-$KCrF_4$	Per.		
	833.5	54	α-$KCrF_4$,K_2CrF_5	Eut.		
	855	58	K_2CrF_5,α-K_3CrF_6	Per.		
	1270	75	α-K_3CrF_6	Comp.		
	840	94	α-K_3CrF_6,KF	Eut.		
	222		$\alpha \rightleftharpoons \beta$ K_3CrF_6	P.T.		
	173		$\beta \rightleftharpoons \gamma$ K_3CrF_6	P.T.		
	786		$\alpha \rightleftharpoons \beta$ $KCrF_4$	P.T.		
	600		K_xCrF_3 in 3 forms: α-hexagonal, β'-hexagonal γ-cubic		3b	430
	850		K_xCrF_3(with CrF_3: $KCrF_3$ of 1:1 ratio)	M.P.		

CrF_3

2nd Component	T°C	Conc. mol %	Phases involved	Def.	Method	Ref.
LiF	790		CrF_3,Li_3CrF_6	Eut.	2d,3b	429
	852		Li_3CrF_6	Comp.		
	740		Li_3CrF_6,LiF	Eut.		
NaF	967		CrF_3,$NaCrF_4$	Per.	3d,3b	429
	860		$NaCrF_4$, $Na_5Cr_3F_{14}$	Eut.		
	872	62.5	$Na_5Cr_3F_{14}$, α-Na_3CrF_6	Per.		
	1140	75	α-Na_3CrF_6	Comp.		
	900		α-Na_3CrF_6,NaF	Eut.		
	650		Na_3CrF_6	P.T.		
PbF_2	672	60.0	CrF_3, $Pb(CrF_6)_2$	Per.		356
	561	74.0	$Pb(CrF_6)_2$, Pb_4CrF_{11}	Eut.		
	592	76.5	Pb_4CrF_{11}, 10% CrF_3 sol. sln. in PbF_2	Per.		
RbF	1135	43	CrF_3, $Rb_2Cr_5F_{17}$	Per.	2d,3b	431
	978	50	$Rb_2Cr_5F_{17}$, $RbCrF_4$	Per.		
	906	56	$RbCrF_4$, $Rb_7Cr_6F_{25}$	Per.		
	879	59	$Rb_7Cr_6F_{25}$, α-Rb_3CrF_6	Eut.		
	1255	75	α-Rb_3CrF_6	Comp.		
	777	97	α-Rb_3CrF_6,RbF	Eut.		
	410		Rb_3CrF_6	P.T.		
TlF	871	51	CrF_3,$TlCrF_4$	Per.	2d,3b	431
	693	61.5	$TlCrF_4$, α-Tl_3CrF_6	Eut.		
	785	75	α-Tl_3CrF_6	Comp.		
	321	98	α-Tl_3CrF_6,TlF	Eut.		
	212		Tl_3CrF_6	P.T.		

CsBr

2nd Component	T°C	Conc. mol %	Phases involved	Def.	Method	Ref.
$GaBr_3$	320		$CsGaBr_4$,CsBr	Per.	1,2a,3b	432
	170		$CsGa_2Br_7$,$CsGaBr_4$	Per.	1f	
	95	80	$GaBr_3$,$CsGa_2Br_7$	Eut.		
$InBr_3$			Fig. 80			433
	444	36	α-Cs_3InBr_6,CsBr	Per.	1,2a,3b	433
	386	42	β-Cs_3InBr_6, α-Cs_3InBr_6	Per.		
	346		α-Cs_2InBr_5, β-Cs_3InBr_6	Per.		
	334		β-Cs_2InBr_5, α-Cs_2InBr_5	Per.		
	290	60	$InBr_3$,β-Cs_2InBr_5	Eut.		
	436	100	$InBr_3$	M.P.		
			$Cs_3InBr_6.H_2O$ and $Cs_2InBr_5.H_2O$ crystallize out of aqueous sln.			

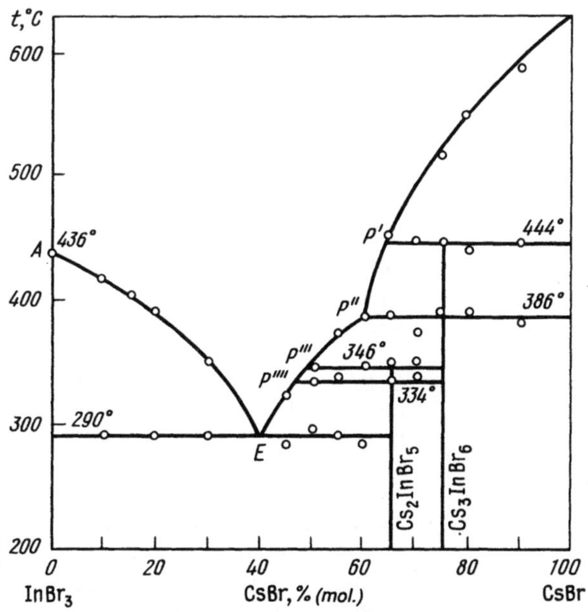

Fig. 80. $InBr_3$-CsBr system.

CsBr

2nd Component	T°C	Conc. mol %	Phases involved	Def.	Method	Ref.
KBr	640	0	CsBr	M.P.	1	360
	586	35	Continuous sol.sln.	Min.		
	636	0	CsBr	M.P.	2b,3a	434
	564	43	Continuous sol.sln.	Min.		
	362		Decomposition of sol.sln.			
	728	100	KBr	M.P.		
	627	0	CsBr	M.P.	1	435
			Continuous sol.sln.			
	732	100	KBr	M.P.		
LiBr	636	0	CsBr	M.P.	2	434
	300	55	CsBr.LiBr,CsBr	Per.		
	262	62.5	LiBr,CsBr.LiBr	Eut.		
	550	100	LiBr	M.P.		
	230	51.5	CsBr,LiBr	Eut.	1	436
NH_4Br			Continuous sol.sln.		6h,3b	437
NaBr	630	0	CsBr	M.P.	2a	438
	460	37.5	CsBr,NaBr	Eut.		
	740	100	NaBr	M.P.		
	482	43	CsBr,NaBr	Eut.	1	439
	647	0	CsBr	M.P.	1,2b,1e	440
	466	42	CsBr,NaBr	Eut.		
	742	100	NaBr	M.P.		
	627	0	CsBr	M.P.	1	441
	484	43	CsBr,NaBr	Eut.		
	747	100	NaBr	M.P.		
$PbBr_2$	490	49	$PbBr_2$.4CsBr,CsBr	Per.	1	442
	465	65	$PbBr_2$.CsBr, $PbBr_2$.4CsBr	Eut.		
	323	95.3	$PbBr_2$,$PbBr_2$.CsBr	Eut.		
	373	0	$PbBr_2$	M.P.		

CsBr

2nd Component	T°C	Conc. mol %	Phases involved	Def.	Method	Ref.
RbBr	636	0	CsBr	M.P.	2b	434
	594	15	Continuous sol.sln.	Min.		
	688	100	RbBr	M.P.		
			Fig. 81			443
	550	20	77% CsBr sol.sln. in RbBr, 15% RbBr sol.sln. in CsBr	Eut.	2d,3b	443
	627	0	CsBr	M.P.	1	441
	627		Continuous sol.sln.	Min.		444
	690	100	RbBr	M.P.		
$SbBr_3$			Fig. 82			445
	520	15	β-3CsBr.2$SbBr_3$, CsBr	Eut.	1,2,3b	445
	625	40	β-3CsBr.2$SbBr_3$	Comp.		
	530		α-3CsBr.2$SbBr_3$, β-3CsBr.2$SbBr_3$	Per.		
	98	~100	$SbBr_3$, α-3CsBr.2$SbBr_3$	Eut.		
			Many phase transformations			

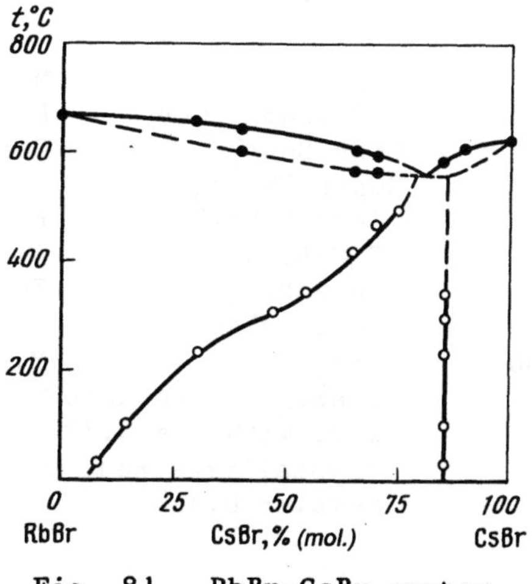

Fig. 81. RbBr-CsBr system.

CsBr

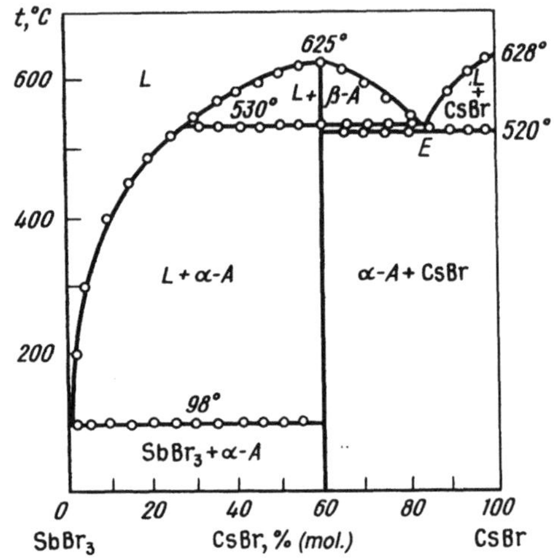

Fig. 82. $SbBr_3$-CsBr system.

2nd Component	T°C	Conc. mol %	Phases involved	Def.	Method	Ref.
$ScBr_3$	596	8.5	CsBr, Cs_3ScBr_6	Eut.		446
	757	25	Cs_3ScBr_6	Comp.		
	706	35.6	Cs_3ScBr_6, $Cs_3Sc_2Br_9$	Eut.		
	728	40	$Cs_3Sc_2Br_9$	Comp.		
	632	54.0	$Cs_3Sc_2Br_9$, $ScBr_3$	Eut.		
$SnBr_2$	640	0	CsBr	M.P.	1,2a,3a	447
	370	36	$CsBr.SnBr_2$,CsBr	Eut.		
	440	50	$CsBr.SnBr_2$	Comp.		
	272	75	$2CsBr.3SnBr_2$, $CsBr.SnBr_2$	Per.		
	190	78	7% CsBr sol.sln. in $SnBr_2$, $2CsBr.3SnBr_2$	Eut.		
	226	100	$SnBr_2$	M.P.		
			$CsSnBr_2$ - Perovskite str. with a = 5.79 ± 0.5Å crystallizes red, room temp. - black.			

CsBr

2nd Component	T°C	Conc. mol %	Phases involved	Def.	Method	Ref.
$SrBr_2$	526	18.5	$CsBr.SrBr_2$,CsBr	Eut.	1	448
	750		$CsBr.SrBr_2$	Comp.		
	572	86.5	$CsBr.SrBr_2$,$SrBr_2$	Eut.		
	539	20	$SrBr_2.CsBr$,CsBr	Eut.	1,2d	449
	759	50	$SrBr_2.CsBr$	Comp.		
	578	86	SBr_2,$SrBr_2.CsBr$	Eut.		
$TeBr_4$			Fig. 83			450
	636	0.0	CsBr	M.P.	1,3b,1e	450
	564	13.0	CsBr,Cs_2TeBr_6	Eut.		
	742	33.3	Cs_2TeBr_6	Comp.		
	344	92.0	Cs_2TeBr_6,$TeBr_4$	Eut.		
	386	100.0	$TeBr_4$	M.P.		
$TiBr_2$			Fig. 84			451
	580	12.5	CsBr,$CsTiBr_3$	Eut.	2a	451
	885	50.0	$CsTiBr_3$	Comp.		
	843	65.0	$CsTiBr_3$,$TiBr_2$	Eut.		
$TiBr_3$	584	10.0	CsBr,Cs_3TiBr_6	Eut.	2a	451
	670	25.0	Cs_3TiBr_6	Comp.		
	658	27.0	Cs_3TiBr_6, $Cs_3Ti_2Br_9$	Eut.		
	697	40.0	$Cs_3Ti_2Br_9$	Comp.		
	593	75.0	$Cs_3Ti_2Br_9$, $TiBr_3$	Eut.		
$TiBr_4$	608	$\leqslant 3$	CsBr,Cs_3TiBr_6	Eut.	2a	451
	720		Cs_3TiBr_6	Comp.		
TlBr	410	72.5	Continuous sol.sln.	Min.	1	452
	425		Continuous sol.sln.	Min.	1	315

CsBr

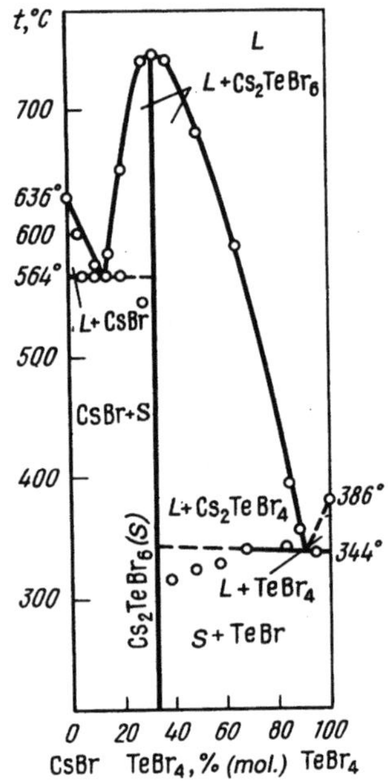

Fig. 83. CsBr–$TeBr_4$ system.

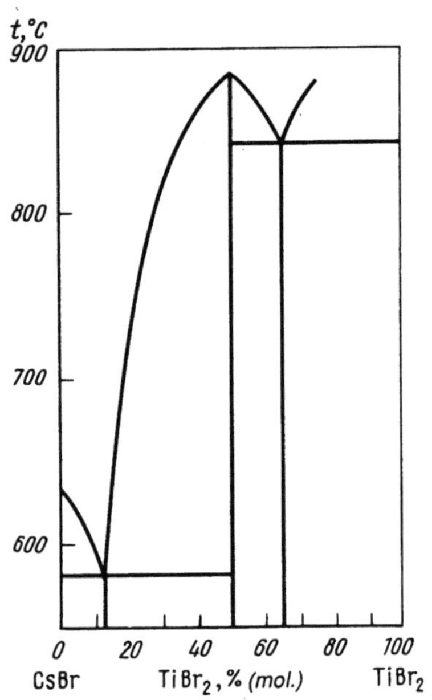

Fig. 84. CsBr–$TiBr_2$ system.

CsBr

2nd Component	T°C	Conc. mol %	Phases involved	Def.	Method	Ref.
CsCl	613	42.5	Continuous sol.sln.	Min.	1	312
			0.4% deviation in density Surface tension isotherm curves towards composition axis.		6i,6j	453
	20-470		Solid solution		3b	454
	627	0	CsBr	M.P.	1,2a	455
	646	100	CsCl	M.P.		456
			Continuous sol.sln.			
			Continuous sol.sln. with a minimum		1 2d,3b	457
CsF	632	0.0	CsBr	M.P.	1,2d	458
	438	48.5	CsF,CsBr			
CsI	578	48	Continuous sol.sln.	Min.	1	459
	640	100	CsI	M.P.		
	250 ± 5	42	Continuous sol.sln. CsBr and CsI separation of sol. sln.	P.T.	3a	460
	800		0.4% deviation in density Surface tension isotherm curves towards composition axis.		6i,6j	453

CsCl

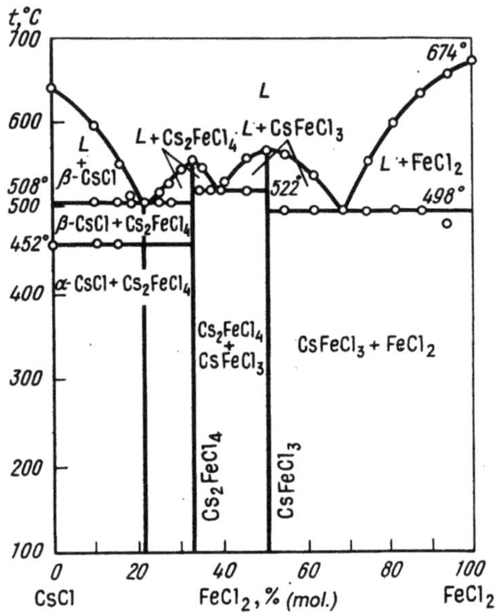

Fig. 85. CsCl-$FeCl_2$ system.

2nd Component	T°C	Conc. mol %	Phases involved	Def.	Method	Ref.
$FeCl_2$			Fig. 85			461
	508	21.4	β-CsCl,Cs_2FeCl_4	Eut.	2c	461
	556	33.3	Cs_2FeCl_4	Comp.		
	522	38.0	Cs_2FeCl_4, $CsFeCl_3$	Eut.		
	558	50	$CsFeCl_3$	Comp.		
	498	69.3	$CsFeCl_3$,$FeCl_2$	Eut.		
	452	0	CsCl	P.T.		
	508	21	CsCl,Cs_3FeCl_5	Eut.	2d,3b	462
	524	25	Cs_3FeCl_5	Comp.		
	522	26.5	Cs_3FeCl_5, Cs_2FeCl_4	Eut.		
	556	33.3	Cs_2FeCl_4	Comp.		
	524	38.5	Cs_2FeCl_4, $CsFeCl_3$	Eut.		
	565	50	$CsFeCl_3$	Comp.		
	499	69	$CsFeCl_3$,$FeCl_2$	Eut.		

CsCl

2nd Component	T°C	Conc. mol %	Phases involved	Def.	Method	Ref.
$GaCl_3$	649	0.0	β-CsCl	M.P.	2a	463
	486		α-CsCl,β-CsCl	Per.		
	378	42	$CsGaCl_4$,α-CsCl	Eut.		
	420	49.4	$CsGaCl_4$	Comp.		
	95	~70	$CsGa_2Cl_7$, $CsGaCl_4$	Per.		
	55	97	$GaCl_3$,$CsGa_2Cl_7$	Eut.		
	52		$CsGaCl_4$	P.T.		
	460	37	α-CsCl,β-CsCl	Per.	2d	464
	350	42.5	β-CsCl,CsCl.$GaCl_3$	Eut.		
	365	50	CsCl.$GaCl_3$	Comp.		
	115	75	CsCl.$GaCl_3$, CsCl.2$GaCl_3$	Per.		
	76	98	CsCl.2$GaCl_3$, $GaCl_3$	Eut.		
	646	0	β-CsCl	M.P.	1,2d	465
	482		α-CsCl,β-CsCl	Per.		
	355	45	$CsGaCl_4$, 1.5% GaCl sol.sln.in α-CsCl	Eut.		
	385	50	$CsGaCl_4$	Comp.		
	105	74	$CsGa_2Cl_7$, $CsGaCl_4$	Per.		
	72	95.5	$GaCl_3$, $CsGa_2Cl_7$	Eut.		
	78	100	$GaCl_3$	M.P.		
$HfCl_4$	302	65.1	$HfCl_4$, Cs_2HfCl_6	Eut.	2	466
	820	33.4	Cs_2HfCl_6	Comp.		
	590	18.4	Cs_2HfCl_6,CsCl	Eut.		
	452		CsCl	P.T.		
$HgCl_2$	642	0	CsCl	M.P.	1,3a	401
	424	27.5	CsCl,Cs_2HgCl_4	Eut.		
	450	33.3	Cs_2HgCl_4	Comp.		
	392	46.5	Cs_2HgCl_4, $CsHgCl_3$	Eut.		
	404	50.0	$CsHgCl_3$	Comp.		
	250	73	$CsHgCl_3$, $CsHg_3Cl_7$	Per.		
	216	84.5	$CsHg_3Cl_7$,$HgCl_2$	Eut.		
	276	100	$HgCl_2$	M.P.		
			$CsHgCl_3$ - perovskite structure cubic parameter a = 5.323Å			

CsCl

2nd Component	T°C	Conc. mol %	Phases involved	Def.	Method	Ref.
$InCl_3$	584	10.6	β-CsCl, β-3CsCl.$InCl_3$	Eut.	2d	464
	632	25	β-3CsCl.$InCl_3$	Comp.		
	428	46.8	β-3CsCl.$InCl_3$, 2CsCl.$InCl_3$	Per.		
	390	51	2CsCl.$InCl_3$, β-CsCl.$InCl_3$	Per.		
	366	55.5	β-CsCl.$InCl_3$, $InCl_3$	Eut.		
	450		CsCl: β ⇌ α	P.T.		
	249		3CsCl.$InCl_3$ β ⇌ α	P.T.		
	176		CsCl.$InCl_3$: β ⇌ α	P.T.		
KCl	606	36	Cont.sol.sln.	Min.	1	467
			Sol. sln.		3b	437
	636	0	CsCl	M.P.	1,3b	
	606	34	Cont. sol.sln.	Min.		
	774	100	KCl	M.P.		

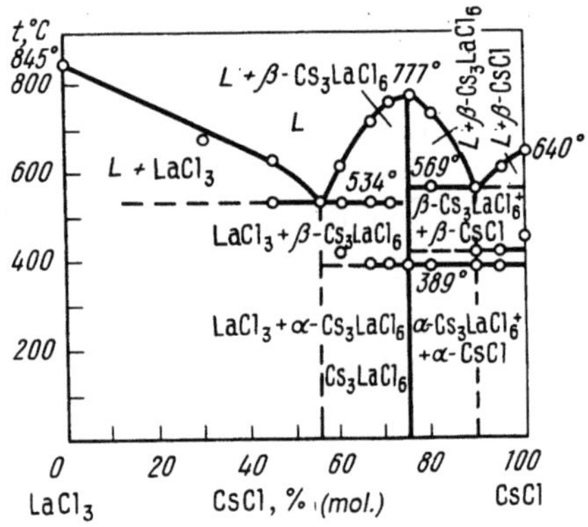

Fig. 86. $LaCl_3$-CsCl system.

CsCl

2nd Component	T°C	Conc. mol %	Phases involved	Def.	Method	Ref.
$LaCl_3$			Fig. 86			469
	569	10.0	β-Cs_3LaCl_6, β-CsCl	Eut.	2c	469
	777	25.0	β-Cs_3LaCl_6	Comp.		
	534	45.0	$LaCl_3$, β-Cs_3LaCl_6	Eut.		
	389		Cs_3LaCl_6	P.T.		
	420		CsCl	P.T.		
LiCl	306	59.5	LiCl,CsCl	Eut.	1	470
	636	0	α-CsCl	M.P.	1,2a	468
	472	30	β-CsCl,α-CsCl	Per.		471
	382	42.5	LiCl.2CsCl,β-CsCl	Per.		
	352	50	LiCl.CsCl, LiCl.2CsCl	Per.		
	332	58.5	β-LiCl,LiCl.CsCl	Eut.		
	566	87.5	α-LiCl,β-LiCl	Per.		
	606	100	α-LiCl	M.P.		
			LiCl.CsCl, LiCl.2CsCl } confirmed		3b	472
$MnCl_2$	499	20.5	$MnCl_2$.3CsCl,CsCl	Eut.	1	473
	511	25.0	$MnCl_2$.3CsCl	Comp.		
	507	27.0	$MnCl_2$.2CsCl, $MnCl_2$.3CsCl	Eut.		
	538	33.3	$MnCl_2$.2CsCl	Comp.		
	526	36.5	$MnCl_2$.CsCl, $MnCl_2$.2CsCl	Eut.		
	593	50.0	$MnCl_2$.CsCl	Comp.		
	517	69.0	CsCl.4$MnCl_2$, $MnCl_2$.CsCl	Eut.		
	537	~79.0	$MnCl_2$, CsCl.4$MnCl_2$	Per.		
$MoCl_5$		<33	2CsCl.$MoCl_4$,CsCl		2b,2c	474
		<33	Components immiscible			

CsCl

2nd Component	T°C	Conc. mol %	Phases involved	Def.	Method	Ref.
NaCl	646	0.0	CsCl	M.P.	2b	475
	490	35.0	CsCl,NaCl	Eut.		
	800	100	NaCl	M.P.		
	454		CsCl $\alpha \rightleftharpoons \beta$			
	493	34	CsCl,NaCl	Eut.	1	467
			Sol.slns.persist		3b,6c	476
	646	0	CsCl	M.P.	1	435
	494	34.5	CsCl,NaCl	Eut.		441
	800	100	NaCl	M.P.		
	640	0	CsCl	M.P.	1	477
	495	35.5	CsCl,NaCl	Eut.		
	800	100	NaCl	M.P.		
	646	0	CsCl	M.P.	1,2d	478
	494	36	CsCl,NaCl	Eut.		
	800	100	NaCl	M.P.		
$NbCl_2$	590	13.0	Cs_2NbCl_4,β-CsCl	Eut.	2a,3b	479
	725	33	Cs_2NbCl_4	Comp.		
	452		CsCl	P.T.		
$NbCl_3$	590	13.0	Cs_2NbCl_5,β-CsCl	Eut.	2a,3b	480
	762	33	Cs_2NbCl_5	Comp.		
	450		CsCl	P.T.		
$NbCl_4$	595	10	Cs_2NbCl_6,CsCl	Eut.	2b	481
	822		Cs_2NbCl_6	Comp.		
	282	57	$NbCl_4$,Cs_2NbCl_6	Eut.		
		>55	Dissociation (two anions: $NbCl_5^-$ & $NbCl_6^{2-}$			
$NdCl_3$			Fig. 87			469
	582	10	β-Cs_3NdCl_6,β-CsCl	Eut.	2c	469
	813	25	β-Cs_3NdCl_6	Comp.		
	495	50	$NdCl_3$, β-Cs_3NdCl_6	Eut.		
	395		Cs_3NdCl_6	P.T.		
	452		CsCl	P.T.		

CsCl

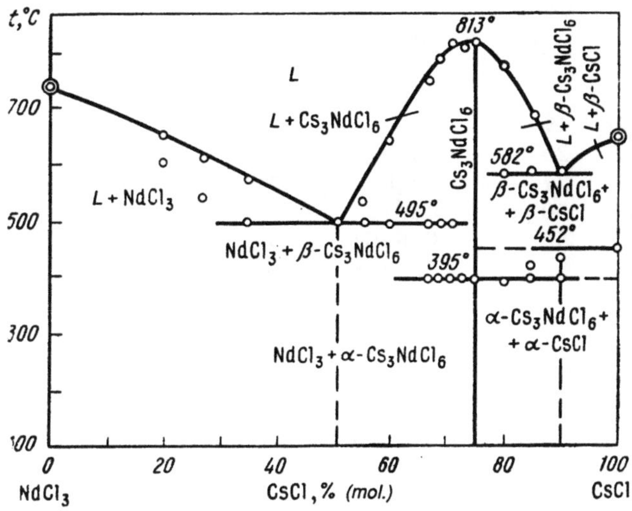

Fig. 87. $NdCl_3$-CsCl system.

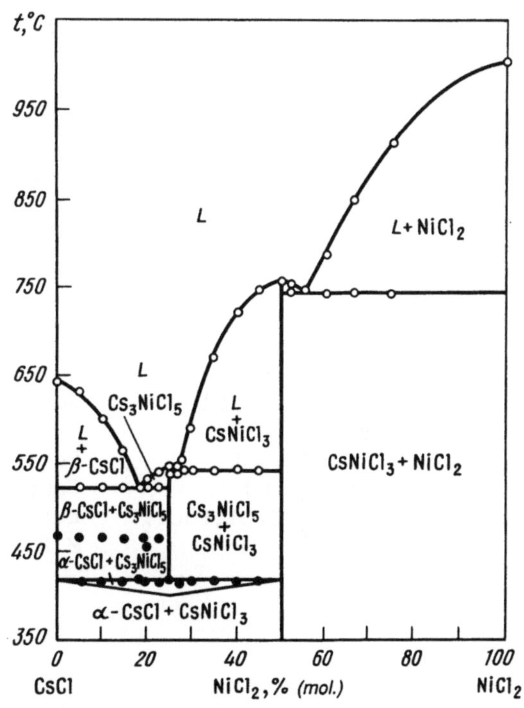

Fig. 88. CsCl-$NiCl_2$ system.

CsCl

2nd Component	T°C	Conc. mol %	Phases involved	Def.	Method	Ref.
$NiCl_2$			Fig. 88			482
	552	19	β-CsCl, Cs_3NiCl_5	Eut.	1,3b	482
	547	25	Cs_3NiCl_5	Comp.		
	541	27	Cs_3NiCl_5, $CsNiCl_3$	Eut.		
	758	50	$CsNiCl_3$	Comp.		
	747	55	$CsNiCl_3$, $NiCl_2$	Eut.		
	466		CsCl − α ⇌ β	P.T.		
	417		Cs_3NiCl_5 (Blue ⇌ yellow)	P.T.		
$PbCl_2$	638	0	α-CsCl	M.P.	1	483
	490	22.0	α-CsCl,β-CsCl	Per.		
	476	28.2	β-CsCl, $CsCl.2PbCl_2$	Eut.		
	605	51	$CsCl.2PbCl_2$	Comp.		
	426	81.8	$CsCl.2PbCl_2$, $PbCl_2$	Eut.		
	498	100	$PbCl_2$	M.P.		
	640	0	CsCl	M.P.		484
	493	25.8	CsCl, $PbCl_2.4CsCl$	Per.		
	488	29.5	$PbCl_2.4CsCl$, $PbCl_2.CsCl$	Eut.		
	615	50	$PbCl_2.CsCl$	Comp.		
	445	81.8	$PbCl_2.CsCl$, $PbCl_2$	Eut.		
	498	100	$PbCl_2$	M.P.		
	640		CsCl	M.P.		485
	493		CsCl, $PbCl_2.4CsCl$	Per.		
	480		$PbCl_2.4CsCl$, $PbCl_2.2CsCl$	Eut.		
	515		$PbCl_2.2CsCl$, $PbCl_2.CsCl$	Per.		
	615		$PbCl_2CsCl$	Comp.		
	440		$PbCl_2.CsCl$, $PbCl_2$	Eut.		
	496		$PbCl_2$	M.P.		

CsCl

2nd Component	T°C	Conc. mol %	Phases involved	Def.	Method	Ref.
$PuCl_3$			Fig. 89			486
	592	10	Cs_3PuCl_6,CsCl	Eut.	2,1f	486
	825±3	25	$CsPuCl_6$	Comp.		
	504	47	$CsPu_2Cl_7$,Cs_3PuCl_6	Eut.		
	616±3	66.6	$CsPu_2Cl_7$	Comp.		
	611	70	$PuCl_3$,$CsPu_2Cl_7$	Eut.		
	410		Cs_3PuCl_6	P.T.		
	465±2		CsCl	P.T.		
$PuCl_4$			Fig. 90			487
	640	0	CsCl	M.P.	2d,3b,5b	488
	510	15-16		Eut.		
			Cs_2PuCl_6	Comp.		
	465	2	CsCl $\alpha \rightleftharpoons \beta$	P.T.		
			Pu content affects acidity			
			Numerous compounds			
RbCl			Continuous sol.sln.		3b	489
			Fig. 91			490
			Continuous sol.sln.		3b,2a	491
			Fig. 91: liquidus B from L. J. Wood (Ref. 492); line A, the authors; dots, L. J. Wood & Krogh-Moc; circles, Refs. 492 & 493			
	336	7.5	Continuous sol.sln.	Min.	1,2d	494
	646	0.0	CsCl	M.P.	1,2a	456
	646		Continuous sol.sln.	Min.		455
	726	100	RbCl	M.P.		435
						495
	640	0	CsCl	M.P.	1	496
	635	15	Continuous sol.sln.	Min.		
	717	100	RbCl	M.P.		
$ReCl_3$			Fig. 92			497
	980		Cs_2ReCl_6	M.P.	2a	497
			Persistence of reaction: $6CsCl + 4ReCl_3 \rightarrow 3Cs_2ReCl_6 + Re$			
			Liquidus approaches c.p. of pure Re			

CsCl

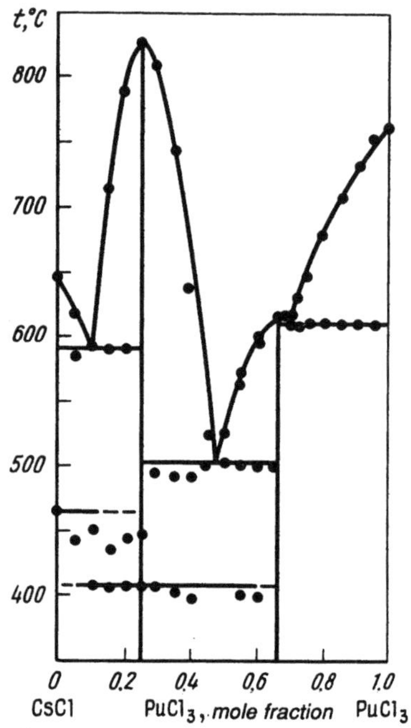

Fig. 89. CsCl-$PuCl_3$ system.

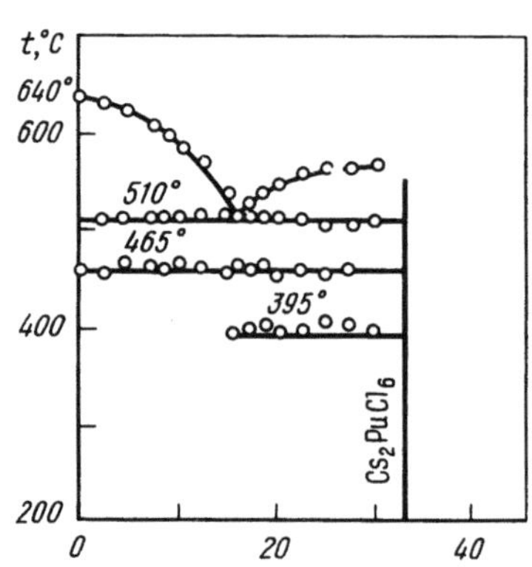

Fig. 90. CsCl-$PuCl_4$ system.

CsCl

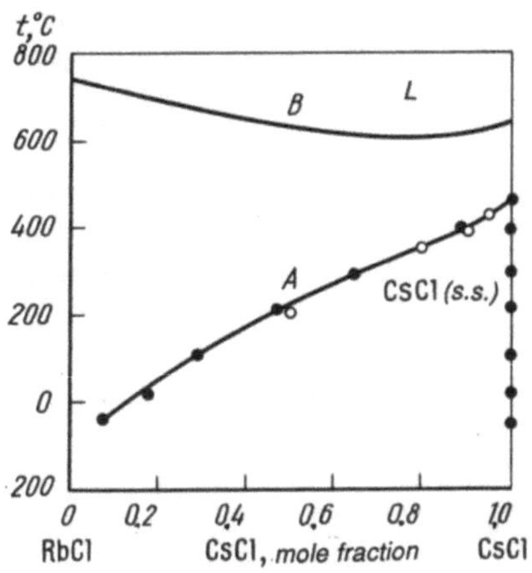

Fig. 91. RbCl-CsCl system.

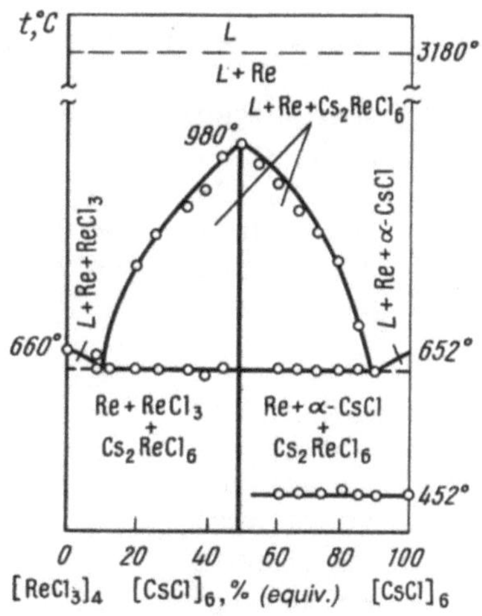

Fig. 92. CsCl-$ReCl_3$ system.

2nd Component	T°C	Conc. mol %	Phases involved	Def.	Method	Ref.
$SbCl_3$			Fig. 93			445
	520	17.5	γ-3CsCl.$SbCl_3$, β-CsCl	Eut.	1,2,3b	445
	550	24	γ-3CsCl.$SbCl_3$	Comp.		
	510	32.5	3CsCl.2$SbCl_3$, γ-3CsCl.$SbCl_3$(B)	Eut.		

CsCl

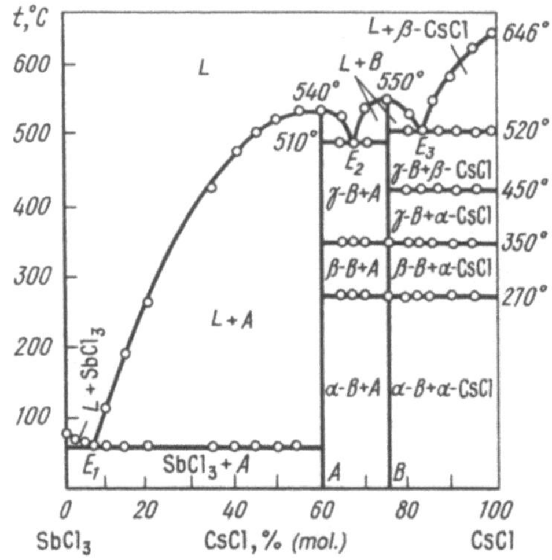

Fig. 93. $SbCl_3$-CsCl system.

2nd Component	T°C	Conc. mol %	Phases involved	Def.	Method	Ref.
$SbCl_3$	540	40	$3CsCl.2SbCl_3$	Comp.		445
	68	92.5	$SbCl_3$, $3CsCl.2SbCl_3$(A)	Eut.		
	270		$3CsCl.SbCl_3$ α ⇌ β	P.T.		
	350		$3CsCl.SbCl_3$ β ⇌ γ	P.T.		
	450		CsCl α ⇌ β	P.T.		
$SbCl_5$	638	0	β-CsCl	M.P.	2d	426
	480	30	β-CsCl, $2CsCl.SbCl_5$	Eut.		
	498		$2CsCl.SbCl_5$, β-$CsSbCl_6$	Per.		
	560	50	β-$CsSbCl_6$	Comp.		
	260		β-$CsSbCl_6$, α-$CsSbCl_6$	Per.		
	3		α-$CsSbCl_6$, $SbCl_5$	Eut.		
	3	100	$SbCl_5$	M.P.		
$ScCl_3$	613	6.5	CsCl, Cs_3ScCl_6	Eut.	2d	498
	814	25	Cs_3ScCl_6	Comp.		
	619	40	Cs_3ScCl_6, $Cs_3Sc_2Cl_9$	Per.		
	569	56.5	$Cs_3Sc_2Cl_9$, $ScCl_3$	Eut.		

CsCl

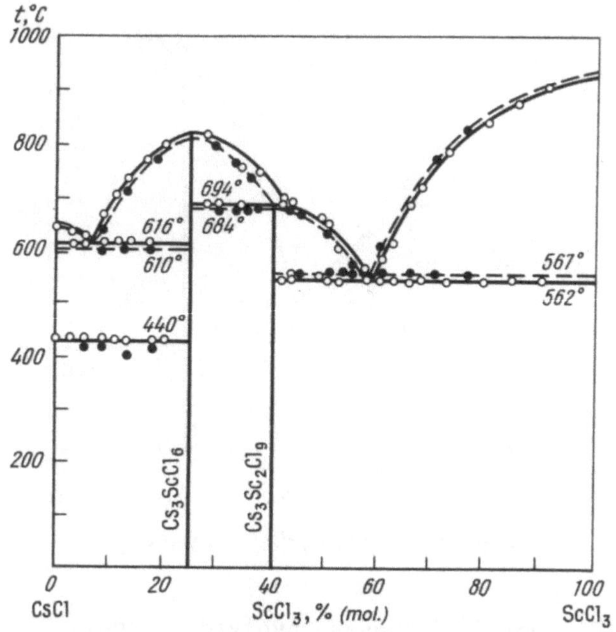

Fig. 94. $CsCl-ScCl_3$ system.

2nd Component	T°C	Conc. mol %	Phases involved	Def.	Method	Ref.
$ScCl_3$			Fig. 94			499
	616	7	$CsCl, 3CsCl.ScCl_3$	Eut.	2d,4a	500
	824	21	$3CsCl.ScCl_3$	Comp.		
	694	43.4	$3CsCl.ScCl_3$, $3CsCl.2ScCl_3$	Per.		
	562	58	$3CsCl.2ScCl_3, ScCl_3$	Eut.		
	440		CsCl	P.T.		
$SmCl_3$			Fig. 95			501
	600	8	β-Cs_3SmCl_6,β-CsCl	Eut.	2a	502
	886	25	β-Cs_3SmCl_6	Comp.		
	510	45	β-$CsSm_2Cl_7$, β-Cs_3SmCl_6	Eut.		
	660	67	β-$CsSm_2Cl_7$	Comp.		
	577	85	$SmCl_3$,β-$CsSm_2Cl_7$	Eut.		
	440		$CsSm_2Cl_7$ $\alpha \rightleftharpoons \beta$	P.T.		
	410		Cs_3SmCl_6 $\alpha \rightleftharpoons \beta$	P.T.		
	452		CsCl $\alpha \rightleftharpoons \beta$	P.T.		

CsCl

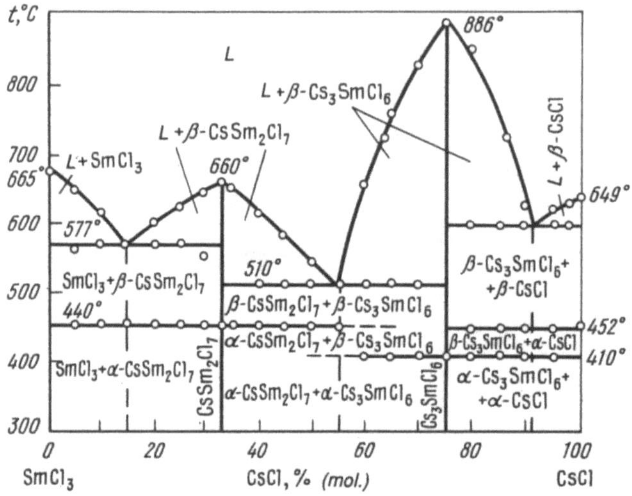

Fig. 95. $SmCl_3$-CsCl system.

2nd Component	T°C	Conc. mol %	Phases involved	Def.	Method	Ref.
$SnCl_2$	337	38.5	CsCl.$SnCl_2$,CsCl	Eut.	1	503
	370	50	CsCl.$SnCl_2$	Comp.		
	224	67.9	CsCl.2$SnCl_2$, CsCl.$SnCl_2$	Per.		
	188	83	$SnCl_2$,CsCl.2$SnCl_2$	Eut.		
			Fig. 96			504
	451	27	α-CsCl,β-CsCl	Per.	2,6c	504
	350	36	$SnCl_2$.CsCl, -CsCl	Eut.		

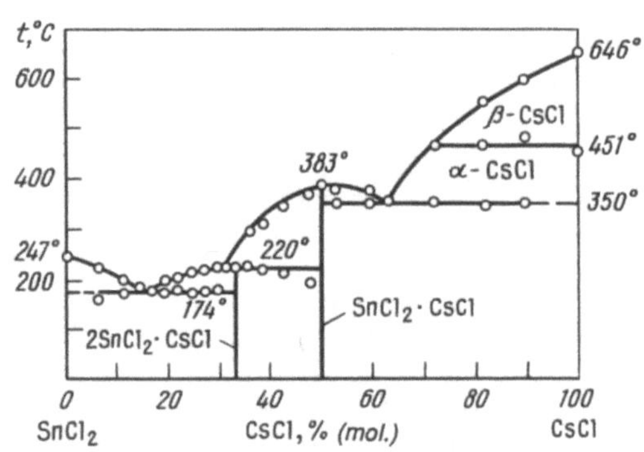

Fig. 96. $SnCl_2$-CsCl system.

CsCl

2nd Component	T°C	Conc. mol %	Phases involved	Def.	Method	Ref.
$SnCl_2$	383	50	$SnCl_2.CsCl$	Comp.		504
	220	70	$2SnCl_2.CsCl$, $SnCl_2.CsCl$	Per.		
	174	84	$SnCl_2$,$2SnCl_2.CsCl$	Eut.		
$SrCl_2$	560	26	CsCl,$CsCl.SrCl_2$	Eut.	1	505
	834		$CsCl.SrCl_2$	Comp.		
	720	89.5	$CsCl.SrCl_2$,$SrCl_2$	Eut.		
	550	25.0	$Cs_2Cl_2.2SrCl_2$,CsCl	Eut.	2a	506
	842	~69.0	$Cs_2Cl_2.2SrCl_2$	Comp.		
	718	87.5	$SrCl_2$, $Cs_2Cl_2.2SrCl_2$	Eut.		
	550	15	$CsCl.SrCl_2$,CsCl	Eut.	1,2d	494
	842	50	$CsCl.SrCl_2$	Comp.		
	718	19	$SrCl_2$,$CsCl_2.SrCl_2$	Eut.		
		85.5	Solid solution	Eut.	8	507
$TaCl_3$	556	20	Cs_2TaCl_5,CsCl	Eut.	2a	508
	710	33	Cs_2TaCl_5	Comp.		
	273	46	$TaCl_3$,Cs_2TaCl_5	Eut.		
	346		Cs_2TaCl_5	P.T.		
	452		CsCl	P.T.		
		60	$TaCl_3$ decomposes			
$TaCl_4$	592	10	β-Cs_2TaCl_6, β-CsCl	Eut.	2b,3a,6a	509
	796	33.3	β-Cs_2TaCl_6	Comp.		
	255	48	$TaCl_4$,α-Cs_2TaCl_6	Eut.		
	345		Cs_2TaCl_6	P.T.		
	452		CsCl	P.T.		
$TaCl_5$	442	36.5	$CsTaCl_6$,CsCl	Eut.	2,6c	510
	548	50	$CsTaCl_6$	Comp.		
	214	~100	$TaCl_5$,$CsTaCl_6$	Eut.		
	451		CsCl	P.T.		
	294		$CsTaCl_6$	P.T.		

CsCl

2nd Component	T°C	Conc. mol %	Phases involved	Def.	Method	Ref.
$TeCl_4$	576	12	β-CsCl,Cs_2TeCl_6	Eut.	2d,3b,3c	511
	719	33	Cs_2TeCl_6	Comp.	6b	
	536	63	Cs_2TeCl_6,$CsTeCl_5$	Per.		
	218	98	$CsTeCl_5$,$TeCl_4$	Eut.		
	452		CsCl	P.T.		
			Existence of Cs_2TeCl_6 & $CsTeCl_6$ confirmed by x-rays			
			Density of Cs_2TeCl_6 = 3.5g/cm^3			
$ThCl_4$			Fig. 97			384
	562	13	CsCl,Cs_3ThCl_7	Eut.	2	384
	720	24	Cs_3ThCl_7	Comp.		
	573	35	Cs_3ThCl_7, Cs_2ThCl_6	Per.		
	460	45.8	Cs_2ThCl_6, $CsThCl_5$	Eut.		
	490	50	$CsThCl_5$	Comp.		
	450	54.5	$CsThCl_5$,$ThCl_4$	Eut.		
$TiCl_2$			Fig. 98			512
	593	10	CsCl,Cs_2TiCl_4	Eut.	2d,3b	512
	747	33	Cs_2TiCl_4, $CsTiCl_3$	Per.		
	932	50	$CsTiCl_3$	Comp.		
	798	68	$CsTiCl_3$,$TiCl_2$	Eut.		
	445		CsCl	P.T.		
$TiCl_3$	591	11.5	CsCl, $TiCl_3$.3CsCl	Eut.	2	513
	770	25	$TiCl_3$.3CsCl	Comp.		
	690	34	$TiCl_3$.3CsCl, $TiCl_3$.CsCl	Eut.		
	781	50	$TiCl_3$.CsCl	Comp.		
	672	65	$TiCl_3$CsCl,$TiCl_3$	Eut.		
	578	7.0	β-CsCl,Cs_3TiCl_6	Eut.	2a	514
	710	25.0	Cs_3TiCl_6	Comp.		
	680	37.0	Cs_3TiCl_6, $Cs_3Ti_2Cl_9$	Eut.		
	705	40.0	$Cs_3Ti_2Cl_9$	Comp.		
	603	58.0	$Cs_3Ti_2Cl_9$, $CsTiCl_4$	Per.		
	551	62.0	$CsTiCl_4$,$TiCl_3$	Eut.		
	452		CsCl	P.T.		

CsCl

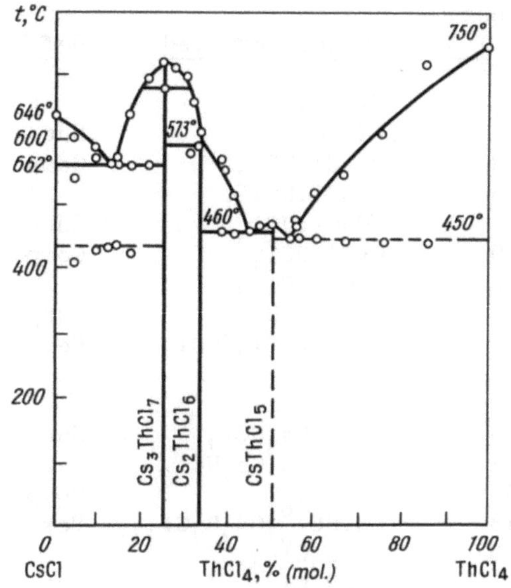

Fig. 97. CsCl-$ThCl_4$ system.

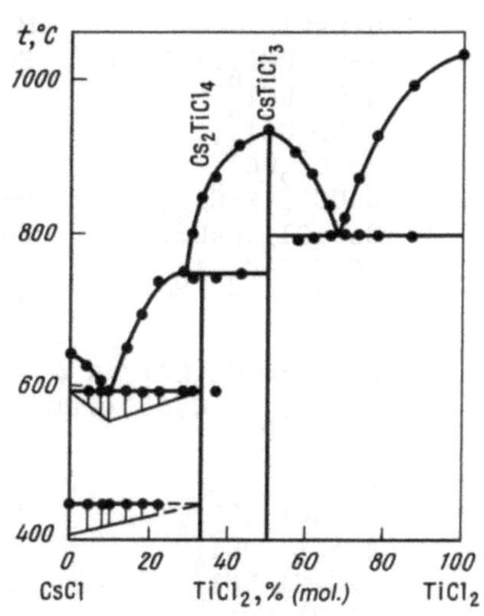

Fig. 98. CsCl-$TiCl_2$ system.

2nd Component	T°C	Conc. mol %	Phases involved	Def.	Method	Ref.
TlCl	390	75		Eut.		305
	393	77	Continuous sol.sln.	Min.	1	515

CsCl

2nd Component	T°C	Conc. mol %	Phases involved	Def.	Method	Ref.
UCl_3	470±2	24	$CsCl_2$, $2CsCl.UCl_3$	Eut.	2d,3b	516
	722±2	33	$2CsCl.UCl_3$	Comp.		
	590±2	45	$2CsCl.UCl_3$, $CsCl.UCl_3$	Eut.		
	602±2	50	$CsCl.UCl_3$	Comp.		
	575±2	56	$CsCl.UCl_3$, $CsCl.3UCl_3$	Eut.		
	625±2		$CsCl.3UCl_3.UCl_3$	Per.		
UCl_4	513.3	20	$CsCl,Cs_3UCl_7$	Eut.	2c,3b	517
	519.1	22	Cs_3UCl_7,Cs_2UCl_6	Per.		
	669.8	33	Cs_2UCl_6	Comp.		
	522.5	48	$Cs_2UCl_6,Cs_3U_2Cl_{11}$	Per.		
	374.5	58	$Cs_3U_2Cl_{11}$, CsU_2Cl_9	Eut.		
	389.6	63	CsU_2Cl_9,UCl_4	Per.		
	642	0	α-CsCl	M.P.	2a,3a,6a	518
	498	19	β-Cs_2UCl_6,α-CsCl	Eut.		
	665	33.4	α-Cs_2UCl_6	Comp.		
	510		β-Cs_2UCl_6, α-Cs_2UCl_6	Per.		
	420	56	$CsUCl_5$,β-Cs_2UCl_6	Per.		
	360	59	$CsU_2Cl_9,CsUCl_5$	Eut.		
	387	64	UCl_4,CsU_2Cl_9	Per.		
	589	100	UCl_4	M.P.		
	445		CsCl	P.T.		
			Density Cs_2UCl_6 = 4.01 (4.04 g/cm³ x-rays) $CsUCl_6$ = 4.04 g/cm³			
			CsCl, $3CsCl.2UCl_4$	Per.	2d	519
	522±2	17.0	$3CsCl.2UCl_4$, $2CsCl.UCl_4$	Eut.		
			$2CsCl.UCl_4$	Comp.		
	356±2	63.0	$2CsCl.UCl_4$, $CsCl.UCl_4$	Eut.		
			$CsCl.UCl_4,UCl_4$	Per.		
VCl_2	642	2.2	$CsCl,CsVCl_3$	Eut.	2d,3b	520
	1084	50	$CsVCl_3$	Comp.		
	1046	66	$CsVCl_3,VCl_2$	Eut.		

CsCl

2nd Component	T°C	Conc. mol %	Phases involved	Def.	Method	Ref.
VCl_3	589	7	Cs_3VCl_6,β-CsCl	Eut.	2a	521
	726	25	Cs_3VCl_6	Comp.		
	676	35	$Cs_3V_2Cl_9$,Cs_3VCl_6	Eut.		
	701	40	$Cs_3V_2Cl_9$	Comp.		
	560	54	VCl_3,$Cs_3V_2Cl_9$	Eut.		
	458		CsCl	P.T.		
WCl_5			Fig. 99			522
	568	11	Cs_2WCl_7,CsCl	Eut.	2b,2c	522
	748	32	Cs_2WCl_7	Comp.		
	510	55	β-$CsWCl_6$,Cs_2WCl_7	Per.		
	~300		α-$CsWCl_6$,β-$CsWCl_6$	Per.		
	275	98	WCl_5,$CsWCl_6$	Eut.		
		60-98	Decomposition (Line GH in Fig. 99)			

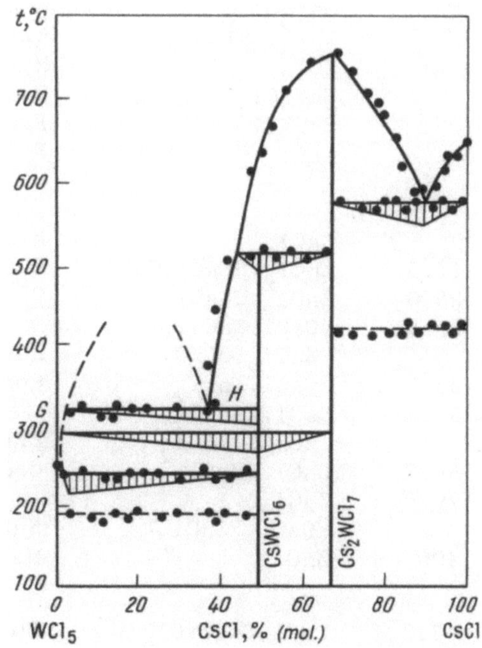

Fig. 99. WCl_5-CsCl system.

CsCl

2nd Component	T°C	Conc. mol %	Phases involved	Def.	Method	Ref.
WCl_6			Fig. 100			522
			Not a binary but a more complex system - as if WCl_6 decomposed to WCl_5 and Cl. Line III shows $CsWCl_6$ $\alpha \rightleftharpoons \beta$ transf. Lines I & II melting pt. and phase transf. of WCl_5		1	522
		0-33	$2CsCl.WCl_5$ exists		2a,5c,6c	523
		33-100	Complete insolubility			
YCl_3			Fig. 101			524
	648	0	CsCl	M.P.	2a	524
	590	8.5	Cs_3YCl_6,CsCl	Eut.		
	870	25	Cs_3YCl_6	Comp.		
	510	52	CsY_3Cl_{10}, Cs_3YCl_6	Eut.		
	640	75	CsY_3Cl_{10}	Comp.		
	584	83.2	YCl_3,CsY_3Cl_{10}	Eut.		
	700	100	YCl_3	M.P.		
	406		CsY_3Cl_{10}	P.T.		
	420		Cs_3YCl_6	P.T.		
	452		CsCl	P.T.		
$ZnCl_2$	647	0	CsCl	M.P.	1,2	525
	535	17.5	$ZnCl_2.3CsCl$,CsCl	Eut.		
	560	25.0	$ZnCl_2.3CsCl$	Comp.		
	556	25.5	$2CsCl.ZnCl_2$, $ZnCl_2.3CsCl$	Eut.		
	601	33.3	$2CsCl.ZnCl_2$	Comp.		
	263	57.5	$CsCl.2ZnCl_2$, $2CsCl.ZnCl_2$	Eut.		
	294	66.7	$CsCl.2ZnCl_2$	Comp.		
	268	79.9	$ZnCl_2$, $CsCl.2ZnCl_2$	Eut.		
	312	100	$ZnCl_2$	M.P.		
$ZrCl_4$	572	15.2	CsCl,Cs_2ZrCl_6	Eut.	2	526
	805	33.4	Cs_2ZrCl_6	Comp.		

CsCl

2nd Component	T°C	Conc. mol %	Phases involved	Def.	Method	Ref.
$ZrCl_4$	286	67.2	Cs_2ZrCl_6, $ZrCl_4$	Eut.		526
	440		CsCl	P.T.		

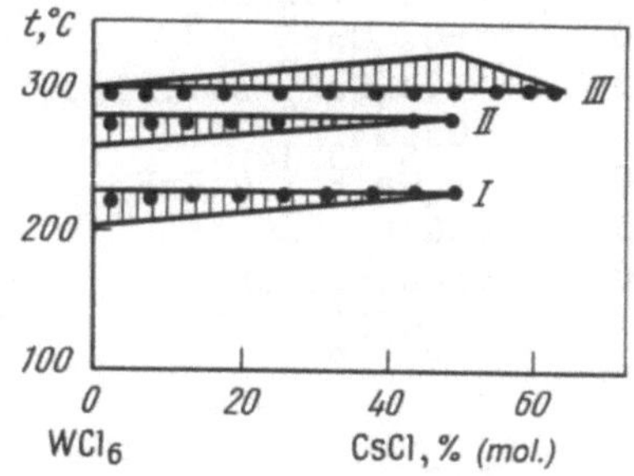

Fig. 100. Part of the WCl_6-CsCl system.

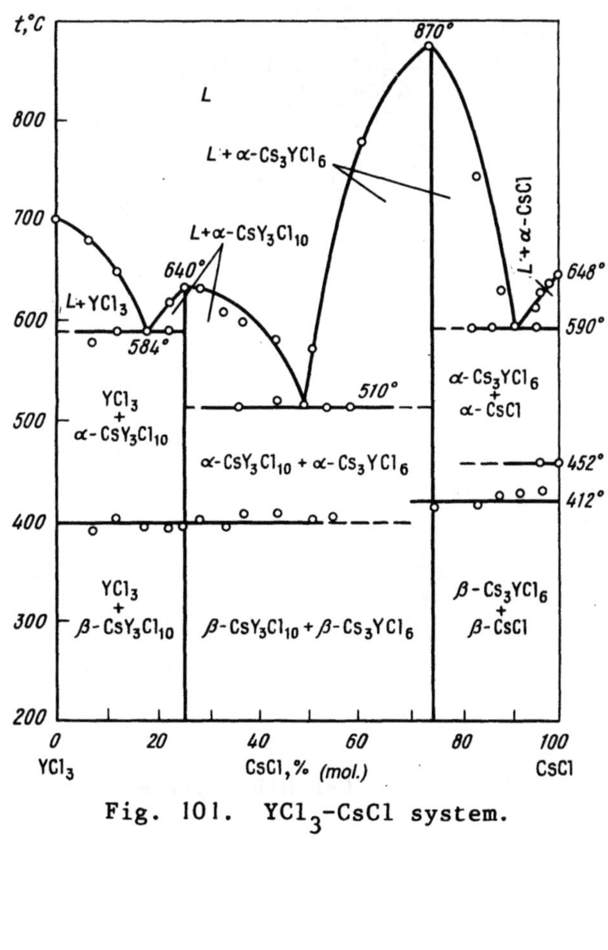

Fig. 101. YCl_3-CsCl system.

CsF

2nd Component	T°C	Conc. mol %	Phases involved	Def.	Method	Ref.
GaF_3	850		$GaCs_3F_6$	M.P.	2d,3b	527
	320		$GaCs_3F_6$	P.T.		
	675	5.2	CsF,α-Cs_3GaF_6	Eut.	2d,3b,2e	528
	851	25.0	α-Cs_3GaF_6	Comp.		
	520	40.5	α-Cs_3GaF_6 sol.sln., $Cs_5Ga_3F_{14}$	Per.		
	500	45.0	$Cs_5Ga_3F_{14}$,$CsGaF_4$	Eut.		
	535	49.5	$CsGaF_4$,$CsGa_2F_7$	Per.		
	590	54.0	$CsGa_2F_7$,GaF_3	Per.		
	316±4		Cs_3GaF_6	P.T.		
HfF_4			Fig. 102			529
	685±5	0.0	CsF	M.P.	2d,3b	529
	620	10.0	CsF,Cs_3HfF_7 sol.sln.	Eut.		
	750	25.0	Cs_3HfF_7	Comp.		
	630	34.0	Cs_3HfF_7 sol.sln., α-Cs_2HfF_6 sol. sln.	Per.		
	570	36.0	α-Cs_2HfF_6 sol.sln., β-Cs_2HfF_6 sol. sln.	Per.		

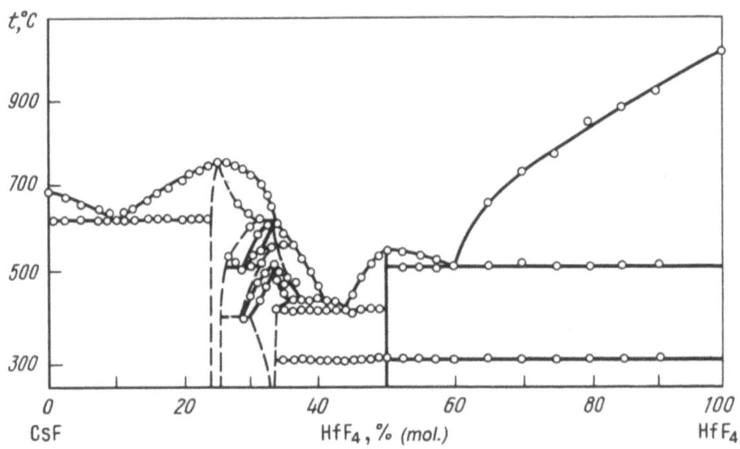

Fig. 102. CsF-HfF_4 system.

CsF

2nd Component	T°C	Conc. mol %	Phases involved	Def.	Method	Ref.
HfF_4	~450	41.0	β-Cs_2HfF_6 sol.sln., γ-Cs_2HfF_6 sol.sln.	Per.		529
	423	43.0	γ-Cs_2HfF_6 sol.sln., α-$CsHfF_5$	Eut.		
	530	50.0	α-$CsHfF_5$	Comp.		
	513	60.0	α-$CsHfF_5$,HfF_4	Eut.		
	1025±5	100.0	HfF_4	M.P.		
	515		Cs_2HfF_6 β ⇌ γ	P.T.		
	320		$CsHfF_5$ α ⇌ β	P.T.		
	395		(Eutectic):β-Cs_2-HfF_6 sol.sln. ⇌ ⇌ γ-Cs_2HfF_6 Cs_3HfF_7 sol. sln.	P.T.		
	501		(Eutectic): α-Cs_2-HfF_6 sol.sln. ⇌ ⇌ Cs_3HfF_7 sol.sln. + β-Cs_2-HfF_6 sol.sln.	P.T.		
KF	685	0.0	CsF	M.P.		530
	625	43	KF, 15% KF sol. sln. in CsF	Eut.	2	
	865	100.0	KF	M.P.		
	627	43	CsF, KF	Eut.	1,2a	531
LiF	470±5	36		Eut.	2c	532
	488±5	46		Per.	1	533
	479	40	CsF,CsF.LiF	Eut.	1,2a	534
	492	50	CsF.LiF	Comp.		535
	490	52	CsF.LiF, LiF	Eut.		536
	480	39	LiF.CsF, CsF	Eut.	1	531
	490	51	LiF, LiF.CsF	Eut.		

CsF

2nd Component	T°C	Conc. mol %	Phases involved	Def.	Method	Ref.
MgF_2	581	15	CsF,2CsF.MgF_2	Eut.	1,2a	537
	740	33.3	2CsF.MgF_2	Comp.		
	699	45	2CsF.MgF_2, CsF.MgF_2	Eut.		
	739	50	CsF.MgF_2	Comp.		
	718	57	CsF.MgF_2,MgF_2	Eut.		
			Fig. 103			538
MnF_2	620	14	CsF,20% CsF sol.sln. in Cs_2MnF_4	Eut.	1,2a	539
	725	33	Cs_2MnF_4	Comp.		
	710	35	Cs_2MnF_4,$CsMnF_3$	Eut.		
	780	50	$CsMnF_3$	Comp.		
	706	71	$CsMnF_3$, CsF	Eut.		
NaF	610±5	24	CsF, NaF	Eut.	1,2c,3b	533
	690	0	CsF	M.P.	2c	540
	615	20	CsF, NaF	Eut.		
	995	100	NaF	M.P.		

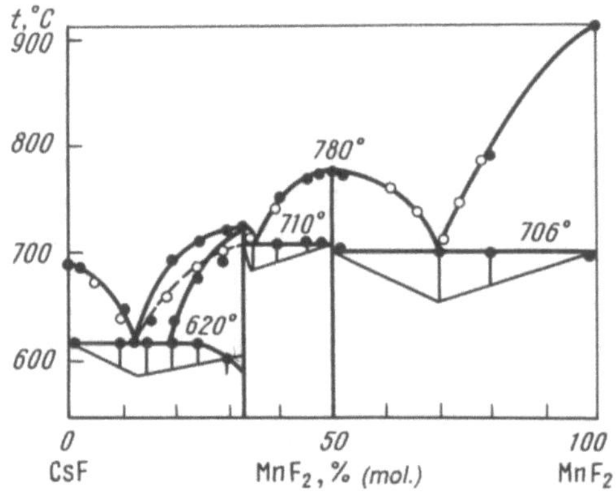

Fig. 103. CsF-MnF_2 system.

CsF

2nd Component	T°C	Conc. mol %	Phases involved	Def.	Method	Ref.
PbF_2	500	28	$CsPbF_3(111_T)CsF$	Eut.	2,3b	541
	615	40	$CsPbF_3(111^T)$, $CsPbF_3(111^n_T)$	Per.		
	725	50	$CsPbF_3(111^T)$	Comp.		
	615	60	$CsPbF_3(111^n_T)$, $CsPbF_3(111^T)$	Per.		
	565	71	PbF_2,$CsPbF_3(^n111_T)$	Eut.		
PrF_3	640	9.5	CsF,$3CsF.PrF_3$	Eut.	1,2d	542
	864	25	$3CsF.PrF_3$	Comp.		
	676	39	$3CsF.PrF_3$,PrF_3	Eut.		
RbF	685	0	CsF	M.P.	1	530
	785	100	RbF	M.P.		
ScF_3	680	4.0	CsF,Cs_3ScF_6	Eut.	1,2a,3b	543
	1084	25.0	Cs_3ScF_6	Comp.		
	798	33.0	Cs_3ScF_6,ScF_3	Eut.		
SrF_2	927		$CsF.SrF_2$ (incongruent melt)	Comp.	1,2a	537
	660	5	CsF,$CsF.SrF_2$	Eut.		
ThF_4			Fig. 104			
	615	9.0	CsF, $3CsF.ThF_4$	Eut.	2,3b	544
	980	25.0	$3CsF.ThF_4$	Comp.		

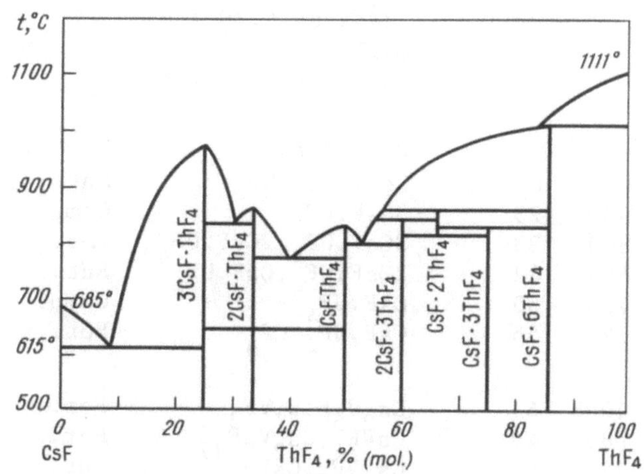

Fig. 104. $CsF-ThF_4$ system.

CsF

2nd Component	T°C	Conc. mol %	Phases involved	Def.	Method	Ref.
ThF_4	838	30.5	$3CsF.ThF_4$, β-$2CsF.ThF_4$	Eut.		544
	869	33.3	β-$2CsF.ThF_4$	Comp.		
	775	40.0	β-$2CsF.ThF_4$, $CsF.ThF_4$	Eut.		
	839	50.0	$CsF.ThF_4$	Comp.		
	797	53.0	$CsF.ThF_4$, $2CsF.3ThF_4$	Eut.		
	842	56.0	$2CsF.3ThF_4$, $CsF.2ThF_4$	Per.		
	860	56.5	$CsF.2ThF_4$, $CsF.6ThF_4$	Per.		
	1010	84.0	$CsF.6ThF_4$, ThF_4	Per.		
	813	66.7	$CsF.2ThF_4 \rightleftharpoons 2CsF.3ThF_4 + CsF.3ThF_4$	P.T.		
	830	75	$CsF.3ThF_4 \rightleftharpoons CsF.6ThF_4 + CsF.2ThF_4$	P.T.		
	645		$\alpha \rightleftharpoons \beta$ $2CsF.ThF_4$	P.T.		
TiF_4			Fig. 105			545
	684	0	CsF	M.P.	2d,3a,6a	545
	596	4.7	CsF sol.sln., Cs_3TiF_7 sol.sln.	Eut.		
	786	16.7	Cs_3TiF_7	Comp.		
	700	23.3	Cs_3TiF_7 sol.sln., Cs_2TiF_6 sol.sln.	Eut.		
	781	33.3	Cs_2TiF_6	Comp.		
			Cs_2TiF_6 - density 4.10g/cm^3.			
			Cs_3TiF_7 - density 3.45g/cm^3.			
UF_4			Fig. 106			
	650	6	CsF, $3CsF.UF_4$	Eut.		546
	980	25	$3CsF.UF_4$	Comp.		546
	800	33	$3CsF.UF_4$, $2CsF.UF_4$	Per.		
	695	42	$2CsF.UF_4$, $CsF.UF_4$	Eut.		
	730	50	$CsF.UF_4$	Comp.		
	720	55	$CsF.UF_4$, UF_4	Eut.		
VF_3	1008	61	$Cs_2V_5F_{17}$, VF_3	Per.	2d,3b	547
	885	49	$CsVF_4$, $Cs_2V_5F_{17}$	Per.		
	790	40	Cs_3VF_6, $CsVF_4$	Eut.		
	1075	-	Cs_3VF_6	Comp.		

CsF

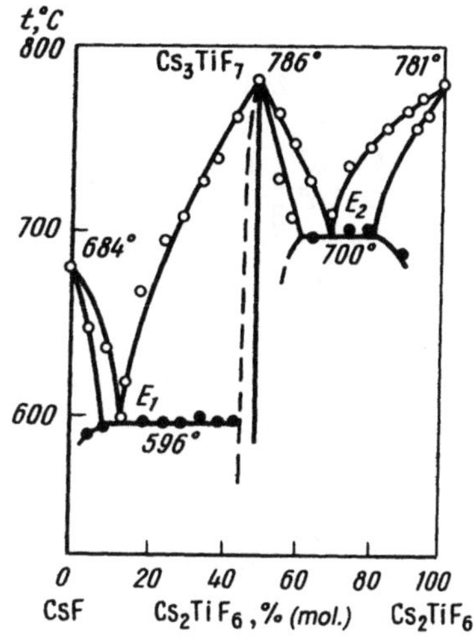

Fig. 105. CsF-Cs_2TiF_6 system.

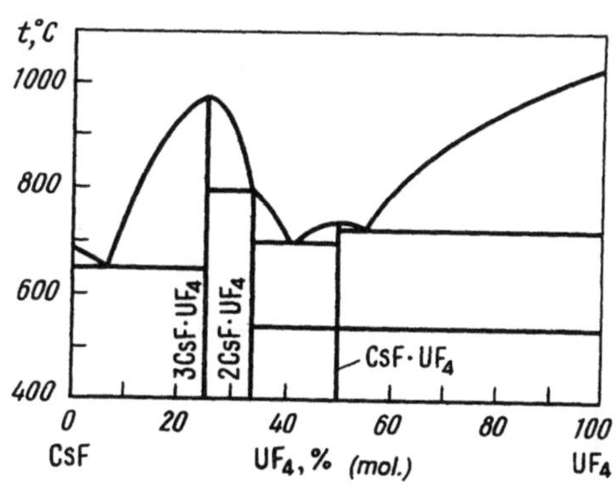

Fig. 106. CsF-UF_4 system.

2nd Component	T°C	Conc. mol %	Phases involved	Def.	Method	Ref.
VF_3	670	5	CsF,Cs_3VF_6	Eut.		547
	345	-	Cs_3VF_6	P.T.		
YF_3	670	4.0	CsF,Cs_3YF_6	Eut.	2a	548
	1074	25.0	Cs_3YF_6	Comp.		
	890	37.0	Cs_3YF_6,YF_3	Eut.		

CsF

2nd Component	T°C	Conc. mol %	Phases involved	Def.	Method	Ref.
ZnF_2			Fig. 107			549
	537	20	β-Cs_2ZnF_4,CsF	Eut.	2,3b	549
	633	34	α-Cs_2ZnF_4	Comp.		
	554	43	$CsZnF_3$,β-Cs_2ZnF_4	Eut.		
	587	50	$CsZnF_3$	Comp.		
	537	56	$CsZn_2F_5$,$CsZnF_3$	Eut.		
	618	67	ZnF_2,$CsZn_2F_5$	Per.		
	563	-	Cs_2ZnF_4	P.T.		
ZrF_4	646	10	CsF,Cs_3ZrF_7 sol.sln.	Eut.	2c,1,3b, 1f	550
	784	25	Cs_3ZrF_7	Comp.		
	530	37	Cs_3ZrF_7 sol.sln., α-Cs_2ZrF_6	Per.		
	465	-	α-Cs_2ZrF_6, β-Cs_2ZrF_6	Per.		
	425	42	β-Cs_2ZrF_6, α-$CsZrF_5$	Eut.		
	518	50	α-$CsZrF_5$	Comp.		
	465	55	α-$CsZrF_5$,ZrF_4	Eut.		
	330		$CsZrF_5$	P.T.		

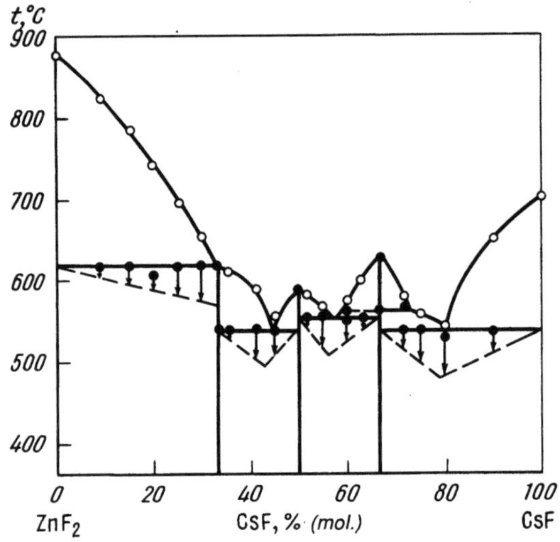

Fig. 107. ZnF_2-CsF system.

CsF

2nd Component	T°C	Conc. mol %	Phases involved	Def.	Method	Ref.
ZrF_4			Fig. 108			551
	760	25	Cs_3ZrF_7	Comp.	2d,3b	551
	630	33.8	Cs_3ZrF_7 sol.sln., α-Cs_2ZrF_6 sol.sln.	Per.		
	583	35.5	α-Cs_2ZrF_6 sol.sln., β-Cs_2ZrF_6 sol.sln.	Per.		
	446	41	β-Cs_2ZrF_6 sol.sln., γ-Cs_2ZrF_6 sol.sln.	Per.		
	420	43	γ-Cs_2ZrF_6 sol.sln., α-$CsZrF_5$	Eut.		
	490	50	α-$CsZrF_5$	Comp.		
	518	31	α-Cs_2ZrF_6 sol.sln. $\rightleftharpoons$ Cs_3ZrF_7 + β-Cs_2ZrF_6 sol. sln.	P.T.		
	467	30.5	β-Cs_2ZrF_6 sol.sln. $\rightleftharpoons$ Cs_3ZrF_7 + γ-Cs_2ZrF_6 sol. sln.	P.T.		
	595		$\alpha \rightleftharpoons \beta$-$Cs_2ZrF_6$	P.T.		
	510		$\beta \rightleftharpoons \gamma$-$Cs_2ZrF_6$	P.T.		
	315		$\alpha \rightleftharpoons \beta$-$CsZrF_5$			

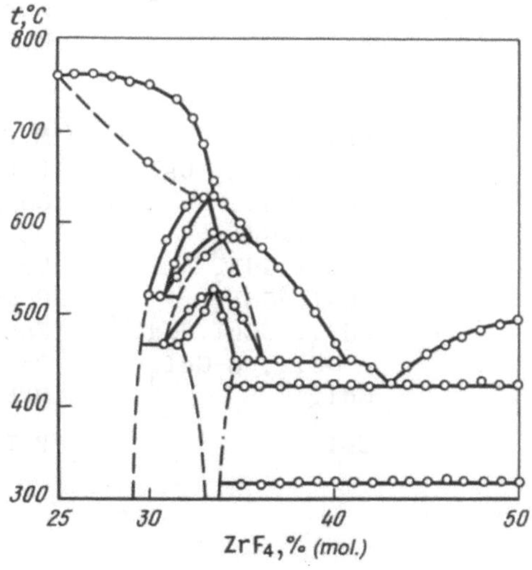

Fig. 108. CsF-ZrF_4 system.

CsI

2nd Component	T°C	Conc. mol %	Phases involved	Def.	Method	Ref.
CsF	699	0	CsF	M.P.	1,2d	552
	430	46.5	CsF, CsI	Eut.		553
	621	100	CsI			
BiI_3	541	22	$3CsI.2BiI_3$, CsI	Eut.	2d	556
	634	40	$3CsI.2BiI_3$	Comp.		
	365	86	BiI_3, $3CsI.2BiI_3$	Eut.		
DyI_3	624		CsI	M.P.	2d	557
	567		$3CsI.DyI_3$,CsI	Eut.		
	715		$3CsI.DyI_3$	Comp.		
	632		$3CsI.2DyI_3$, $3CsI.DyI_3$	Eut.		
	643		$3CsI.2DyI_3$	Comp.		
	607		DyI_3, $3CsI.2DyI_3$	Eut.		
	977		DyI_3	M.P.		
ErI_3	624		CsI	M.P.	2d	557
	569		$3CsI.ErI_3$, CsI	Eut.		
	717		$3CsI.ErI_3$	Comp.		
	653		$3CsI.2ErI_3$, $3CsI.ErI_3$	Eut.		
	667		$3CsI.2ErI_3$	Comp.		
	637		$3CsI.2ErI_3$,ErI_3	Eut.		
	1004		ErI_3	M.P.		
GdI_3	624		CsI	M.P.	2d	557
	564		$3CsI.GdI_3$, CsI	Eut.		
	712		$3CsI.GdI_3$	Comp.		
	590		$3CsI.2GdI_3$, $3CsI.GdI_3$	Eut.		
	597		$3CsI.2GdI_3$	Comp.		
	562		GdI_3, $3CsI.2GdI_3$	Eut.		
	748		α-GdI_3, β-GdI_3	Per.		
	940		GdI_3	M.P.		
	483) 430)		CsI	P.T.		

CsI

2nd Component	T°C	Conc. mol %	Phases involved	Def.	Method	Ref.
HgI_2	641	0.0	CsI	M.P.	2a	558
	380	-	$2CsI.HgI_2$, CsI	Per.		
	184	52.5	$CsI.2HgI_2$, $2CsI.HgI_2$	Eut.		
	233	66.7	$CsI.2HgI_2$	Comp.		
	212	75	α-HgI_2, $CsI.2HgI_2$	Eut.		
	273	100	α-HgI_2	M.P.		
	134		HgI_2	P.T.		
InI	330	75	CsI, InI	Eut.	2d,3b	559
	365	100	InI	M.P.		
InI_3			Fig. 109			560
	290	40	Cs_3InI_6, CsI	Per.	2	560
	122	82	InI_3, Cs_3InI_6	Eut.		
KI	634	0.0	CsI	M.P.	2e	561
	516	45.0	Continuous sol.sln.	Min.		
	694	100.0	KI			
	640	0.0	CsI	M.P.	1	562
	494	21.0	Continuous sol.sln.	Min.		

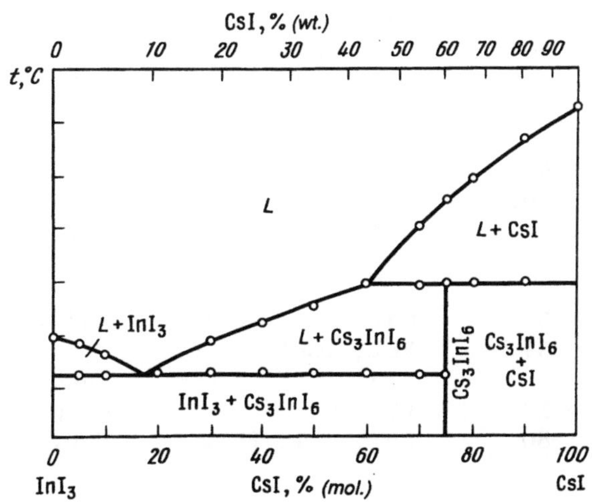

Fig. 109. InI_3-CsI system.

CsI

2nd Component	T°C	Conc. mol %	Phases involved	Def.	Method	Ref.
KI	559	30.0	Continuous sol.sln.	Min.	1	563
LaI_3	624	0.0	CsI	M.P.	2d	564
	539	14.0	$3CsI.LaI_3$, CsI	Eut.		
	668	25.0	$3CsI.LaI_3$	Comp.		
	467	44.0	$CsI.9LaI_3$, $3CsI.LaI_3$	Eut.		
	542	56	LaI_3, $CsI.9LaI_3$	Per.		
	768	100	LaI_3	M.P.		
	478		CsI	P.T.		
	471		$3CsI.LaI_3$	P.T.		
NaI	625	0.0	CsI	M.P.	2b,2c	565
	435	45	CsI, NaI	Eut.		
	667	100	NaI			
	428	48.5	CsI, NaI	Eut.	1	566
	661	100	NaI	M.P.		
NdI_3	624	0	CsI	M.P.	2d	557
	547	12	CsI, $3CsI.NdI_3$	Eut.		
	693	25	$3CsI.NdI_3$	Comp.		
	500	40	$3CsI.NdI_3$, $3CsI.2NdI_3$	Per.		
	478	50	$3CsI.2NdI_3$, $CsI.4NdI_3$	Eut.		
	523	58	$CsI.4NdI_3$, β-NdI_3	Per.		
	563	62	β-NdI_3, α-NdI_3	Per.		
	780	100	α-NdI_3	M.P.		
	471		$3CsI.NdI_3$	P.T.		
PbI_2	464	31	CsI, $PbI_2.4CsI$	Per.	1,3b	567
	434	39	$PbI_2.4CsI$, $PbI_2.CsI$	Eut.		568
	481	~57	$PbI_2.CsI$	Comp.		
	341	80	$PbI_2.CsI$, PbI_2	Eut.		

CsI

2nd Component	T°C	Conc. mol %	Phases involved	Def.	Method	Ref.
PbI_2	641	0.0	CsI	M.P.	2a	569
	468	30	$4CsI.PbI_2$, CsI	Per.		
	432	40	α-$PbI_2.CsI$, $4CsI.PbI_2$	Eut.		
	476	50	α-$PbI_2.CsI$	Comp.		
	342	80	PbI_2, α-$PbI_2.CsI$	Eut.		
	404	100	PbI_2	M.P.		
	327		$PbI_2.CsI$	P.T.		
PrI_3	624	0.0	CsI	M.P.	2d	570
	545	13	$3CsI.PrI_3$, CsI	Eut.		
	685	25	$3CsI.PrI_3$	Comp.		
	470	40	$3CsI.2PrI_3$, $3CsI.PrI_3$	Per.		
	460	43	$CsI.9PrI_3$, $3CsI.2PrI_3$	Eut.		
	534	58	PrI_3, $CsI.9PrI_3$	Per.		
	730	100	PrI_3	M.P.		
	480		CsI	P.T.		
RbI	634	0.0	CsI	M.P.	2a	571
	566	35	Continuous sol.sln.	Min.		
	656	100	RbI	M.P.		
	640	0.0	CsI	M.P.	1	572
	578	35	Continuous sol.sln.	Min.		
	640	100	RbI	M.P.		
SbI_3	621	0.0	CsI	M.P.	1,2d,3a	573
	580	22	$3CsI.2SbI_3$, CsI	Eut.		
	619	40	$3CsI.2SbI_3$	Comp.		
	166	82	SbI_3, $3CsI.2SbI_3$	Eut.		
	170	100	SbI_3	M.P.		
SmI_3	624		CsI	M.P.		574
	557		$3CsI.SmI_3$, CsI	Eut.		
	705		$3CsI.SmI_3$	Comp.		
	540		$3CsI.2SmI_3$, $3CsI.SmI_3$	Eut.		
	546		$3CsI.2SmI_3$	Comp.		
	506		$3CsI.2SmI_3$, SmI_3	Eut.		

CsI

2nd Component	T°C	Conc. mol %	Phases involved	Def.	Method	Ref.
SmI_3	670		α-SmI_3, β-SmI_3	Per.		574
	489		$\alpha \rightleftharpoons \beta$ CsI	P.T.		
	382		$\beta \rightleftharpoons \gamma$ SmI_3	P.T.		
	546		$\alpha \rightleftharpoons \beta$ 3CsI.SmI_3	P.T.		
SnI_2	626	0	CsI	M.P.	2a,3b	575
	393	36	CsI.$CsSnI_3$	Eut.		
	452	50	$CsSnI_3$	Comp.		
	269	82	$CsSnI_3$, SnI_2	Eut.		
	320	100	SnI_2	M.P.		
	641	0	CsI	M.P.	2a	576
	417	36	α-CsI.SnI_2, CsI	Eut.		
	467	50	α-CsI.SnI_2	Comp.		
	295	83	SnI_2, α-CsI.SnI_2	Eut.		
	330	100	SnI_2	M.P.		
	172		CsI.SnI_2	P.T.		
TeI_4	621	0	CsI	M.P.	2,3b,1e	577
	444	13	CsI, α-Cs_2TeI_6	Eut.		
	558	33	α-Cs_2TeI_6	Comp.		
	400		$\alpha \rightleftharpoons \beta$-$Cs_2TeI_6$	Per.		
	224	74	β-Cs_2TeI_6, TeI_4	Eut.		
	280	100	TeI_4	M.P.		
TlI	397±1	76	9% CsI sol. sln. in TlI; 20% TlI sol.sln. in CsI	Eut.	2	578
	170		TlI	P.T.		
			Rate of heating:			
	162	99.1	2°C/min } Cub.sol. sln. $\rightleftharpoons$ rhombic sol.sln. + CsI	P.T.		
	175	99.1	4°C/min }			
	406	76	Continuous sol.sln.	Min.	1	579
			Limited sol.sln. of 2-3.5% CsI in TlI		2,6d	580

CuBr

2nd Component	T°C	Conc. mol %	Phases involved	Def.	Method	Ref.
$TeBr_4$	484	0.0	CuBr	M.P.	2a,2e,3b	581
	320	82	β-CuBr, $TeBr_4$	Eut.		
	388	100	$TeBr_4$	M.P.		
	360	23-66	Decomposition of $TeBr_4$			
	382		CuBr	P.T.		
CuCl			Fig. 110			582
			All forms of CuBr form sol. slns. with CuCl		1	582

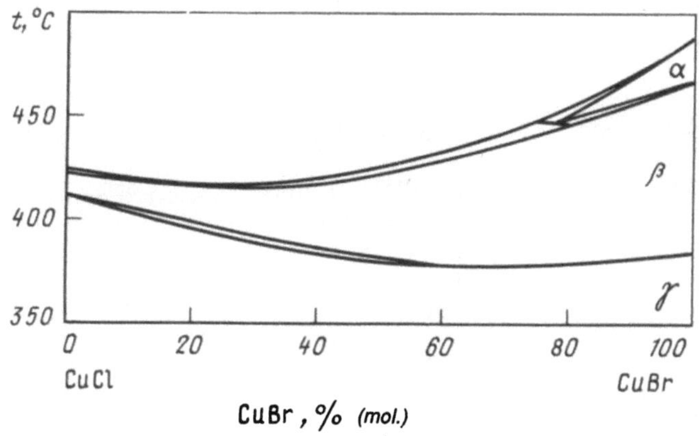

Fig. 110. CuCl-CuBr system.

2nd Component	T°C	Conc. mol %	Phases involved	Def.	Method	Ref.
CuI			Continuous sol. sln.		2a	583

CuCl

2nd Component	T°C	Conc. mol %	Phases involved	Def.	Method	Ref.
$GaCl_3$	250	57	$CuCl.GaCl_3$,CuCl	Per.	1,2b	584
	60	95	$GaCl_3$,$CuCl.GaCl_3$	Eut.		
$InCl_3$	423	0.0	CuCl	M.P.	2b,2c,1	585
	365	4.8	$InCl_3$,CuCl	Eut.		

CuCl

2nd Component	T°C	Conc. mol %	Phases involved	Def.	Method	Ref.
KCl	150	35	-	Eut.		586
LiCl	408	20	-			586
$MgCl_2$	412	25	CuCl, $MgCl_2$		2,1f	587
NaCl	325	23		Eut.		586
	330	24	12% NaCl sol.sln. in CuCl, 11% CuCl in sol.sln. in NaCl		2,1f	588
$NbCl_5$	202	>77	CuCl, $NbCl_5$	Eut.	2,6c	589
	255		Cu[$NbCl_6$] non-homogeneous melting	Comp.	1,2d	
		100	Very close to pure $NbCl_5$	Eut.		
$PbCl_2$	280	42		Eut.		586
	290	46	CuCl, $PbCl_2$		2,1d	591
$SnCl_2$	172	66				568
$TaCl_5$	215	<30	CuCl, $TaCl_5$		2,6c	589
			Fig. 111			590
	295	99	$CuTeCl_6$, CuCl	Per.	2d,1	590
	214	100	$TaCl_5$, $CuTaCl_5$	Eut.	2	
$ZnCl_2$	283	79	3% CuCl sol.sln. $ZnCl_2$ 4% ZnCl sol.sln.in CuCl	Eut.	2	592

CuCl

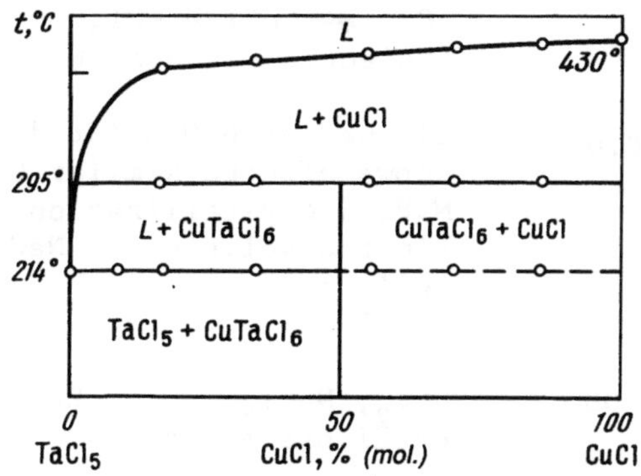

Fig. 111. $TaCl_5$-CuCl system.

$CuCl_2$

2nd Component	T°C	Conc. mol %	Phases involved	Def.	Method	Ref.
KCl	598±2	0.0	β-$CuCl_2$	M.P.	2a,3b	593
	486		α-$CuCl_2$,β-$CuCl_2$	Per.		
	407		Disordered struct. of $CuCl_2$	Per.		
	355	45±1	$KCuCl_3$, Disordered struc. of $CuCl_2$,α-$CuCl_2$	Eut		
	365	50	$KCuCl_3$	Comp.		
	310	64±1	KCl, $KCuCl_3$ In the non-isolated melt a reaction: $CuCl_2 \rightleftharpoons CuCl + \frac{1}{2}Cl_2$	Eut.		
	280)		Compound + existing	Eut.		
	246)		components	Eut.		
NaCl	598±2	0.0	β-$CuCl_2$	M.P.	2a,2d,3b	593
	486		α-$CuCl_2$, β-$CuCl_2$	Per.	5c	
	386	54-55	NaCl, α-$CuCl_2$	Eut.		
	800	100	NaCl	M.P.		

$CuCl_2$

2nd Component	T°C	Conc. mol %	Phases involved	Def.	Method	Ref.
NaCl	285		$CuCl_2 \rightleftharpoons CuCl + \frac{1}{2}Cl_2$ (Eutectic mixture) P.T.			593
	358 353	0.0	Calcined at 620°C for 1 hr. shows effects similar to M.P. & crystallization. Limited sol.sln. at NaCl & $CuCl_2$ ends.			
$NbCl_5$	200 >450	~100	$CuCl_2$, $NbCl_5$ Containers explode	Eut.	2d, 1	594
$SbCl_5$	~3 3	~100 100	$CuCl_2$, $SbCl_5$ $SbCl_5$	Eut.	2d	595
$TaCl_5$	208 >450	~100	$CuCl_2$, $TaCl_5$ Containers explode	Eut.	2d, 1	594
TlCl	312	55	$3TlCl.2CuCl_2$, $CuCl_2$	Eut.	1	596
	322	60	$3TlCl.2CuCl_2$	Comp.		
	298	70	$4TlCl.2CuCl_2$, $3TlCl.2CuCl_2$	Eut.		
	312	80	$4TlCl.CuCl_2$	Comp.		
	308	84	TlCl, $4TlCl.CuCl_2$	Eut.		
	428	100	TlCl	M.P.		

CuF_2

2nd Component	T°C	Conc. mol %	Phases involved	Def.	Method	Ref.
SrF_2			Sr_2CuF_6, $SrCuF_4$ Exist. confirmed	Comp. Comp.	3b	602

CuI

2nd Component	T°C	Conc. mol %	Phases involved	Def.	Method	Ref.
GaI_3			Fig. 112			597
	296	~87	α-CuI, $CuGaI_4$	Per.	2,1e,3b	597
	200	87.5	$CuGaI_4$, GaI_3	Eut.		
	392	0.0	γ ⇌ β CuI	P.T.		
	358	0.0	β ⇌ α CuI	P.T.		
InI_3			Fig. 113			598
	500	7.5	InI_3.3CuI, CuI	Eut.	2a	598
	-	25	InI_3.3CuI	Comp.		
	300	53	InI_3.CuI, InI_3.3CuI	Per.		
	170	95	InI_3,InI_3.CuI	Eut.		
	500	12.5-38	Decomposition (line CGD in the fig.)			
	370	0.0	CuI	P.T.		
KI	400	23	α-CuI, β-CuI	Per.	2,2d,3b	599
	367	25	β-CuI, γ-CuI	Per.		
	332	30	γ-CuI, KCu_4I_5	Per.		
	272	42	KCu_4I_5, KI	Eut.		
Nb_3I_8	589	0.0	CuI	M.P.	2d	600
	500	28.4	CuI, Phase x	Eut.		
	402		CuI	P.T.		
	368		CuI	P.T.		
RbI	400	22	α-CuI, β-CuI	Per.	2,2d,3b	599
	367	26	β-CuI, γ-CuI	Per.		
	285	35	γ-CuI, $RbCu_2I_3$	Per.		
	270	48	$RbCu_2I_3$,$RbCu_2I_3$	Eut.		
	297	56	Rb_2CuI_3, RbI	Per.		
TeI_4	590	0.0	CuI	M.P.	2a,2e,6c 3b	601
	260	77	γ-CuI, TeI_4	Eut.		
	440		CuI α ⇌ β	P.T.		
	400		CuI β ⇌ γ	P.T.		

CuI

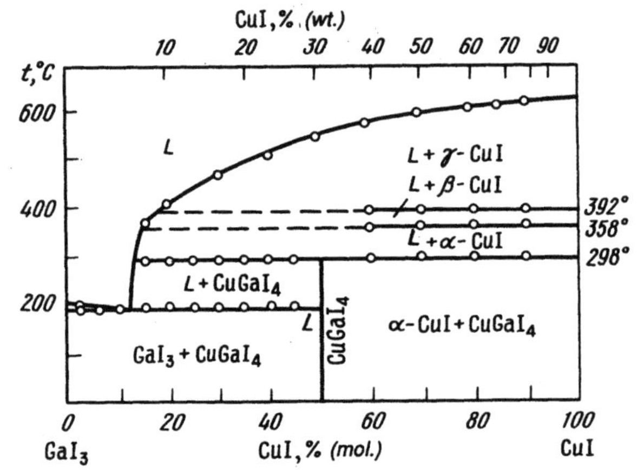

Fig. 112. CuI-GaI$_3$ system.

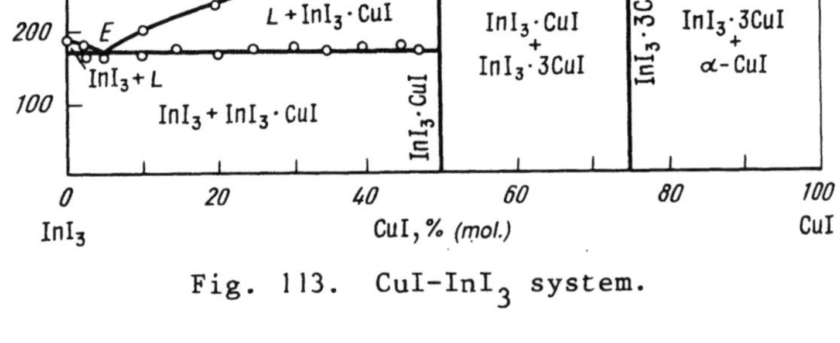

Fig. 113. CuI-InI$_3$ system.

$DyCl_3$

2nd Component	T°C	Conc. mol %	Phases involved	Def.	Method	Ref.
KCl			Fig. 114			603
	510	11.5	$DyCl_3$,KDy_3Cl_{10}	Eut.	2a	604
	600	25	KDy_3Cl_{10}	Comp.		
	415	50	KDy_3Cl_{10}, β-K_3DyCl_6	Eut.		
	790	75	β-K_3DyCl_6	Comp.		
	685	85	β-K_3DyCl_6,KCl	Eut.		
	390		K_3DyCl_6	P.T.		
NaCl			Fig. 115			603
	420	30	$NaDy_3Cl_{10}$, $DyCl_3$	Per. Per.	2a	604
	378	45	Na_3DyCl_6, $NaDy_3Cl_{10}$	Eut.		
	520	74	NaCl, Na_3DyCl_6	Per.		

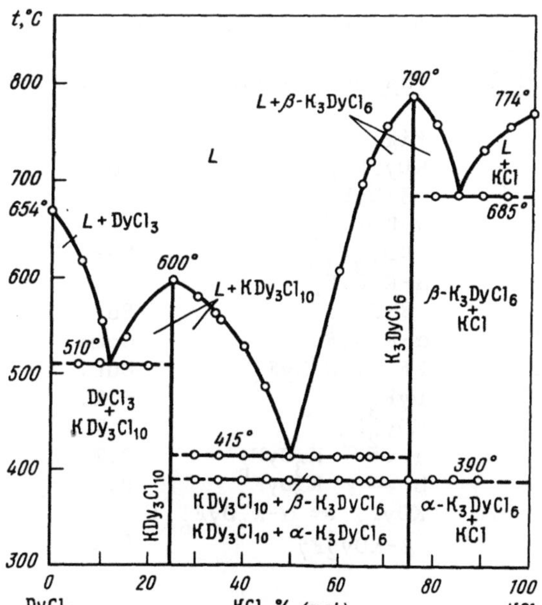

Fig. 114. $DyCl_3$-KCl system.

DyF_3

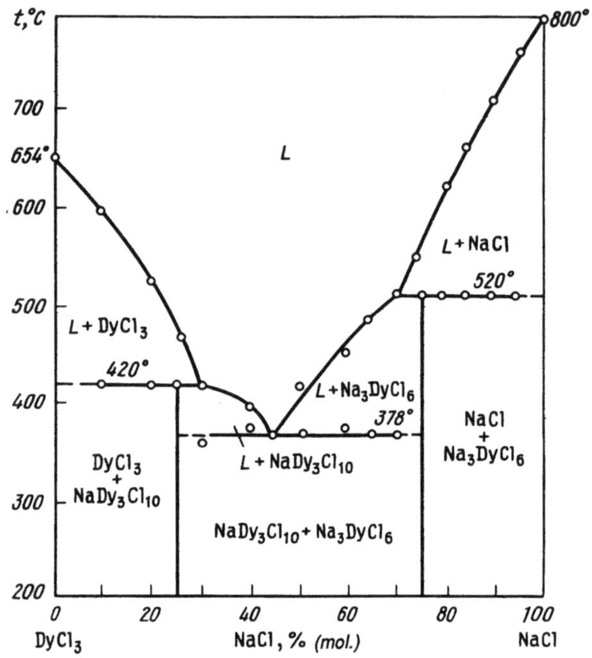

Fig. 115. $DyCl_3$-NaCl system.

2nd Component	T°C	Conc. mol %	Phases involved	Def.	Method	Ref.
KF			Fig. 116			605
	1000	41	KDy_3F_{10}, β-DyF_3	Per.	2a,3b,1e	605
	915	52	α-KDy_2F_7, KDy_3F_{10}	Per.		
	765	67	α-K_3DyF_6, β-KDy_2F_7	Eut.		
	950	75	α-K_3DyF_6	Comp.		
	775	83	KF, α-K_3DyF_6	Eut.		
	1040		DyF_3	P.T.		
	805		KDy_2F_7	P.T.		
	395		α ⇌ β K_3DyF_6	P.T.		
	300		β ⇌ γ K_3DyF_6	P.T.		
	625		$KDyF_4$ ⇌ α-K_3DyF_6 + β-KDy_2F_7	P.T.		

DyF_3

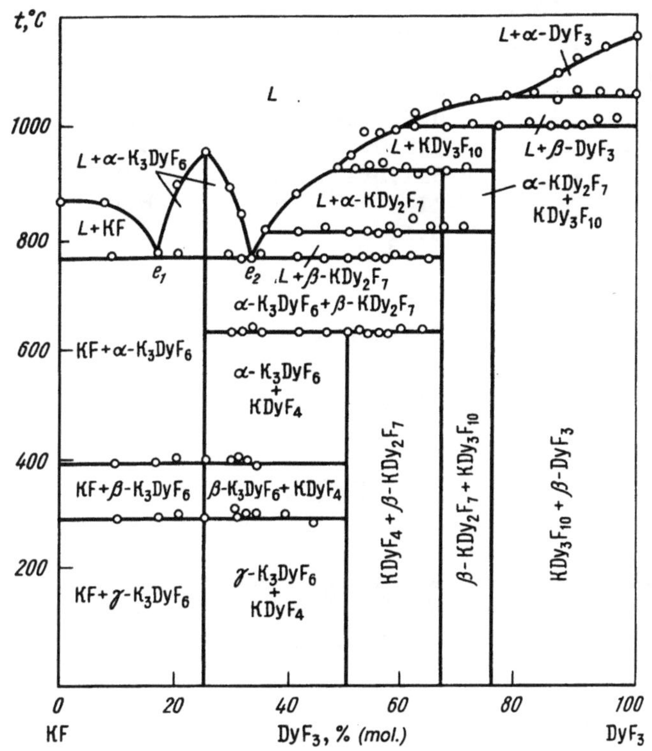

Fig. 116. KF-DyF_3 system.

2nd Component	T°C	Conc. mol %	Phases involved	Def.	Method	Ref.
LiF	1090	15	α-, β-DyF_3	Per.	2	606
	820	53	LiF.DyF_3, DyF_3	Per.		
	700	71	LiF, LiF.DyF_3	Eut.		
NaF	970	30	5NaF.9DyF_3, DyF_3	Eut.	2,1f,3b	607
	1000	36	5NaF.9DyF_3	Comp.		
	820	60	NaF.DyF_3, 5NaF.9DyF_3	Per.		
	676	72	NaF, NaF.DyF_3	Eut.		
			NaF.DyF_3 exists in two modifications: cubic (fluorite) at very high temp. and hexagonal. Cubic NaF.DyF_3 form sol.sln. with 5NaF.9DyF_3 which decomposes at lower temp. according to a "eutectic scheme."			

DyI_3

2nd Component	T°C	Conc. mol %	Phases involved	Def.	Method	Ref.
KI	977		DyI_3	M.P.	2d	608
	462		DyI_3, $3KI.DyF_3$	Eut.		
	582		$3KI.DyI_3$	Comp.		
	536		$3KI.DyI_3$, KI	Eut.		
	677		KI	M.P.		
	432		DyI_3	P.T.		
NaI	977		DyI_3	M.P.	2d	608
	480		DyI_3, NaI	Eut.		
	654		NaI	M.P.		
	427		DyI_3	P.T.		

$ErCl_3$

2nd Component	T°C	Conc. mol %	Phases involved	Def.	Method	Ref.
KCl			Fig. 117			604
	490	34	KEr_3Cl_{10}, $ErCl_3$	Per.		609
	416	50	K_3ErCl_6, KEr_3Cl_{10}	Eut.		

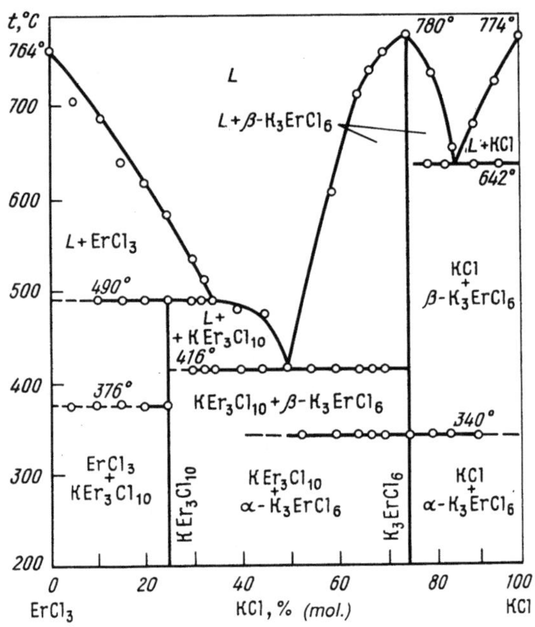

Fig. 117. $ErCl_3$-KCl system.

$ErCl_3$

2nd Component	T°C	Conc. mol %	Phases involved	Def.	Method	Ref.
KCl	780	75	K_3ErCl_6	Comp.		609
	642	88	KCl, K_3ErCl_6	Eut.		
	340		K_3ErCl_6	P.T.		
	376		Thermal effect of unknown form			
	547	34	$ErCl_3$, $KCl.4ErCl_3$	Per.	2e	610
	426	56	β-K_3ErCl_6, $KCl.4ErCl_3$	Eut.		
	808	75	β-K_3ErCl_6	Comp.		
	647	87	KCl, β-K_3ErCl_6	Eut.		
	364		K_3ErCl_6	P.T.		
NaCl	768	0	$ErCl_3$	M.P.	2a	611
	394	48	$ErCl_3$, Na_3ErCl_6	Eut.		604
	542	73	Na_3ErCl_6, $NaCl$	Per.		
	800	100	$NaCl$	M.P.		

ErF_3

2nd Component	T°C	Conc. mol %	Phases involved	Def.	Method	Ref.
LiF	800	43	$LiF.ErF_3$, ErF_3	Eut.	2	606
	820	50	$LiF.ErF_3$	Comp.		
	670	70	LiF, $LiF.ErF_3$	Eut.		
NaF			Fig. 118			612
	920	28	$5NaF.9ErF_3$, ErF_3	Eut.	2,1f,3b	613
	960	35	$5NaF.9ErF_3$	Comp.		

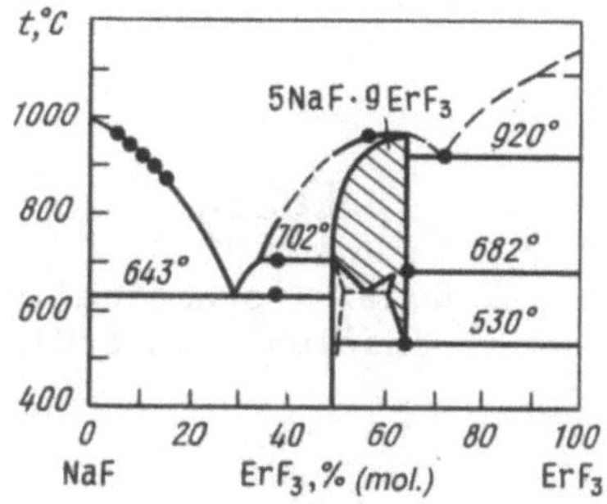

Fig. 118. NaF-ErF_3 system.

ErF_3

2nd Component	T°C	Conc. mol %	Phases involved	Def.	Method	Ref.
NaF	702	65	$NaF.ErF_3$, $5NaF.9ErF_3$	Per.		613
	643	70	NaF, $NaF.ErF_3$	Eut.		
			$NaF.ErF_3$ is in two modifications: cubic (fluorite) at high temp. and hexagonal. $5NaF.9ErF_3$ - two modifications: cubic (fluorite) at high temp. and orthorhombic at low temp. The cubic modifications form a sol. sln. with themselves, which develops at lower temp.			
SrF_2			Sol. sln. of fluorite type and limited sol.sln. of "tisonite" (hexagonal) type.		3b	614

ErI_3

2nd Component	T°C	Conc. mol %	Phases involved	Def.	Method	Ref.
KI	1004		ErI_3	M.P.	2d	608
	477		ErI_3, $3KI.ErI_3$	Eut.		
	586		$3KI.ErI_3$	Comp.		
	539		$3KI.ErI_3$, KI	Eut.		
	677		KI	M.P.		
	433		ErI_3	P.T.		
NaI	1004		ErI_3	M.P.	2d	608
	477		ErI_3, NaI	Eut.		
	654		NaI	M.P.		
	437		ErI_3	P.T.		

$EuCl_2$

2nd Component	T°C	Conc. mol %	Phases involved	Def.	Method	Ref.
$SrCl_2$			$EuCl_2$ - practically insoluble in $SrCl_2$		3b,1f	615

$EuCl_3$

2nd Component	T°C	Conc. mol %	Phases involved	Def.	Method	Ref..
NaCl	420	52	$EuCl_3$, Na_2EuCl_5	Eut.		616
	464	63.5	Na_2EuCl_5, Na_3EuCl_6	Per.		
	489	73	Na_3EuCl_6, NaCl	Per.		
	418		Na_3EuCl_6	P.T.		
$SrCl_2$			Limited sol.sln. At 400°C 23% of $EuCl_3$ dissolves in $SrCl_2$		3b,1f	615

EuF_3

2nd Component	T°C	Conc. mol %	Phases involved	Def.	Method	Ref..
LiF	690	55	$LiF.EuF_3$, EuF_3	Per.	2	606
	600	75	LiF, $LiF.EuF_3$	Eut.		
			EuF_3.LiF - existence confirmed			
	760	50	EuF_3 - possible transformation			
NaF			Fig. 119			607
	1060	30	$5NaF.9EuF_3$,EuF_3	Eut.	2,1f,3a	613
	1091	35.7	$5NaF.9EuF_3$	Comp.		
	825	61	$NaF.EuF_3$, $5NaF.9EuF_3$	Per.		
	718	72	NaF, $NaF.EuF_3$	Eut.		

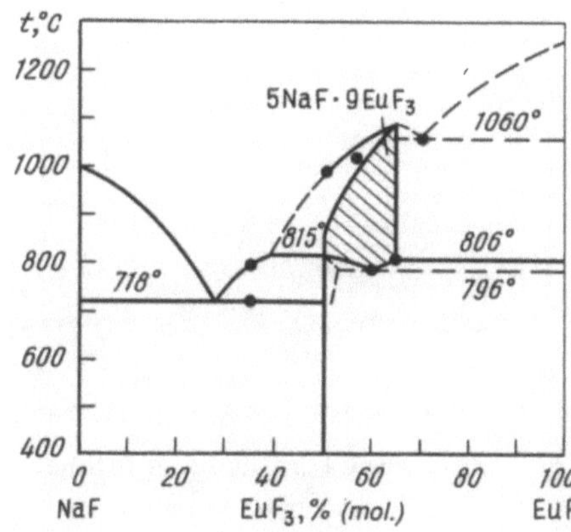

Fig. 119. NaF-EuF_3 system.

$EuCl_3$

2nd Component	T°C	Conc. mol %	Phases involved	Def.	Method	Ref.
NaF			$NaF.EuF_3$ - two phases: cubic (fluorite) at high temp. and hexagonal. The cubic phase forms sol. sln. with $5NaF.9EuF_3$, decomposing at lower temp.			613
	796	60	$5NaF.9EuF_3$ sol. sln. $\rightleftharpoons$ $EuF_3.NaF$ + EuF_3			
	806		$\alpha \rightleftharpoons \beta$ EuF_3			

$FeCl_2$

2nd Component	T°C	Conc. mol %	Phases involved	Def.	Method	Ref.
$GaCl_3$			Fig. 120			617
	65	97	$FeGaCl_5$, $GaCl_3$	Eut.	2a	617
	185	62	$FeCl_2$, $FeGaCl_5$	Per.		
	672	0-54	Decomposition			
$InCl_3$	520	65	$InCl_3$, 4% $InCl_3$ sol. sln. in $FeCl_2$		2a	618

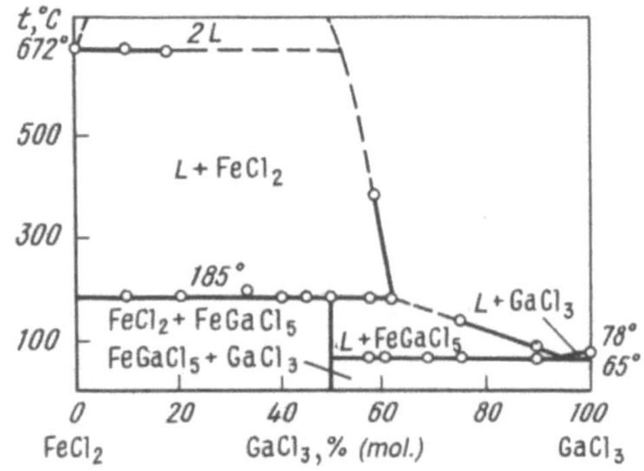

Fig. 120. $FeCl_2$-$GaCl_3$ system.

$FeCl_2$

2nd Component	T°C	Conc. mol %	Phases involved	Def.	Method	Ref.
KCl			Fig. 121			619
	393	45.8	$KFeCl_3$, $FeCl_2$	Eut.	2c	
	399	50.0	$KFeCl_3$	Comp.		
	351	61.8	K_2FeCl_4, $KFeCl_3$	Eut.		
	380	66.7	KCl, K_2FeCl_4	Per.		
	260±5			P.T.		
	293±5			P.T.		
			Fig. 122			620
	380	47.8	$KFeCl_3$, $FeCl_2$	Eut.	2c	620
	385	50	$KFeCl_3$	Comp.		
	340	60.2	K_2FeCl_4, $KFeCl_3$	Eut.		
	367	67	KCl, K_2FeCl_4	Per.		
	240		$\alpha \rightleftharpoons \beta$ K_2FeCl_4	P.T.		
	322		$\beta \rightleftharpoons \gamma$ K_2FeCl_4	P.T.		
	305		$\alpha \rightleftharpoons \beta$ $KFeCl_3$			
			Fig. 123			621
	390	47	β-$KFeCl_3$, $FeCl_2$	Eut.	2a	621
	406	50	β-$KFeCl_3$	Comp.		
	350	59.3	β-K_2FeCl_4, β-$KFeCl_3$	Eut.		
	374	66.7	KCl, β-K_2FeCl_4	Per.		
	268		K_2FeCl_4	P.T.		
	308		$KFeCl_3$	P.T.		
$LaCl_3$	620	27.7	$FeCl_2$, $LaCl_3$	Eut.	2c	622
	641	39.1	$FeCl_2$, $LaCl_3$	Eut.	2a,6c	623
$MgCl_2$			Continuous sol. sln.		2a	624
$MnCl_2$			Continuous sol. sln.		3b	625
NaCl	370	56	$FeCl_2$, NaCl	Eut.	2c	620
	374	56	$FeCl_3$, Na_2FeCl_4	Eut.		621
	400	~59	Na_2FeCl_4, NaCl	Per.		
	370	55	NaCl, $FeCl_2$	Eut.	2a,6c	626

$FeCl_2$

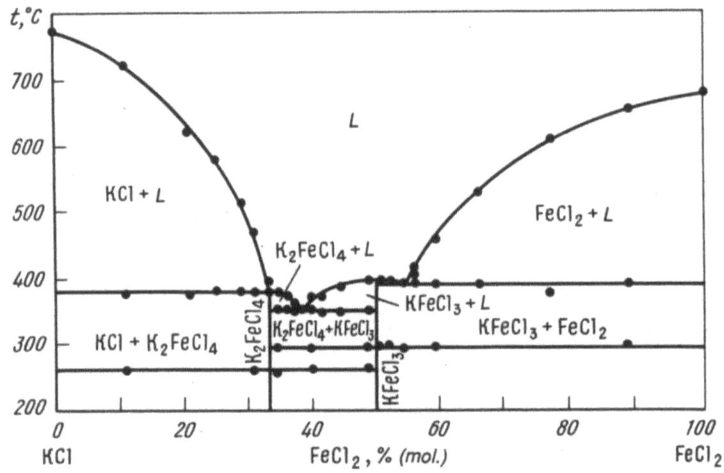

Fig. 121. KCl-$FeCl_2$ system.

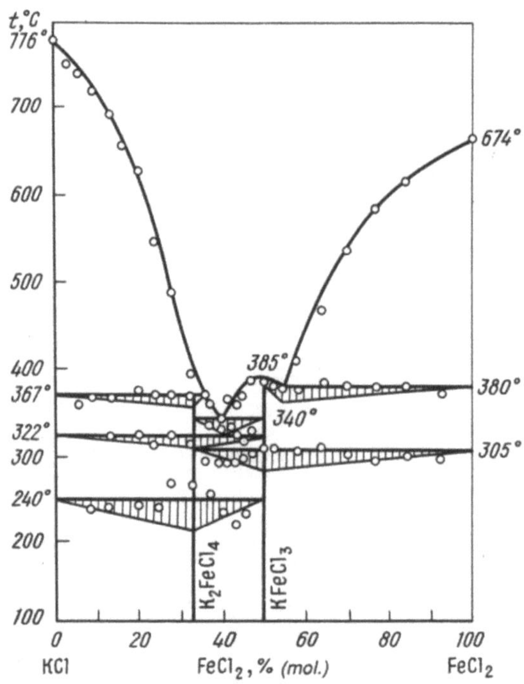

Fig. 122. KCl-$FeCl_2$ system.

$FeCl_2$

2nd Component	T°C	Conc. mol %	Phases involved	Def.	Method	Ref.
$NbCl_5$			Fig. 124			627
			Slight mutual solubility no new phases. Solidus temp. concides with melting temp. of $NbCl_5$		2a,3a	627
			Eutectic temp. and comp. very close to pure $NbCl_5$		2d,1	628
$NdCl_3$	608	40.2	$NdCl_3$, $FeCl_2$	Eut.	2c	620
	610	45	$NdCl_3$, $FeCl_2$	Eut.	2a,6c	623
$NiCl_2$	672	0.7	$FeCl_2$, $NiCl_2$	Eut.	2d	629
			Eutectic exists		3b	630
$PrCl_3$	463	43	$FeCl_2$, $PrCl_3$	Eut.	2a,6c	623
RbCl	643	36	$RbFeCl_3$, $FeCl_2$	Eut.	2d,3b	631
	501	50	$RbFeCl_3$	Comp.		
	458	59.5	Rb_2FeCl_4, $RbFeCl_3$	Eut.		
	489	66.7	Rb_2FeCl_4	Comp.		
	461	70	Rb_3FeCl_5, Rb_2FeCl_4	Eut.		
	467	72	RbCl, Rb_3FeCl_5	Per.		
$SnCl_2$	230	94	$FeCl_2$, $SnCl_3$		2c	622
$TaCl_5$			Slight mutual solubility Solidus at melting temp. of $TaCl_5$		2a,3b	627
			Eutectic close $TaCl_5$ comp. & temp.		2d,1	628

FeCl$_2$

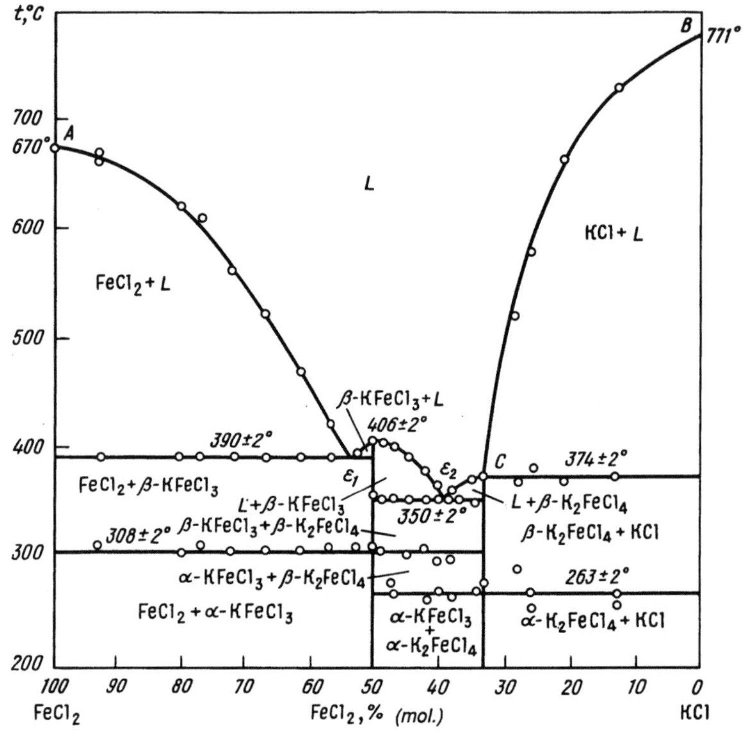

Fig. 123. FeCl$_2$-KCl system.

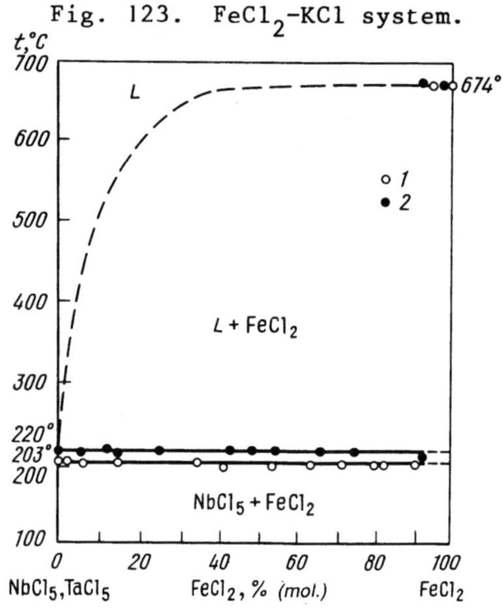

Fig. 124. NbCl$_5$, TaCl$_5$-FeCl$_2$ system.

$FeCl_2$

2nd Component	T°C	Conc. mol %	Phases involved	Def.	Method	Ref.
YCl_3	674	0.0	$FeCl_2$	M.P.	2	632
	550	58.4	Sol.slns.	Eut.		
	700	100	YCl_3	M.P.		
	420		$FeCl_2.YCl_3$	P.T.		
	360		$\alpha \rightleftharpoons \beta$ $FeCl_2.YCl_3$	P.T.		

$FeCl_3$

2nd Component	T°C	Conc. mol %	Phases involved	Def.	Method	Ref.
$GaCl_3$			Figs. 125 and 126			633
	305	0.0	$FeCl_3$	M.P.	2,3b,6k	633
	74.7	99.3	Sol.sln. with 99.7 and 60% $GaCl_3$	Eut.		
$GeCl_4$			Eutectic very close to pure $GeCl_4$ - comp. & temp.	Eut.	2d,1e	634

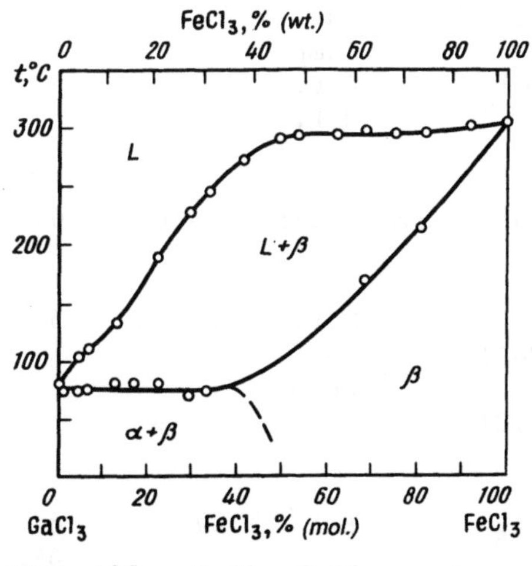

Fig. 125. $GaCl_3$-$FeCl_3$ system.

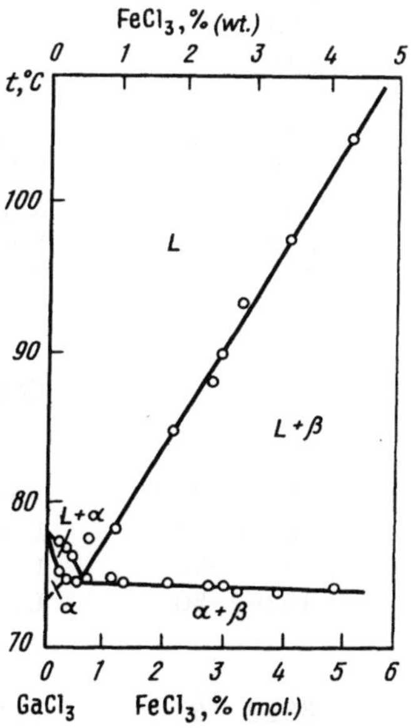

Fig. 126. $GaCl_3$-$FeCl_3$ system.

$FeCl_3$

2nd Component	T°C	Conc. mol %	Phases involved	Def.	Method	Ref.
$HfCl_4$	304	0.0	$FeCl_3$	M.P.	2a	635
	300±2	38.5	β-$HfCl_4$,$FeCl_3$	Eut.		
	432	100	α-$HfCl_4$	M.P.		
	416		$HfCl_4$	P.T.		
$InCl_3$	312	0.0	$FeCl_3$	M.P.	2a	636
	~312	~0.0	3% $FeCl_3$ sol.sln. in $InCl_3$, $FeCl_3$	Eut.		
KCl	303	0.0	$FeCl_3$	M.P.	2a	637
	202	47	$FeCl_3$.$KFeCl_4$	Eut.		
	243	50	$KFeCl_4$	Comp.		
	248	54	$KFeCl_4$, KCl	Eut.		
	242	52	$KFeCl_4$, KCl	Eut.	2,3b	638
	249	50	$KFeCl_4$	Comp.		
	206	45	$FeCl_3$, $KFeCl_4$	Eut.		
	530	55	KCl, K_2FeCl_5	Per.	2a	639
	256	67	K_2FeCl_5,$KFeCl_4$	Per.		
	220	39	$KFeCl_4$,$FeCl_3$	Eut.		
	23±3		K_2FeCl_5	P.T.		
	776	0.0	KCl	M.P.	1	640
	242	47	$KFeCl_4$, KCl	Eut.		
	251	50	$KFeCl_4$	Comp.		
	202	54	$FeCl_3$, $KFeCl_4$	Eut.		
	309	100	$FeCl_3$	M.P.		
$LaCl_3$			Complete mutual solubility		2a,6c	623
LiCl	303	0.0	$FeCl_3$	M.P.	2d,3b	641
	146-8	48	$FeCl_3$, $LiFeCl_4$	Eut.		
	~151-2	~50	$LiFeCl_4$	Comp.		
	151-2	50	$LiFeCl_4$, LiCl			
$MoCl_5$	88±2	63	$FeCl_3$, $MoCl_5$	Eut.	2c	642, 643
	185	93		Eut.	2a,6c	644

$FeCl_3$

2nd Component	T°C	Conc. mol %	Phases involved	Def.	Method	Ref.
NaCl	156	48	$FeCl_3$, NaCl	Eut.	2	645
			$NaFeCl_4$ confirmed (Latent heat of fusion 17.8 kcal/mol)	Comp.	6c	646
	157	48	$FeCl_3$, $NaFeCl_4$	Eut.	2,3b	638
	163	50	$NaFeCl_4$	Comp.		
	162	51	$NaFeCl_4$, NaCl	Eut.		
$NbCl_5$	198	80	$FeCl_3$, $NbCl_5$	Eut.	2,6c	647
	191	81	$FeCl_3$, $NbCl_5$	Eut.	1	648
$NdCl_3$			Mutual insolubility		2a	623
$ReCl_5$	236	55	$ReCl_5$, $FeCl_3$	Eut.	2a	649
$SbCl_3$	309	0.0	$FeCl_3$	M.P.	2d	650
	66	~98	$FeCl_3$, $SbCl_3$	Eut.		
	72	100	$SbCl_3$	M.P.		
$SbCl_5$	304	0.0	$FeCl_3$	M.P.	2d	651
	~3	~100	$SbCl_5$, $FeCl_3$	Eut.		
	3	100	$SbCl_5$	M.P.		
$SnCl_4$	303	0.0	$FeCl_3$	M.P.	2c	652
	-33	~100	$FeCl_3$, $SnCl_4$	Eut.		
	-33	100	$SnCl_4$	M.P.		
$TaCl_5$	200	76	$FeCl_3$, $TaCl_5$	Eut.	2,6c	653
	203	85.5	$FeCl_3$, $TaCl_5$	Eut.	1	648
$TeCl_4$	142	34.7	$FeCl_3$, $FeCl_3.TeCl_4$	Eut.	2a,3a	654
	155	50	$FeCl_3.TeCl_4$	Comp.		
	119	62.9	$FeCl_3.TeCl_4$, $TeCl_4$	Eut.		

$FeCl_3$

2nd Component	T°C	Conc. mol %	Phases involved	Def.	Method	Ref.
$ThCl_4$			Fig. 124			655
			Almost complete mutual insolubility		2,6c	655
	303	~0.0	$FeCl_3$, $ThCl_4$	Eut.		

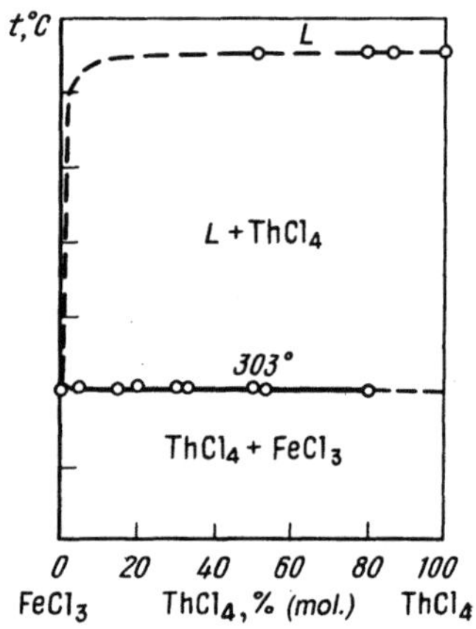

Fig. 127. $FeCl_3$-$ThCl_4$ system.

2nd Component	T°C	Conc. mol %	Phases involved	Def.	Method	Ref.
$TiCl_4$	-24	~0.0		Eut.	2,2d,1	645
	204	99.5)				
	212	99.0)	Solubility of $FeCl_3$ in $TiCl_4$			
	212	98.0)				
	215	97.1				
UCl_4	309	0.0	$FeCl_3$	M.P.	2d	656
	298	4.0	$FeCl_3$, UCl_4	Eut.		
	587	100	UCl_4	M.P.		
WCl_5	304	0.0	$FeCl_3$	M.P.	2d,3b	657
	180	30.0	$FeCl_3$, α-WCl_5	Eut.		
	240	100.0	β-WCl_5	M.P.		
	180		WCl_5	P.T.		
$ZrCl_4$	300	15	$FeCl_3$, $ZrCl_4$	Eut.	2	658

$FeCl_3$

2nd Component	T°C	Conc. mol %	Phases involved	Def.	Method	Ref.
$ZrCl_4$	304	0.0	$FeCl_3$	M.P.	2a	635
	300±2	6.0	$FeCl_3$, $ZrCl_4$	Eut.		
	437	100	$ZrCl_4$	M.P.		
	426		$\alpha \rightleftharpoons \beta$ $ZrCl_4$	P.T.		

FeF_2

2nd Component	T°C	Conc. mol %	Phases involved	Def.	Method	Ref.
LiF	<1000		Do not react		3b	659
NaF	788 ±10		$NaFeF_3$ - decomposes	P.T.	3b	659
RbF	730 ±10		Rb_2FeF_4	Comp.	3b	659
	"		$RbFeF_3$	Comp.		
SnF_2			Compounds at molar ratios: 14:1, 8:1, 4:1, 3:2, 2:1, 1:1 contain anions: SnF_5^-, SnF_3^-		2c	660
TlF	665 ±10		$TlFeF_3$	Comp.	3b	659

FeF_3

2nd Component	T°C	Conc. mol %	Phases involved	Def.	Method	Ref.
PbF_2	569±8		$Pb_3(FeF_6)_2$ (Quadratic modification)	Comp.	3b, 1e, 2d	661
			Sol. slns. of orthorhombic form PbF_2 transforms in solid state.			
SrF_2	775±10		$SrFeF_5$	Comp.		661a
	788±12		$Sr_3(FeF_6)_2$	Comp.		

FeI_2

2nd Component	T°C	Conc. mol %	Phases involved	Def.	Method	Ref.
GaI_3	188	75	GaI_3, 16% GaI_3 sol.sln. in FeI_2	Eut.	2,3b	662
InI_3	570	0.0	FeI_3	M.P.	2d	663
	~570	~100	FeI_3, InI_3	Eut.		
	208	100	InI_3	M.P.		
	130		$InI_3 + 4FeI_2 \rightleftharpoons InI_3.4FeI_2$	P.T.		
TeI_4	576	0.0	FeI_2	M.P.	2,3b	664
	253	73	FeI_2, TeI_4	Eut.		
	280	100	TeI_4	M.P.		
	570	0.0	FeI_2	M.P.	2d	665
	260	72.5	FeI_2, TeI_4	Eut.		
	272	100	TeI_4	M.P.		

GaBr

2nd Component	T°C	Conc. mol %	Phases involved	Def.	Method	Ref.
$GaBr_2$	152	98.25	Ga_2Br_4,GaBr	Eut.	2c	666

$GaBr_3$

2nd Component	T°C	Conc. mol %	Phases involved	Def.	Method	Ref.
NaBr	123	0	$GaBr_3$	M.P.	2,3b	668
	87	30	$GaBr_3$,$NaGaBr_4$	Eut.		
	180	44.5	$NaGaBr_4$, NaBr	Per.		
	755	100	NaBr	M.P.		
RbBr	80	20	$GaBr_3$, $RbGa_2Br_7$	Eut.	1,2,3b, 1e	669
	125		$RbGa_2Br_7$, $RbGaBr_4$	Per.		
	252		$RbGaBr_4$, RbBr	Per.		
TlBr	114	30	$GaBr_3$, α-$TlGaBr_4$	Eut.	1,2e,3b, 1e	669

$GaBr_3$

2nd Component	T°C	Conc. mol %	Phases involved	Def.	Method	Ref.
TlBr	120		α-$TlGaBr_4$, β-$TlGaBr_4$	Per.		669
	196	50	β-$TlGaBr_4$	Comp.		
	190	55	β-$TlGaBr_4$, TlBr	Eut.		
GaI_3			Fig. 128			670
	122	0.0	$GaBr_3$	M.P.	2d,3b,4a	670
	100	30	Continuous sol. sln. (GaI_3base)	Min.		
	40		Sol.sln. decomposes $3GaBr_3.GaI_3$ appears	P.T.		

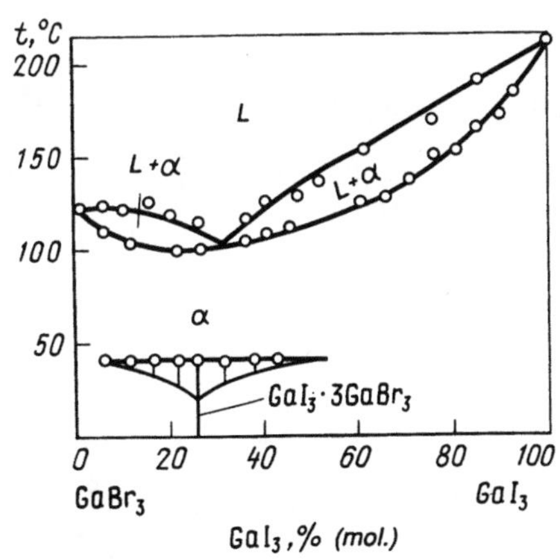

Fig. 128. $GaBr_3$-GaI_3 system.

$GaCl_2$

2nd Component	T°C	Conc mol %	Phases involved	Def.	Method	Ref.
$GaCl_3$	84	47	β-$GaCl_2$, Ga_3Cl_7	Per.	2,4a	667
	65	59	Ga_3Cl_7	Eut.		
	55-200		Existence of Ga_3Cl_7 (or $Ga.Ga_2Cl_7$) shown by electrical conductivity measurements	P.T.		
$InCl_2$			Fig. 129			671
	161	2.0	Continuous sol.sln. $GaCl_2$-$InGaCl_4$	Min.	2d	671
	262		$InGaCl_4$	Comp.		
	224	88.5	Continuous sol.sln. $InGaCl_4$-$InCl_2$			
$MgCl_2$	170	~0.0	Ga_2Cl_4, $Ga_2Cl_4.MgCl_2$	Eut.	2a,3b,5	672
	210	<2.0	$Ga_2Cl_4.MgCl_2$, $MgCl_2$	Per.		
	58		Ga_2Cl_4	P.T.		

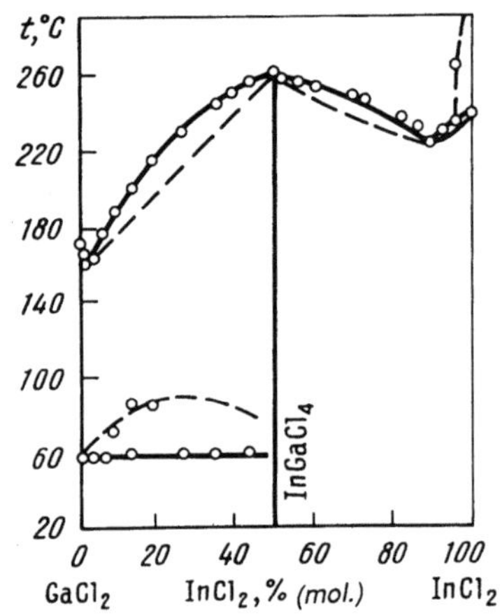

Fig. 129. $GaCl_2$-$InCl_2$ system.

$GaCl_2$

2nd Component	T°C	Conc. mol %	Phases involved	Def.	Method	Ref.
$MnCl_2$	170	~0.0		Eut.	2a,5	672
			Melts containing 80, 44.5 and 20.5% of Ga_2Cl_4 partly decompose (2.7-3.4% $GaCl_2$) to: $2GaCl_3$ and Ga			
	58		Ga_2Cl_4	P.T.		
NaCl			On melting the mixture a metallic phase is liberated and the system becomes multiphase		2a	671
$SrCl_2$	170	~0.0	Ga_2Cl_4, $Ga_2Cl_4.SrCl_2$	Eut.	2a,3b,5	672
	300	8	$Ga_2Cl_4.SrCl_2$, $SrCl_2$	Per.		
	58		Ga_2Cl_4	P.T.		
$ZnCl_2$			Fig. 130 Almost complete mutual insolubility in either solid or liquid state			673

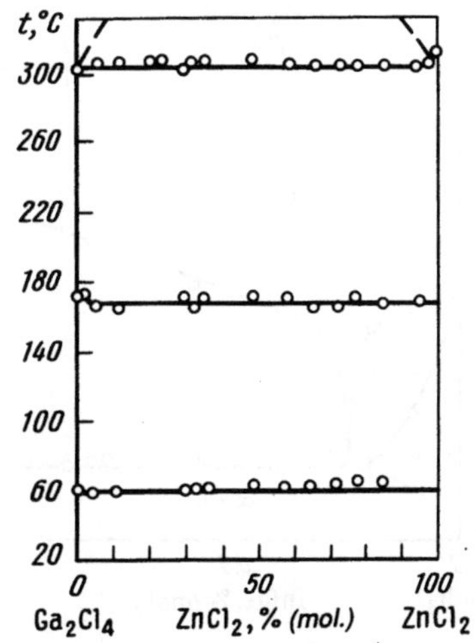

Fig. 130. Ga_2Cl_4-$ZnCl_2$ System.

$GaCl_3$

2nd Component	T°C	Conc mol %	Phases involved	Def.	Method	Ref.
$GeCl_4$	54	~93	$GaCl_3$, $GeCl_4$	Eut.	2a	674
$HgCl_2$	55	8	$GaCl_3$, $Hg(GaCl_4)_2$	Eut.	2a	675
	72	40.1	$Hg(GaCl_4)_2$, $HgCl_2$	Eut.		
	277	100	$HgCl_2$	M.P.		
InCl	240	40	$InGaCl_4$, $In_5Ga_3Cl_{14}$	Per.	2,5,3b	676
	208		$\beta \rightleftharpoons \alpha$ $In_5Ga_3Cl_{14}$	Per.		
	190	75	α-$In_5Ga_3Cl_{14}$, InCl	Eut.		
$InCl_3$			Fig. 131			677
	78	0.0	$GaCl_3$	M.P.	2d	677
	75	~1.3	$GaCl_3$, $InCl_3$	Eut.		
	580	100	$InCl_3$	M.P.		
	75.9	~0.25	~0.2% $InCl_3$ sol. sln. in $GaCl_3$		2d,3a	678

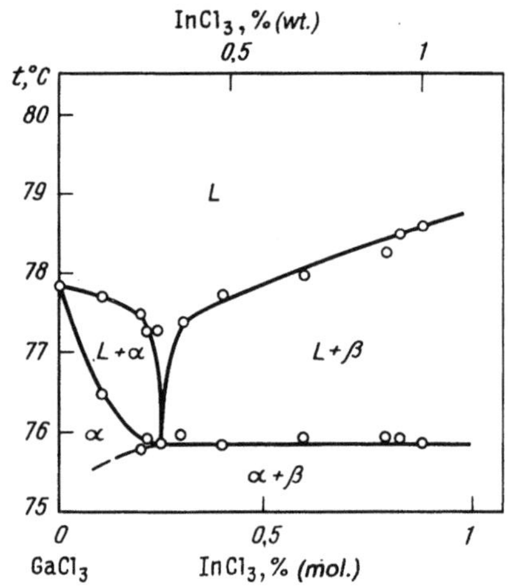

Fig. 131. $GaCl_3$-$InCl_3$ system.

$GaCl_3$

2nd Component	T°C	Conc. mol %	Phases involved	Def.	Method	Ref.
KCl	63	14.9	$GaCl_3$,KGa_2Cl_7	Eut.	2a	679
	128	33.7	KGa_2Cl_7	Comp.		
	94	37.1	KGa_2Cl_7,$KGaCl_4$	Eut.		
	228	50	$KGaCl_4$	Comp.		
	210	51	$KGaCl_4$,KCl	Eut.		
	90		$KGaCl_4$	P.T.		
	74	17	KCl.2$GaCl_3$,$GaCl_3$	Eut.		680
	106	33	KCl.$GaCl_3$, KCl.2$GaCl_3$	Per.		
	248	50	KCl,KCl.$GaCl_3$	Per.		
LiCl	78	0	$GaCl_3$	M.P.	2a	679
	57	18	$GaCl_3$,$LiGaCl_4$	Eut.		
	166	45.2	$LiGaCl_4$,LiCl			
			Fig. 132			681
	67	20	2LiCl.3$GaCl_3$,$GaCl_3$	Eut.		681

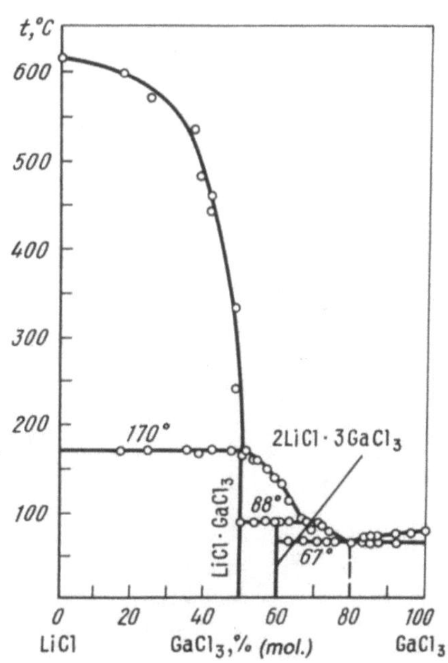

Fig. 132. LiCl-$GaCl_3$ system.

$GaCl_3$

2nd Component	T°C	Conc. mol %	Phases involved	Def.	Method	Ref.
LiCl	88	32	$LiCl.GaCl_3$, $2LiCl.3GaCl_3$	Per.		681
	170	48.5	$LiCl, LiCl.GaCl_3$	Per.		
			Tensometric analysis suggests limited sol.sln.			
$MgCl_2$			Fig. 133			682
	73.6	1.15	0.7% $MgCl_2$ sol.sln. in $GaCl_3$, $Mg(GaCl_4)_2$		2a	682
	205		$Mg(GaCl_4)_2$ decomposes	P.T.		
$MnCl_2$	76	<0.3	$GaCl_3$, $MnCl_2.2GaCl_3$	Eut.	2,3b	683
	210	13.5	$MnCl_2.2GaCl_3$, $MnCl_2$	Per.		

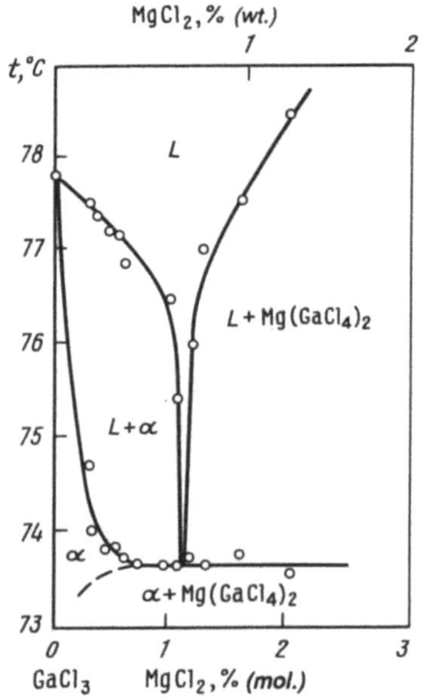

Fig. 133. $GaCl_3$-$MgCl_2$ system.

$GaCl_3$

2nd Component	T°C	Conc. mol %	Phases involved	Def.	Method	Ref.
$MoCl_5$	67	~7	$GaCl_3$,$GaCl_3.MoCl_5$	Eut.	2	684
	109	~47	$GaCl_3.MoCl_5$,$MoCl_5$	Per.		
			Probable sol.sln.			
NH_4Cl	54	15.8	$GaCl_3$,$NH_4Ga_2Cl_7$	Eut.		676
	308	49.7	NH_4GaCl_4	Comp.		
	225	65.0	NH_4GaCl_4,NH_4Cl	Eut.		
	92		$NH_4Ga_2Cl_7$ - Melts and decomposes	M.P.,T.P.		
NaCl	62	2.9	$GaCl_3$,$NaCl.GaCl_3$	Eut.	2b,1	685
	238	18.0	$NaCl.GaCl_3$,NaCl	Per.		
	62	3.0	$GaCl_3$,$NaGaCl_4$	Eut.	2a,6c	686
	169	50	$NaGaCl_4$			
	167	52	$NaGaCl_4$,NaCl	Eut.		
$NbCl_5$	127	40		Per.	2	684
			Eut.temp. coincides with M.P. of $GaCl_3$			
	78		$GaNbCl_8$,$GaCl_3$	Eut.	2a	687
	126		$GaNbCl_8$,$NbCl_5$	Per.		
			Eutectoid composition close to $GaCl_3$			
$NiCl_2$		<0.5	Solubility of $NiCl_2$		2	683
	77			Eut.		
RbCl	75	3	$RbCl.2GaCl_3$,$GaCl_3$	Eut.	2d	688
	110	25	$RbCl.GaCl_3$, $RbCl.2GaCl_3$	Per.		
	330	50	$RbCl.GaCl_3$	Comp.		
	310	54.5	RbCl,$RbCl.GaCl_3$	Eut.		
	70	8	$3GaCl_3.2RbCl$,$GaCl_3$	Eut.		689
	102	28	$RbCl.GaCl_3$, $3GaCl_3.2RbCl$	Per.		
	345	50	$RbCl.GaCl_3$	Comp.		
	328	54	RbCl,$RbCl.GaCl_3$	Eut.		

$GaCl_3$

2nd Component	T°C	Conc. mol %	Phases involved	Def.	Method	Ref.
$SbCl_3$	50.5	30.6	$SbCl_3.GaCl_3$,$GaCl_3$	Eut.		690
	60.5	50.0	$SbCl_3.GaCl_3$	Comp.		
	51.8	67.7	$SbCl_3$,$SbCl_3.GaCl_3$	Eut.		
$SbCl_5$	78	0.0	$GaCl_3$	M.P.	2d	691
	-6	84	$GaCl_3$,$SbCl_5$	Eut.		
	3	100	$SbCl_5$	M.P.		
$SeCl_4$			Fig. 134			692
	30.0	21	$GaCl_3$, β-$2GaCl_3.SeCl_4$	Eut.	2a	692
	46.0	~28	β-$2GaCl_3.SeCl_4$, $GaCl_3.SeCl_4$	Per.		
	170	~49	$GaCl_3.SeCl_4$, $SeCl_4$	Per.		
	24		$2GaCl_3.SeCl_4$	P.T.		

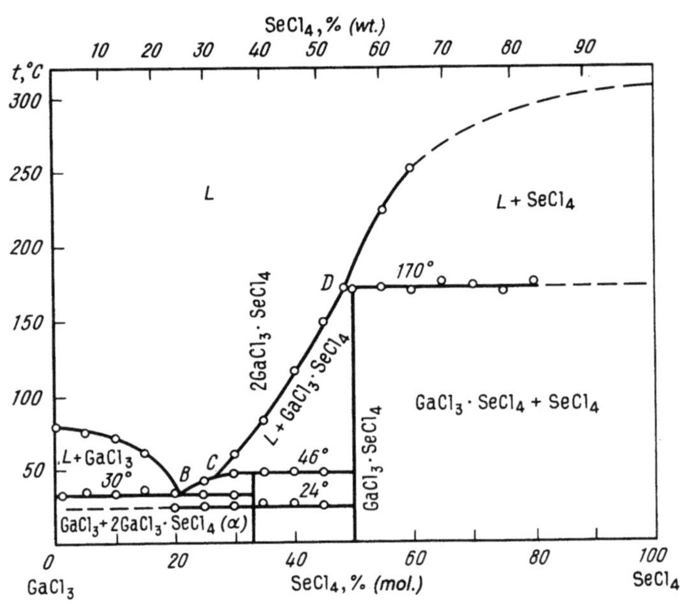

Fig. 134. $GaCl_3$-$SeCl_4$ system.

$GaCl_3$

2nd Component	T°C	Conc. mol %	Phases involved	Def.	Method	Ref.
$SrCl_2$	78	0.0	$GaCl_3$, $Sr(GaCl_4)_2$	Eut.		693
	198	14.2	$Sr(GaCl_4)_2$, $SrGaCl_5$	Per.		
	297	35.7	$SrGaCl_5$, $SrCl_2$	Per.		
			X-ray analysis confirmed the existence of $SrGaCl_5$ and failed to confirm $Sr(GaCl_4)_2$ due to strong hygroscopy of melts.			
$TaCl_5$	78	0.0	$GaCl_3$	M.P.	2a	683
	~78	~0		Eut.		
			$GaTaCl_8$ - melts incongruously	Comp.		
$TeCl_4$			Fig. 135			692
	45	20	$GaCl_3$, $2GaCl_3.TeCl_4$	Eut.	2a	692
	57	24	$2GaCl_3.TeCl_4$, $GaCl_3.TeCl_4$	Per.		
	158	50	$GaCl_3.TeCl_4$	Comp.		
	140	60	$GaCl_3.TeCl_4$, $TeCl_4$	Eut.		

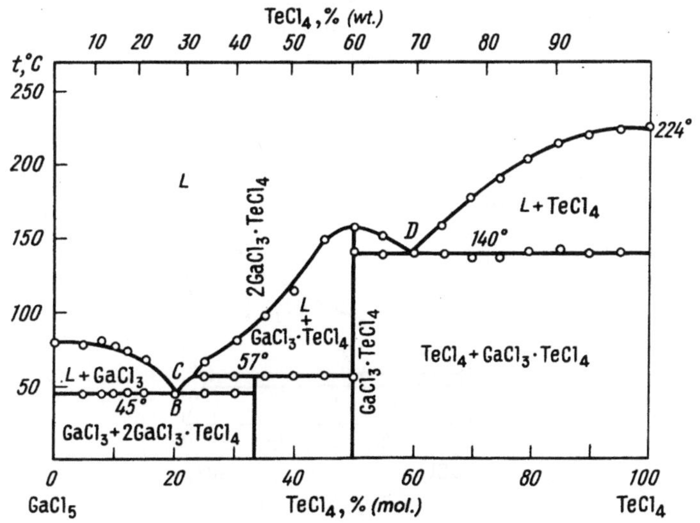

Fig. 135. $GaCl_5$-$TeCl_4$ system.

$GaCl_3$

2nd Component	T°C	Conc. mol %	Phases involved	Def.	Method	Ref.
$TiCl_4$	-25	~90	$GaCl_3$,$TiCl_4$	Eut.	2a	674
TlCl	130	25	$TlGa_2Cl_7$, $TlGaCl_4$	Per.	2a	679
	378	50	$TlGaCl_4$	Comp.		
	290	67.8	$TlGaCl_4$,TlCl	Eut.		
			Liquidus line corresponding to cryst. of $GaCl_3$ is not marked on the diagram.			
WCl_6	75		WCl_6 sol.sln. $GaCl_3$	Eut.	2d	694
	224	100	WCl_6 $\alpha \rightleftharpoons \beta$	P.T.		
	174	100	WCl_6 $\beta \rightleftharpoons \gamma$	P.T.		
$ZnCl_2$	78	0.0	$GaCl_3$	M.P.	2a	675
	~30	~32	$GaCl_3$, $ZnCl_2$	Eut.		
		>25	Glass			
		25-35	Glass-semiliquid at ambient temp.			
GaI_3			Fig. 136			725
	43	9.5	2% GaL_3 sol.sln. in $GaCl_3(\alpha)$	Eut.	2d,3b,4a	725

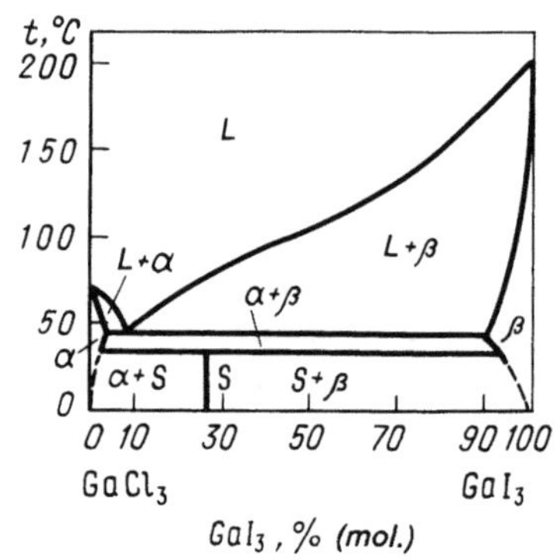

Fig. 136. $GaCl_3$-GaI_3 system.

$GaCl_3$

2nd Component	T°C	Conc. mol %	Phases involved	Def.	Method	Ref.
GaI_3	212	100	10% $GaCl_3$ sol. sln. in GaI_3(β) GaI_3	M.P.		725
	32		GaI_3 sol.sln. (β) + $GaCl_3$ sol.sln. (α) ⇌ $3GaCl_3.GaI_3$	P.T.		

GaF_3

2nd Component	T°C	Conc. mol %	Phases involved	Def.	Method	Ref.
KF	615-620	47-48	$K_2Ga_5F_{17}$,GaF_3	Per.		695
	588	50.5	α-$KGaF_4$, $K_2Ga_5F_{17}$	Per.		
	540	54.5	K_2GaF_5,α-$KGaF_4$	Eut.		
	600	58.5	γ-K_2GaF_6,K_2GaF_5	Per.		
	1002	75.0	γ-K_3GaF_6	Comp.		
	810	92.5	KF, γ-K_3GaF_6	Eut.		
	172±1		α ⇌ β } K_3GaF_6	P.T.		
	239±1		β ⇌ γ } K_3GaF_6			
	540		α ⇌ β $KGaF_4$	P.T.		
LiF	600	61.2	α-$GaLi_3F_6$,GaF_3	Eut.		695
	740	75.0	α-$GaLi_3F_6$	Comp.		696
	690	82.0	LiF, α-$GaLi_3F_6$	Eut.		
	525		$GaLi_3F_6$	P.T.		
NaF	598	50	$NaGaF_4$, GaF_3	Per.	2a,3b	695
	595	56	$Na_5Ga_3F_{14}$, $NaGaF_4$	Eut.		
	665	61.5	α-Na_3GaF_6, $Na_5Ga_3F_{14}$	Per.		
	960	75	α-Na_3GaF_6	Comp.		
	855	82	NaF,α-Na_3GaF_6	Eut.		
	593-605		Na_3GaF_6	P.T.		

GaF_3

2nd Component	T°C	Conc. mol %	Phases involved	Def.	Method	Ref.
PbF_2	630	52-53	$PbGaF_5$, Phase of unknown composition	Per.	2d,3b	697
	608	55.5	$Pb_3(GaF_6)_2$, $PbGaF_5$	Eut. Eut.		
	626	60.0	$Pb_3(GaF_6)_2$	Comp.		
	550	72.3	$Pb_9Ga_2F_{24}$, $Pb_3(GaF_6)_2$	Eut.		
	571	74.5	$Pb_9Ga_2F_{24}$, 12%GaF_3 sol.sln. in PbF_2	Per.		
	510±3		$PbGaF_5$	P.T.		
	395±10		$Pb_3(GaF_6)_2$	P.T.		
	500		α-phase - sol.sln. of cubic form, β-phase - rhombic form, $Pb_3(GaF_6)_2$, $PbGaF_5$ } hexagonal		3b	698
RbF	620	47.0	$Rb_2Ga_5F_{17}$, GaF_3	Per.	2e,3b	695
	595	50.5	$RbGaF_4$, $Rb_2Ga_5F_{17}$	Per.		
	548	57.0	Rb_2GaF_5, $RbGaF_4$	Eut.		
	565	58.8	δ-Rb_3GaF_6, Rb_3GaF_5	Per.		
	952	75.0	δ-Rb_3GaF_6	Comp.		
	765	94.0	RbF, δ-Rb_3GaF_6	Eut.		
	91±2		α ⇌ β } Rb_3GaF_6	P.T.		
	106±2		β ⇌ γ } Rb_3GaF_6	P.T.		
	385±4		γ ⇌ δ } Rb_3GaF_6	P.T.		

GaI_3

2nd Component	T°C	Conc. mol %	Phases involved	Def.	Method	Ref.
GeI_4	118	26	GaI_3, GeI_4	Eut.	2,3b	699
HgI_2			Fig. 137			700
	140	30	GaI_3, HgI_2 The inflection on the HgI_2 liquidus is probably due to the density of gallium ions which do not form a separate phase.	Eut.	2a,3b	700
InI_3			Continuous sol. sln. with a flat minimum.		2,3b	699
KI	140	35	GaI_3, $KGaI_4$	Eut.	2,3b,1e	701
	206	50	$KGaI_4$	Comp.		
	200	53	$KGaI_4$, KI	Eut.		
MgI_2			Fig. 138			700
	200	17	GaI_3,$MgGaI_5$	Eut.	2,3b	700
	230	~24	$MgGaI_5$, MgI_2	Per.		
NaI	150	33	GaI_3, β-$NaGaI_4$	Eut.	2,3b,1e	701
	214	50	β-$NaGaI_4$	Comp.		
	210	53	β-$NaGaI_4$, NaI	Eut.		
	58		$NaGaI_4$	P.T.		

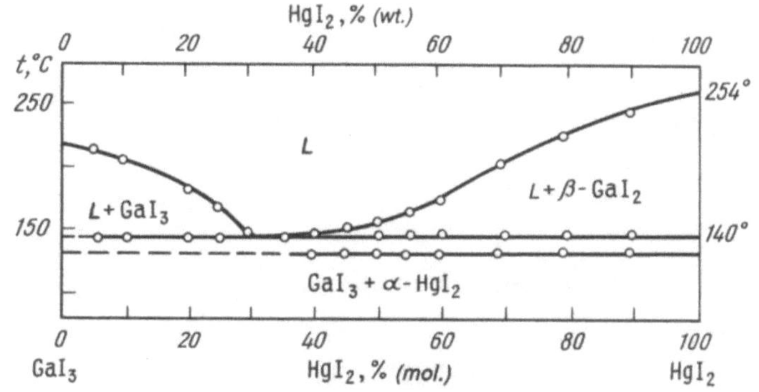

Fig. 137. GaI_3-HgI_2 system.

GaI_3

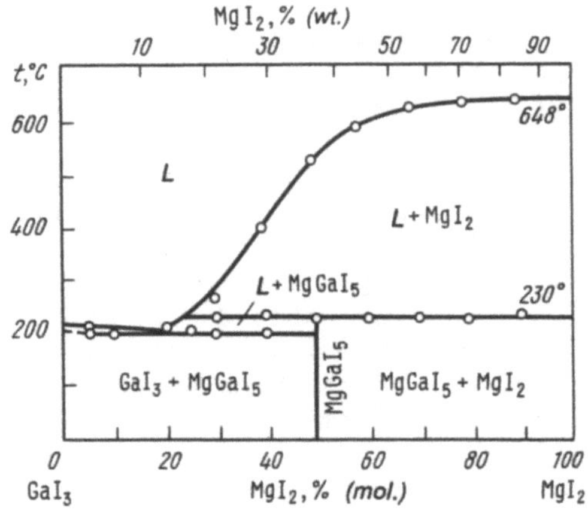

Fig. 138. GaI_3-MgI_2 system.

2nd Component	T°C	Conc. mol %	Phases involved	Def.	Method	Ref.
NiI_2			Fig. 139			700
	188	29	GaI_3, 30% GaI_3 sol.sln. in NiI_2	Eut.	2,3b	700
			Fig. 140			700
PbI_2	180	15	$Pb(GaI_4)_2$, 10% PbI_2 sol.sln. in GaI_3	Eut.	2,3b	700

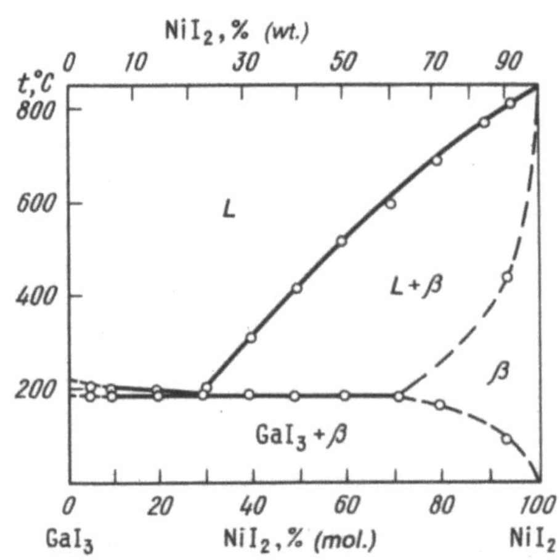

Fig. 139. GaI_3-NiI_2 system.

GaI_3

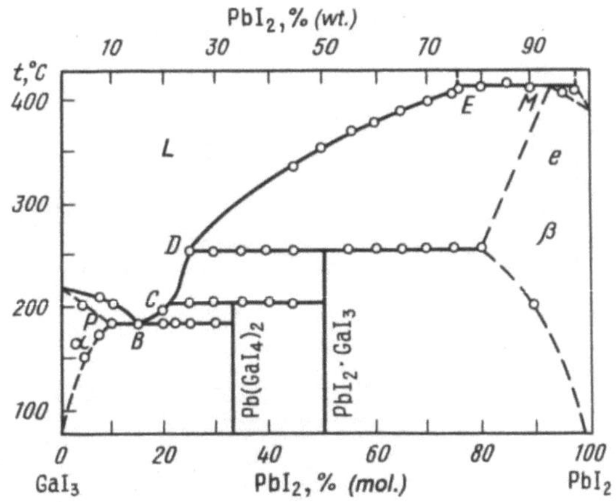

Fig. 140. GaI_3-PbI_2 system.

2nd Component	T°C	Conc. mol %	Phases involved	Def.	Method	Ref.
PbI_2	200	~22	$Pb(GaI_4)I, Pb(GaI_4)_2$	Per.		700
	250	~25	20% GaI_3 sol.sln. in PbI_2, $Pb(GaI_4)I$	Per.		
		75-95	Separation at temperature slightly higher than M.P. of PbI_2			
SbI_3	120	54		Eut.	2,3b	699
SnI_2			Fig. 141			700
	160	18	$Sn(GaI_4)_2$, 6% SnI_2 sol.sln. in GaI_3	Eut.	2,3b	700
	230	~30	20% sol.sln. GaI_3 in β-SnI_2, $Sn(GaI_4)_2$	Per.		
	112		$Sn(GaI_4)_2$	P.T.		
	142		SnI_2	P.T.		
SnI_4	120	30	GaI_3, SnI_4	Eut.	2,3b	699
TeI_4	175	73	GaI_3, TeI_4	Eut.	2,3b	699

GaI_3

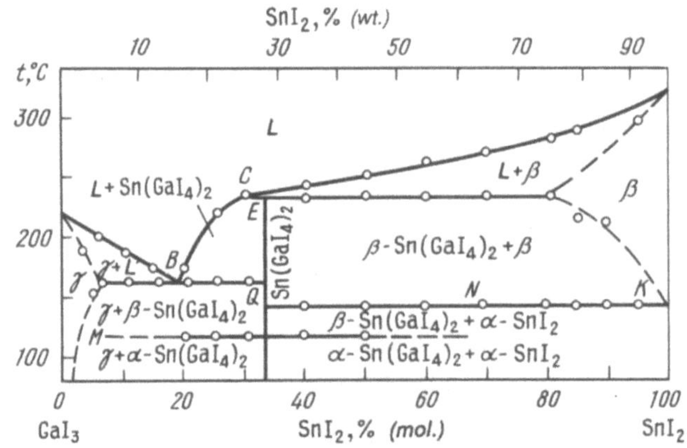

Fig. 141. GaI_3-SnI_2 system.

2nd Component	T°C	Conc. mol %	Phases involved	Def.	Method	Ref.
TlI	140	35	$TlGaI_4$, GaI_3	Eut.	2,3b,1e	701
	220	50	$TlGaI_4$	Comp.		
	208	55	β-TlI, $TlGaI_4$	Eut.		
	178		TlI	P.T.		
ZnI_2	140	38	GaI_3, ZnI_2	Eut.	2,3b	700

$GdCl_3$

2nd Component	T°C	Conc. mol %	Phases involved	Def.	Method	Ref.
KCl			Fig. 142			702
	540	15	$GdCl_3$,KGd_3Cl_{10}	Eut.		703
	580	25	KGd_3Cl_{10}	Comp.		
	494	45	KGd_3Cl_{10},K_2GdCl_5	Eut.		
	542	56	K_2GdCl_5,β-K_3GdCl_6	Per.		
	825	77	β-K_3GdCl_6	Comp.		
	686	93	β-K_3GdCl_6,KCl	Eut.		
	380		K_3GdCl_6	P.T.		

$GdCl_3$

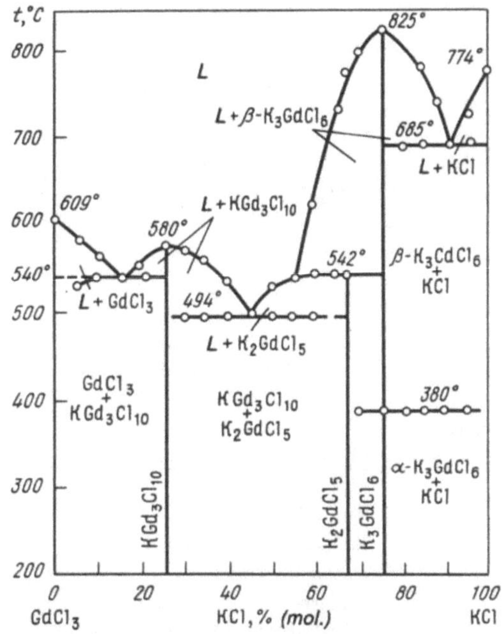

Fig. 142. $GdCl_3$-KCl system.

2nd Component	T°C	Conc. mol %	Phases involved	Def.	Method	Ref.
NaCl	396	40	$GdCl_3$, Na_3GdCl_6	Eut.	2a	702 703
	486	70	Na_3GdCl_6,NaCl	Per.		
YCl_3	552	25	Continuous sol.sln.	Min.	2	704

GdF_3

2nd Component	T°C	Conc. mol %	Phases involved	Def.	Method	Ref.
LiF	750	57	LiF.GdF_3,GdF_3	Per.	2	705
	630	75	LiF,LiF.GdF_3	Eut.		
	880		Probable p.t.	P.T.		
NaF			Fig. 143			706
	1050	30	5NaF.9GdF_3, GdF_3	Eut.	2,1f,3b	707
	1090	34	5NaF.9GdF_3	Comp.		
	840	56	NaF.GdF_3, 5NaF.9GdF_3	Per.		

$GdCl_3$

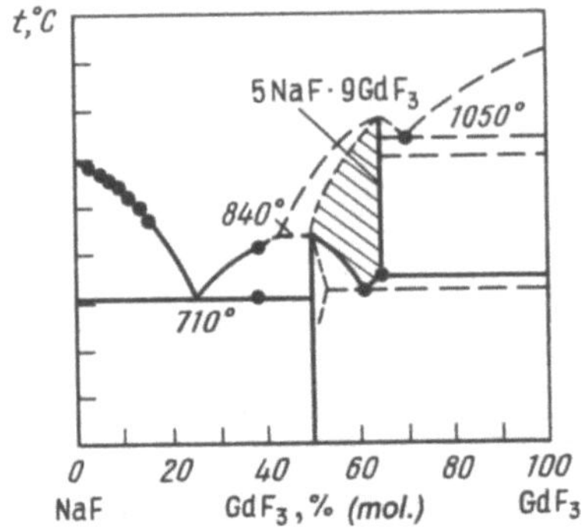

Fig. 143. NaF-GdF_3 system.

2nd Component	T°C	Conc. mol %	Phases involved	Def.	Method	Ref.
NaF	710	75	NaF,NaF.GdF_3 NaF.GdF_3 exhibits two forms: high temp.- cubic (fluorite), and low temp.- hexagonal. The cubic phase forms sol.sln. with 5NaF.9GdF_3 decomposing at lower temp.	Eut.		707
SrF_2			Limited sol.slns. of hexagonal modification.		3b	708

GdI_3

2nd Component	T°C	Conc. mol %	Phases involved	Def.	Method	Ref.
KI	940		GdI_3	M.P.	2d	709
	748		α-GdI_3, β-GdI_3	Per.		
	450		GdI_3,3KI.GdI_3	Eut.		
	579		3KI.GdI_3	Comp.		
	535		3KI.GdI_3, KI	Eut.		
	677		KI	M.P.		
NaI	940		GdI_3	M.P.	2d	709
	460		GdI_3,3NaI.GdI_3	Eut.		
	463		3NaI.GdI_3,NaI	Per.		
	654		NaI	M.P.		
	381		3NaI.GdI_3	P.T.		

$GeBr_4$

2nd Component	T°C	Conc. mol %	Phases involved	Def.	Method	Ref.
$CuBr_2$		~0.0	$GeBr_4$, $CuBr_2$	Eut.	2d	710
		4-90	Decomposition			
$SbBr_3$	25	1.0	$SbBr_3$, $GeBr_4$		2d	710
$SnBr_4$	19.4	56	Continuous sol.sln.	Min.	2d	710
$TiBr_4$			Continuous sol.sln. with no invariants		2d	710

$GeCl_4$

2nd Component	T°C	Conc. mol %	Phases involved	Def.	Method	Ref.
$MoCl_5$	194	100	$MoCl_5$	M.P.	2d,1	711
			Eutectic			
$SnCl_4$	-49.7	0.0	$GeCl_4$	M.P.	2d	712
	-51.3	9.0	Continuous sol.sln.	Min.		
	-33	100.0	$SnCl_4$	M.P.		
$TeCl_4$	-51.4	0.0	$GeCl_4$	M.P.	2d	713
	-52.0	~0.0	$GeCl_4$, $TeCl_4$	Eut.		
	224	100	$TeCl_4$	M.P.		
	217	10-80	Decomposition			
$TiCl_4$	-49.7	0.0	$GeCl_4$	M.P.	2d	712
	-50.8	~3.3	Continuous sol.sln.	Min.		
	-24	100	$TiCl_4$	M.P.		
VCl_4	-51.3	2.4	Continuous sol.sln.	Min.	2d	714
WCl_6		10-80	WCl_6 sol.sln.		2d	715
$ZrCl_4$			Eutectic temp. of composition coincides with pure $GeCl_4$.		2d	714

GeI_4

2nd Component	T°C	Conc. mol %	Phases involved	Def.	Method	Ref.
InI_3	144	0.0	GeI_4	M.P.	2d	716
	131	20	GeI_4, 7% GeI_4 sol.sln. in InI_3	Eut.		
	210	100	InI_3	M.P.		

GeI_4

2nd Component	T°C	Conc. mol %	Phases involved	Def.	Method	Ref.
SnI_4	134	40	Continuous sol.sln.	Min.	2d,3b	717
TiI_4	136.5	15	Continuous sol.sln.	Min.	2d,3b	717
ZrI_4		~0.0	GeI_4, ZrI_4 (temp. close to M.P.of GeI_4)	Eut.	2d,3b	717

$HfBr_4$

2nd Component	T°C	Conc. mol %	Phases involved	Def.	Method	Ref.
$ZrBr_4$			Unlimited sln. in both liquid and solid. Liquidus almost a straight line.		1	718

$HfCl_4$

2nd Component	T°C	Conc. mol %	Phases involved	Def.	Method	Ref.
KCl	432	0.0	$HfCl_4$	M.P.	2	719
	242	38	$HfCl_4$,K_2HfCl_6	Eut.		
	802	66.7	K_2HfCl_6	Comp.		
	550	77.6	K_2HfCl_6, KCl	Eut.		
	550		K_3HfCl_7 - Decomposition			
NaCl	548	73	$HfCl_4$, NaCl	Eut.	2	720
			Fig. 144			719
	336	40.6	Na_2HfCl_6,$HfCl_6$	Eut.	2	719
	660	66.7	Na_2HfCI_6	Comp.		
	540	73.4	NaCl,Na_2HfCl_6	Eut.		
	384		$\alpha \rightleftharpoons \beta$ Na_2HfCl_6	P.T.		
	440		$\beta \rightleftharpoons \gamma$ Na_2HfCl_6	P.T.		
	484		$\gamma \rightleftharpoons \delta$ Na_2HfCl_6	P.T.		
$SnCl_2$	432	0.0	$HfCl_4$	M.P.	1	721
	342	42.3	$SnCl_2.2HfCl_4$, $HfCl_4$	Eut.		
	218	81.0	$SnCl_2$, $SnCl_2.2HfCl_4$	Eut.		
	253	100	$SnCl_2$	M.P.		

$HfCl_4$

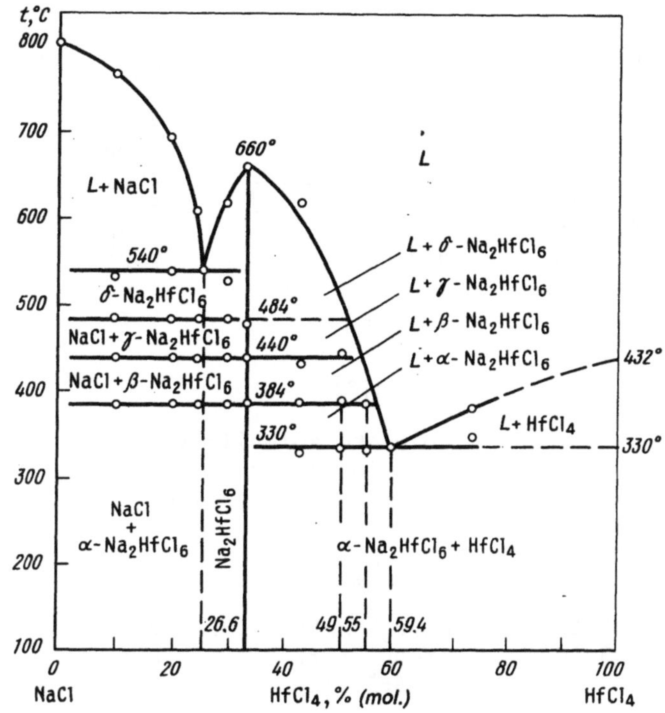

Fig. 144. $NaCl-HfCl_4$ system.

2nd Component	T°C	Conc. mol %	Phases involved	Def.	Method	Ref.
$ZnCl_2$	432	0.0	$HfCl_4$	M.P.	1	721
	305	93.5	$ZnCl_2$, $HfCl_4$	Eut.		
	315	100	$ZnCl_2$	M.P.		
$ZrCl_4$			Complete solubility in liquid and solid. Liquidus line is almost straight			

HfF_4

KF	420	43	$KHfF_5$, HfF_4	Eut.	1,2a,3b	722
	435	50	$KHfF_5$	Comp.		
	408	60	K_2HfF_6, $KHfF_5$	Eut.		
	608	65	K_3HfF_7, K_2HfF_6	Per.		
	923	75	K_3HfF_7	Comp.		
	772	87	KF, K_3HfF_7	Eut.		
	850	100	KF	M.P.		

HfF_4

2nd Component	T°C	Conc. mol %	Phases involved	Def.	Method	Ref.
NaF	538	48	$NaHfF_5$, HfF_4	Eut.	1,2a,3b	723
	540	50	$NaHfF_5$	Comp.		
	500	58	Na_2HfF_6,$NaHfF_5$	Eut.		
	582	65	Na_3HfF_7,Na_2HfF_6	Per.		
	860	75	Na_3HfF_7	Comp.		
	762	79	NaF,Na_3HfF_7	Eut.		
	750		NaF + Na_3HfF_7 ⇌ Na_4HfF_8	P.T.		

HfI_4

2nd Component	T°C	Conc. mol %	Phases involved	Def.	Method	Ref.
ZrI_4			Complete solubility in liquid and solid. Liquidus line almost straight		1	718

$HgBr_2$

2nd Component	T°C	Conc. mol %	Phases involved	Def.	Method	Ref.
TlBr	201±3	23	$HgBr_2$,$5HgBr_2$.TlBr	Per.	2d,3b	724
	164±3	40	$5HgBr_2$.TlBr $HgBr_2$.TlBr	Eut.		
	188±3	50	$HgBr_2$.TlBr	Comp.		
	175±3	52	$HgBr_2$.TlBr, $HgBr_2$.4TlBr	Eut. Eut.		
	302±3	73.5	$HgBr_2$.4TlBr,TlBr	Per.		
HgI_2		42	Continuous sol.sln. βHgI_2-$HgBr_2$	Min.	2d,3b,1g	726
	127		$\beta \rightleftharpoons \alpha$ HgI_2			
	<127		β-HgI_2-$HgBr_2$ ⇌ sol.sln. α-HgI_2-$HgBr_2$			

HgCl

2nd Component	T°C	Conc. mol %	Phases involved	Def.	Method	Ref.
$HgCl_2$	270	5	HgCl, $HgCl_2$	Eut.	1	727
$NbCl_5$	230		$Hg(NbCl_6)$ - incongruent melting. Eutectic comp. and temp. close to pure $NbCl_5$.	Comp.	2d,1	728
$TaCl_5$	270		$Hg(TaCl_6)$ - incongruent melting. Eutectic comp. and temp. close to pure $TaCl_5$.	Comp.	2d,1	728

Hg_2Cl_2

2nd Component	T°C	Conc. mol %	Phases involved	Def.	Method	Ref.
Hg_2I_2			Fig. 145			729
	216	90	Continuous sol.sln.	Min.	2d,3b,1d	729
			Two immiscible regions exist above liquidus:			

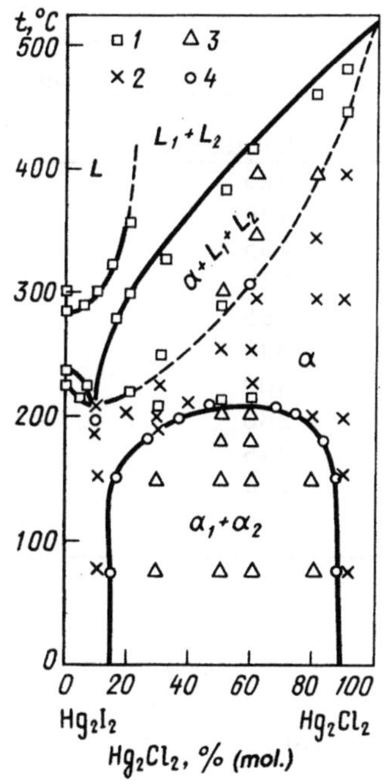

Fig. 145. Hg_2I_2-Hg_2Cl_2 system.

2nd Component	T°C	Conc. mol %	Phases involved	Def.	Method	Ref.
	360 → 288	80 → 100	Critical separation line			
			The immiscibility region probably closes at higher temp.			
		<90	Melting accompanied by decomposition: liquid enriched with metallic mercury and liquid enriched with the compound whose composition is shown in the diagram			

Hg_2Cl_2

2nd Component	T°C	Conc. mol %	Phases involved	Def.	Method	Ref.
Hg_2I_2	<212		Decomposition of solid solution			729
	Amb.	11-86	Limited sol.sln. Table V			

TABLE V

% Hg_2I_2	Temp. °C	
	Liquidus	Solidus
0.0	525	525
10.0	490	450
20.0	465	-
40.0	420	312
50.0	385	292
70.0	330	248
80.0	300	220
85.0	280	217
90.0	216	-
95.0	224	216
100.0	246	241

$HgCl_2$

2nd Component	T°C	Conc. mol %	Phases involved	Def.	Method	Ref.
$InCl_3$	270	2.0	$InCl_3$, $HgCl_2$		2a	730
$MoCl_5$	194	85	$MoCl_5$, $HgCl_2$	Eut.	2d,1	711
	194	100	$MoCl_5$	M.P.		
$NbCl_5$	196	82.3	$NbCl_5$, $HgCl_2$	Eut.	2d,1	731
RbCl	276	0.0	$HgCl_2$	M.P.	1,3b	732
	204	37	$RbHgCl_3$, $HgCl_2$	Eut.		
	275	50	$RbHgCl_3$	Comp.		
	256	59	Rb_2HgCl_4,$RbHgCl_3$	Eut.		
	328	66	RbCl,Rb_2HgCl_4	Per.		
	716	100	RbCl	M.P.		
			$RbHgCl_3$ is of fluorite cubic structure with a = 5.219Å			

$HgCl_2$

2nd Component	T°C	Conc. mol %	Phases involved	Def.	Method	Ref.
$SbCl_5$	280	0.0	$HgCl_2$	M.P.	2d	733
	~3	~100	$HgCl_2$,$SbCl_5$	Eut.		
	3	100	$SbCl_5$	M.P.		
	>268	30-80	Separation			
$TaCl_5$	205	86.3	$HgCl_2$, $TeCl_5$	Eut.	2d,1	731
TlCl	210±3	28.0	$HgCl_2$,$5HgCl_2$.TlCl	Per.	2d,3b	724
	189±3	33.3	$5HgCl_2$.TlCl, $HgCl_2$.TlCl	Eut.		
	224±3	50	$HgCl_2$.TlCl	Comp.		
	209±3	60	$HgCl_2$.TlCl, $3HgCl_2$.10TlCl	Eut.		
	258±3	73	$3HgCl_2$.10TlCl,TlCl	Per.		
WCl_6	276	0.0	$HgCl_2$	M.P.	2d	734
	224	95	α-WCl_6, $HgCl_2$	Eut.		
	256	100	WCl_6	M.P.		
	174		β ⇌ γ WCl_6	P.T.		

HgI_2

2nd Component	T°C	Conc. mol %	Phases involved	Def.	Method	Ref.
InI_3	148	45.0	InI_3.$2HgI_2$, β-HgI_2	Per.		735
	100	75	InI_3,InI_3.$2HgI_2$	Eut.		
	90	0.0	HgI_2	P.T.		
KI	257	0.0	HgI_2	M.P.	2a	736
	132	37	HgI_2,2KI.HgI_2	Eut.		
	207	52	2KI.HgI_2, KI	Per.		
	680	100	KI	M.P.		
Nb_3I_8	250	0.0	HgI_2	M.P.	2d	737
			Complete mutual insolubility in liquid and lack of reaction in solid.			
RbI	273	0	α-HgI_2	M.P.	2a	736
	138	42	α-HgI_2,3RbI.$2HgI_2$	Eut.		
	230	55	3RbI.$2HgI_2$, 4RbI.HgI_2	Per.		
	260	60	4RbI.HgI_2, RbI	Per.		
	642	100	RbI	M.P.		
	134		HgI_2	P.T.		

HgI_2

2nd Component	T°C	Conc. mol %	Phases involved	Def.	Method	Ref.
SbI_3	149	76	42% SbI_3 soln. in α-HgI_2,HgI_2	Eut.	2d	738
	130.5		$\alpha \rightleftharpoons \beta$-$HgI_2$			
	132		α-HgI_2 sol.sln. $\rightleftharpoons$ β-HgI_2 with 7% SbI_3			
TeI_4	253	0.0	HgI_2	M.P.	2a	739
	202	28	HgI_2, TeI_4	Eut.		
	280	100	TeI_4	M.P.		
	128		HgI_2	P.T.		
TlI	149±3	40	HgI_2,HgI_2.4Tl	Eut.	2d,3b	724
	340±3	80	HgI_2.4TlI, TlI	Per.		
	127		HgI_2 (ortho-rhombic $\rightleftharpoons$ quadratic)	P.T.		

$HoCl_3$

2nd Component	T°C	Conc. mol %	Phases involved	Def.	Method	Ref.
KCl			Fig. 146			740
	560	20	$HoCl_3$,KHo_2Cl_7	Eut.	2a,3b	740
	587	33	KHo_2Cl_7	Comp.		
	454	55	KHo_2Cl_7, β-K_3HoCl_6	Eut.		
	860	76	β-K_3HoCl_6 (temp. different in text and diagram)	Comp.		
	664	90	β-K_3HoCl_6, KCl	Eut.		
	400		K_3HoCl_6	P.T.		
NaCl	720	0	$HoCl_3$	M.P.	2	741
	438	55	$HoCl_3$,Na_3HoCl_6	Eut.		
	550	69	Na_3HoCl_6,NaCl	Per.		
	800	100	NaCl	M.P.		

HoF_3

2nd Component	T°C	Conc. mol %	Phases involved	Def.	Method	Ref.
KF			Fig. 147			742
	1145	0	α-HoF_3	M.P.	2d,3b	742
	1070	81	α-HoF_3,β-HoF_3	Per.		
	990	33	KHo_7F_{22},β-HoF_3	Per.		
	880	45	α-KHo_2F_7, KHo_7F_{22}	Per.		
	800	54	β-KHo_2F_7, α-KHo_2F_7	Per.		

$HoCl_3$

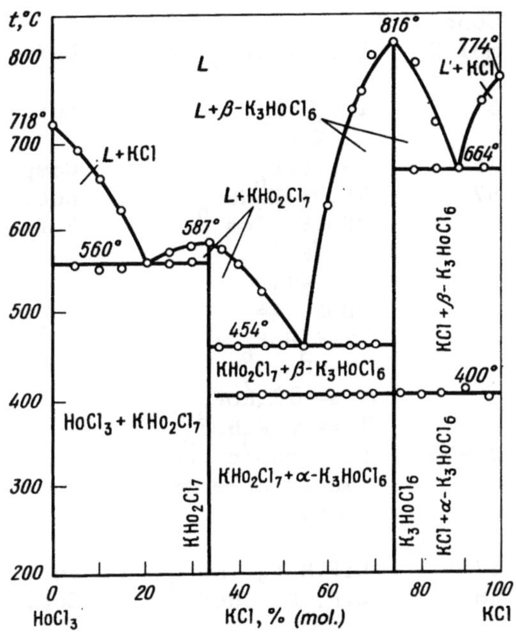

Fig. 146. $HoCl_3$-KCl system.

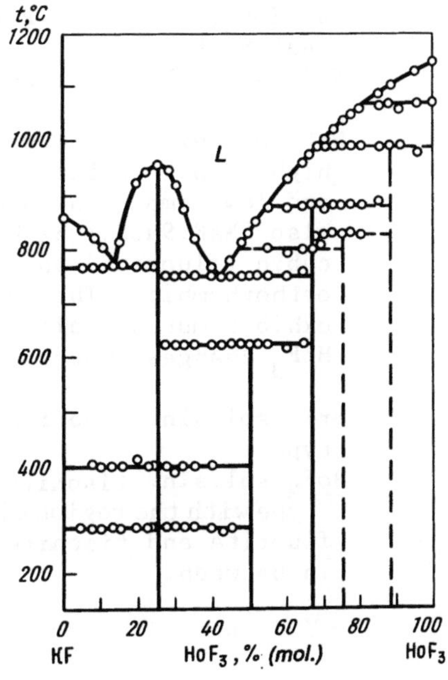

Fig. 147. KF-HoF_3 system.

HoF_3

2nd Component	T°C	Conc. mol %	Phases involved	Def.	Method	Ref.
KF	750	57.5	α-K_3HoF_6, β-KHo_2F_7	Eut.	2d,3b	742
	960	75	α-K_3HoF_6	Comp.		
	770	87	KF, α-K_3HoF_6	Eut.		
	825		KHo_3F_{10} ⇌ α-KHo_2F_7 + KHo_7F_{22}	P.T.		
	620		$KHoF_4$ ⇌ α-K_3HoF_6 + β-KHo_2F_7	P.T.		
	400		α ⇌ β K_3HoF_6			
	285		β ⇌ γ K_3HoF_6			
		20-45	With rapid cooling a metastable sol.sln. obtained.			
LiF	800	52	$LiF.HoF_3$, HoF_3	Per.	2	705
	680	68	LiF, $LiF.HoF_3$	Eut.		
NaF			Fig. 148			706
	962	29	$5NaF.9HoF_3$,HoF_3	Eut.	2,1f,3b	707
	1010	35	$5NaF.9HoF_3$	Comp.		
	745	65	$NaF.HoF_3$, $5NaF.9HoF_3$	Per.		
	665	73	NaF, $NaF.HoF_3$	Eut.		
			$NaF.HoF_3$ exists in two forms: high temp. - cubic (fluorite) and low temp. - hexagonal, also $5NaF.9HoF_3$: high temp.- cubic (fluorite) and low temp.- orthorhombic. The cubic forms exhibit mutual solid solutions. HoF_3 changes phase.			
SrF_2			SrF_2 sol.sln. fluorite type HoF_3 sol.sln. tisonite type with the region of fluorite and tisonite in between.		3b	743

$InBr_3$

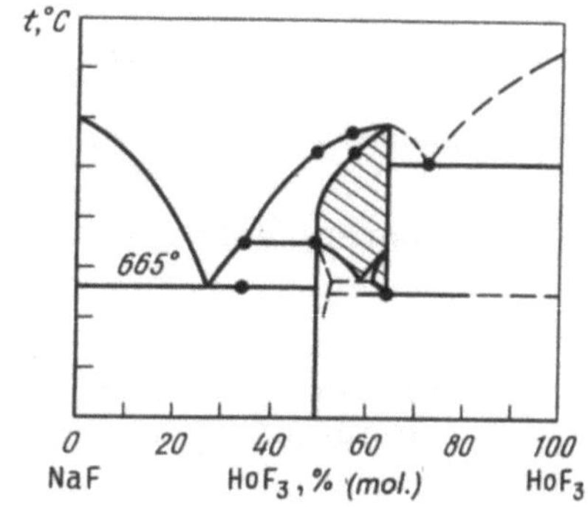

Fig. 148. $NaF-HoF_3$ system.

2nd Component	T°C	Conc. mol %	Phases involved	Def.	Method	Ref.
KBr			Fig. 149			744
	435	0	$InBr_3$	M.P.	1,2d,3b,	744
	232	50	$InBr_3$,K_3InBr_6	Eut.	1e,6a	
	320	52	K_3InBr_6,KBr	Per.		
	735	100	KBr	M.P.		

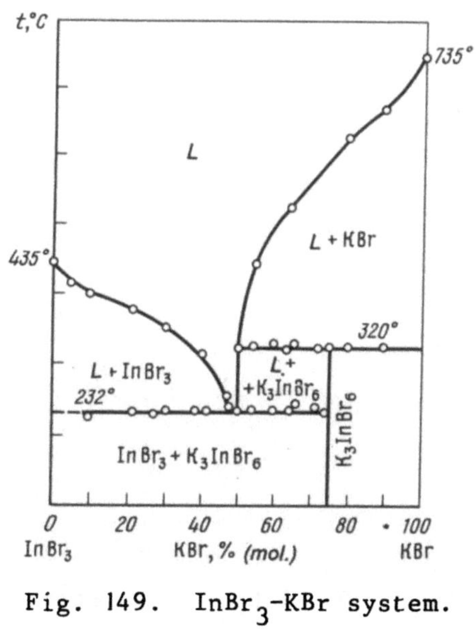

Fig. 149. $InBr_3$-KBr system.

$InBr_3$

2nd Component	T°C	Conc. mol %	Phases involved	Def.	Method	Ref.
KBr			Existence of K_3InBr_6 confirmed by optical crystallography (n = 1.704, d = 5.61 g/cm³)			744
	140	50, 55, 60, 65	Thermal effects			
			X-ray analysis confirms existence of penta- or tetrabromide of potassium			
LiBr			Fig. 150			744
	435	0	$InBr_3$	M.P.	1,2d,3b, 1e,6a	744
	220	52	18% LiBr sol.sln. in InBr,Li_3InBr_6	Eut.		
	237	~54	Li_3InBr_6,LiBr	Per.		
			Optical crystallography confirms existence of Li_3InBr_6 (n = 1.648, d = 5.69 g/cm³)			

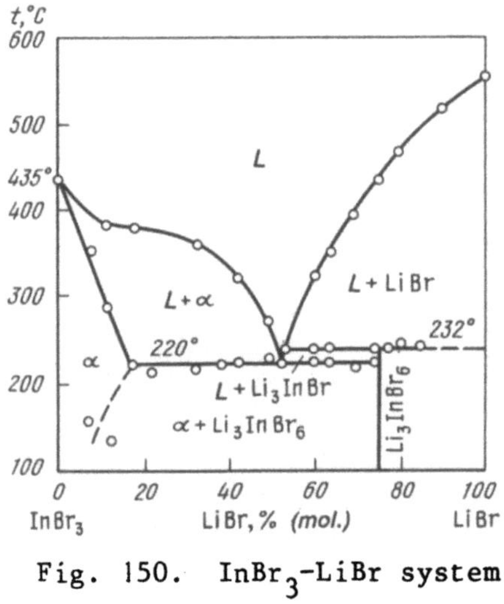

Fig. 150. $InBr_3$-LiBr system.

$InBr_3$

2nd Component	T°C	Conc. mol %	Phases involved	Def.	Method	Ref.
NH_4Br			Fig. 151			744
	435	0	$InBr_3$	M.P.	1,2d,3b	744
	198	58	$InBr_3$, $(NH_4)_3InBr_6$	Eut.	1e,6a	
	328	71	$(NH_4)_3InBr_6$, NH_4Br	Per.		
	160		Decomposition of NH_4InBr_4 Optical crystallograhy confirms the existence of $(NH_4)_3InBr_6$ (n = 1.760 d = 4.19g/cm^3) and NH_4InBr_4 (n = 1.76 d = 3.21g/cm^3)			
NaBr	436	0.0	$InBr_3$	M.P.	2d, 1h	745
	200	53	$InBr_3Na_2InBr_5$	Eut.		
	230	54	Na_2InBr_5, β-Na_3InBr_6	Per.		
	266	57	β-Na_3InBr_6, α-Na_3InBr_6	Per.		
	330	60	α-Na_3InBr_6,NaBr	Per.		
	755	100	NaBr	M.P.		

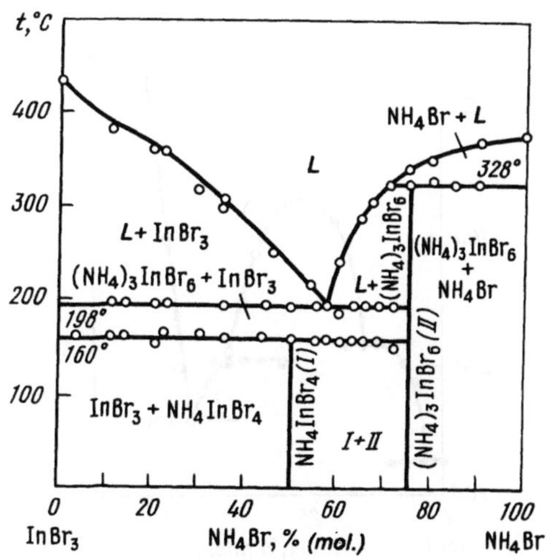

Fig. 151. $InBr_3$-NH_4Br system.

$InBr_3$

2nd Component	T°C	Conc. mol %	Phases involved	Def.	Method	Ref.
$PbBr_2$	435	0.0	$InBr_3$	M.P.	2a, 1	746
	278	60	$InBr_3$,$PbBr_3$	Eut.		
$ZnBr_2$	436	0.0	$InBr_3$	M.P.	2a,3b,1	746
	340	40	α – 24% $ZnBr_2$ sol. in $InBr_3$, β – 22% $InBr_3$ sol. in $ZnBr_2$	Eut.		
	394	100	$ZnBr_2$	M.P.		

InCl

2nd Component	T°C	Conc. mol %	Phases involved	Def.	Method	Ref.
$InCl_3$			Fig. 152			747
	230		$InCl_3$,Compound of undetermined composition	Eut.	2	747
	252		In_2Cl_3,Compound of undetermined composition	Per.		
	314		In_2Cl_3	Comp.		
	260		In_4Cl_5,In_2Cl_3	Per.		
	214.5		InCl, In_4Cl_5	Eut.		

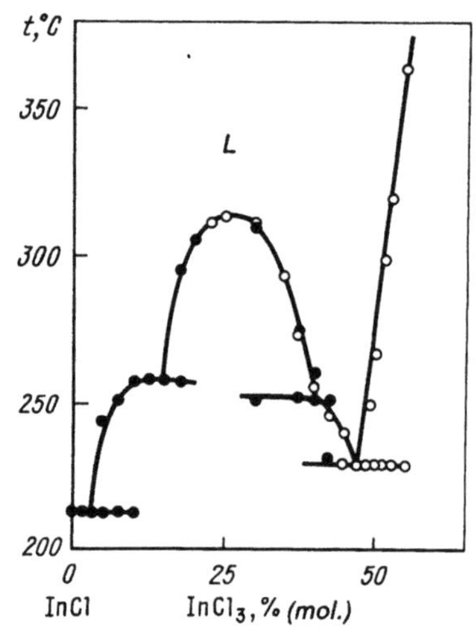

Fig. 152. InCl-$InCl_3$ system.

InCl

2nd Component	T°C	Conc. mol %	Phases involved	Def.	Method	Ref.
LiCl	220	~0.0	InCl, LiCl	Eut.	2a,5,3b	748
		<25	Liquidus line is approximate and does not show melting (crystallization) effects.			
		35-85	Tendency of separation			
$PbCl_2$			Fig. 153			749
	222	0.0	InCl, α-$InPbCl_3$	Eut.	2d,3b,1	749
	239		α-$InPbCl_3$, β-$InPbCl_3$			
	358	50	β-$InPbCl_3$	Comp.		
	353	56	β-$InPbCl_3$,$PbCl_2$	Eut.		
	287		$InPb_2Cl_5$	P.T.		
$SnCl_2$			Existence of compound 1:1 ratio Complex ion $[SnCl_3]^-$ in melt		4c	750
TlCl			Fig. 154			751
	216	0.0	InCl	M.P.	2d,1	751
	210	~0.0	InCl,InCl.3TlCl	Eut.		
	290	~52	InCl.3TlCl, 6% InCl sol.sln. in β-TlCl	Per.		
	430	100	TlCl	M.P.		
			Fig. 155			752
	222	33	40 TlCl sol.sln. and 83% TlCl sol.sln.	Per.	2,3b	752
	210	3	Liquidus of InCl sol.sln.	Min.		
$ZnCl_2$	224	0.0	InCl	M.P.	2,3b,1a	753
	170	13	$ZnCl_2$.2InCl,InCl	Eut.		
	211	31	$ZnCl_2$.2InCl	Comp.		
	146	43	$2ZnCl_2$.InCl, $ZnCl_2$.2InCl	Eut.		
	227	59	$ZnCl_2$,$2ZnCl_2$.InCl	Per.		
	318	100	$ZnCl_2$	M.P.		
	166	15	InCl,In_2ZnCl_4	Eut.	2a,1,3b	
	200	33.3	In_2ZnCl_4	Comp.		

InCl

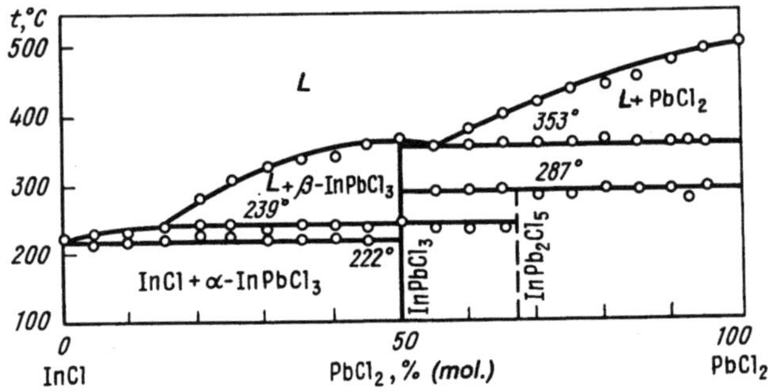

Fig. 153. InCl-$PbCl_2$ system.

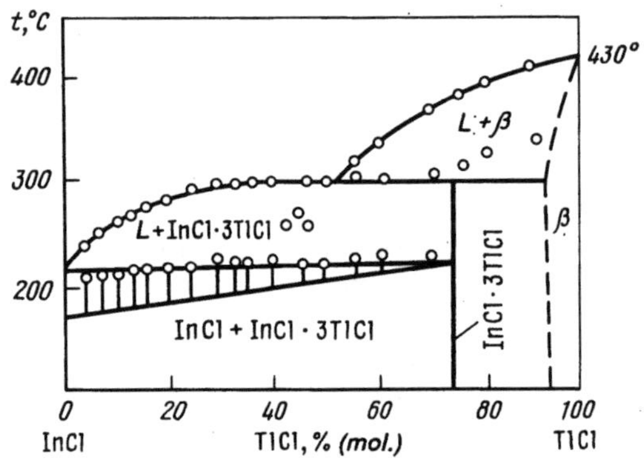

Fig. 154. InCl-TiCl system.

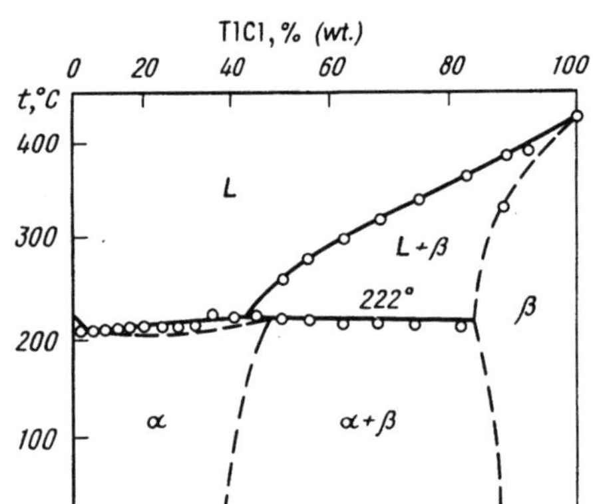

Fig. 155. InCl-TiCl system.

InCl

2nd Component	T°C	Conc. mol %	Phases involved	Def.	Method	Ref.
$ZnCl_2$	137	45	In_2ZnCl_4, $InZn_2Cl_5$	Eut.	2a,1,3b	754
	208	60	$InZn_2Cl_5$, $ZnCl_2$	Per.		
	124		$InZnCl_3$	P.T.		
	224	0.0	InCl	M.P.	1,3b,1f	755
	170	14	$ZnCl_2$.2InCl,InCl	Eut.		
	211	31	$ZnCl_2$.2InCl	Comp.		
	145	43	$2ZnCl_2$.InCl, $ZnCl_2$.2InCl	Eut. Eut.		
	227	64.5	$ZnCl_2$, $2ZnCl_2$.InCl	Per.		
	318	100	$ZnCl_2$	M.P.		

$InCl_3$

2nd Component	T°C	Conc. mol %	Phases involved	Def.	Method	Ref.
KCl	316	52.5	$InCl_3$, K_2InCl_5	Eut.	2b,2c,1	756
	355	53.5	K_2InCl_5, α-K_3InCl_6	Per.,		
	400	54.5	α-K_3InCl_6, β-K_3InCl_6	Per.		
	598	66	β-K_3InCl_6,KCl	Per.		
	288	49	K_2InCl_5, $InCl_3$	Eut.	2d	757
	337	53	K_3InCl_6	Comp.		
	618	75	K_3InCl_6, K_2InCl_5	Per.		
	606	81.5	KCl, K_3InCl_6	Eut.		
			K_3InCl_6 decomposes on melting.			
	420) 369)		K_3InCl_6 - Thermal effects of unidentified nature (probably P.T.)			
$LaCl_3$	568	10	$InCl_3$, $LaCl_3$	Eut.	2b,2c	758
LiCl	381	52	Li_3InCl_6, $InCl_3$	Eut.	2d	757
	462	62	LiCl, Li_3InCl_6	Per.		
$MgCl_2$			Fig. 156			759
	554	42	35% $MgCl_2$ sol. sln. in $InCl_3$, 16% $InCl_3$ sol. sln. in $MgCl_2$.	Eut.	2a	759

$InCl_3$

2nd Component	T°C	Conc. mol %	Phases involved	Def.	Method	Ref.
$MnCl_2$			Fig. 157			760
	495	35	4% $MnCl_2$ sol.sln. in $InCl_3$, 4% $InCl_3$ sol.sln. in $MnCl_2$	Eut.	2b,2c	760
NaCl	416	37	NaCl,Na_3InCl_6	Per.	2d	761
	270	49	Na_3InCl_6,$InCl_3$	Eut.		

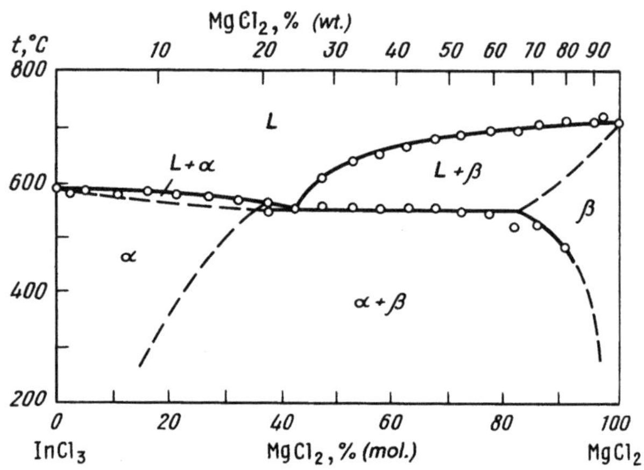

Fig. 156. $InCl_3$-$MgCl_2$ system.

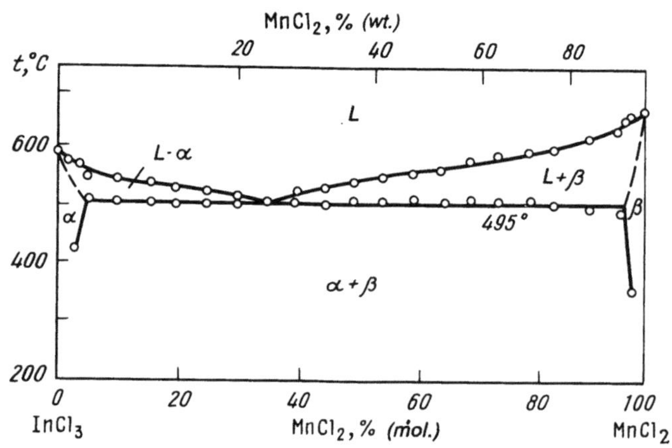

Fig. 157. $InCl_3$-$MnCl_2$ system.

$InCl_3$

2nd Component	T°C	Conc. mol %	Phases involved	Def.	Method	Ref.
NaCl	272	51	$InCl_3$,3NaCl.$InCl_3$	Eut.	2b	762
	410	58.5	3NaCl.$InCl_3$,NaCl	Per.		
	272	51.0	$InCl_3$, Na_3InCl_6	Eut.	2a,6c	763
	408	~57	Na_3InCl_6,NaCl	Per.		
$PbCl_2$	400	56	$InCl_3$,$PbCl_2$ Slight possibility of solid slns.	Eut.	2b	764
RbCl	326	47.8	2RbCl.$InCl_3$, $InCl_3$	Eut.	2d	765
	389	52	3RbCl.$InCl_3$, 2RbCl.$InCl_3$	Per.		
	666	75	3RbCl.$InCl_3$	Comp.		
	633	89.6	RbCl,3RbCl.$InCl_3$	Eut.		
$SbCl_5$	~3	~100	$InCl_3$,$SbCl_5$	Eut.	2d	733
	3	100	$SbCl_5$	M.P.		
$SmCl_3$	550	20	5% $SmCl_3$,sol. sln. in $InCl_3$ 12% $InCl_3$ sol. sln. in $SmCl_3$	Eut.	2b,2c,2	758
$SnCl_2$	218	91	6% $SnCl_2$ sol. sln. in $InCl_2$, 4% $InCl_3$ sol. sln. in $SnCl_2$	Eut.	2b,1	764
$TeCl_4$	216	85	15% $TeCl_4$ sol. sln. in $InCl_3$, 3% $InCl_3$ sol. sln. in $TeCl_4$	Eut.	2b,2c	760
TlCl	260	52	II-$InCl_3$.2TlCl,	Eut.	2a,1	766
	320	53	I-$InCl_3$.2TlCl, II-$InCl_3$.2TlCl	Per.		
	350	54	$InCl_3$.3TlCl, I-$InCl_3$.2TlCl	Per.		
	480	75	$InCl_2$.3TlCl	Comp.		
	390	93.6	TlCl, $InCl_3$.3TlCl	Eut.		

$InCl_3$

2nd Component	T°C	Conc. mol %	Phases involved	Def.	Method	Ref.
TlCl	280	50	$InCl_3$, Tl_2InCl_5	Eut.	2b,2c,1	756
	474	75	Tl_2InCl_5	Comp.		
	400	92.8	Tl_2InCl_5,TlCl	Eut.		
	360		Tl_2InCl_5 decomposes			
	325		Tl_2InCl	P.T.		
	260	52	$InCl_3$, $InCl_3$.2TlCl	Eut.	2a	767
	350	54	$InCl_3$.2TlCl, $InCl_3$.3TlCl	Per.		
	480	76	$InCl_3$.3TlCl	Comp.		
	390	73.6	$InCl_3$.3TlCl,TlCl	Eut.		
	320		$InCl_3$.2TlCl	P.T.		
$YbCl_3$			Seemingly continuous sol.sln.		2b,2c	768
$ZnCl_2$			Fig. 158			759
	260	91	18% ZnCl sol.sln. in $InCl_3$, 3% $InCl_3$ sol. sln. in $ZnCl_2$		2a	759

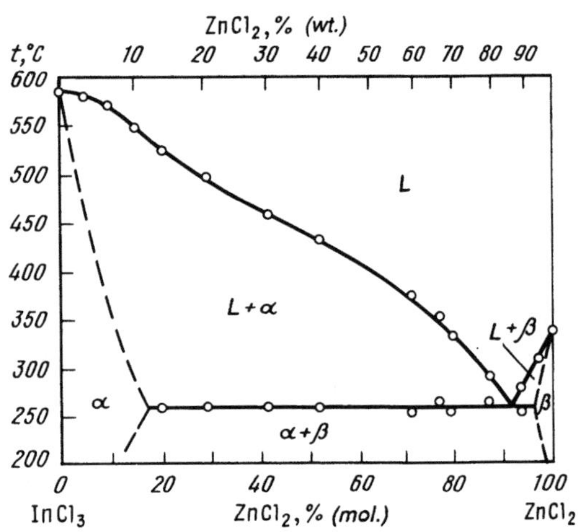

Fig. 158. $InCl_3$-$ZnCl_2$ system.

InF_3

2nd Component	T°C	Conc. mol %	Phases involved	Def.	Method	Ref.
KF			The following compounds exist:		2d,3b	
			KIn_2F_7,			
			$KInF_4$ - rhombic with {a = 7.93Å, b = 7.76Å, c = 12.57Å}			
			K_3InF_6			
			$K_5In_3F_{14}$ - tetragonal with {a = 7.933Å, c = 11.882Å}			
LiF	627	64	InF_3,Li_3InF_6	Eut.	2,1f,3b	770
	867	75	Li_3InF_6	Comp.		
	760	86	Li_3InF_6,LiF	Eut.		
			Li_3InF_6 } compounds confirmed	Comp.	2d,3b	
			$LiInF_4$ }	Comp.		
	415		α ⇌ β Li_3InF_6	P.T.		769
	505		β ⇌ α Li_3InF_6	P.T.		
	505		α ⇌ β $LiInF_4$	P.T.		
NaF			Na_3InF_6 } compounds confirmed	Comp.	2d, 3b	769
			$NaInF_4$ }	Comp.		
	705		α ⇌ β-Na_3InF_6	P.T.		
	780		β ⇌ γ-Na_3InF_6	P.T.		
			α-Na_3InF_6 - monoclinic with {a = 5.643Å, b = 5.844Å, c = 8.190Å, β = 90°52'}			
			$NaInF_4$ - hexagonal with {a = 13.18Å, c = 9.39Å}			
PbF_2			$Pb_5In_4F_{22}$ confirmed	Comp.	3b	771
	500	73.5-100	Phase with a structure of β-PbF_2			
RbF			Rb_3InF_6	Comp.	2d,3b	772
			Rb_2InF_5	Comp.		
			$Rb_4In_3F_{13}$	Comp.		
			$RbInF_4$	Comp.		

InF_3

2nd Component	T°C	Conc. mol %	Phases involved	Def.	Method	Ref.
RbF			$Rb_2In_3F_{11}$	Comp.	2d,3b	772
			$RbIn_2F_7$	Comp.		
			$RbIn_3F_{10}$	Comp.		
	165		$\alpha \rightleftharpoons \beta$ Rb_3InF_6	P.T.		
	220		$\beta \rightleftharpoons \gamma$ Rb_3InF_6	P.T.		
	725		$Rb_4In_3F_{13}$ - low temp. α = high temp. β.	P.T.		
	540		$RbInF_4$ - low temp. $\alpha \rightleftharpoons$ high temp. β	P.T.		
	625		RbI_2F_7 - low temp. $\alpha \rightleftharpoons$ high temp. β.	P.T.		
			$Rb_4In_3F_{13}$, $RbInF_4$ } do not stabilize by hardening			
			$RbInF_4$ - hydrolyzing			
	>650		$5Rb_2InF_5 \rightarrow 2Rb_3InF_6 + Rb_4In_3F_{13}$	Decomposition		
	>815		$3RbIn_2F_7 \rightarrow Rb_2In_3F_{11} + RbIn_3F_{10}$	Decomposition		
			γ-Rb_3InF_6 - tetragonal: a = 9.249Å			
			$RbInF_4$ - rhombic: a = 8.268Å, b = 8.082Å, c = 13.105Å, z = 8			
SrF_2	826 ±15		Sr_2InF_7 (incongruent melting)	Comp.	3b	773
TlF			The following compounds confirmed:			772
	130		Tl_3InF_6	P.T.	2d,3b	
			$Tl_4In_3F_{13}$			
	480		$TlInF_4$ $\alpha \rightleftharpoons \beta$; β- tendency to hydrolyze	P.T.		
	<480		$Tl_2In_3F_{11}$			
	>480		$2Tl_2In_3F_{11}$ - $3TlInF_4 + TlIn_3F_{10}$			

InF_3

2nd Component	T°C	Conc. mol %	Phases involved	Def.	Method	Ref.
TiF			$TlIn_3F_{10}$ α-Tl_3InF_6 - tetragonal a = 6.45Å, c = 9.2Å β-Tl_3InF_6 - cubic a = 9.193Å			772
ZrF_4	720		Zr_3InF_{15} decomposes: $Zr_3InF_{15} \rightarrow ZrF_4$ + cubic sol.sln. of ReO_3 type	Comp.	3d,3e, 6a,j	774
	700	$.25 \leqslant x \leqslant .50$	$Zr_xIn_{1-x}F_{3+x}$ sol.sln. range			
	800	$.17 \leqslant x \leqslant .65$	As above With lower values of x transformations into rhombic InF sol.sln.			

InI

2nd Component	T°C	Conc. mol %	Phases involved	Def.	Method	Ref.
KI	365	0.0	InI	M.P.	2d,3b	775
	350	17.5	InI, KI	Eut.		
MnI_2	365	0.0	InI	M.P.	2,3b	776
	340	30	MnI_2, InI	Eut.		
	620	100	MnI_2	M.P.		
	236		$InI.MnI_2$ (solid)	Comp.		
NaI	365	0.0	InI	M.P.	2d,3b	775
	345	7.5	InI, NaI	Eut.		
PbI_2	365	0.0	InI	M.P.	2,3b	776
	300	17.5	InI, $InI.PbI_2$	Eut.		
	318	43	$InI.PbI_2$, PbI_2	Per.		
	386	100	PbI_2	M.P.		
		85-100	PbI_2 dissociates			
RbI	365	0.0	InI	M.P.	2d,3b	775
	340	22.5	InI, RbI	Eut.		
SnI_2	365	0.0	InI	M.P.	2d,3b	777
	270	30	InI, In_4SnI_6	Per.		
	253	40	In_4SnI_6, $InSn_2I_5$	Eut.		
	290	65	$InSn_2I_5$	Comp.		

InI

2nd Component	T°C	Conc. mol %	Phases involved	Def.	Method	Ref.
SnI_2	280	80	$InSn_2I_5$,SnI_2	Eut.		777
	320	100	SnI_2	M.P.		
	246		In_4SnI_6	P.T.		
TlI			Fig. 159			778
	351	0.0	InI	M.P.	2d	778
	391	64.4	25% and 30% InI sol.sln. in TlI	Per.		
	175		α ⇌ β TlI			
	208		α-sol.sln. dissociates			
ZnI_2	365	0.0	InI	M.P.	2,3b	776
	250	30	InI,α-2InI.ZnI_2	Eut.		
	260	33	α-2InI.ZnI_2	Comp.		
	236	46	α-2InI.ZnI_2,ZnI_2	Eut.		
	440	100	ZnI_2	M.P.		
	210		2InI.MnI_2	P.T.		

InI_2

2nd Component	T°C	Conc. mol %	Phases involved	Def.	Method	Ref.
MnI_2			Fig. 160			779
	226	0.0	InI_2	M.P.	2a	779
	~226	~0.0	InI_2,MnI_2	Eut.		
	620	100	MnI_2	M.P.		
	161	0.0	InI_2	P.T.		

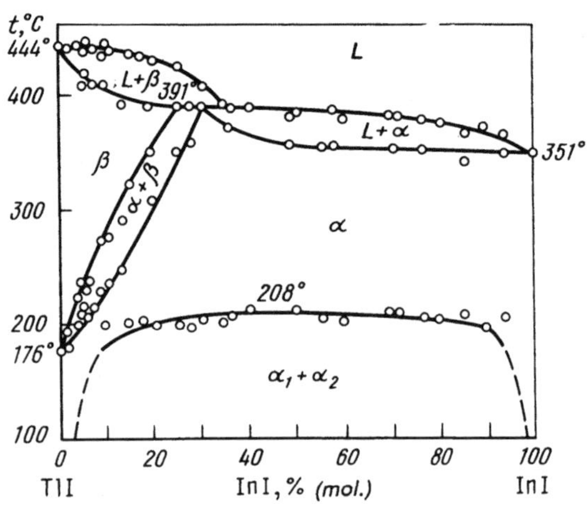

Fig. 159. TlI-InI system.

InI_2

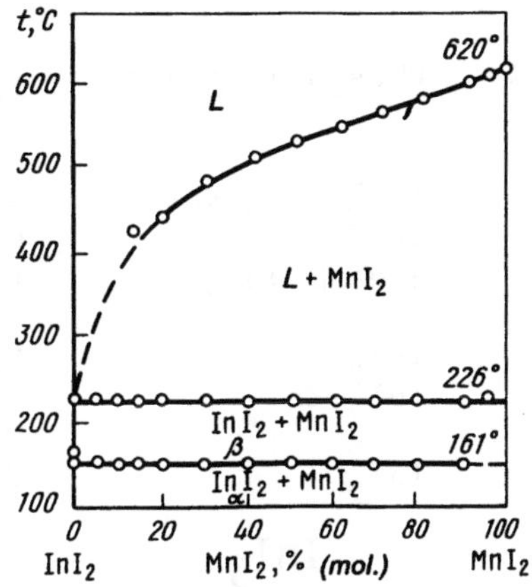

Fig. 160. InI_2-MnI_2 system.

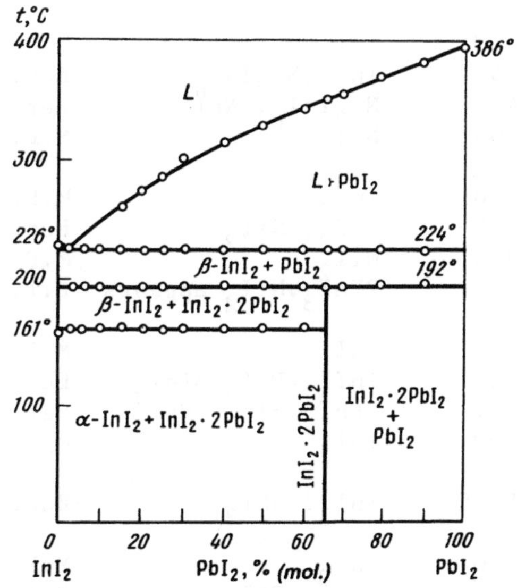

Fig. 161. InI_2-PbI_2 system.

2nd Component	T°C	Conc. mol %	Phases involved	Def.	Method	Ref.
PbI_2			Fig. 161			779
	226	0.0	InI_2	M.P.	2d,3b	779
	224	2.5	α-InI_2,PbI_2	Eut.		
	386	100	PbI_2	M.P.		
	161		α ⇌ β InI	P.T.		
	192		$InI_2.2PbI_2$	Comp.		

InI_2

2nd Component	T°C	Conc. mol %	Phases involved	Def.	Method	Ref.
SnI_2	226	0.0	InI_2	M.P.	2a	779
	210	25.5	InI_2,SnI_2 sol.sln.	Eut.		
	320	100	SnI_2	M.P.		
	161		$\alpha \rightleftharpoons \beta$ InI_2	P.T.		
ZnI_2	226	0.0	InI_2	M.P.	2a	779
	220	3.0	InI_2,ZnI_2	Eut.		
	440	100	ZnI_2	M.P.		
	161		InI_2	P.T.		

InI_3

2nd Component	T°C	Conc. mol %	Phases involved	Def.	Method	Ref.
KI	100	15	InI_3,K_3InI_6	Eut.	2	780
	170	18	K_3InI_6,KI	Per.		
NaI	210	0.0	InI_3	M.P.	2,3b	781
	85	25	InI_3,Na_3InI_6	Eut.		
	193	47	Na_3InI_6, NaI	Per.		
	655	100	NaI	M.P.		
NiI_2	208	0.0	InI_3	M.P.	2d	782
	~208	~0.0	InI_3, NiI_2	Eut.		
	760	100	NiI_2	M.P.		
	140	~33	$2InI_3.NiI_2$ appears	P.T.		
PbI_2	198	0.0	InI_3	M.P.	2c	784
	140	22.5	InI_3,$2PbI_2.3InI_3$	Eut.		
	186	36.9	$2PbI_2.3InI_3$,PbI_2	Per.		
	398	100	PbI_2	M.P.		
SbI_3	127	52.5	InI_3, SbI_3	Eut.	2a	783
SnI_2	198	0.0	InI_3	M.P.	2c	784
	110	15	InI_3, SnI_2	Eut.		
	328	100	SnI_2	M.P.		
TlI			Fig. 162			780
	118	15	InI_3, $TlInI_4$	Eut.	2	780
	193	~65	$TlInI_4$, TlI	Per.		
ZnI_2	210	0.0	InI_3	M.P.	2	785
	182	~25	10% ZnI_2 sol.sln. in InI_3, 30% InI_3 sol.sln. in ZnI_2	Eut.		

InI_3

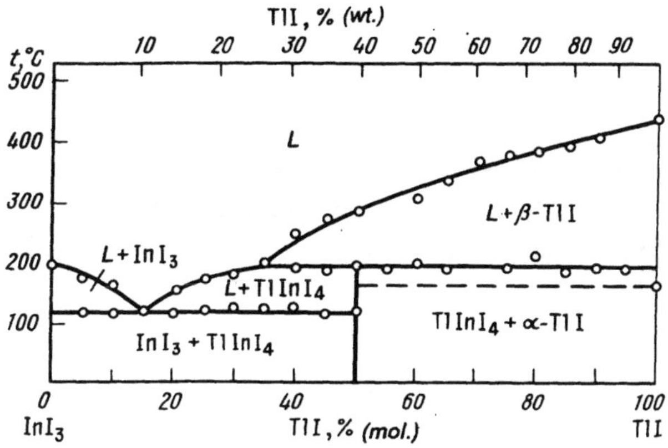

Fig. 162. InI_3-TlI system.

2nd Component	T°C	Conc. mol %	Phases involved	Def.	Method	Ref.
ZnI_2	446	100	ZnI_2	M.P.		785
	<130		Sol.slns. on the InI_3 side decompose	P.T.		
	<140		Sol.slns. on the ZnI_2 side decompose	P.T.		

KBF_4

2nd Component	T°C	Conc. mol %	Phases involved	Def.	Method	Ref.
KCl	508	3	KBF_4,KCl.11KBF_4	Eut.	2	786
	590	10	KCl.11KBF_4	Comp.	1	
	471	19	KCl.11KBF_4,KCl	Eut.		
KF	514	0	KBF_4	M.P.	2c,1	787
	410	34	KBF_4, KF	Eut.		
	858	100	KF	M.P.		
	441	24.0	KBF_4,KF.2KBF_4	Eut.	1	788
	531	33.3	KF.2KBF_4,KF.KBF_4	Per.		
	669	50.0	KF.KBF_4,KF	Per.		

KBF_4

TABLE VI

	** t°C				*** t°C	
			Transformation			
KBF_4%*	Liquidus	Eutectic	1	2	Liquidus	Eutectic
10	828	459	286	-	811 ±5	-
25	778	-	-	-	748 ±4	-
30	-	-	-	-	724 ±4	-
33.3	745	462	285	-	712 ±2	-
40	715	-	-	-	679 ±3	-
50	654	462	-	346	624 ±3	460±9
60	589	463	285	345	552 ±3	460±2
70	505	461	284	346	479 ±2	454±2
80	489	457	281	345	485 ±2	460±2
85	512	460	282	345	504 ±2	455±3
90	534	459	282	345	522 ±2	-
95	551	456	281	345,414	536 ±2	-
100	570	-	283	-	-	-

* Crystallizing phases: KF 10-70%, KBF_4 80-100%.
** According to thermographic data.
*** According to "differential hardness" data.

2nd Component	T°C	Conc. mol %	Phases involved	Def.	Method	Ref.
KF	460	26		Eut.		789
	283±2	0.0	KBF_4	P.T.		
	345		Effect ascribed to traces of KBF_3OH which melted at 355. Compounds $KF.KBF_4$ and $KF.2KBF_4$ confirmed.			
	441	25	KF, KBF_4	Eut.	3b,1	790

KBr

2nd Component	T°C	Conc. mol %	Phases involved	Def.	Method	Ref.
LiBr	328	60.0	KBr, LiBr	Eut.	2,1,6e	791
	315	60.5	KBr, LiBr	Eut.	1	792
	326	62.5	KBr, β-LiBr	Eut.	1	793
	515	100	LiBr	P.T.		
	334	62	KBr, LiBr	Eut.	1	794
	334	62	KBr, LiBr (Small deviations from ideality)	Eut.	2a,6b,61	795
	318	60.5	KBr, LiBr	Eut.	1	796
NaBr	638	50	Continuous sol.sln.	Min.	1	797 798
	638	48	Continuous sol.sln.	Min.	1	799
	395	65	Limit of sol.sln.	P.T.	36	800
	627	50	Continuous sol.sln.	Min.	1,3b	801 802
$PbBr_2$			Fig. 163			803
	426	42	$2KBr.PbBr_2$, KBr	Per.	1	803
	355	54	$KBr.2PbBr_2$, $2KBr.PbBr_2$	Eut.		
	335	90	$PbBr_2$, $KBr.2PbBr_2$	Eut.		
	417	42.3	$2KBr.PbBr_2$, KBr	Per.	1	804
	354	55.6	$KBr.2PbBr_2$, $2KBr.PbBr_2$	Eut.		
	380	66.7	$KBr.2PbBr_2$	Comp.		
	330	84	$PbBr_2$, $KBr.2PbBr_2$	Eut.		

KBr

2nd Component	T°C	Conc. mol %	Phases involved	Def.	Method	Ref.
$PbBr_2$	425	42.9	$PbBr_2.2KBr$, KBr	Per.	2a,3b,	795
	367	54.4	$2PbBr_2.KBr$, $PbBr_2.2KBr$	Eut.	6b, 61	
	348	37.8	$PbBr_2.2KBr$, KBr	Eut.	61	
			4,10,20,35 and 70% equiv. Br confirmed by x-ray-graphic method to contain $KBr.2PbBr_2$ and $2KBr.PbBr_2$. The isotherms of molar volume show positive deviation from ideality but the isotherms of viscosity a negative deviation.			
			A number of compounds confirmed		3b	805

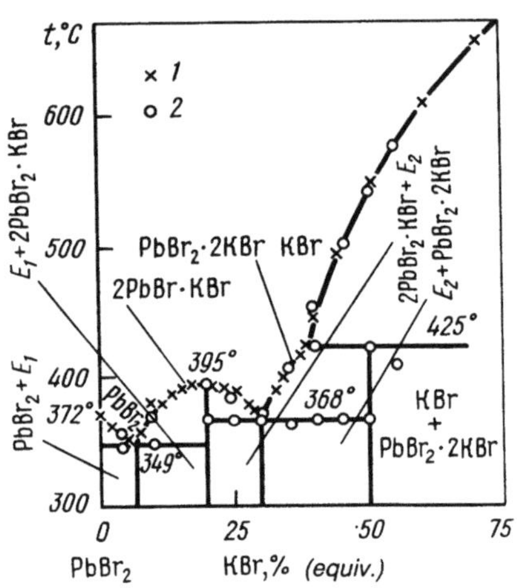

Fig. 163. $PbBr_2$-KBr system.

1 - Visual
2 - Thermographic

KBr

2nd Component	T°C	Conc. mol %	Phases involved	Def.	Method	Ref.
RbBr	728	0.0	KBr	M.P.	1	806
	675	85	Continuous sol.sln.	Min.		
	680	100	RbBr	M.P.		
	732	0.0	KBr	M.P.	1	807
	690	100	RbBr	M.P.		808
$SbBr_3$			Fig. 164			809
	350	40	KBr,2Kbr.$SbBr_3$(B)	Per.	2	809
	300	66	3KBr.$SbBr_3$, 3KBr.2$SbBr_3$(A)	Per.	1	
	96	~100	3KBr.2$SbBr_3$, $SbBr_3$	Eut.		
$ScBr_3$	637	15.8	KBr, K_3ScBr_6	Eut.	2d,3b	810
	697	25	K_3ScBr_6	Comp.		
	525	40.0	K_3ScBr_6, $K_3Sc_2Br_9$	Per.		
	512	45.2	K_3ScBr_9,$ScBr_3$	Eut.		
	440		K_3ScBr_9	P.T.		

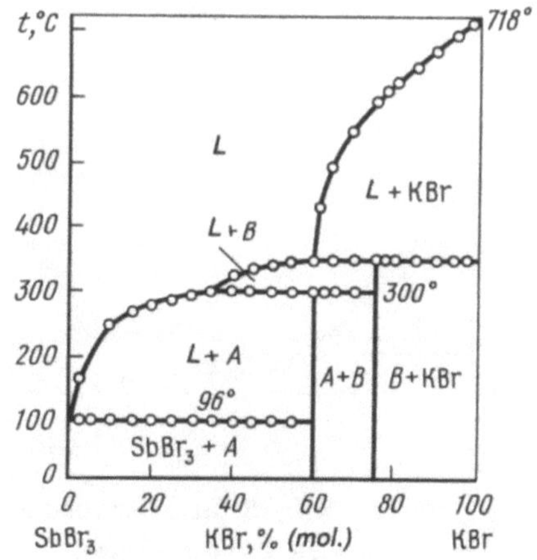

Fig. 164. $SbBr_3$-KBr System.

KBr

2nd Component	T°C	Conc. mol %	Phases involved	Def.	Method	Ref.
$SnBr_2$	730	0	KBr	M.P.	1,2a	811
	264	50	$3KBr.SnBr_2$, KBr	Per.		
	238	52.5	$KBr.2SnBr_2$, $3KBr.SnBr_2$	Eut.		
	272	66.7	$KBr.2SnBr_2$	Comp.		
	212	87.5	5%KBr sol.sln. in $SnBr_2$, $KBr.2SnBr_2$	Eut.		
	226	200	$SnBr_2$	M.P.		
	>600		$SnBr_2$ decomposes			
$SrBr_2$			Fig. 165			812
	563	33	KBr, $2KBr.SnBr_2$	Eut.	1	812
	519	52	$2KBr.SrBr_2$, $KBr.2SrBr_2$	Eut.		
	562	82	$KBr.2SrBr_2$, $SrBr_2$	Eut.		
	554	-	$SrBr_2.4KBr$, KBr	Per.	1,3b,2d	813
	522	45	$2SrBr_2.KBr$, $SrBr_2.4KBr$	Eut.		
	572	72	$2SrBr_2.KBr$	Comp.		
	556	82	$SrBr_2$, $2SrBr_2.KBr$	Eut.		

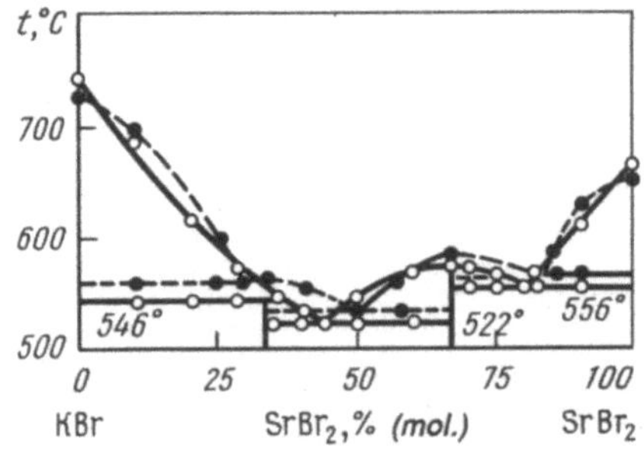

Fig. 165. KBr-$SrBr_2$ System.
——— Ref. 814
– – – Ref. 815

KBr

2nd Component	T°C	Conc. mol %	Phases involved	Def.	Method	Ref.
$SrBr_2$	740	0.00	KBr	M.P.	1, 3a	814
	550	34.00	KBr, K_2SrBr_4	Per.		
	552	45.00	K_2SrBr_4, KSr_2Br_5	Eut.		
	570	66.67	KSr_2Br_5	Comp.		
	556	80.00	KSr_2Br_5, β-$SrBr_2$	Eut.		
	642	-	β-$SrBr_2$, α-$SrBr_2$	Per.		
	660	100	α-$SrBr_2$	M.P.		
$TeBr_4$			Fig. 166			816
	732	0	KBr	M.P.	2,	816
	580	20	KBr,K_2TeBr_5	Eut.	3b	
	636	33	K_2TeBr_6	Comp.	1e	
	350	80	K_2TeBr_6,$TeBr_4$	Eut.		
	386	100	$TeBr_4$	M.P.		

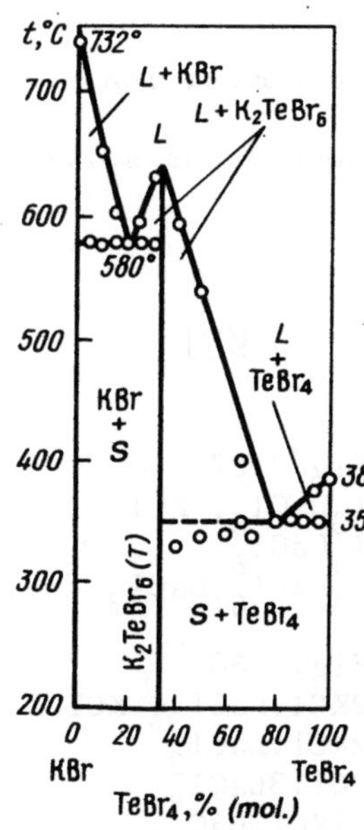

Fig. 166. KBr–$TeBr_4$ system.

KBr

2nd Component	T°C	Conc. mol %	Phases involved	Def.	Method	Ref.
$TiBr_3$	635	20	KBr, K_3TiBr_6	Eut.	2d	817
	662	25	K_3TiBr_6	Comp.	3b	
	575	40	$K_3TiBr_6, TiBr_3$	Eut.		
TlBr	459	13.5	KBr, TlBr	Eut.	1	818
	460	100	TlBr	M.P.		819
						820
KCl	724	40	Continuous sol.sln.	Min.	1	812
			Continuous sol. sln. KCl-KBr		3b	821
			Continuous sol.sln.		6h,3b	822
	732	0.0	KBr	M.P.	1	823
			Continuous sol.sln.		3b	
	755	100.0	KCl	M.P.		
KI	205±5	33	Decomposition	D.P.	3b	824
	664		Continuous sol.sln.	Min.	1	825

KCl

2nd Component	T°C	Conc. mol %	Phases involved	Def.	Method	Ref.
$LaCl_3$			Fig. 167			826
	580	20.0	K_3LaCl_6, KCl	Eut.	2c	826
	625	25.0	K_3LaCl_6	Comp.		
	554	53.0	$K_3LaCl_6, LaCl_3$	Eut.		
			Fig. 168			827
	585	22	$2KCl.LaCl_3$, KCl	Eut.	2a	827
	645	33.3	$2KCl.LaCl_3$	Comp.		
	580	48	$KCl.3LaCl_3$, $2KCl.LaCl_3$	Eut.		
	620	70	$LaCl_3$, $KCl.3LaCl_3$	Per.		

KCl

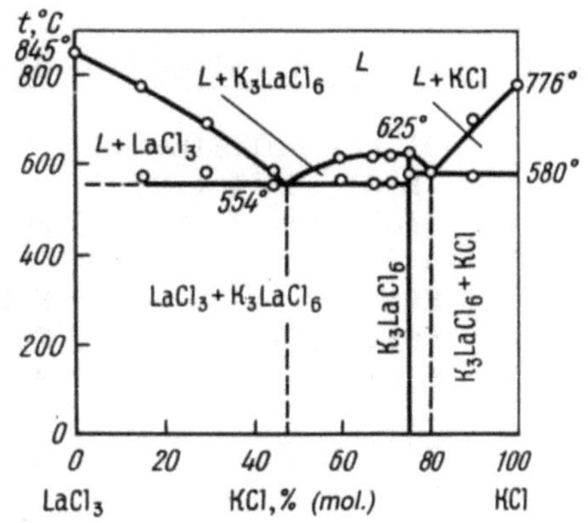

Fig. 167. $LaCl_3$-KCl System.

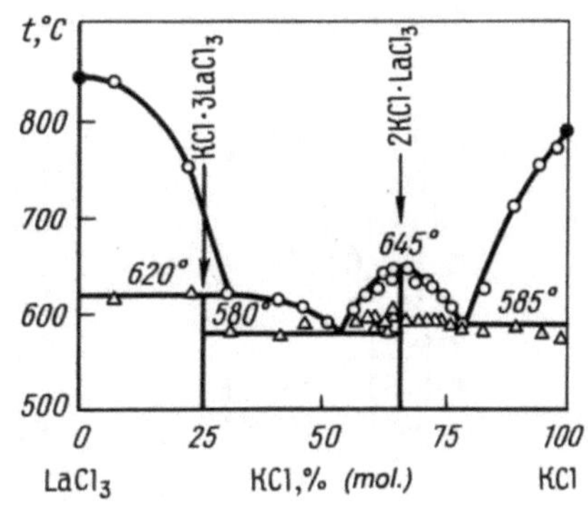

Fig. 168. $LaCl_3$-KCl System.

2nd Component	T°C	Conc. mol %	Phases involved	Def.	Method	Ref.
LiCl	352	57	KCl, LiCl	Eut.	1	828
	354	58.5	KCl, LiCl	Eut.	1,2,6e	829
	450	60		Eut.		830
	348	58	β-LiCl, KCl	Eut.	1	831
	565	92	LiCl transf., KCl	P.T.		
	346	56	KCl, LiCl	Eut.	1	832
	354	58	KCl, LiCl	Eut.	1	833
	352	59	KCl, LiCl	Eut.	2c	834

KCl

2nd Component	T°C	Conc. mol %	Phases involved	Def.	Method	Ref.
LiCl	352	59	3% KCl sol.sln. in LiCl, 8.5% LiCl sol.sln. in KCl	Eut.	2.3b,2f	835
	774	0.0	KCl	M.P.	1,2a	836
	348	58	KCl, LiCl	Eut.		
	606	100	LiCl	M.P.		
	358	61.5	KCl, LiCl	Eut.	-	837
$MgCl_2$	430	-	KCl, $KMgCl_3$	Eut.	2a	838
	488	-	$KMgCl_3$	Comp.		
	470	-	$KMgCl_3$, $MgCl_2$	Eut.		
$MnCl_2$	448	30.5	KCl, K_4MnCl_6	Per.	2d,3b	839
	420	33	K_4MnCl_6, $K_3Mn_2Cl_7$	Eut.		
	437	37	$K_3Mn_2Cl_7$, $KMnCl_3$	Per.		
	490	50	$KMnCl_3$	Comp.		
	448	64	$KMnCl_3$, $MnCl_2$	Eut.		
	386		$\alpha \rightleftharpoons \beta$ $KMnCl_3$	P.T.		
	775	0.0	KCl	M.P.	2a	840
	445	30.0	$4KCl.MnCl_2$, KCl	Per.		
	428	35.0	$KCl.MnCl_2$, $4KCl.MnCl_2$	Eut.		
	495	50.0	$KCl.MnCl_2$	Comp.		
	449	65.0	$MnCl_2$,$KCl.MnCl_2$	Eut.		
	650	100.0	$MnCl_2$	M.P.		
	772	0	KCl	M.P.	2a,3b, 5c	841 842
	452	33	K_2MnCl_4, KCl	Per.		
	420	35	$K_3Mn_2Cl_7$, K_2MnCl_4	Eut.		
	438	37	$KMnCl_3$, $K_3Mn_2Cl_7$	Per.		
	490	50	$KMnCl_3$	Comp.		
	450	64	$MnCl_2$,$KMnCl_3$	Eut.		
	652	100	$MnCl_2$	M.P.		

KCl

2nd Component	T°C	Conc. mol %	Phases involved	Def.	Method	Ref.
$MoCl_3$	693	10	KCl, K_3MoCl_6	Eut.	2d,3b	845
	873	25	K_3MoCl_6	Comp.		
	745	43	K_3MoCl_6, $K_3Mo_2Cl_9$	Per.		
	200		$KMoCl_4$ appears:			
			Refractive indices:			
			K_3MoCl_6: N'g = 1.702 N'p = 1.698			
			$KMoCl_4$: N'g = 1.925 N'p = 1.897			
	774	0.0	KCl	M.P.	2d	846
	698	12.0	KCl, β-K_3MoCl_6	Eut.		
	865	-	β-K_3MoCl_6	Comp.		
	650	40.0	β-K_3MoCl_6, $MoCl_3$	Eut.		
	720	50-85	Decomposes	D		
	460		α-β K_3MoCl_6	P.T.		
	290		Partial decomposition of $MoCl_3$ with presence of $KMoCl_4$	D		
$MoCl_5$	194	~100	Slight mutual solubility	Eut.	2,6c	847
		<33	Appearance of compound $2KCl.MoCl_4$		2b,2c,5	848
		>33	Complete mutual insolubility			
NH_4Cl	-	-	26% NH_4Cl sol. sln. in KCl	-	3e	849
NaCl	658		Continuous sol.sln.	Min.	1	850
	<500		Sol.sln. dissociates	P.T.		

KCl

2nd Component	T°C	Conc. mol %	Phases involved	Def.	Method	Ref.
NaCl	666	40	Continuous sol.sln.	Min.	1	851
	772	0.0	KCl	M.P.		
	800	100.0	NaCl	M.P.		
	660	50	Continuous sol.sln.	Min	1	852
	770	0.0	KCl	M.P.		
	800	100.0	NaCl	M.P.		
			Limited sol.sln.		3b	853
	<500	70.0	Sol.sln. dissoci- ates	P.T.		
	502±5	70.0	The dissociation limits determined	P.T.	3b	854
	665	51.0	Continuous sol.sln.	Min.	2d,1d, 3b,5	855
			Sol.sln. dissoci- ates at lower temp.			
	515	66-75	Minimum on dissoci- ation curve	Min.		
	645±2	50	30% KCl sol.sln. in NaCl, 40% NaCl sol.sln. in KCl			856
			Continuous sol.sln.		1	857
			Fig. 169			858
	664	50	Continuous sol.sln.	Min.	2d,5c, 3b	858
			Limited solubility in solid state.			
	500±5		Critical point of dissociation.	P.T.		
	650	56	Continuous sol.sln.	Min.	2,3b	859
			Limited range of solubility in solid state.			
	520±2	25	Critical pt. of sol. state disso- ciation.	P.T.		
	658	50	Continuous sol.sln.	Min.	1,3b	860 861

KCl

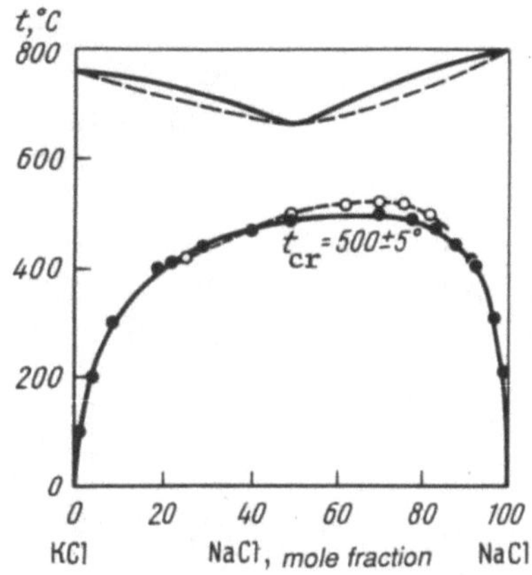

Fig. 169. KCl-NaCl system.

2nd Component	T°C	Conc. mol %	Phases involved	Def.	Method	Ref.
$NbCl_2$	658	27	β-K_2NbCl_4, KCl	Eut.	2a,3b	862
	695	33	β-K_2NbCl_4	Comp.		
	634		K_2NbCl_4	P.T.		
Nb_3Cl_8	663		KCl, α-$K_7Nb_3Cl_{15}$	Eut.	2	863
	725		α-$K_7Nb_3Cl_{15}$	Comp.		
	630		α-$K_7Nb_3Cl_{15}$, $K_4Nb_3Cl_{12}$	Per.		
	357		$K_7Nb_3Cl_{15}$	P.T.		
$NbCl_3$			Fig. 170			864
	644	20	β-K_2NbCl_5, KCl	Eut.	2e	864
	718	32	β-K_2NbCl_5	Comp.		
	590	-	K_2NbCl_5	P.T.		
$NbCl_4$	615	22	K_2NbCl_6, KCl	Eut.	2c	865
	782	33	K_2NbCl_6	Comp.		
	298	60	$NbCl_4$, K_2NbCl_6	Eut.		
$NbCl_5$			Fig. 171			866
	376	48	KCl, $KNbCl_6$	Eut.	1,2a	866
	396	50	$KNbCl_6$	Comp.		

KCl

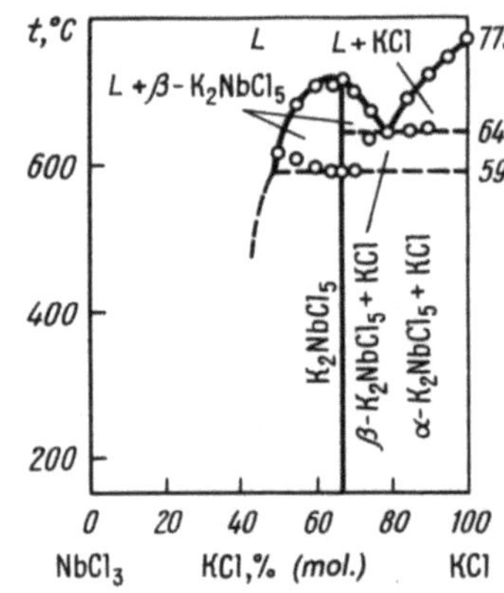

Fig. 170. $NbCl_3$-KCl system.

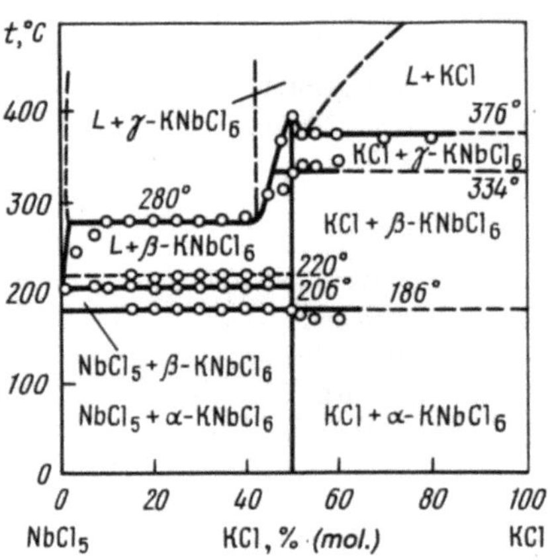

Fig. 171. $NbCl_5$-KCl system.

2nd Component	T°C	Conc. mol %	Phases involved	Def.	Method	Ref.
$NbCl_5$	206	~100	$KNbCl_6$, $NbCl_5$	Eut.		866
	220	57-99	Dissociation	D.		
	186	-	$\alpha \rightleftharpoons \beta$ $KNbCl_6$	P.T.		
	334	-	$\alpha \rightleftharpoons \beta$ $KNbCl_6$	P.T.		

KCl

2nd Component	T°C	Conc. mol %	Phases involved	Def.	Method	Ref.
$NbCl_5$			Fig. 172			867
	368.0	44	$KNbCl_6$, KCl	Eut.	1	867
	394.2	50	$KNbCl_6$	Comp.		
	204.5	~100	$NbCl_5$, $KNbCl_6$	Eut.		
	285	61-99	Dissociation	D.		
			Existence of $KNbCl_6$ confirmed			868
	95-100		$KNbCl_6$ sol.sln. in $NbCl_5$		3b	
		41-50	KCl sol.sln. in $KNbCl_6$			
$NdCl_3$			Fig. 173			869
	615	16.5	β-K_3NdCl_6, KCl	Eut.	2c	869
	682	26	β-K_3NdCl_6	Comp.		
	590	41	K_2NdCl_5, β-K_3NdCl_6	Per.		

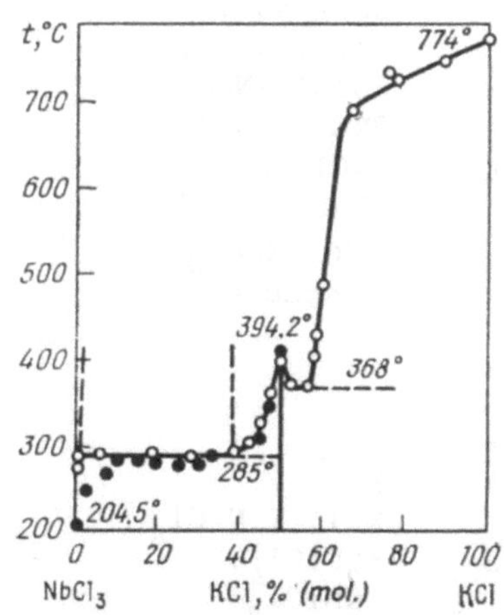

Fig. 172. $NbCl_3$-KCl system.

2nd Component	T°C	Conc. mol %	Phases involved	Def.	Method	Ref.
$NdCl_3$	480	50	$NdCl_3$,K_2NdCl_5	Eut.	2c	869
	345		K_3NdCl_6	P.T.		
	395		K_2NdCl_5	P.T.		

KCl

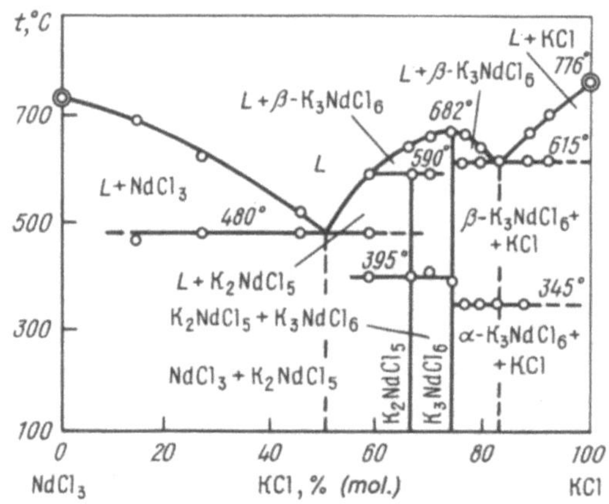

Fig. 173. $NdCl_3$-KCl system.

2nd Component	T°C	Conc. mol %	Phases involved	Def.	Method	Ref.
			Fig. 174			870
	625	21	KCl, $3KCl.NdCl_3$	Eut.	2a	870
	696	26	$3KCl.NdCl_3$	Comp.		
	590	36	$3KCl.NdCl_3$, $3KCl.2NdCl_3$	Eut.		
	615	40	$3KCl.2NdCl_3$	Comp.		
	500	50.2	$3KCl.2NdCl_3$, $NdCl_3$	Eut.		
	450		$2KCl.NdCl_3$	P.T.		
	435		detected	P.T.		
$NiCl_2$	770	0	KCl	M.P.	1,2a,3a	871
	494	22	KCl, K_2NiCl_4	Eut.		
	502	35.5	K_2NiCl_4, $KNiCl_3$	Per.		
	644	50	$KNiCl_3$	Comp.		
	638	58.5	$KNiCl_3$, $NiCl_2$	Eut.		
	480		$KNiCl_3$ (The structure similar to $CsNiCl_3$ - hexagonal: a = 6.37Å, c = 5.608Å	P.T.		

KCl

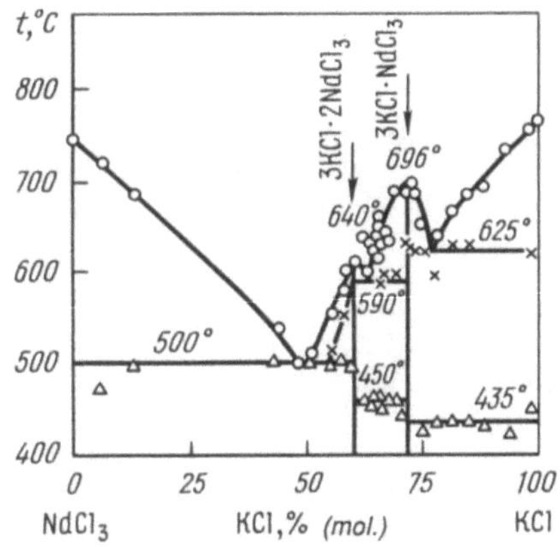

Fig. 174. $NdCl_3$-KCl system.

2nd Component	T°C	Conc. mol %	Phases involved	Def.	Method	Ref.
$PbCl_2$	490	36.5	$2KCl.PbCl_2$, KCl	Per.	1	872
	410	51	$KCl.2PbCl_2$,			873
			$2KCl.PbCl_2$	Eut.		874
	440	67	$KCl.2PbCl_2$	Comp.		
	421	77	$PbCl_2$,			
			$KCl.2PbCl_2$	Eut.		
	496	100	$PbCl_2$	M.P.		
	386		$KCl.2PbCl_2$	P.T.		
	411	67	-	Eut.		875
	428	88		Eut.		
			Fig. 175			876
	462	45	$2KCl.PbCl_2$, β-KCl	Per.	2b,2c	876
	390-426	45-74	$2KCl.PbCl_2$ (E_2FD)			
			C_3B_1)			
			C_2B_1) β-KCl (Metastable equilibrium)			
		33	C_3			
	482	38	α-KCl, β-KCl (C_3)			
	482-778	0-33	α-KCl (C_3B)			
	482-778	0-35	α-KCl (C_2B)			
	390-482		β-KCl (E_4C_3)			
	400-482	38-54	β-KCl (E_3C_2)			
	400	54	$KCl.2PbCl_2$, β-KCl (E_3)	Eut.		

KCl

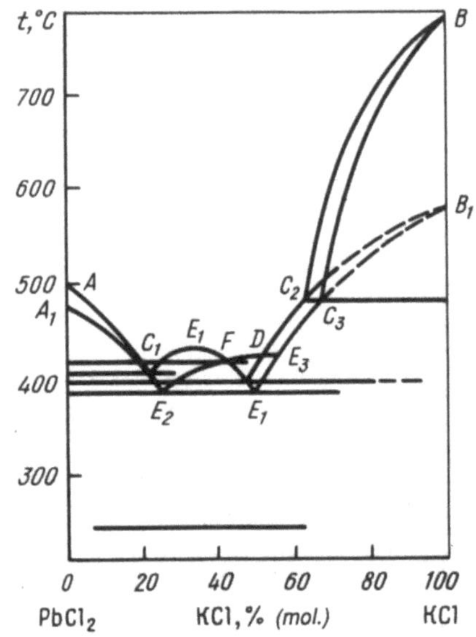

Fig. 175. $PbCl_2$-KCl system.

2nd Component	T°C	Conc. mol %	Phases involved	Def.	Method	Ref.
$PbCl_2$	251	40-92.5	α-KCl.2$PbCl_2$, β-KCl.2$PbCl_2$	-		876
	-	-	KCl.2$PbCl_2$ (E_1E_3)	-		
	422	58-80	Dissociation range (C_1)	-		
	390	74	β-$PbCl_2$, $PbCl_2$.2KCl (E_2-curve BC_3)	Eut.		
	410	78	β-$PbCl_2$, KCl.2$PbCl_2$(E_1)	Eut.		
	410-422	78-80	β-$PbCl_2$(C_1E_1)			
	422	80	α-$PbCl_2$, β-$PbCl_2$(C_1)	Per.		
	422-480	80-100	β-$PbCl_2$ (Metastable equilibr.- A_1C_1)			
	422-501	80-100	α-$PbCl_2$			
	400		System dissociates and eutectic (E_2) crystallizes (the curve $AC_1FE_3DC_2B$)	Eut.		
			Liquidus curve BC_2 according to M. Usanovich suggests equimolar compound.			

KCl

2nd Component	T°C	Conc. mol %	Phases involved	Def.	Method	Ref.
$PbCl_2$			Thermographic records suggest compounds: $KCl.2PbCl_2$, $2KCl.PbCl_2$ and crystallization of eutectics:			876
	410		E_1	Eut.		
	390		E_2	Eut:		
	410		E_4	Eut.		
			Existence of compounds confirmed		3b	877
	493	40.0	KCl, $2KCl.PbCl_2$	Per.	1,2	878
	426	53.0	$2KCl.PbCl_2$, $KCl.PbCl_2$	Per.		
	420	56.5	$KCl.PbCl_2$, $KCl.2PbCl_2$	Eut.		
	448	66.7	$KCl.2PbCl_2$	Comp.		
	424	79.0	$KCl.2PbCl_2$, $PbCl_2$	Eut.		
	T liqs. +50	0.0, 6.3, 25.4, 35.0, 45.0, 60.5, 72.5, 85.0, 100.0	Existence of compounds K_2PbCl_4 and KPb_2Cl_5 confirmed		2g	879
		52	-	Eut.		
$PrCl_3$	610	19	KCl, β-K_3PrCl_6	Eut.	2	880
	658	25	β-K_3PrCl_6	Comp.		
	628	36	K_2PrCl_5, β-K_3PrCl_6	Per.		
	483	57	$PrCl_3$, K_2PrCl_5	Eut.		
	450		K_3PrCl_6	P.T.		
	610	22.8	KCl, $3KCl.PrCl_3$	Eut.	2a	881
	675	28	$3KCl.PrCl_3$	Comp.		
	620	34	$3KCl.PrCl_3$, $2KCl.PrCl_3$	Per.		
	580	35	$2KCl.PrCl_3$, $3KCl.2PrCl_3$	Eut.		

KCl

2nd Component	T°C	Conc. mol %	Phases involved	Def.	Method	Ref.
$PrCl_3$	615	38	$3KCl.2PrCl_3$	Comp.		881
	500	53.8	$3KCl.2PrCl_3$, $PrCl_3$	Eut.		
$PuCl_3$	621±3	17	K_3PuCl_6, KCl	Eut.	2c	882
	685±3		K_3PuCl_6	Comp.		
	611±3	35	K_2PuCl_5, K_3PuCl_6	Per.		
	486 3	57	$PuCl_3$, K_2PuCl_5	Eut.		
$PuCl_4$			Fig. 176			883
	770	0.0	KCl	M.P.	2d,3b,1g	884
	582	18-19		Eut.		
			K_2PuCl_6 congruently melting but unstable	Comp.		
			X-ray analysis confirms various compounds.			
	435		Equilibrium between Pu(III) & Pu(IV)	P.T.		

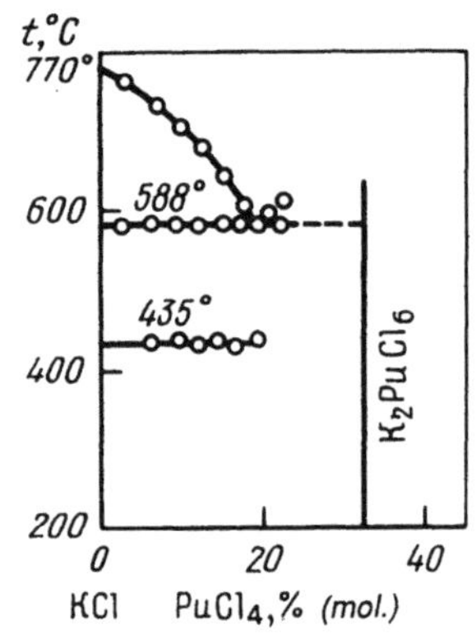

Fig. 176. $KCl-PuCl_4$ system.

KCl

2nd Component	T°C	Conc. mol %	Phases involved	Def.	Method	Ref.
$RbCl$			Continuous sol.sln.		1,2a	885
	775	0.0	KCl	M.P.	1,3b	886
			Continuous sol.sln.			887
	726	100.0	$RbCl_2$	M.P.		
$SbCl_5$	768	0	KCl	M.P.	2d	888
	415	50	KCl, $3KCl.2SbCl_5$	Per.		
	402	54	$3KCl.2SbCl_5$, β-$KSbCl_6$	Per.		
	250		α-$KSbCl_6$, β-$KSbCl_6$	Per.		
	~3	~100	α-$KSbCl_6$, $SbCl_5$	Eut.		
	3	100	$SbCl_5$	M.P.		
	316	70-97	Decomposition range	D		
$ScCl_3$	692	16	KCl, $3KCl.ScCl_3$	Eut.	2d,4a	889
	818	25	$3KCl.ScCl_3$	Comp.		
	473	39	$3KCl.ScCl_3$, $2KCl.ScCl_3$	Per.		
	445	49	$2KCl.ScCl_3$, $ScCl_3$	Eut.		
$SmCl_2$	553	35	KCl, $KCl.2SmCl_2$	Eut.	2a	890
	620	65	$KCl.2SmCl_2$, $SmCl_2$	Per.		
$SmCl_3$			Fig. 177			891
	645	15.5	KCl, $3KCl.SmCl_3$	Eut.	2a	891
	732	24	$3KCl.SmCl_3$	Comp.		
	570	40	$3KCl.SmCl_3$, $2KCl.SmCl_3$	Per.		
	490	52	$2KCl.SmCl_3$, $SmCl_3$	Eut.		
	365		K_3SmCl_6	P.T.		

KCl

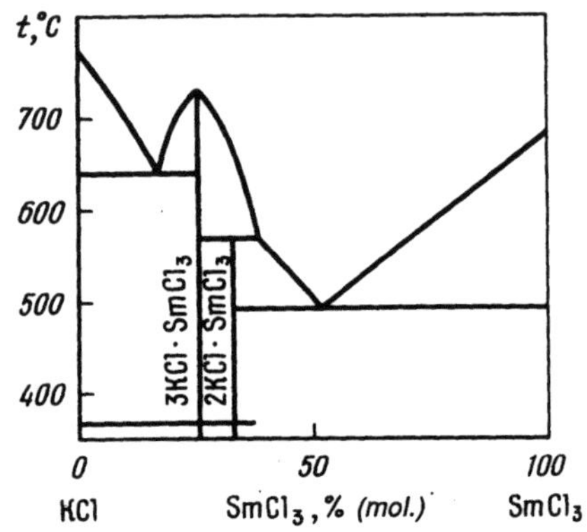

Fig. 177. $KCl-SmCl_3$ system.

2nd Component	T°C	Conc. mol %	Phases involved	Def.	Method	Ref.
$SmCl_3$	643	17	$3KCl.SmCl_3$, KCl	Eut.	2a	892
	574	42	$2KCl.SmCl_3$, $3KCl.SmCl_3$	Per.		
	482	56	$KCl.2SmCl_3$, $2KCl.SmCl_3$	Eut.		
	530	68	$SmCl_3$,$KCl.2SmCl_3$	Per.		
	360	-	$3KCl.SmCl_3$	P.T.		
	644	17	K_3SmCl_6, KCl	Eut.	2c	893
	745	25	K_3SmCl_6	Comp.		894
	570	36-38	K_2SmCl_5, K_3SmCl_6	Per.		
	460	60	KSm_2Cl_7,K_2SmCl_5	Eut.		
	545	66.7	KSm_2Cl_7	Comp.		
	490	75	$SmCl_3$,KSm_2Cl_7	Eut.		
	350	-	KSm_2Cl_7	P.T.		
	360	K_3SmCl_6	P.T.			
$SnCl_2$	224	48	$SnCl_2.KCl$, KCl	Eut.	2a	895
	225	50	$SnCl_2.KCl$	Comp.	6c	
	176	60	$3SnCl_2.KCl$, $SnCl_2.KCl$	Eut.		
	218	75	$3SnCl_2.KCl$	Comp.		
	200	80	$SnCl_2$,$3SnCl_2.KCl$	Eut.		

KCl

2nd Component	T°C	Conc. mol %	Phases involved	Def.	Method	Ref.
$SrCl_2$	588	30.5	$SrCl_2.K_2Cl_2$, K_2Cl_2	Eut.	1	896
	597	33.3	$SrCl_2.K_2Cl_2$	Comp.		
	575	45	$SrCl_2.K_2Cl_2$, $4SrCl_2.K_2Cl_2$	Eut.		
	642	66.7	$SrCl_2$, $4SrCl_2.K_2Cl_2$	Per.		
	596	30.3	KCl, $2KCl.SrCl_2$	Eut.	1	897
	572	45	$2KCl.SrCl_2$, $KCl.2SrCl_2$	Eut.		
	637	67	$KCl.2SrCl_2$, $SrCl_2$	Eut.		
	580	30.3	$(KCl)_2.SrCl_2$, KCl	Eut.	2a	898
	595	33.3	$(KCl)_2.SrCl_2$	Comp.		899
	560	45.5	$(KCl)_2.3SrCl_2$, $(KCl)_2.SrCl_2$	Eut.		
	620	60	$(KCl)_2.3SrCl_2$	Comp.		
	612	68	$SrCl_2$, $(KCl)_2.3SrCl_2$	Eut.		
	594	32.5	KCl,$2KCl.SrCl_2$	Eut.	1	900
	603	33.3	$2KCl.SrCl_2$	Comp.		
	577	44.8	$2KCl.SrCl_2$, $KCl.2SrCl_2$	Eut.		
	642	66.3	$KCl.2SrCl_2$,$SrCl_2$	Per		
		71	Solid solutions on the KCl side	Eut.	8	901
			Fig. 178			902
	770	0	KCl	M.P.	2a	902
	590	29.00	KCl,K_2SrCl_4	Eut.	3a	
	602	33.33	K_2SrCl_4	Comp.		
	570	44.00	K_2SrCl_4, KSr_2Cl_5	Eut.		
	675	66.67	KSr_2Cl_5	Comp.		
	643	68.00	KSr_2Cl_5, $SrCl_2$	Eut.		
	878	100	$SrCl_2$	M.P.		
			In figure: 1. The authors 2. Refs. 898, 899			
$TaCl_3$	520	32	K_2TaCl_5, KCl	Eut.	2a	903

KCl

2nd Component	T°C	Conc. mol %	Phases involved	Def.	Method	Ref.
$TaCl_3$	560	33	K_2TaCl_5	Comp.		903
	230	50	$TaCl_3$, K_2TaCl_5	Eut.		
	263		K_2TaCl_5	P.T.		

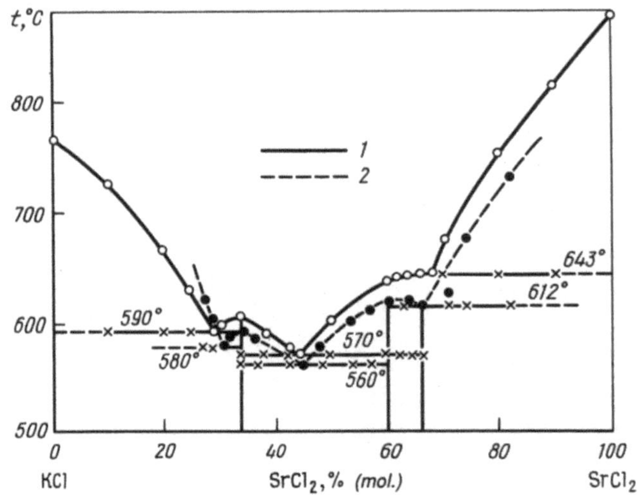

Fig. 178. KCl-$SrCl_2$ system.

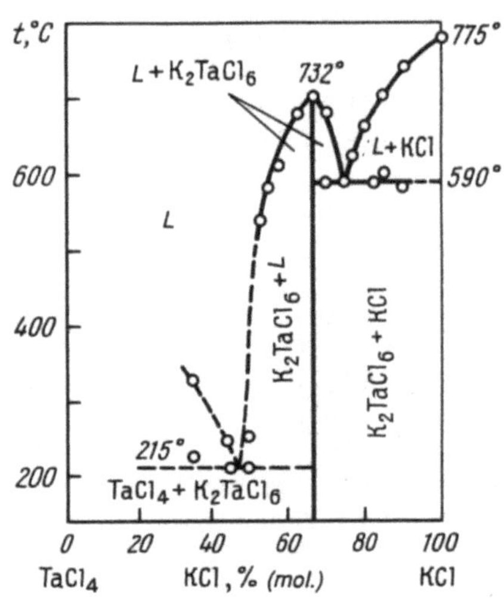

Fig. 179. $TaCl_4$-KCl system.

KCl

2nd Component	T°C	Conc. mol %	Phases involved	Def.	Method	Ref.
$TaCl_4$			Fig. 179			904
	590	25	K_2TaCl_6, KCl	Eut.	2a,3b	904
	732	33.3	K_2TaCl_6	Comp.		
	215	51	K_2TaCl_6, $TaCl_4$	Eut.		
			K_2TaCl_6 {Refractive index = 1.789; Density = 3.017 g/cm³}			
$TaCl_5$			Table VII			905
	370	48	KCl, $KTaCl_6$	Eut.	2d,1,4a	905
	~216	~100	$TaCl_5$, $KTaCl_6$	Eut.		
			Phase transformations of $KTaCl_6$ associated with sharp increase of conductivity			
			$KTaCl_6$ exists	Comp.	3b	906
		41-50	KCl sol.sln. in $KTaCl_6$			
		95-100	$KTaCl_6$ sol.sln. in KCl			

Table VII

$TaCl_5$, (mol %)	Beginning of crystn.	Transformn. $\alpha \rightleftarrows \beta$ $KTaCl_6$	End of crystn.	trans. $\gamma \rightleftarrows \beta$ $KTaCl_6$ According to heating curve	trans. $\gamma \rightleftarrows \beta$ $KTaCl_6$ According to cooling curve	Solid Phase	Transformation
100	216	-	-	-	-	$TaCl_5$	One liq. layer
99	-	-	215	-	-		
95	300	-	215	181	160	α-$KTaCl_5$	Two liq. layer
90	309	-	215	181	160		
80	309	-	215	182	160		
70	310	308	215	182	160		
60	316	309	215	185	161		
58	320	310	214	185	161		
55	351	310	214	185	162		
52	390	310	214	185	161		
50	410	312	-	185	160		
49	400	314	-	185	135		
48	370	314	370	185	135	β-$KTaCl_6$, KCl	Single liq. layer
47	443	314	370	185	135	KCl	
45	525	314	368	185	127		
40	-	314	370	183	126		
20	-	314	369	181	-		
0	775	-	-	-	-		

KCl

2nd Component	T°C	Conc. mol %	Phases involved	Def.	Method	Ref.
$TeCl_4$			Fig. 180			907
	559	29.0	KCl, K_2TeCl_6	Eut.	2d, le, 3b	907
	570	33.3	K_2TeCl_6	Comp.	6a	
	208	85.0	K_2TeCl_6, $TeCl_4$	Eut.		
			K_2TeCl_6 - optically anisotropic density 2.68 g/cm^3			
$ThCl_4$	620	11.7	Fig. 181			908
			KCl, K_3ThCl_7	Eut.	2	908
	705	24.0	K_3ThCl_7	Comp.		
	406	39.0	K_3ThCl_7, K_2ThCl_6	Per.		
	384	42.8	K_2ThCl_6, $KThCl_5$	Eut.		
	428	50.0	$KThCl_5$	Comp.		
	403	52.5	$KThCl_5$, $ThCl_4$	Eut.		
	622	18	KCl, K_2ThCl_6	Eut.	2	909
	715	-	K_2ThCl_6	Comp.		
	432	52	K_2ThCl_6, $ThCl_4$	Eut.		
	617	33-34	KCl, K_2ThCl_6	Per.	2a	910
	376	44-45	K_2ThCl_6, $KThCl_5$	Eut.		
	457	50	$KThCl_5$	Comp.		
	416	56-58	$KThCl_5$, $ThCl_4$	Eut.		

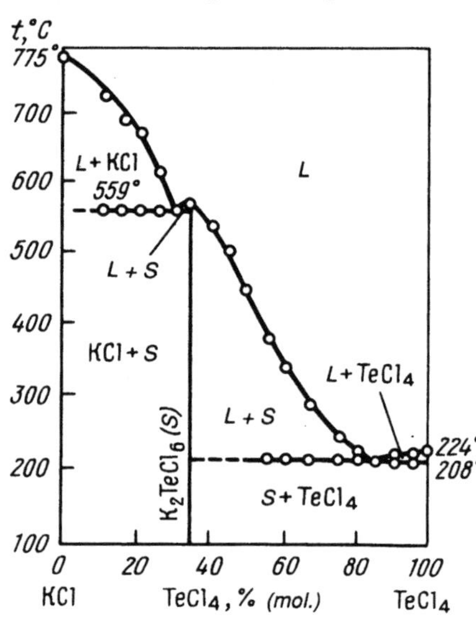

Fig. 180. KCl-$TeCl_4$ system.

KCl

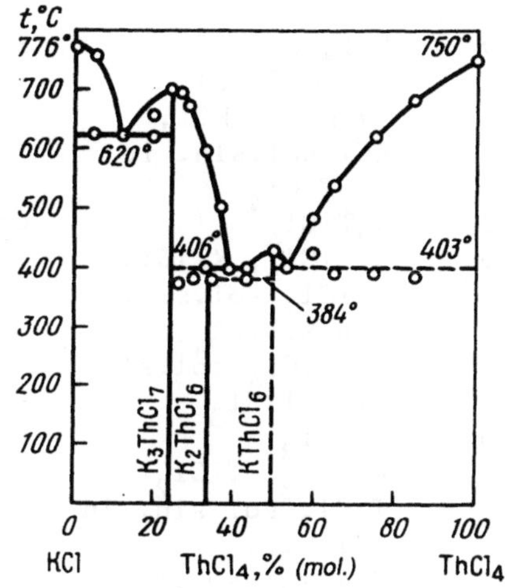

Fig. 181. $KCl-ThCl_4$ system.

2nd Component	T°C	Conc. mol %	Phases involved	Def.	Method	Ref.
$TiCl_2$	632	27	KCl, K_2TiCl_4	Eut.	2,3b	911
	671	33	K_2TiCl_4, $KTiCl_3$	Per.		
	762	50	$KTiCl_3$	Comp.		
	730	62	$KTiCl_3$, $TiCl_2$	Eut.		
	770	0	KCl	M.P.	2a	912
	650	16.2	KCl,K_3TiCl_5	Eut.		
	766	25	K_3TiCl_5	Comp.		
	650	16.2	KCl, K_3TiCl_5	Eut.	2	913
	766	25.0	K_3TiCl_5	Comp.		
	620	36.5	K_3TiCl_5, $KTiCl_3$	Eut.		
	804	50.0	$KTiCl_3$	Comp.		
	688	56.5	$KTiCl_3$, $TiCl_2$	Eut.		
		0-6	K_2TiCl_4 sol.sln. in KCl		2a,3b	914
		27-33.3	KCl sol.sln. in K_2TiCl_4			
	671		K_2TiCl_4	Comp.		
		45-50	K_2TiCl_4 sol. sln. in $KTiCl_3$			
	762		$KTiCl_3$	Comp.		
		50-62	$TiCl_2$ sol.sln. in $KTiCl_3$			

KCl

2nd Component	T°C	Conc. mol %	Phases involved	Def.	Method	Ref.
$TiCl_2$	615	21.0	KCl, K_2TiCl_4 K_2TiCl_4 sol. sln. in KCl, KCl sol.sln. in K_2TiCl_4	Eut.	2a,3b	915
	705	33.3	K_2TiCl_4	Comp.		
	648	39.0	K_2TiCl_4, $KTiCl_3$ $KTiCl_3$ sol.sln. in K_2TiCl_4, K_2TiCl_4 sol.sln. in $KTiCl_3$	Eut.		
	747	50.0	$KTiCl_3$	Comp.		
	694	56.6	$KTiCl_3$, $TiCl_2$ $TiCl_2$ sol.sln. in $KTiCl_3$, $KTiCl_3$, sol.sln. in $TiCl_2$	Eut.		
$TiCl_3$	676	14.8	KCl, $3KCl.TiCl_3$	Eut.	2c	916
	760	24.9	$3KCl.TiCl_3$	Comp.		917
	595	33.6 (nominal)	$3KCl.TiCl_3$, $TiCl_2$	Eut. (nominal)		
			Fig. 182 Letters referring to figure			918
	673	17	E_1 - KCl, $TiCl_3.3KCl$	Eut.	2	918
	760	25	S_1 - $TiCl_3.3KCl$	Comp.		
	581	37.5	E_2 - $TiCl_3.3KCl$, $TiCl_3.KCl$	Eut.		
	654	50	S_2 - $TiCl_3KCl$	Comp.		
	534	75	E_3 - $TiCl_3.KCl$, $TiCl_3$	Eut.		

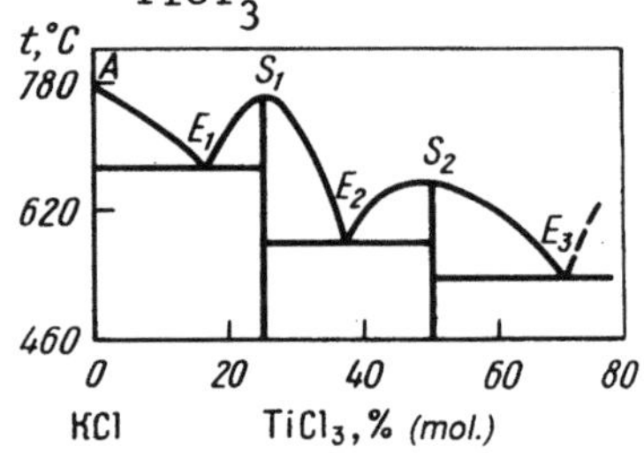

Fig. 182. KCl-$TiCl_3$ system.

KCl

2nd Component	T°C	Conc. mol %	Phases involved	Def.	Method	Ref.
$TiCl_3$	676	15	KCl, $3KCl.TiCl_3$	Eut.	2c	919
	760	25	$3KCl.TiCl_3$	Comp.		
	580	40	$3KCl.TiCl_3$, $2KCl.TiCl_3$	Per.		
	682	15	KCl, K_3TiCl_6	Eut.	2d,3e, 1d	920
	783	25	K_3TiCl_6	Comp.		
	605	40	K_3TiCl_6, K_2TiCl_5	Per.		
	535	50	K_2TiCl_5, $TiCl_3$	Eut.		
	420	-	K_3TiCl_6	P.T.		
		0-6	K_2TiCl_4, KCl		3b	921
		27-33.3	K_2TiCl_4, $KTiCl_3$			
		33.3-45	$KTiCl_3$ base sol.sln., K_2TiCl_4			
		45-62	$KTiCl_3$ base sol.sln.			
			Fig. 183			922
	675	15	KCl with 10% $TiCl_3$ base sol.sln., K_3TiCl_6	Eut.	2a,3a 6a	922
	762	25	K_3TiCl_6	Comp.		
	574	47	K_3TiCl_3, α-$K_3Ti_2Cl_9$	Per.		
	550	52	α-K_3TiCl_9, β-K_3TiCl_9	Per.		
	510	58	β-K_3TiCl_9, $TiCl_3$	Eut.		
			$K_3Ti_2Cl_9$ base sol.slns.present			
			Unit cell sizes a = 7.06Å c = 17.80Å			
			Density = 2.33 g/cm^3			

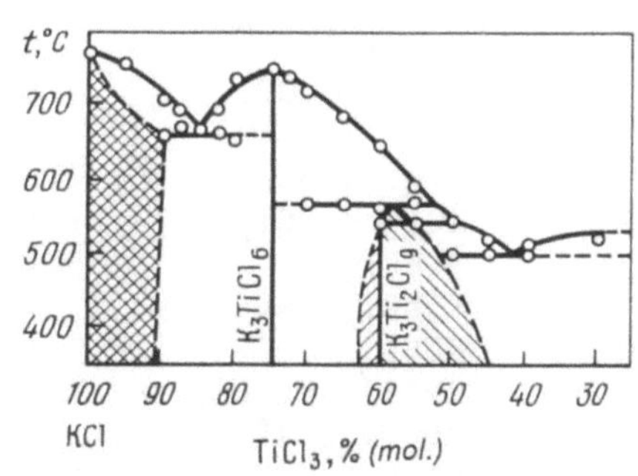

Fig. 183. KCl-$TiCl_3$ system.

KCl

2nd Component	T°C	Conc. mol %	Phases involved	Def.	Method	Ref.
$TiCl_3$			K_3TiCl_6 in solid state partially decomposes to KCl and $K_3Ti_2Cl_9$			922
UCl_3			Fig. 184			923
	595	20	KCl, K_2UCl_5	Eut.		923
	625	33	K_2UCl_5	Comp.		
	550	50	K_2UCl_5, KCl′	Eut.		
	525		KCl + K_2UCl_5 K_3UCl_6	P.T.		
	590	16.0	KCl, 2KCl.UCl_3	Eut.	2d,3b	924
	660	33.3	2KCl.UCl_3	Comp.		
	543	50.0	2KCl.UCl_3, UCl_3	Eut.		

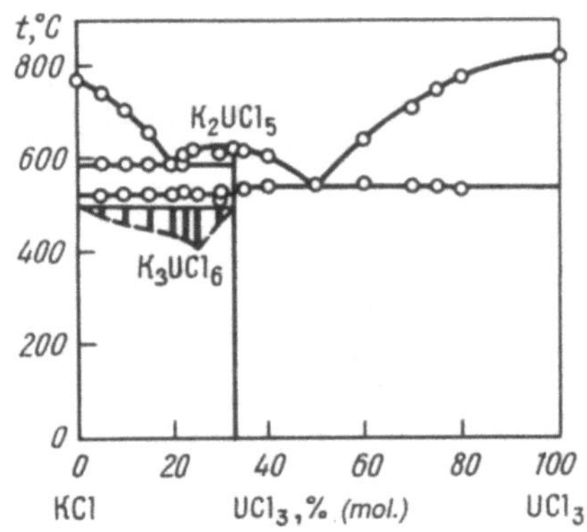

Fig. 184. KCl-UCl_3 system.

2nd Component	T°C	Conc. mol %	Phases involved	Def.	Method	Ref.
UCl_4	555	25	KCl, K_2UCl_6	Eut.	2c	925
	654	~33	K_2UCl_6	Comp.		
	334	51	K_2UCl_6, UCl_4	Eut.		
			Fig. 185			926
	552	27	KCl, K_2UCl_6	Eut.	2c,3b	926
	642.8	33	K_2UCl_6	Comp.		
	334.9	43	K_2UCl_6, $KUCl_5$	Eut.		
	349.0	50	$KUCl_5$	Comp.		
	341.1	52	$KUCl_5$, KU_2Cl_9	Eut.		
	350	56	KU_2Cl_9, KU_3Cl_{13}	Per.		
	452.0	60	KU_3Cl_{13}, UCl_4	Per.		

KCl

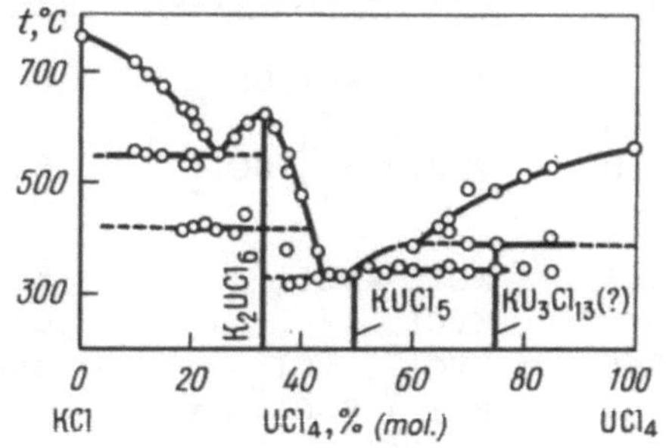

Fig. 185. KCl-UCl_4 system.

2nd Component	T°C	Conc. mol %	Phases involved	Def.	Method	Ref.
UCl_4	529	24	KCl, K_2UCl_6	Eut.	2a	927
	605	33.3	K_2UCl_6	Comp.		
	418	41	α-K_2UCl_6, β-K_2UCl_6	Per.		
	300	43	β-K_2UCl_6, β-KU_2Cl_9 (composition conditional)	Eut.		
	320	52	β-KU_2Cl_9, α-KU_2Cl_9	Per.		
	384	61	α-KU_2Cl_9,α-UCl_4	Per.		
	485	100	α-UCl_4, β-UCl_4	Per. ?		
	540	100	β-UCl_4, γ-UCl_4	Per. ?		
	551	25.0	KCl, K_2UCl_6	Eut.	1, le	928
	622	33.3	K_2UCl_6	Comp.		
	335	51.0	K_2UCl_6, KU_2Cl_9	Eut.		
	410	58.0	KU_2Cl_9,UCl_4	Per.		
	188		K_2UCl_6	P.T.		
	312		K_2UCl_6	P.T.		
			Existence of K_2UCl_4 & KU_2Cl_9 confirmed by "immersion analysis"			
			For K_2UCl_6 N_p = 1.640 ± 0.003 $N_\%$ = 1.680 ± 0.003			
			For KU_2Cl_9 N_p = 1.744 ± 0.003 $N_\%$ = 1.780 ± 0.003			
			For KCl N_p = 1.488 ± 0.002			
			For UCl_4 N_p = 1.780 $N_\%$ = 1.780			

KCl

2nd Component	T°C	Conc. mol %	Phases involved	Def.	Method	Ref.
UCl_4	562±2	26.0	KCl, $2KCl.UCl_4$	Eut.	2d	929
		33.3	$2KCl.UCl_4$	Comp.		
	335±2	51.0	$2KU.UCl_4$, UCl_4	Eut.		
VCl_2	687	21.5	KCl, $KVCl_3$	Eut.	2d,3e	930
	946	50	$KVCl_3$	Comp.		
	930	52.5	$KVCl_3$, VCl_2	Eut.		
	645		KCl + $KVCl_3$ ⇌ K_2VCl_4	P.T.		
			Solubilities			931
		26-45	KCl in K_2VCl_4		3b	931
		45-50	K_2VCl_4 in $KVCl_3$			
		50-63	VCl_2 in $KVCl_3$			
			Existence of compounds: $KVCl_3$ in liquid phase K_2VCl_4 in solid phase			
	687	21.5	KCl, $KVCl_3$	Eut.	2,3b	932
	946	50	$KVCl_3$	Comp.		
	930	52.5	$KVCl_3$, VCl_2	Eut.		
VCl_3	655±5	16.00	KCl, $3KCl.VCl_3$	Eut.	2c	933
	743	25.5	$3KCl.VCl_3$	Comp.		
	676	13	KCl, K_3VCl_6	Eut.	2c	934
	722	25	K_3VCl_6	Comp.		
	628		K_3VCl_6, $K_3V_2Cl_9$	Per.		
	580	52	$K_3V_2Cl_9$, phase comp. unknown	Eut.		
	<628		K_3VCl_6 transforms	P.T.		
	61	13	KCl, β-K_3VCl_6	Eut.	2	935
	436	44	β-K_3VCl_6, $KCl.3VCl_3$	Eut.		
	550	65.9	$KCl.3VCl_3$, VCl_3	Per.		
	366	-		P.T.		
	774	0	KCl	M.P.	2b	936
	675	13	K_3VCl_6, KCl	Eut.		
	830	25	K_3VCl_6	Comp.		
	430	50	KV_3Cl_{10}, K_3VCl_6	Eut.		
	535	60	VCl_3, KV_3Cl_{10}	Per.		
	700	100	VCl_3	M.P.		
	340		K_3VCl_6	P.T.		

KCl

2nd Component	T°C	Conc. mol %	Phases involved	Def.	Method	Ref.
VCl_3			$3KCl.VCl_3$ - congruently melting compound, $\Delta H = -14.4 \pm 1.0$ kcal/mol		2a,6c, 2h	937
			$3KCl.2VCl_3$ - incongruently melting compound			
			Solid solution range:			
		24.7-25	KCl in K_3VCl_6		3b	938
		25-29.4	VCl_3 in K_3VCl_6			
	668		KCl, K_3VCl_6	Eut.	2d	939
	720		K_3VCl_6	Comp.		
	624		K_3VCl_6, $K_3V_2Cl_9$	Per.		
	553		K_3VCl_9, VCl_3	Eut.		
WCl_5			Fig. 186			
	386±2	~40	K_2WCl_7, KCl	Per.	2a	940
	360	44	$KWCl_6$, K_2WCl_7	Eut.		
	406±2	50	$KWCl_6$	Comp.		
	240±2	98	WCl_5, $KWCl_6$	Eut.		
		58-98	Decomposition (Line EF in the fig.) Line AB & CD - sol.sln. ranges.			

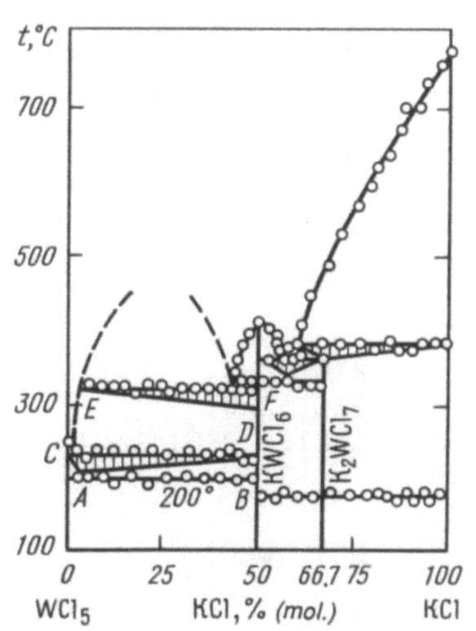

Fig. 186. WCl_5-KCl system.

KCl

2nd Component	T°C	Conc. mol %	Phases involved	Def.	Method	Ref.
WCl_6	260	86	KCl, γ-WCl_6	Eut.	2,6c	941
	174	100	$\alpha \rightleftharpoons \beta$ WCl_6	P.T.		
	222	100	$\beta \rightleftharpoons \gamma$ WCl_6	P.T.		
	406		$KWCl_7$	Comp.	2a	942
	227		WCl_6			
		0-33	Complete insolubility		2a,6c	943
		>33	Compound exists (supposedly) $2KCl.WCl_6$			
YCl_3	675	10	K_3YCl_6, KCl	Eut.	2a	944
	830	25	K_3YCl_6	Comp.		
	430	50	KY_3Cl_{10}, K_3YCl_6	Eut.		
	535	60	YCl_3, KY_3Cl_{10}	Per.		
	340		K_3YCl_6	P.T.		
	672	15.0	KCl, $3KCl.YCl_2$	Eut.	2	945
	780	25.0	$3KCl.YCl_3$	Comp.		
	440	46.5	$3KCl.YCl_3$, $KCl.3YCl_3$	Eut.		
	550	62	$KCl.3YCl_3$, YCl_3	Per.		
	368			P.T.		
$YbCl_2$	560	33	$KCl.YbCl_2$, KCl	Eut.	2a	946
	558	73	$YbCl_2$, $KCl.YbCl_2$	Eut.		
$YbCl_3$	665	15	$YbCl_3$, $3KCl.YbCl_3$	Eut.	2a	946
	423	45	$3KCl.YbCl_3$, KCl	Eut.		
	385		$\alpha \rightleftharpoons \beta$ $3KCl.YbCl_3$	P.T.		
$ZnCl_2$	433	31	$2KCl.ZnCl_2$, KCl	Eut.	2c	947
	450	33	$2KCl.ZnCl_2$	Comp.		
	230	49	$ZnCl_2$, $2KCl.ZnCl_2$	Eut.		
$ZrCl_2$	698±1	53.5	KCl, $ZrCl_2$	Eut.	2c,3b	948

KCl

2nd Component	T°C	Conc. mol %	Phases involved	Def.	Method	Ref.
$ZrCl_2$			In molten state a dissociation reaction takes place: $2ZrCl_2 \rightleftharpoons ZrCl_4 + Zr$, and the system segregates into top layer containing a soln.of $ZrCl_4$ and bottom containing metallic Zr. Below 500°C $ZrCl_2$ does not coexist with KCl.		2c	949
$ZrCl_3$	581	98	$ZrCl_3$, KCl	Eut.	2c,3b	948
$ZrCl_4$	235	65	KCl, $ZrCl_4$	Eut.	1,2c	950
			Fig. 187			951
	596	24.2	K_2ZrCl_6, KCl	Eut.	2	951
	798	33.3	K_2ZrCl_6	Comp.		
	220	57.8	$ZrCl_4$, K_2ZrCl_6	Eut.		
	437	100	$ZrCl_4$	M.P.		
	532		Dissociation of K_3ZrCl_7	P.T.		
	594	13.3	KCl, K_2ZrCl_6	Eut.	2c,6g,3b	952
KF	606	45	KF, KCl	Eut.	1	953
	20-900		Presence of KF does not influence the phase transformation of KCl (5-45 kbar press.)		6b,3b 6d	954
	770	0	KCl	M.P.	1,2a	955
	605	45.2	KF, KCl	Eut.		
	850	100	KF	M.P.		
KI	584	52.5	Continuous sol.sln. range	Min.	1	956
	770	0.0	KCl	M.P.	1	957
	598	51.0	KCl, KI	Eut.		
	680	100.0	KI	M.P.		

KCl

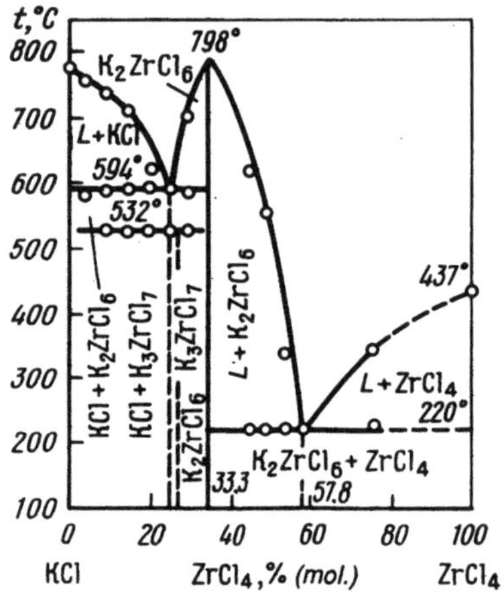

Fig. 187. $KCl-ZrCl_4$ system.

KF

2nd Component	T°C	Conc. mol %	Phases involved	Def.	Method	Ref.
BaF_2	729	26.5	KF, BaF_2	Eut.	2a	958
LaF_3	630±5	20	KF, $KLaF_4$	Eut.	2a	959
	770		Peritectic reaction	Per.		
	550-590		Unspecified effect			
	625	22.0	KF, $KLaF_4$	Eut.	1,2a,3b	960
	746	34.5	$KLaF_4$, LaF_3	Per.		
	540		$2KF + KLaF_4 \rightleftharpoons K_2LaF_6$	P.T.		
			X-ray analysis reveals many compounds			
LiF	852	0.0	KF	M.P.	1	961
	487	50.0	γ-LiF, KF	Eut.		
	730	79.5	βLiF, γ-LiF	Per.		
	812	92.5	α-LiF, β-LiF	Per.		
	848	100.0	α-LiF	M.P.		

KF

2nd Component	T°C	Conc. mol %	Phases involved	Def.	Method	Ref.
LiF	476	50.0	KF, LiF	Eut.	1,2a	962
	493	52	KF, LiF	Eut.	2,6e,1	963
	492	49.9	KF, LiF	Eut.	1	964
	856	0.0	KF	M.P.	1,2a	965
	492	50.0	KF, LiF	Eut.		
	842	100.0	LiF	M.P.		
	852	0	KF	M.P.	1	966
	487	-	KF, KF.LiF	Per.		
	482	53	KF.LiF, LiF	Eut.		
	850	100	LiF	M.P.		
MgF_2		21.5-50	Confirmed existence of: $KMgF_3$ - perovskite structure K_2MgF_4		3b	967
	850	0	KF	M.P.	1,2a	968
	785	21.7	KF, $2KF.MgF_2$*	Eut.		
	872	35.9	$2KF.MgF_2$* $KF.MgF_2$	Per.		
	1090	50.0	$KF.MgF_2$	Comp.		
	1028	81.2	$KF.MgF_2$	Eut.		
	1263	100	MgF_2	M.P.		
			* Limited sol. sln. ranges in the compounds.			
MnF_2			Fig. 188			
	743	17	KF, $KF.MnF_2$	Eut.	1	969
	1032	50	$KF.MnF_2$	Comp.		
	816	84	$KF.MnF_2$, MnF_2	Eut.		
	743	17	KF, K_2MnF_4	Eut.	1,2a	970
	795	20	K_2MnF_4, $KMnF_3$	Per.		
	1032	50	$KMnF_3$	Comp.		
	814	84	$KMnF_3$, MnF_2	Eut.		

KF

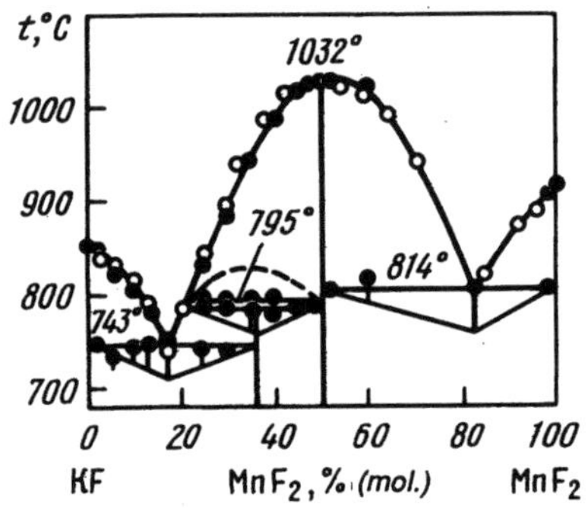

Fig. 188. KF-MnF_2 system.

2nd Component	T°C	Conc. mol %	Phases involved	Def.	Method	Ref.
NaF	714	40	NaF, KF	Eut.	1	971
			Limited sol.sln.in KF		3B	972
	721	40	NaF, 5%NaF sol.sln. in KF	Eut.	2,2d	973
	716	40	NaF, KF	Eut.	1,2a	974
	710	40	NaF, KF	Eut.	1	975
NbF_5	723	16.5	KF, KF.K_2NbF_7	Eut.	2	976
	760	25.0	KF.K_2NbF_7	Comp.		
	706	30.0	KF.K_2NbF_7, K_2NbF_7	Eut.		
	719	17.3	KF, KF.K_2NbF_7	M.P.	1,2a,3b	977
	768	25.0	KF.K_2NbF_7	Comp.	le	
	703	30.9	KF.K_2NbF_7,K_2NbF_7	Eut.		
NdF_3	625	23	KF, $KNdF_4$	Eut.	2a,3b,le	978
	740	~37	$KNdF_4$, Phase A	Per.		
	550		2KF + $KNdF_4$ ⇌ K_3NdF_6	P.T.		
PbF_2	460	40.4	KF, PbF_2	Eut.	1	979

KF

2nd Component	T°C	Conc. mol %	Phases involved	Def.	Method	Ref.
PbF_2			Fig. 189			
	475		K_4PbF_6, KF	Per.	1,3b	980
	465	42	56% KF sol.sln. in phase III_H, K_4PbF_6	Eut.		
	645	70	12% KF sol.sln. in PbF_2 (with modification II_H), unidentified comp. sol.sln. in phase III_H	Per.		
	420		Sol.sln. in PbF_2 (modification II_H) $\rightleftharpoons$ PbF_2 (modification II_T) + + sol.sln. in phase III_H,	P.T.(j;)		
	300		Sol.sln. in phase III_H $\rightleftharpoons$ PbF_2+ +K_4PbF_6	P.T.(c)		
			Line AB - decomposition of sol.sln. of phase III_H			

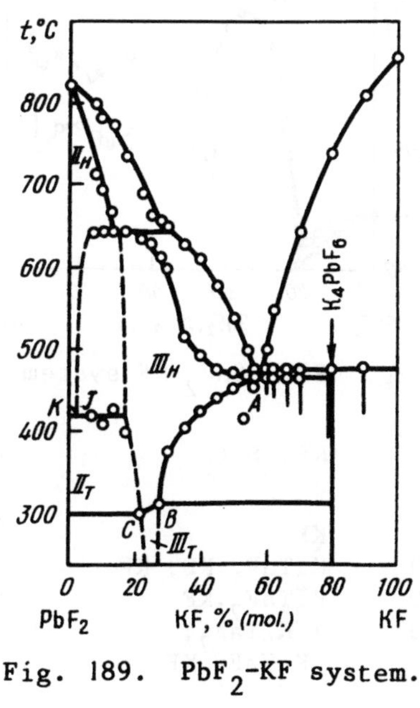

Fig. 189. PbF_2-KF system.

KF

2nd Component	T°C	Conc. mol %	Phases involved	Def.	Method	Ref.
PrF_3	620	22.5	KF, $KPrF_4$	Eut.	1,2d	981
	740	37.5	$KPrF_4$, PrF_3	Per.		
	520		α ⇌ β K_3PrF_6	P.T.		
	580		K_3PrF_6 ⇌ $KPrF_4$ + 2KF	P.T.		
RbF			Complete range of sol.sln.		1	982
ScF_3			Congruently melting $3KF.ScF_3$ and $KF.ScF_3$ (solid)		5b	983
SnF_2			Fig. 190			
	354		$KF.2SnF_2$	Comp.	2c,3b	984
	310		$KF.SnF_2$	Comp.		
	323		$2KF.SnF_2$	Comp.		
	325		$6KF.SnF_2$	Comp.		

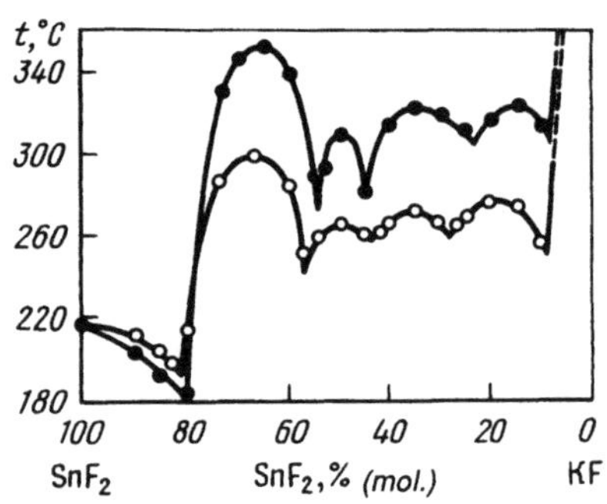

Fig. 190. SnF_2-KF system.

2nd Component	T°C	Conc. mol %	Phases involved	Def.	Method	Ref.
SrF_2	744	22	KF, SrF_2	Eut.	1	985
TaF_5	856	0.0	KF	M.P.	1,2c,3b	986
	727	16.9	$K_2TaF_7.KF$, KF	Eut.	1e	
	776	25.0	$K_2TaF_7.KF$	Comp.		
	717	30.5	α-K_2TaF_7, $K_2TaF_7.KF$	Eut.		

KF

2nd Component	T°C	Conc. mol %	Phases involved	Def.	Method	Ref.
TaF_5	740	31.5	β-K_2TaF_7, α-K_2TaF_7	Per.		986
	775	33.3	β-K_2TaF_7	M.P.		
	858	0.0	KF	M.P.	1,2a	987
	718	15.3	$KF.K_2TaF_7$, KF	Eut.		
	780	25.0	$KF.K_2TaF_7$	Comp.		
	697	29.2	K_2TaF_7, $KF.K_2TaF_7$	Eut.		
	730	33.3	K_2TaF_7	M.P.		
	850	0.0	KF	M.P.	1	988
	738	18.3	KF, $K_2TaF_7.KF$	Eut.		
	776	25.0	$K_2TaF_7.KF$	Comp.		
	697	30.2	$K_2TaF_7.KF$, K_2TaF_7	Eut.		
	771	33.3	K_2TaF_7	M.P.		

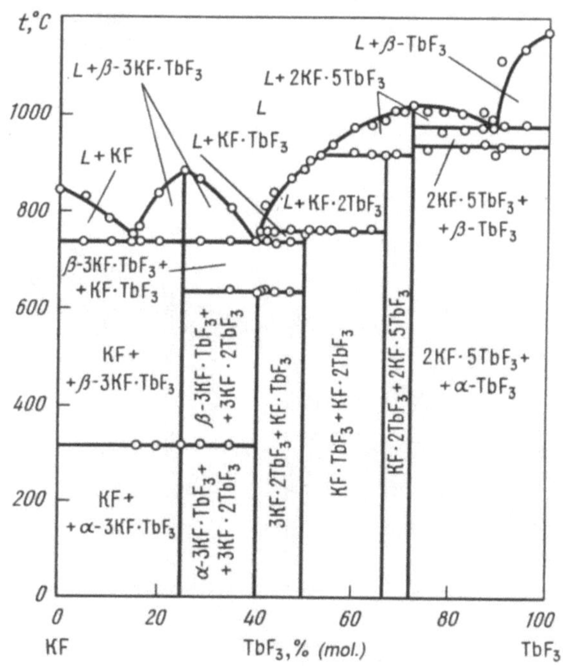

Fig. 191. KF-TbF_3 system.

KF

2nd Component	T°C	Conc. mol %	Phases involved	Def.	Method	Ref.
TbF_3	740	15.0	KF, β-3KF.TbF_3	Eut.	2d,3b, 1e	989
	885	25.0	β-3KF.TbF_3	Comp.		
	740	39.9	β-3KF.TbF_3, KF.TbF_3	Eut.		
	760	40.0	KF.TbF_3, KF.2TbF_3	Per.		
	920	52.0	KF.2TbF_3, 2KF.5TbF_3	Per.		
	1015	71.5	2KF.5TbF_3	Comp.		
	980	89.0	2KF.5TbF_3, β-TbF_3	Eut.		
	315		α ⇌ β 3KF.TbF_3	P.T.		
	940		α ⇌ β TbF_3	P.T.		
	635		β-3KF.TbF_3 + 3(KF.TbF_3) ⇌ ⇌ 2(3KF.2TbF_3)	P.T.		
ThF_4			Fig. 192			
	662	18	KF, K_5ThF_9	Eut.	2,3b	990
	693	~18	K_5ThF_9, K_3ThF_7	Comp.		
	741	34	K_3ThF_7, $K_3Th_2F_{11}$	Eut.		
	856	25	K_3ThF_7	Eut.		
	887	40	$K_3Th_2F_{11}$	Comp.		

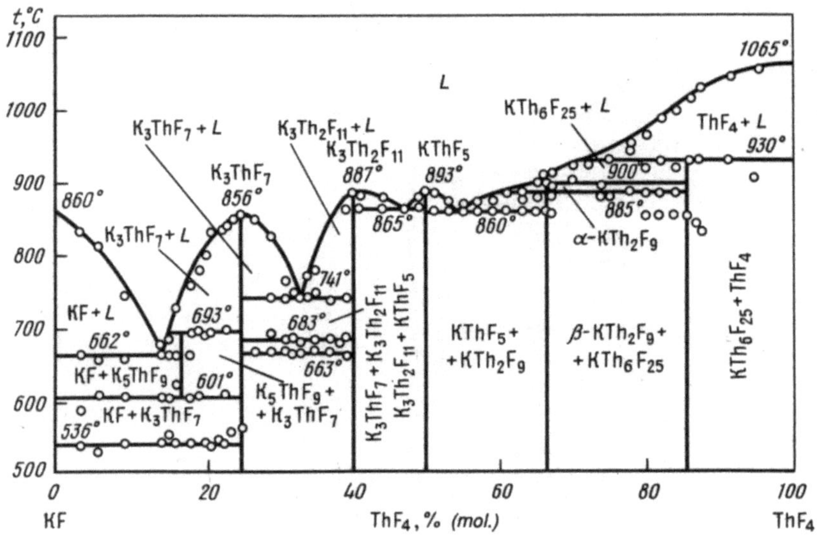

Fig. 192. KF-ThF_4 system.

KF

2nd Component	T°C	Conc. mol %	Phases involved	Def.	Method	Ref.
ThF_4	865	47	$K_3Th_2F_{11}$, $KThF_5$	Eut.	2,3b	
	893	50	$KThF_5$	Comp.		
	860	54	$KThF_5$, α-KTh_2F_9	Eut.		
	885	67	α-KTh_2F_9, β-KTh_2F_9	Per.		
	900	~67	KTh_2F_9, KTh_6F_{25}	Per.		
	930	75	KTh_6F_{25}, ThF_4	Per.		
	536		K_3ThF_7	P.T.		
	601		$K_5ThF_9 \rightleftharpoons K_3ThF_7$ + KF	P.T.		
	683		Transformation not described			
	663		" " "			
TiF_4			Fig. 193			
	830	14.3	KF, K_3TiF_7	Eut.	2d, 1d,	991
	890	25.0	K_3TiF_7	Comp.	3b, 1e	
	820	30.0	K_3TiF_7, K_2TiF_6	Eut.		

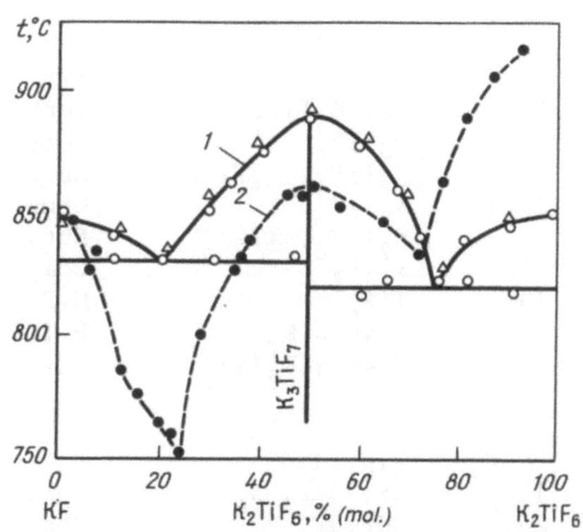

Fig. 193. KF-K_2TiF_6 system.

T°C	Conc. mol %	Phases involved	Def.	Method	Ref.
		Fig. 194			
830	15.3	KF, K_3TiF_7	Eut.	2a,3b	992
890	25.0	K_3TiF_7	Comp.	1e,3e,	993
820	30.3	K_3TiF_7, K_2TiF_6	Eut.		

KF

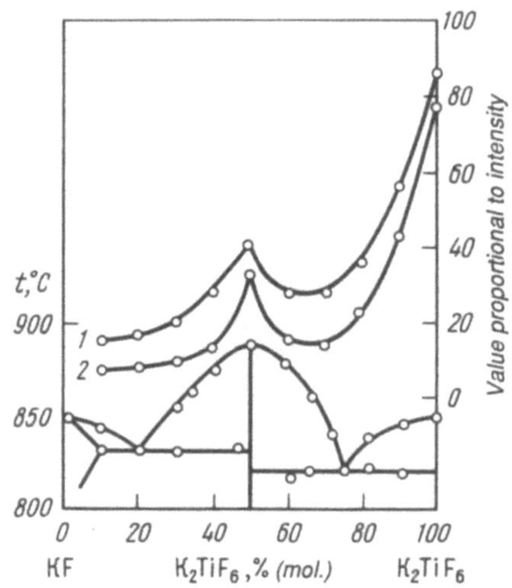

Fig. 194. KF-K_2TiF_6 system.

2nd Component	T°C	Conc. mol %	Phases involved	Def.	Method	Ref.
TiF_4			Fig. 195			
	854	0	KF	M.P.	2d,2a,1	994
	710	16	KF, K_3TiF_7	Eut.		
	775	25	K_3TiF_7	Comp.		
	740	29	K_3TiF_7, K_2TiF_6	Eut.		
	899	33.3	K_2TiF_6	Comp.		
	420	47	K_2TiF_6, $KTiF_5$	Eut.		
	480	50	$KTiF_5$	Comp.		
	430	100	TiF_4	M.P.		
	395		$\alpha \rightleftharpoons \beta$ K_2TiF_6	P.T.		
	580		$\beta \rightleftharpoons \gamma$ K_2TiF_6			
UF_4			Fig. 196			
	735	15	KF, $3KF.UF_4$	Eut.	1,2,1f	995
	957	25	$3KF.UF_4$	Comp.	3b	
	755	35	$2KF.UF_4$, $3KF.UF_4$	Per.		
	740	38.5	$2KF.UF_4$, $7KF.6UF_4$	Eut.		
	789	46.2	$7KF.6UF_4$	Comp.		
	735	54.5	$7KF.6UF_4$, $KF.2UF_4$	Eut.		
	765	65	$6KF.2UF_4$, UF_4	Per.		
	608		$2KF.UF_4 \rightleftharpoons$ $3KF.UF_4 + 7KF.6UF_4$			

KF

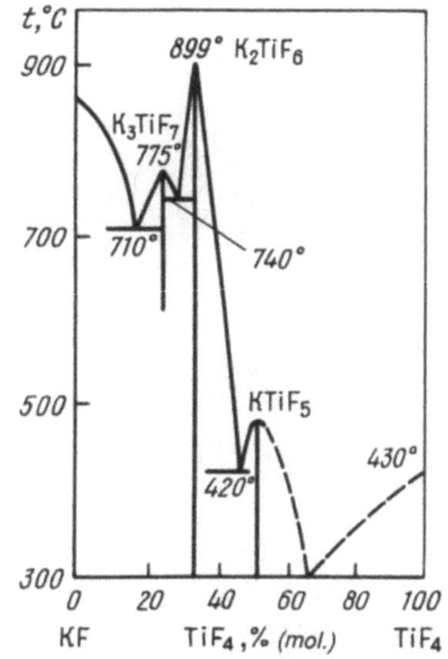

Fig. 195. $KF-TiF_4$ system.

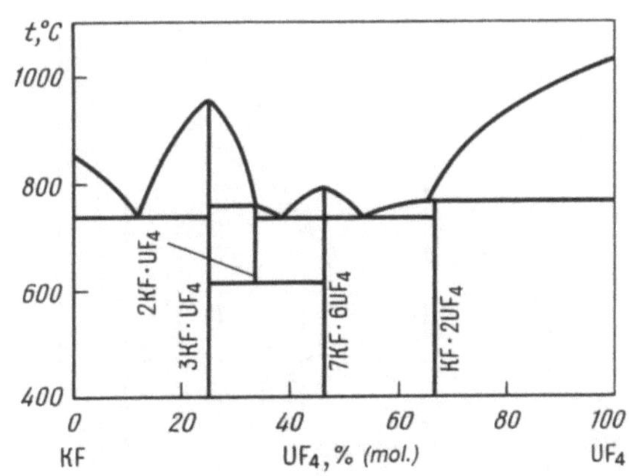

Fig. 196. $KF-UF_4$ system.

KF

2nd Component	T°C	Conc. mol %	Phases involved	Def.	Method	Ref.
VF_3	806	4.5	KF, K_3VF_6	Eut.	2d,3b	996
	1175	-	K_3VF_6	Comp.		
	795	40.5	K_3VF_6, K_2VF_5	Per.		
	840	49	KVF_4, $K_2V_3F_{11}$	Per.		
	860	53.5	$K_2V_3F_{11}$, $K_2V_5F_{17}$	Per.		
	920	59.5	$K_2V_5F_{17}$, VF_3	Per.		
	200		K_3VF_6	P.T.		
	775		KVF_4	P.T.		
YF_3	760	13	KF, K_3YF_6	Eut.	2a,1	997
	980	25	K_3YF_6	Comp.		
	752	43	K_3YF_6, YF_6	Eut.		
	702	-	KYF_4	P.T.		
			Fig. 197			
	770	11	KF, α-K_3YF_6	Eut.	1,2a,3b	998
	980	25	α-K_3YF_6	Comp.		
	750	45	α-K_3YF_6, $KF.2YF_3$	Eut.		
	845±5	50.0	$KF.2YF_3$, $KF.6YF_3$	Per.		
	965	63	$KF.6YF_3$, β-YF_3	Per.		
	1052	85	β-YF_3, α-YF_3	Per.		

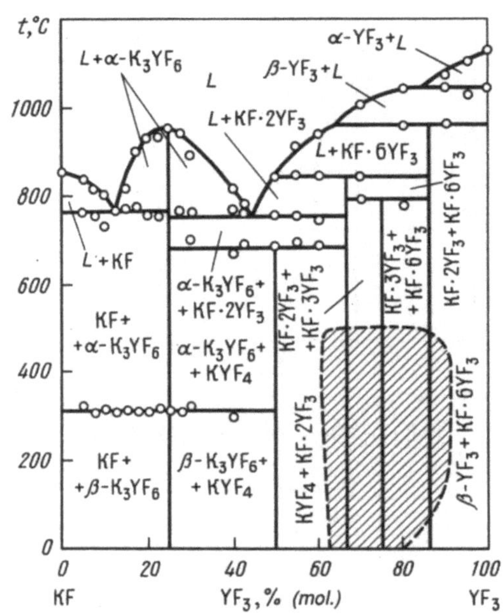

Fig. 197. KF-YF_3 system.

KF

2nd Component	T°C	Conc. mol %	Phases involved	Def.	Method	Ref.
YF_3	800±5		$KY_3F_{10} \rightleftharpoons KF.2YF_3 + KF.6YF_3$	P.T.		998
	320		K_3YF_6	P.T.		
		62-89	Metastable cubic (fluorite) phase (crosshatched in the fig.)			
			Fig. 198			
	780	11	KF, α-K_3YF_6	Eut.	2	999
	980	25	α-K_3YF_6	Comp.	3b	
	770	45	α-K_3YF_6, $KF.2YF_3$	Eut.	1	
	890±5	53	$KF.2YF_3$, $KF.7YF_3$	Per.		
	965±5	64	$KF.7YF_3$, β-YF_3	Per.		
	1035	81	β-YF_3, α-YF_3	Per.		
	795		$KF.2YF_3 + KF.7YF_3 \rightleftharpoons KF.3YF_3$	P.T.		
	700		α-$K_3YF_6 + KF.2YF_3 \rightleftharpoons KYF_4$	P.T.		
	320		$\alpha \rightleftharpoons \beta$ K_3YF_6			

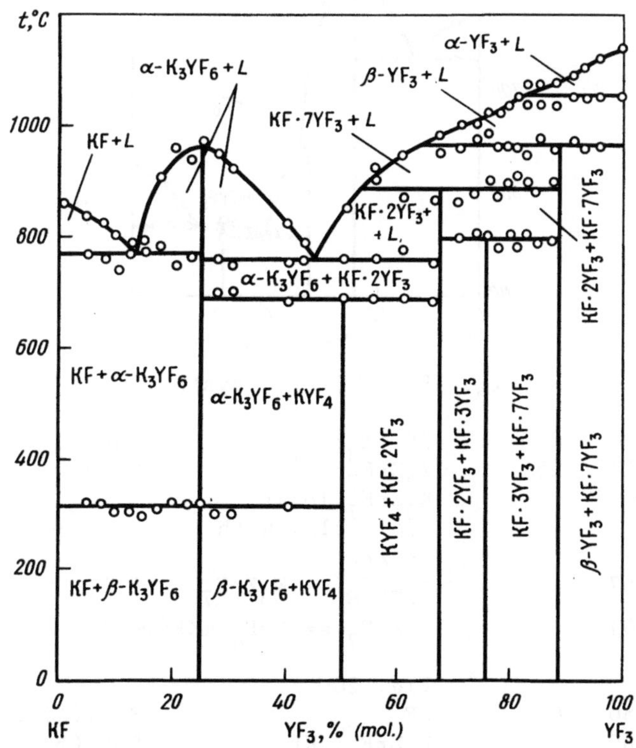

Fig. 198. KF-YF_3 system.

KF

2nd Component	T°C	Conc. mol %	Phases involved	Def.	Method	Ref.
ZnF_2	683	24	K_2ZnF_4, KF	Eut.	2,3b	1000
	737	34	$KZnF_3$, K_2ZnF_4	Per.		
	870	50	$KZnF_3$	Comp.		
	748	80	ZnF_2, $KZnF_3$	Eut.		
ZrF_4			Fig. 199			1001
	760	13	KF, K_3ZrF_7	Eut.	2,3b	1001
	900	25	K_3ZrF_7	Comp.		
	600	35	K_3ZrF_7, K_2ZrF_6	Per.		
	410	42	K_2ZrF_6, $KZrF_5$	Eut.		
	440	55	$KZrF_5$, ZrF_4	Eut.		

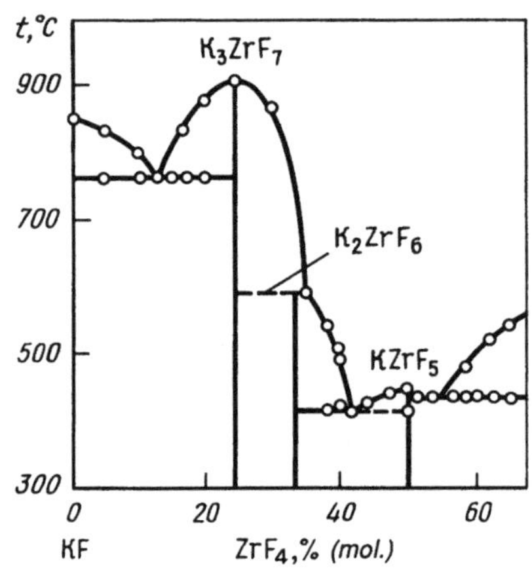

Fig. 199. KF-ZrF_4 system.

2nd Component	T°C	Conc. mol %	Phases involved	Def.	Method	Ref.
	776	8.5	KF, K_3ZrF_7	Eut.	1,2c	1002
	921	25	K_3ZrF_7	Comp.		
			K_3ZrF_7 forms sol. sln. with K_2ZrF_6			
	757		$K_2ZrF_6 \rightleftharpoons K_3ZrF_7$	P.T.		
	591		$K_3ZrF_7 \rightleftharpoons ZrF_4.mKF(m<3)$			
			Fig. 200			1003
	850	0	KF	M.P.(A)	2a,3b	1003
	760	13	K_3ZrF_7, KF	Eut.(B)		
	923	25	K_3ZrF_7	Comp.(C)		

KF

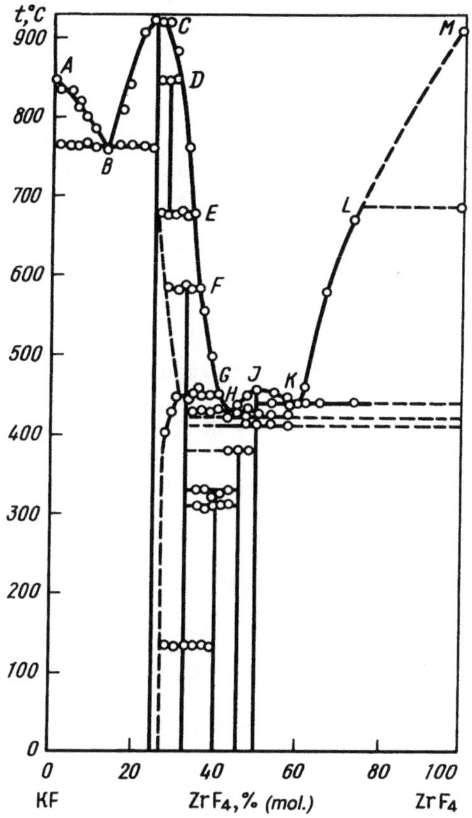

Fig. 200. KF-ZrF_4 system.

2nd Component	T°C	Conc. mol %	Phases involved	Def.	Method	Ref.
ZrF_4	848	31	K_3ZrF_7, $K_5Zr_2F_{13}$	Per.(D)		1003
	680	34	$K_5Zr_2F_{13}$, K_3ZrF_7 base sol. sln.	Per.(E)		
	585	36	K_3ZrF_7 base sol. sln., δ-K_2ZrF_6	Per.(F)		
	445	41	δ-K_2ZrF_6, γ-K_2ZrF_6	Per.(G)		
	430	47	γ-K_2ZrF_6, γ-$KZrF_5$	Eut.(H)		
	455	50	γ-$KZrF_5$	Comp.(J)		
	440	60	γ-$KZrF_5$, α-ZrF_4	Eut.(K)		
	685	73	α-ZrF_4, β-ZrF_4	Per.		
	903±5	100	β-ZrF_4	M.P.		
	424		γ ⇌ β $KZrF_5$	P.T.		
	400		β ⇌ α $KZrF_5$	P.T.		

KF

2nd Component	T°C	Conc. mol %	Phases involved	Def.	Method	Ref.
ZrF_4	380		$K_7Zr_6F_{31}$	P.T.		
	327		$K_3Zr_2F_{11}$	P.T.		
	313		$\alpha \rightleftharpoons \beta$ $K_3Zr_2F_{11}$	P.T.		
	260		$\alpha \rightleftharpoons \beta$ K_2ZrF_6	P.T.		
	298		$\beta \rightleftharpoons \gamma$ K_2ZrF_6	P.T.		
	130		$\gamma \rightleftharpoons \alpha$ K_2ZrF_6			
			$K_2ZrF_6 \rightarrow K_3ZrF_7$ (incongruent melting)	M.P.	1,2c	1004
	766	6	KF, K_3ZrF_7	Eut.		
	921	25	K_3ZrF_7	Comp.		
	757		Limit of Liquidus Sol. sln. on the K_2ZrF_6 side			
	591		$K_2ZrF_6 \rightarrow K_3ZrF_7 + KZrF_5$ (incongruent melting)	M.P.		

KI

2nd Component	T°C	Conc. mol %	Phases involved	Def.	Method	Ref.
LaI_3	667	0	KI	M.P.	2d	1005
	489	20	$2KI.LaI_3$, KI	Eut.		
	527	33.33	$2KI.LaI_3$	Comp.		
	500	47	LaI_3, $2KI.LaI_3$	Eut.		
	768	100	LaI_3	M.P.		
LiI	260	47	LiI, KI (possibly 7% KI sol. sln. in LiI and 11% LI sol.sln. in KI)	Eut.	2	1006
	274	64	LiI, KI	Eut.	2d	1007
	286.4	63.2	KI, LiI	Eut.	1,2	1008
	285	63.5	LiI, KI	Eut.	2	1009
NaI	575	-	Continuous sol.sln.	Min.	1,2	1010
	580	60	Continuous sol. sln. - These frequently decompose at ambient temp.	Min.	1,2	1011

KI

2nd Component	T°C	Conc. mol %	Phases involved	Def.	Method	Ref.
NaI	240	62	Limited range of sol. slns.	Min.	3b	1012
NdI_3	677	0	KI	M.P.	2d	1013
	500	19	$3KI.NdI_3$, KI	Eut.		
	531	25	$3KI.NdI_3$	Comp.		
	448	34	$2KI.NdI_3$, $3KI.NdI_3$	Per.		
	430	44	NdI_3, $2KI.NdI_3$	Eut.		
	563	56	α-NdI_3, β-NdI_3	Per.		
	780	100	NdI_3	M.P.		
	410		$3KI.NdI_3$	P.T.		
	356		$3KI.NdI_3$	P.T.		
PbI_2	357	48	$KI.PbI_2$, KI	Eut.	1,2	1014
	358	50	$KI.PbI_2$	Comp.		
	330	69	PbI_2, $KI.PbI_2$	Eut.		
	358	47	$KI.PbI_2$, KI	Eut.	1,2	1015
	321	68	PbI_2, $KI.PbI_2$	Eut.		
			$KI.PbI_2$ confirmed		3b	1016
	680	0.0	KI	M.P.	2a	1017
	370	50.0	$2KI.PbI_2$, KI	Per.		
	342	70.0	PbI_2, $2KI.PbI_2$	Eut,		
	408	100	PbI_2	M.P.		
PrI_3	677	0	KI	M.P.	2d	1018
	482	20	$3KI.PrI_3$, KI	Eut.		
	503	25	$3KI.PrI_3$	Comp.		
	472	32	$2KI.PrI_3$, $3KI.PrI_3$	Eut.		
	475	33.3	$2KI.PrI_3$	Comp.		
	455	45	PrI_3, $2KI.PrI_3$	Eut.		
	730	100	PrI_3	M.P.		
	407		-	P.T.		
	373		-	P.T.		
RbI	694	0	KI	M.P.	2b	1019

KI

2nd Component	T°C	Conc. mol %	Phases involved	Def.	Method	Ref.
RbI		0-100	Continuous sol. sln. range		2b	1019
	656	100	RbI	M.P.		
	514	63	Continuous sol. sln. range	Min.	1,2	1020
	642	100	RbI	M.P.		
SbI_3	632	0.0	KI	M.P.	2a,1,3a	1021
	390	37	$2KI.SbI_3$, KI	Per.		
	164	84	SbI_2, $2KI.SbI_3$	Eut.		
	170	100	SbI_3	M.P.		
SmI_3	677	0	KI	M.P.	2d	1022
	509	18	α-$3KI.SmI_3$, KI	Eut.		
	544	25	α-$3KI.SmI_3$	Comp.		
	415	38	α-$3KI.SmI_3$, β-$3KI.SmI_3$	Per.		
	403	40	β-$3KI.SmI_3$, β-SmI_3	Eut.		
	670	60	α-SmI_3, β-SmI_3	Per.		
	339		$\gamma \rightleftharpoons \beta$ $3KI.SmI_3$	P.T.		
SnI_2	681	0	KI	M.P.	2a,3b	1023
	249	50	KI, $KSnI_3$	Eut.		
	260	74.9	$KSnI_3$, SnI_2(?)	Per.(?)		
	320	100	SnI_2	M.P.		
			Possibility of $KSnI_3$ confirmed by solution of Schrödinger eqn.			
		60-70	Possibility of another compound			
	680	0	KI	M.P.	2a	1024
	294	65	$KI.SnI_2$	Eut.		
	330	100	SnI_2	M.P.		
TeI_4	682	0	KI	M.P.	2,3b,1e	1025
	364	17	KI, α-K_2TeI_6	Eut.		
	410	33	α-K_2TeI_6	Comp.		

KI

2nd Component	T°C	Conc. mol %	Phases involved	Def.	Method	Ref.
TeI_4	296	-	α-K_2TeI_6, β-K_2TeI_6	Per.	2,3b,1e	1025
	224	90	β-K_2TeI_6, TeI_4	Eut.		
	280	100	TeI_4	M.P.		
TlI	436	90	KI, TlI	Eut.	1,2	1026

$LaCl_3$

2nd Component	T°C	Conc. mol %	Phases involved	Def.	Method	Ref.
NaCl	850	0	$LaCl_3$	M.P.	2b	1027
	525	78	$LaCl_3$, NaCl	Eut.		
	800	100	NaCl			
$SbCl_3$	72	100	$SbCl_3$ With slight addition of $LaCl_3$ temperature of crystallization rises sharply	M.P.	2d	1028
$SbCl_5$	3	~100	$LaCl_3$, $SbCl_5$	Eut.	2d	1029
	3	100	$SbCl_5$	M.P.		
$SmCl_3$	860	12	Continuous sol.sln.	Max.	2a	1030
	665	100	$SmCl_3$	M.P.		
$SnCl_2$	273	97.7	$LaCl_3$, $SnCl_2$	Eut.	2c	1031
YCl_3	852	0	$LaCl_3$	M.P.	2	1032
	650	75	$LaCl_3$, YCl_3	Eut.		
	700	100	YCl_3	M.P.		

LaF_3

2nd Component	T°C	Conc. mol %	Phases involved	Def.	Method	Ref.
LiF	770	79		Eut.	2d,1e	1033

LaF_3

2nd Component	T°C	Conc. mol %	Phases involved	Def.	Method	Ref.
LiF	750	86	LiF, LaF_3	Eut.	1,2	1034
	780	80	LiF, LaF_3	Eut.	2	1035
	755-7	84.5	LiF, LaF_3	Eut.	2a	1036
NaF	777	65	$NaLaF_4$, LaF_3	Per.	2,3b	1037
	734.6	72.8	NaF, $NaLaF_4$	Eut.		
	763	58	LaF_3.NaF, LaF_3	Per.	2,1f,3b	1038
	620	72	NaF, LaF_3.NaF	Eut.		
	780	64	$NaLaF_4$, LaF_3	Per.	1,2a	1039
	725	71	NaF, $NaLaF_4$	Eut.		
	810	57	NaF.LaF_3 (hexagonal), LaF_3	Per.	2,3b,5, le	1040
	730	71	NaF, NaF.LaF_3 (hexagonal)	Eut.		
SrF_2			Limited sol. slns: SrF_2 base (fluorite) and LaF_3 (tisonite), with mixture of both between the limits		3b	1041

LaI_3

2nd Component	T°C	Conc. mol %	Phases involved	Def.	Method	Ref.
NaI	768	0	LaI_3	M.P.	2d	1042
	454	70	LaI_3, NaI	Eut.		
	654	100	NaI	M.P.		

LiBr

2nd Component	T°C	Conc. mol %	Phases involved	Def.	Method	Ref.
NaBr	522	20	Continuous range of sol.sln.	Min.	1	1044
	510		Continuous range of sol. sln.	Min.	1	1045

LiBr

2nd Component	T°C	Conc mol %	Phases involved	Def.	Method	Ref.
NaBr	544	0	LiBr	M.P.	1	1046
	507	17	LiBr, LiBr.NaBr	Eut.		1047
	512	29	LiBr.NaBr, LiBr.2NaBr	Per.		
	537	40.5	LiBr.2NaBr, NaBr	Per.		
	510	25	Continuous range of sol. sln.	Min.	1	1048
$PbBr_2$	322	79	$PbBr_2$, LiBr	Eut.	2a,6b	1049
			Molar volume isotherms are close to "additive" Viscosity isotherms slightly depart from "additive"		6f,6ℓ	
RbBr	380		Continuous sol.sln.	Min.	1	1050
	549	0	LiBr	M.P.	1	1051
	292	39	LiBr, LiBr.RbBr	Eut.		
	306	47.5	LiBr.RbBr, RbBr	Per.		
	680	100	RbBr	M.P.		
$ScBr_3$	536	2.7	LiBr,3LiBr.$ScBr_3$	Eut.	2d,3b	1052
	574	8.1	3LiBr.$ScBr_3$,$ScBr_3$	Per.		
$SnBr_2$	545	0	LiBr	M.P.	1,2a	1053
	188	87.5	$SnBr_2$,LiBr	Eut.		
	226	100	$SnBr_2$	M.P.		
$SrBr_2$	438	38	LiBr,LiBr.2$SrBr_2$	Eut.	1	1054
	488	63	LiBr.2$SrBr_2$,$SrBr_2$	Per.		
	497		$SrBr_2$, 3$SrBr_2$.LiBr	Per.	1,2d,3b	1055
	432	34	3$SrBr_2$.LiBr,LiBr	Eut.		
TlBr	342	58.5	TlBr,LiBr	Eut.	1	1056

LiBr

2nd Component	T°C	Conc. mol %	Phases involved	Def.	Method	Ref.
LiCl	519	33	Continuous range of sol.sln.	Min.	1	1057
	526	35	Continuous range sol.sln.	Min.	1	1058
LiF	448	25	LiBr, LiF	Eut.	1	1059

LiCl

2nd Component	T°C	Conc. mol %	Phases involved	Def.	Method	Ref.
$MgCl_2$			Fig. 201			1060
	590	50	α and β sol.slns. of compounds with 55 and 80% $MgCl_2$ sol. sln. crystallizing range 10-12°C	Per.	2d, le, 3b	1060
		50-80	Breaking up of sol.slns. (optical investigation)			
		20-80	Breaking up of sol.slns. (x-ray investigation)			

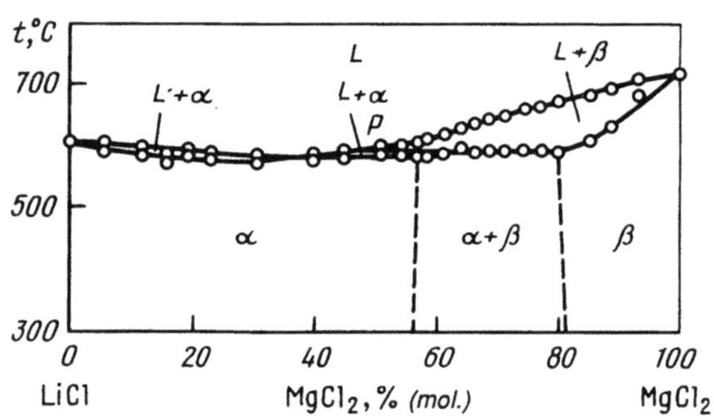

Fig. 201. $LiCl-MgCl_2$ system.

LiCl

2nd Component	T°C	Conc. mol %	Phases involved	Def.	Method	Ref.
$MgCl_2$	580	35	Continuous sol. sln. Breaking up of sol.sln.	Min.	2d,3b	1061
$MnCl_2$	573	~23	α-$LiCl_2$, β-LiCl	Per.	1	1062
	545	38	β-LiCl, $2LiCl.MnCl_2$	Eut.		
	553	43	$2LiCl.MnCl_2$	Comp.		
	536	63	$2LiCl.MnCl_2$, $MnCl_2$	Eut.		
NaCl	553	21.5	LiCl,LiCl.NaCl	Eut.	1	1063
	575	33	LiCl.NaCl,NaCl Data according to text, according to phase diagram: Eut. - 27.5% NaCl and Per. - 38% NaCl.	Per.		
	570		α-LiCl, β-LiCl	Per.	1	1064
	545	28	β-LiCl, LiCl.NaCl	Eut.		
	575	38.5	LiCl.NaCl, NaCl	Per.		
	550	28	Continuous range sol.sln.	Min.	1	1065
	606	0	α-LiCl	M.P.	1	1066
	570		α-LiCl, β-LiCl	Per.		
	551	24.5	β-LiCl, LiCl.NaCl	Eut.		
	572	36	LiCl.NaCl, LiCl.2NaCl	Per.		
	622	50	LiCl.2NaCl, NaCl	Per.	1	1066
	570	10.5	α-LiCl, β-LiCl	Per.	1	1067
	546	30	β-LiCl, LiCl.NaCl	Eut.		
	576	40	LiCl.NaCl, 2NaCl.LiCl	Per.		
	620	50	2NaCl.LiCl, NaCl	Per.		

LiCl

2nd Component	T°C	Conc. mol %	Phases involved	Def.	Method	Ref.
NaCl			Fig. 202			1068
	565	21	α-LiCl, β-LiCl	Per.	2a, 1	1068
	556	28.5	LiCl base sol. sln., LiCl.NaCl sol. sln. in NaCl (E)	Eut.		
	575	37	LiCl.NaCl sol. sln. in NaCl, LiCl.2NaCl sol. sln. in NaCl (P_1)	Per.		1068
	610	47	LiCl.2NaCl sol. sln. in NaCl, LiCl sol.sln. in NaCl (P_2)	Per.		

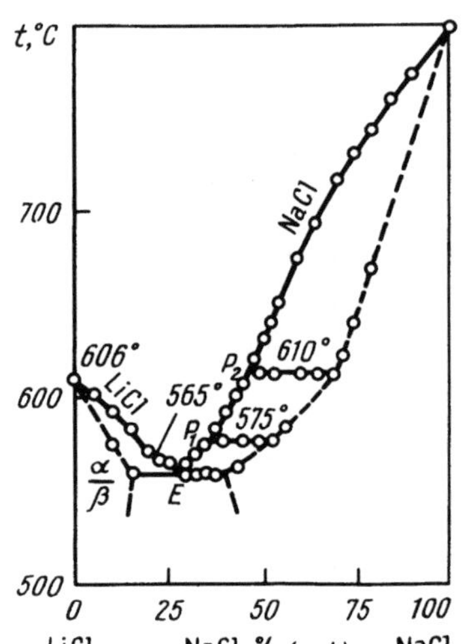

Fig. 202. LiCl-NaCl system.

	602	0	LiCl	M.P.	1,2a,1e	1069
	540	30	Continuous sol. sln.	Min.		
	800	100	NaCl	M.P.		
			Continuous sol. sln.		2a	1070

LiCl

2nd Component	T°C	Conc. mol %	Phases involved	Def.	Method	Ref.
NaCl			Fig. 203.			1071
	606	0	LiCl	M.P.	2a,3b	1071
	555	25	LiCl base sol. sln.	Min.		
	575	37	Sol.slns.with 58% and 78% NaCl	Per.		
	801	100	NaCl	M.P.		
	<580		Sol.slns. break up			1071
	500	20-83	Heterogeneous range			
			LiCl based sol.sln. - unstable			
			Breaking up process begins after 6-8 hrs. and ends after 24-48 hrs.			

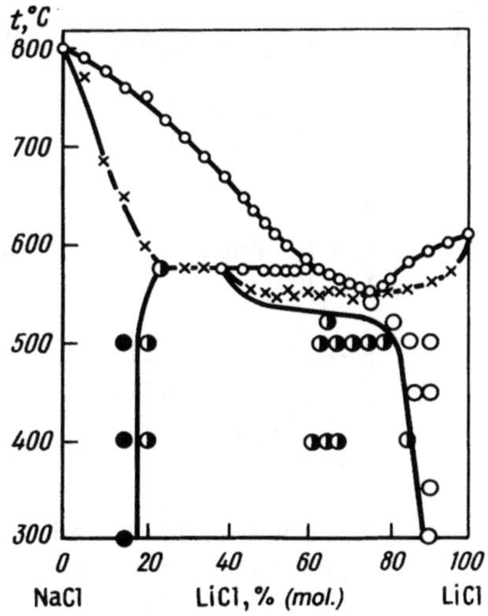

Fig. 203. NaCl-LiCl system.

2nd Component	T°C	Conc. mol %	Phases involved	Def.	Method	Ref.
$NbCl_5$			Fig. 204			1072
	253		$LiNbCl_6$ (under pressure)	Comp.	2,6c	1072
	195	99	$NbCl_5$, $LiNbCl_6$	Eut.		

LiCl

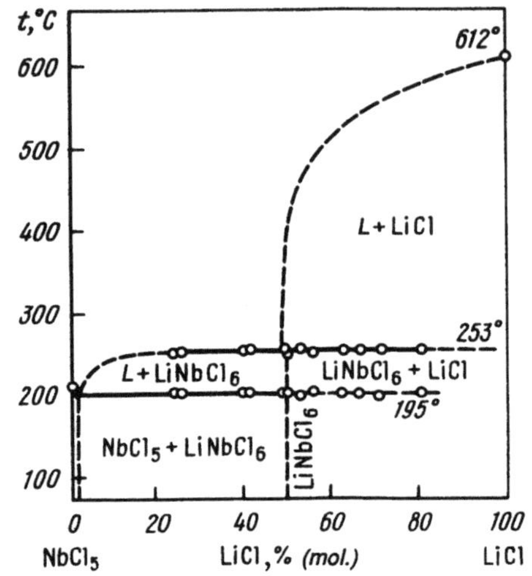

Fig. 204. $NbCl_5$-LiCl system.

2nd Component	T°C	Conc. mol %	Phases involved	Def.	Method	Ref.
$NiCl_2$	620	27.5	$NiCl_2$ & $LiCl_2$ base sol.sln.	Eut.	1,2	1073
$PbCl_2$	404	63.3	LiCl, $PbCl_2$	Eut.	1,2	1074
	566	100	LiCl	P.T.		
	563	12	β-LiCl, α-LiCl	Per.	1,2	1075
	406	64	β-LiCl, β-$PbCl_2$	Eut.		
	425	72	α-$PbCl_2$, β-$PbCl_2$	Per.		
	626		No compounds found Deviation of heat of mixing from ideality caused by destruction of $PbCl_2$ and by mutual solution of components.		2a	1076

LiCl

2nd Component	T°C	Conc. mol %	Phases involved	Def.	Method	Ref.
$PuCl_3$	607	0	LiCl	M.P.	2c	1077
	464	28	LiCl, $PuCl_3$	Eut.		
	767	100	$PuCl_3$	M.P.		
RbCl	608	0	LiCl	M.P.	1,2	1078
	314	41.5	LiCl.RbCl, LiCl	Eut.	3b	1079
	324	50	LiCl.RbCl	Comp.		
	322	53	RbCl, LiCl.RbCl	Eut.		
	716	100	RbCl	M.P.		
$SbCl_5$	~3	~100	LiCl, $SbCl_5$	Eut.	2d	1080
	3	100	$SbCl_5$	M.P.		
$ScCl_3$	562	13	LiCl, $3LiCl.ScCl_3$	Eut.	2a,4a	1081
	585	18	$3LiCl.ScCl_3$, $ScCl_3$	Per.		1082
	470		$3LiCl.ScCl_3$	P.T.		
$SnCl_2$	215	85		Eut.		1083
$SrCl_2$	473	48		Eut.		1083
	478	47	LiCl, $SrCl_2$	Eut.	1,2	1084
	474	35.5	LiCl, $SrCl_2$	Eut.	1,2	1085, 1086
$TaCl_5$			Fig. 205			1087
	337		β-$LiTaCl_6$, LiCl	Per.	2,6c	1087
	295		α-$LiTaCl_6$, β-$LiTaCl_6$	Per.		
	205		$TaCl_5$, α-$LiTaCl_6$	Eut.		
$TeCl_4$	220	96	LiCl, $TeCl_4$	Eut.	2d	1088
	599	8-69	Decomposition			

LiCl

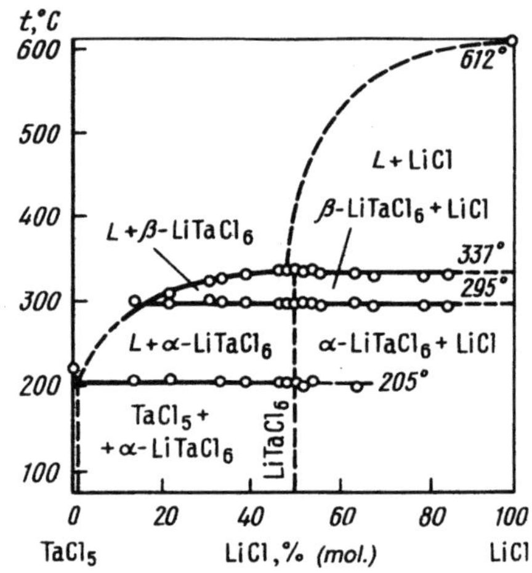

Fig. 205. $TaCl_5$-LiCl system.

2nd Component	T°C	Conc. mol %	Phases involved	Def.	Method	Ref.
$ThCl_4$	417	35	$ThCl_4$, Li_4ThCl_8	Eut.	2	1089
	452		Li_4ThCl_8, LiCl	Per.		
$TiCl_2$			Continuous sol.sln.		2d,3b	1090
$TiCl_3$			Fig. 206			1091
	554			Eut.	2,1e,3a	1091
	560	78.0	Li_3TiCl_6	Per.		
	454	830	LiCl, Li_3TiCl_6	Eut.		
	23-70		Decomposition range			
	200		$LiTiCl_4$	P.T.		
			Li_3TiCl_6 crystals are colored with refr. indices: Ng = 1.765 Np = 1.761			
			$LiTiCl_4$ crystals - polychromic in the range yellow to red Ng =1.945, Np = 1.876			

LiCl

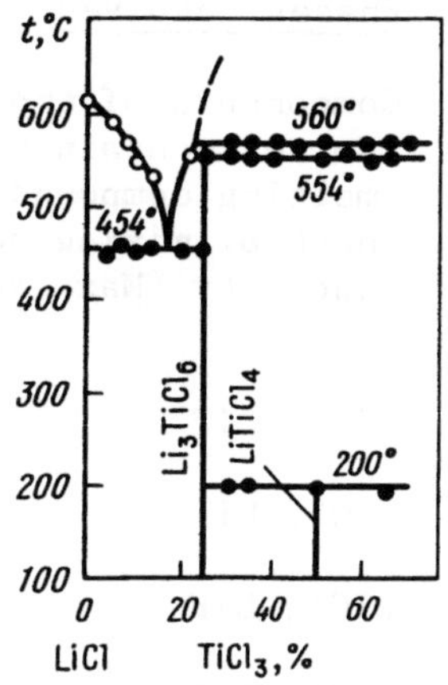

Fig. 206. LiCl-$TiCl_3$ system.

2nd Component	T°C	Conc. mol %	Phases involved	Def.	Method	Ref.
UCl_3	495		UCl_3, LiCl	Eut.		1092
	490±2	26	UCl_3, LiCl		2d,3b	1093 1094
UCl_4	429	28	LiCl,Li_2UCl_6	Eut.	2c,3b	1095
	448	33	Li_2UCl_6	Comp.		
	415±3	43	Li_2UCl_6, UCl_4	Eut.		
	415±2	31.0	LiCl, 2LiCl.UCl_4	Eut.	2d	1096
		33.3	2LiCl.UCl_4	Comp.		
	400±2	43.0	2LiCl.UCl_4, UCl_4	Eut.		
	415	31.0	LiCl, 2LiCl.UCl_4	Eut.	2d	1097
	430±2	33.3	2LiCl.UCl_4	Comp.		
	400±2	44.0	2LiCl.UCl_4, 30% LiCl sol.sln. in UCl_4	Eut.		
VCl_2	612	0.09	LiCl	M.P.	3b	1098
			Indefinite interactions between the components			

LiCl

2nd Component	T°C	Conc. mol %	Phases involved	Def.	Method	Ref.
$ZnCl_2$	626		Components of the system do not contain congruently melting compounds heat of mixing deviates from ideality (Max. at 66% $ZnCl_2$)		2a	1099
LiF	501	30.5	LiCl, LiF	Eut.	2,3b	1100
	496	30	LiCl, LiF	Eut.	1,2	1101
	488	30	LiCl, LiF	Eut.	1,2	1102
	498	30	LiCl, LiF	Eut.	1,2	1103
	570	9	$\alpha \rightleftharpoons \beta$ LiCl	P.T.		1104
	501	30.5	LiCl, LiF	Eut.	2a	1105
LiH	495.6	34	LiCl, LiH	Eut.	2	1106 1107
LiI	607		LiCl	M.P.	2a	1108
	368		LiCl, LiI	Eut.		
	469		LiI	M.P.		
	368	65.4	LiCl, LiI	Eut.	2a,3b	1109
		0.63.5	LiCl			
		70-100	LiI			
		5-95	No sol.sln. observed			
			Table VII			

LiF

2nd Component	T°C	Conc. mol %	Phases involved	Def.	Method	Ref.
LuF_3	680	30	LiF, $LiF.LuF_3$	Eut.	2	1110
	800	50	$LiF.LuF_3$	Comp.		
	780	58	$LiF.LuF_3$, LuF_3	Eut.		
	940		LuF_3	P.T.		
MnF_2	854	0.0	LiF	M.P.	1,2	1111
	610	43	LiF, MnF_2	Eut.		1112
	600	37	LiF, MnF_2	Eut.	6k	1113

LiF

TABLE VII

LiI (mol %)	Temp. °C Liquidus	Eutectic	LiI (mol %)	Temp. °C Liquidus	Eutectic
0	606.8	-	63.5	379.1	368.7
5	592.4	367.6	65.4	368.2	368.2
15	560.7	367.9	70	388.2	367.7
25	532.0	368.7	74.9	399.3	367.8
35	498.9	369.3	80	414.7	368.0
45	461.6	369.4	90	442.3	366.7
55	421.9	369.1	94.9	456.1	367.5
60	395.7	369.0	100	468.9	-

2nd Component	T°C	Conc. mol %	Phases involved	Def.	Method	Ref.
NaF	632	39	LiF, NaF	Eut.	1,2,2a	1114
	652	39	LiF, NaF	Eut.	1,2	1115
	625	40	LiF, NaF	Eut.	-	1116
	649	39	LiF, 5% LiF sol. sln. in NaF	Eut.	2d	1117
	659±2	39.6	LiF, 3% LiF sol. sln. in NaF	Eut.	2,3b	1118
	652	40	LiF, NaF	Eut.	2d	1119
NdF_2	720	20	LiF, NdF_3	Eut.	2	1120
PbF_2	595	67	LiF, PbF_2		1,2	1121
PmF_3	690	27	LiF, PmF	Eut.	2	1122

LiF

2nd Component	T°C	Conc. mol %	Phases involved	Def.	Method	Ref.
PrF_3	720	23	LiF, PrF_3	Eut.	2	1122
	730	21	LiF, PrF_3	Eut.	1,2d	1123
PuF_3	743±2	19.5	LiF, PuF	Eut.	2c	1124
RbF	465	53	LiF.RbF, LiF	Per.		1125
	460	55	RbF, LiF.RbF	Eut.		
	850	0.0	LiF	M.P.	1,2	1126
	465	48	LiF.RbF, LiF	Eut.		
	469	50	LiF.RbF	Comp.		
	464	52.5	RbF, LiF.RbF	Eut.		
	790	100	RbF	M.P.		
ScF_3	594	25	LiF, ScF_3	Eut.	1,2	1127
	606	27	LiF, ScF_3	Eut.	1,2	1127
	606	27	LiF, ScF_3	Eut.	1,2,2a	1128
	440-445		$LiScF_4$	Comp.		
SmF_3	660	33	LiF, SmF_3	Eut.	2	1122
SrF_2	761	20	LiF, SrF_2	Eut.	1,2	1129
TbF_3	660	27	LiF, LiF.TbF_3	Eut.	2	1122
	790	48	LiF.TbF_3, TbF_3	Per.		
	1000	100	TbF_3	P.T.		
ThF_4			Fig. 207			
	568	23	LiF, 3LiF.ThF_4	Eut.	2b,2c, 3b	1130
	573	25	3LiF.ThF_4	Comp.		
	565	29	3LiF.ThF_4, 7LiF.6ThF_4	Eut.		
	597	30.5	7LiF.6ThF_4, LiF.2ThF_4	Per.		
	762	42	LiF.2ThF_4, LiF.4ThF_4	Per.		
	897	58	LiF.4ThF_4, ThF_4	Per.		

LiF

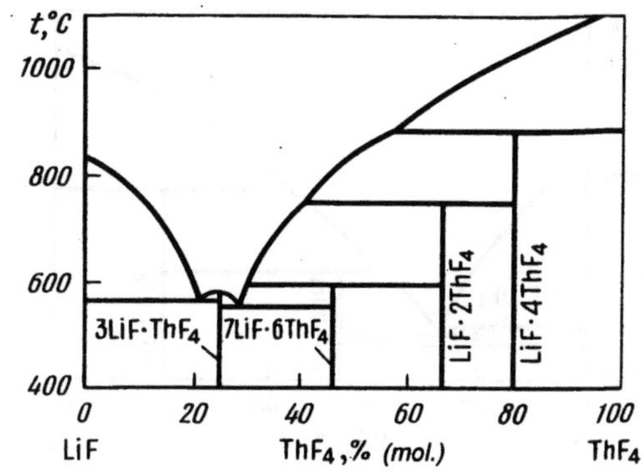

Fig. 207. LiF-ThF_4 system.

2nd Component	T°C	Conc. mol %	Phases involved	Def.	Method	Ref.
TiF_4	846	0.0	LiF	M.P.	1,2,2d,	1131
	596	17.5	LiF, Li_2TiF_6	Eut.	3b,5	
	680	50	Li_2TiF_6	Comp.		
	520		Li_2TiF_6	P.T.		
TmF_3	670	30	LiF, LiF.TmF_3	Eut.	2	1122
	800	50	LiF.TmF_3	Comp.		
	770	59	LiF.TmF_3, TmF_3	Eut.		
	1050	85	α-TmF_3, β-TmF_3	Per.		
UF_3	770	28	UF_3, LiF	Eut.		1132
UF_4			Fig. 208			
	500	26	LiF, 4LiF.UF_4	Per.	2,2d,	1133
	490	27	4LiF.UF_4, 7LiF.6UF_4	Eut.	3b,1	
	610	40	7LiF.6UF_4, LiF,4UF_4	Per.		
	775	57	LiF.4UF_4, UF_4	Per.		
	470		Eut. metastable, LiF, 7LiF.6UF_4	Eut.		
YF_3	695	19	LiF, LiF.YF_3	Eut.	2b,2c,	1134
	815	49	LiF.YF_3, YF_3 Table VIII	Per.	3b	

LiF

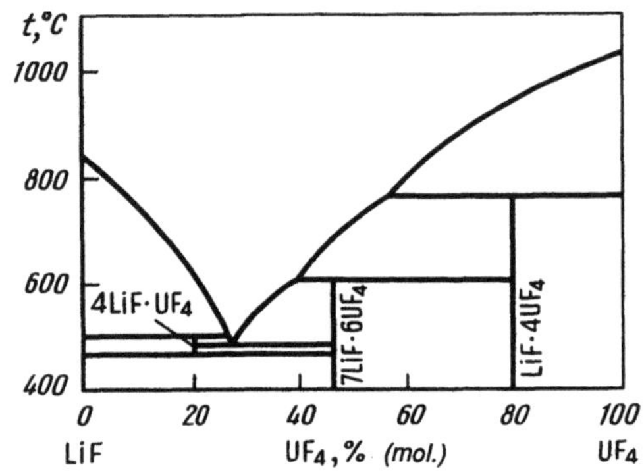

Fig. 208. LiF-UF_4 system.

TABLE VIII

YF_3	Temp. °C Beginning of crystallization	Eutectic	Peritectic	Inversion	Phases involved
10	786	690	-	-	LiF
20	720	690	-	-	$LiF.YF_3$
30	755	697	-	-	
40	803	697	-	-	
50	883	-	825	-	YF_3
60	925	-	829	-	
65	955	-	830	-	
85	-	-	807	1050	
100	1148	-	-	1055	

2nd Component	T°C	Conc. mol %	Phases involved	Def.	Method	Ref.
YF_3	695	19	LiF, $LiYF_4$	Eut.	2a	1135
	819	49	$LiYF_4$, YF_3	Per.		

LiF

2nd Component	T°C	Conc. mol %	Phases involved	Def.	Method	Ref.
YbF_3	680	30	LiF, $LiF.YbF_3$	Eut.	2	1122
	800	50	$LiF.YbF_3$	Comp.		
	780	56	$LiF.YbF_3$, YbF_3	Eut.		
	940	74	α-YbF_3, β-YbF_3	Per.		
	690	30	YbF_3, LiF	Eut.	2a	1136
ZnF_2			Fig. 209			
	620	36	50%LiF sol.sln. in ZnF_2, 35%ZnF_2 sol.sln. in LiF	Eut.	2,3b	1137
		19-90	Breakup of sol.sln. (pts. A & B in the fig.)			

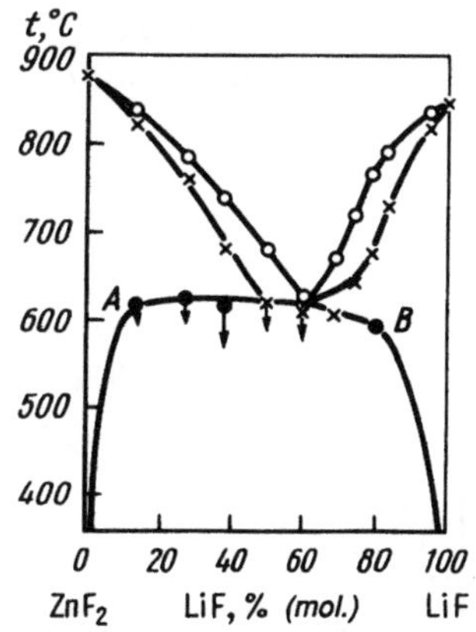

Fig. 209. ZnF_2-LiF system

2nd Component	T°C	Conc. mol %	Phases involved	Def.	Method	Ref.
ZrF_4	585	21	LiF, β-Li_3ZrF_7	Eut.	2d,3b	1138
	640	24	β-Li_3ZrF_7	Comp.		
	592		β-Li_3ZrF_7, α-Li_3ZrF_7	Per.		
	570	33	α-Li_3ZrF_7,Li_2ZrF_6	Per.		
	502	50	Li_2ZrF_6, α-ZrF_4	Eut.		
	522	53	α-ZrF_4, $LiZr_4F_{17}$(?)	Per.		
	556	57	$LiZr_4F_{17}$(?), α-ZrF_4	Per.		

LiF

2nd Component	T°C	Conc. mol %	Phases involved	Def.	Method	Ref.
ZrF_4	681	63	α-ZrF_4, β-ZrF_4	Per.	2d,3b	1138
	464		$2LiF + Li_2ZrF_6 \rightleftharpoons Li_4ZrF_8$	P.T.		
	474		$LiF + Li_2ZrF_6 \rightleftharpoons Li_3ZrF_7$	P.T.		
	465		$\alpha \rightleftharpoons \beta$-$Li_2ZrF_6$ (probably)	P.T.		
			Fig. 210			1139
	598	21.0	α-$3LiF.ZrF_4$, LiF	Eut.	2,1f,3b	1139
	662	125.0	α-$3LiF.ZrF_4$	Comp.		1140
	570	29.5	$2LiF.ZrF_4$, α-$3LiF.ZrF_4$	Eut.		
	596	33.3	$2LiF.ZrF_4$	Comp.		
	507	49.0	$3LiF.4ZrF_4$, $2LiF.ZrF_4$	Eut.		
	520	51.5	ZrF_4, $3LiF.4ZrF_4$	Per.		
	470		β-$3LiF.ZrF_4 \rightleftharpoons LiF + 2LiF.ZrF_4$	P.T.		
	475		$\alpha \rightleftharpoons \beta$ $3LiF.ZrF_4$	P.T.		

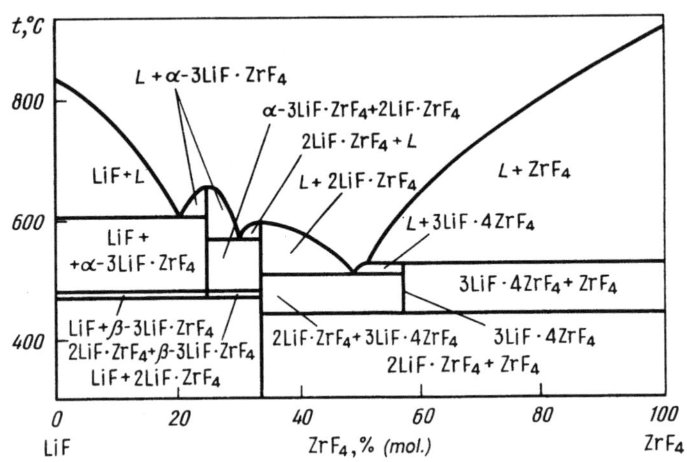

Fig. 210. LiF-ZrF_4 system.

2nd Component	T°C	Conc. mol %	Phases involved	Def.	Method	Ref.
LiH	682±0.5	86±0.5	Sol.slns. (10-95%) Table IX	Min.	2,1e	1141

LiF

TABLE IX

LiH%	Temp. °C Liquidus	Solidus	LiH%	Temp. Liquidus	Solidus
0.00	848.0	848.0	68.54	693.4	685.0
9.91	832.4	786.8	74.97	688.6	683.5
19.99	802.0	741.0	79.95	685.6	683.1
29.94	778.5	715.6	84.58	684.1	682.6
40.01	772.4	698.4	90.56	684.2	683.1
49.99	704.4	691.7	94.94	685.0	683.8
60.05	698.0	686.7	100.00	686.4	686.4

2nd Component	T°C	Conc. mol %	Phases involved	Def.	Method	Ref.
LiI	-	0-80	LiF		2a	1142
	410	83.5	LiI	Eut.		
	-	84.9	LiI			
			Table X			

TABLE X

LiI	Temp. °C Liquidus	Solidus	LiI	Temp. °C Liquidus	Solidus
0.000	848.0	-	0.700	547.5	411.0
0.100	822.2	411.3	0.800	448.1	411.1
0.200	798.7	410.2	0.835	410.9	-
0.275	779.9	411.5	0.849	416.2	411.5
0.350	755.3	411.7	0.900	436.6	410.4
0.450	709.8	411.3	0.950	455.2	409.4
0.550	652.0	411.2	1.000	469.2	-
0.650	583.9	410.8			

LiH

2nd Component	T°C	Conc. mol %	Phases involved	Def.	Method	Ref.
LiI	390.8	70.3	LiH, LiI	Eut.	2	1143
TeI_4	449	0.0	LiI	M.P.	2a,2e,	1144
	266	~97.0	LiI, TeI_4	Eut.	6c,3b	
	280	100.0	TeI_4	M.P.		
	380	20-80	Decomposition			

$LuCl_3$

2nd Component	T°C	Conc. mol %	Phases involved	Def.	Method	Ref.
NaCl	450	60.0	$LuCl_3$, Na_3LuCl_6		2a,3b	1145
	560	70.0	Na_3LuCl_6, NaCl	Per.		

LuF_3

2nd Component	T°C	Conc. mol %	Phases involved	Def.	Method	Ref.
NaF			Fig. 211			
	930	27	α-LuF_3, β-LuF_3	Per.	2,1f,3b	1146
	900	29	5NaF.9LuF_3, LuF_3	Eut.		
	940	36	5NaF.9LuF_3	Comp.		
	650	74	NaF.LuF_3, 5NaF.9LuF_3	Per.		
	595	79	NaF, NaF.LuF_3	Eut.		
			5NaF.9LuF_3 exists in two allotropes: cubic fluorite (high temp.) and orthorhombic. The fluorite modification forms sol. slns. with NaF.LuF_3			

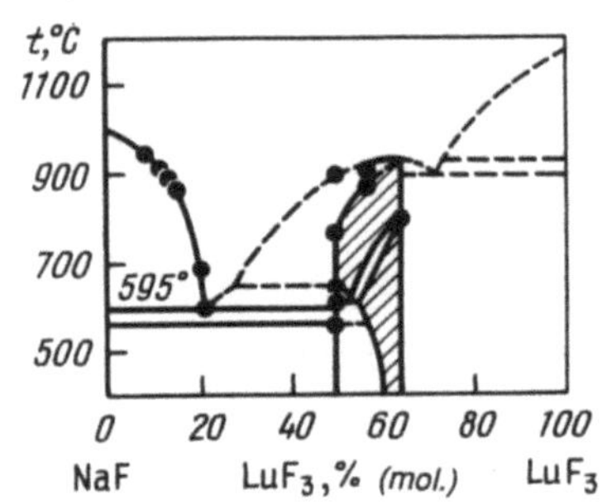

Fig. 211. NaF-LuF_3 system.

LuF_3

2nd Component	T°C	Conc. mol %	Phases involved	Def.	Method	Ref.
NaF			Fig. 212			1147
	873	28.0	$5NaF.9LuF_3$ (cubic) LuF_3	Eut.	2,3b,5, le	1147
	930	35.7	$5NaF.9LuF_3$ (cubic)	Comp.		
	595	71	NaF, $5NaF.9LuF_3$ (or sol.sln. on $5NaF.9LuF_3$ base)	Eut.		
			Limited sol. slns. in $NaF.LuF_3$ and $5NaF.9LuF_3$			

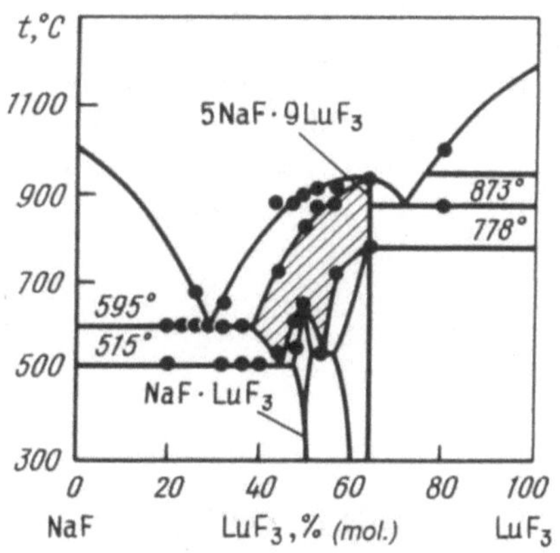

Fig. 212. $NaF-LuF_3$ system.

2nd Component	T°C	Conc. mol %	Phases involved	Def.	Method	Ref.
SrF_2			Limited sol. slns.: SrF_2 fluorite and LuF_3 tisonite, and mixture of those above between them.		3b	1148

$MgCl_2$

2nd Component	T°C	Conc. mol %	Phases involved	Def.	Method	Ref.
NaCl	466	51.0	$NaMgCl_3$, $MgCl_2$	Per.	2,2d,3b	1149
	451	56.1	Na_2MgCl_4, $NaMgCl_3$	Eut.		
	488	56.3	NaCl, Na_2MgCl_4	Per.		

$MgCl_2$

2nd Component	T°C	Conc. mol %	Phases involved	Def.	Method	Ref.
NaCl	442	56.1	$MgCl_2$.NaCl, $MgCl_2$.2NaCl	Eut.	2d, le, 3b	1150 1151
			$MgCl_2$.NaCl - decomposes on melting	Comp.		
			$MgCl_2$.2NaCl - decomposes on melting	Comp.		
$NbCl_5$	720	0.0	$MgCl_2$	M.P.	2,6c	1152
	192	92	$MgCl_2$, $NbCl_5$	Eut.		
	207	100	$NbCl_5$	M.P.		
		61-100	High vapor pressure of $NbCl_5$ prevents melting - no liquidus			
		0-34	Solidus traced			
$NdCl_3$	625	40	-	Eut.	2	1153
$PbCl_2$	459	83.1	-	Eut.	-	1154
	440	80.5	$MgCl_2$, $PbCl_2$	Eut.	2,1f	1155
$PrCl_3$	645	38	$MgCl_2$, $PrCl_3$	Eut.	2	1156
$PuCl_3$	714	0	$MgCl_2$	M.P.	2,1d	1157
	650	38	$MgCl_2$, $PuCl_3$	Eut.		
	767	100	$PuCl_3$	M.P.		
$SbCl_5$	~3	~100	$MgCl_2$, $SbCl_5$	Eut.	2d	1158
	3	100	$SbCl_5$	M.P.		
$ThCl_4$	610	55	$MgCl_2$, $ThCl_4$	Eut.	2,6c	1159
	714 ±	0	$MgCl_2$	M.P.	2d	1160
	604 ±	60	$MgCl_2$, $ThCl_4$	Eut.		
	768 ±	100	$ThCl_4$	M.P.		

$MgCl_2$

2nd Component	T°C	Conc. mol %	Phases involved	Def.	Method	Ref.
$TiCl_2$	716	0.29	$MgCl_2$ base sol. sln., $TiCl_2$	Per.	2	1161
TlCl	362	74	-	Eut.	0	1154
UCl_3	670	35	$MgCl_2$, UCl_3	Eut.	2d	1162
UCl_4	717	0	$MgCl_2$	M.P.	2a,3b,1	1163
	540	69.3	$MgCl_2$, UCl_4	Eut.	1,5	
	589	100	UCl_4	M.P.		
	540±2	74	$MgCl_2$, UCl_4	Eut.	2d,3b	1164 1165
YCl_3	577	60	$MgCl_2$, YCl_3	Eut.	2	1166
	476		$3MgCl_2.2YCl_3$			
$ZrCl_4$	720	0	$MgCl_2$	M.P.	2a,6c	1167
	426	96.4	$MgCl_2$, $ZrCl_4$	Eut.		
	437	100	$ZrCl_4$	M.P.		
		<75	Mixtures melt due to high vapor pressure of $ZrCl_4$ vapor			
MgF_2	628±2	22±0.5	$MgCl_2$, MgF_2	Eut.	2d,2, 6k,3b	1168
	714	0.0	$MgCl_2$	M.P.	2d	1169
	596	16.5	$MgCl_2$, MgF_2	Eut.		
	996	100.0	MgF_2	M.P.		

MgF_2

2nd Component	T°C	Conc. mol %	Phases involved	Def.	Method	Ref.
NaF	1265	0.0	MgF_2	M.P.	2,1f	1170
	994	36.3	$NaMgF_3$, MgF_2	Eut.		
	1023	50	$NaMgF_3$	Comp.		
	824	87.4	NaF, $NaMgF_3$	Eut.		
	997	100.0	NaF	M.P.		

MgF_2

2nd Component	T°C	Conc. mol %	Phases involved	Def.	Method	Ref.
ScF_3	1255	0.0	MgF_2	M.P.	2,3b	1171
	1985	34	10% ScF_3 sol.sln. in MgF_2, ScF_3	Eut.		
	1530	100	ScF_3	M.P.		
	840		MgF_2	P.T.		
	1350		ScF_3	P.T.		
SrF_2	890	50	- (composition not confirmed experimentally)	Eut.	1,2	1172
ThF_4			Fig. 213			1173
	912	60	$MgTh_2F_{10}$,MgF_2	Eut.		1173
	920	67	$MgTh_2F_{10}$	Comp.		
	910	77	ThF_4, $MgTh_2F_{10}$	Eut.		

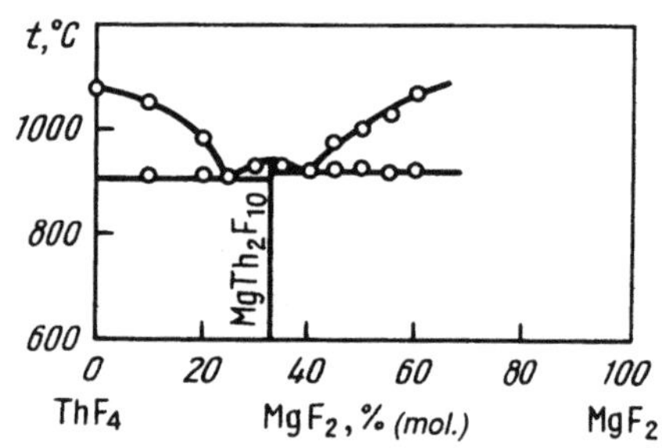

Fig. 213. ThF_4-MgF_2 system.

$MnCl_2$

2nd Component	T°C	Conc. mol %	Phases involved	Def.	Method	Ref.
NaCl	448	45	$NaMn_2Cl_5$, $MnCl_2$	Per.	2d,3b	1174
	428	~50	$NaMnCl_3$,$NaMn_2Cl_5$	Eut.		
	428	50	$NaMnCl_3$	Comp.		
	420	53	Na_2MnCl_4, $NaMnCl_3$	Eut.		
	437	57.5	Na_4MnCl_6, Na_2MnCl_4	Eut.		
	458	61	NaCl, Na_4MnCl_6	Per.		
	318	-	$NaMn_2Cl_5$	P.T.		

$MnCl_2$

2nd Component	T°C	Conc. mol %	Phases involved	Def.	Method	Ref.
NaCl	650	0	$MnCl_2$	M.P.	2a	1175
	442	45.0	$MnCl_2$,NaCl.2$MnCl_2$	Per.		
	425	50.0	NaCl.2$MnCl_2$, 4NaCl.$MnCl_2$	Eut.		
	445	55.0	4NaCl.$MnCl_2$,NaCl	Per.		
	800	100	NaCl	M.P.		
			Fig. 214			1176
	652	0.0	$MnCl_2$	M.P.	2d,3b	1176
	445	45	$MnCl_2$ sol.sln., $NaMn_2Cl_5$	Per.		
	428	50	$NaMn_2Cl_5$, $NaMnCl_3$	Per.		
	420	~53	$NaMnCl_3$, β-Na_2MnCl_4	Eut.		
	~420	~53	α-Na_2MnCl_4, NaCl	Per.		
	~420	~53	α-Na_2MnCl_4	-		
	446	58	α-Na_2MnCl_4, NaCl	Per.		
	800	100	NaCl	M.P.		
			$NaMnCl_3$ unit cell parameters a = 9.69Å, c = 9.60Å			

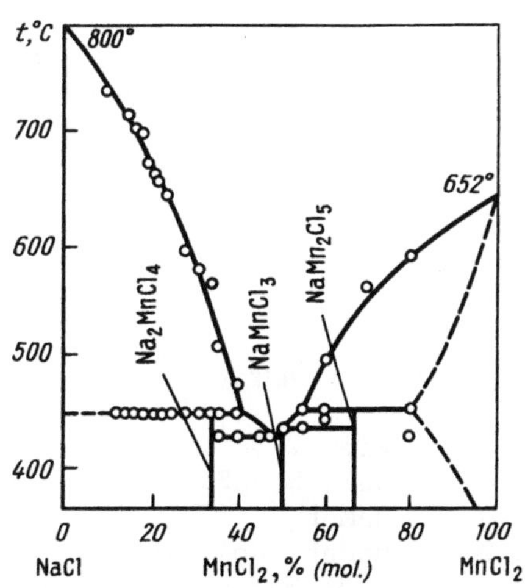

Fig. 214. NaCl-$MnCl_2$ system.

$NbCl_5$	-	~100	Temp.and composition close to pure $NbCl_5$	Eut.	2d	1177

$MnCl_2$

2nd Component	T°C	Conc. mol %	Phases involved	Def.	Method	Ref.
$NiCl_2$	650	0.0	$MnCl_2$	M.P.	2a,2b,	1178
	648	0.5	$MnCl_2$, $NiCl_2$	Eut.	2c	
	1003	100	$NiCl_2$	M.P.		
RbCl	479	30	$MnCl_2$, $MnCl_2$.RbCl	Eut.	1,2	1179
	552	50	$MnCl_2$.RbCl	Comp.		
	475	64	$MnCl_2$.RbCl, $2MnCl_2$.3RbCl	Per.		
	462	69	$2MnCl_2$.3RbCl, $MnCl_2$.2RbCl	Per.		
	446	73	$MnCl_2$.2RbCl,RbCl	Eut.		
$SbCl_5$	3	~100	$MnCl_2$, $SbCl_5$	Eut.	2d	1180
	3	100	$SbCl_5$	M.P.		
$TaCl_5$	-	~100	Temp.& composition close to pure $TaCl_5$	Eut.	2d,1	1181
$TiCl_2$	650	0.0	$MnCl_2$	M.P.	2d	1182
		90-100	Solubility range of $MnCl_2$ in $TiCl_2$			
MnF_2	520	37.5	$MnCl_2$, MnF_2	Eut.	1,2,2a	1183

MnF_2

2nd Component	T°C	Conc. mol %	Phases involved	Def.	Method	Ref.
NaF			Fig. 215			1184
	692	34.0	NaF, $NaMnF_3$	Eut.	1,2,2a	1185
	672	50.0	$NaMnF_3$	Comp.		
	738	66.0	$NaMnF_3$ sol.sln., 23% NaF sol.sln. in MnF_2	Eut.		
	610-140		Dissociaion of sol. slns. with appearance of $NaMn_2F_5$	P.T.		

MnF_2

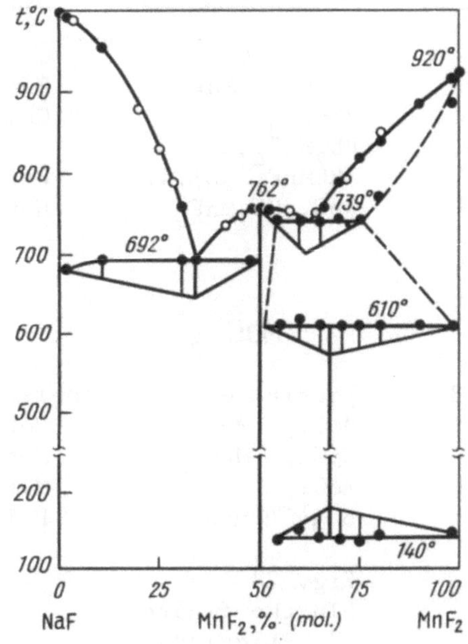

Fig. 215. NaF-MnF_2 system.

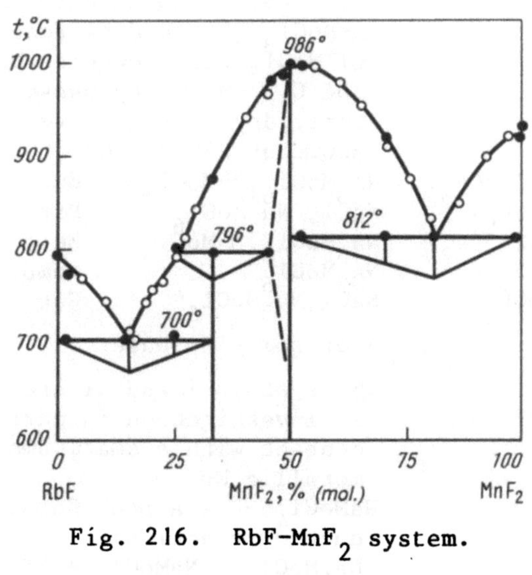

Fig. 216. RbF-MnF_2 system.

MnF_2

2nd Component	T°C	Conc. mol %	Phases involved	Def.	Method	Ref.
RbF			Fig. 216			1184
	812	19	$RbMnF_3$, MnF_3	Eut.	1,2,2a	1185
	986	50	$RbMnF_3$	Comp.		
	796	76	Rb_2MnF_4, $RbMnF_3$ sol.sln.	Per.		
	700	84	RbF, Rb_2MnF_4	Eut.		

$MoCl_3$

2nd Component	T°C	Conc. mol %	Phases involved	Def.	Method	Ref.
$MoCl_5$	300	5-35	Thermal effects undefined		2	1186
	272	-	$MoCl_4$, $MoCl_3$	Per.		
	180	95	$MoCl_5$, $MoCl_4$	Eut.		
	194	100	$MoCl_5$	M.P.		
	98	-	$\alpha \rightleftharpoons \beta$ $MoCl_5$	P.T.		
NaCl			Fig. 217			1187
			[In the figure 1 - Liquidus 2 - solidus]		2d, le, 3b	
			The system has two types of diagrams: (ii) NaCl-α-$MoCl_3$ stable, (i) NaCl-β-$MoCl_3$ metastable (α-$MoCl_3$ obtained by reaction of $MoCl_5$ with powdered Mo at 354°C, β-$MoCl_3$ by passing pure, dry hydrogen over $MoCl_5$ at 270-300°C)			
	448	47.5(e_3)	Na_3MoCl_6, β-$MoCl_3$*	Eut.		
	586	74(p)	NaCl, Na_3MoCl_6	Per.		
	548	62.5(e_2)	Na_3MoCl_6, α-$MoCl_3$*	Eut.		
	608	75(D)	Na_3MoCl_6	Comp.		
	590	80(e_1)	NaCl, Na_3MoCl_6*	Eut.		
			* or sol.sln. base			
			The systems i and ii are identical if investigation is carried out in contact with a small amount of metallic Mo			
	~200	-	$NaMoCl_4$ - as a product of reaction: $Na_3MoCl_6 \rightarrow NaMoCl_4 + 2NaCl$ Table XI	Comp.		

TABLE XI

OPTICAL PROPERTIES

Phase	Light	Refraction index		Formula, habitus	Cleavage	Comment
		N'_g	N'_f			
α-$MoCl_3$	Pleachroism from blood-red to black	>2.1	>2.1	Isometric grains	-	Polysynthetic doublets
α-$MoCl_3$	Red-Magenta	1.932	1.820	Long fibrous crystals	Diagonal	-
Na_3MoCl_6	Rosy	1.662	1.646	Isometric grains	-	-
$NaMoCl_4$	Yellow-Green	2.010	1.980	Long fibrous crystals	Diagonal	Primary extinction
Na_3MoCl_6 sol. sln.	Rosy-Magenta	>1.830	<1.640	Isometric grains	-	-
α-$MoCl_3$ sol. sln.	Pleachroism from yellow to dark red	>1.990	<1.834	Isometric grains	-	-

$MoCl_3$

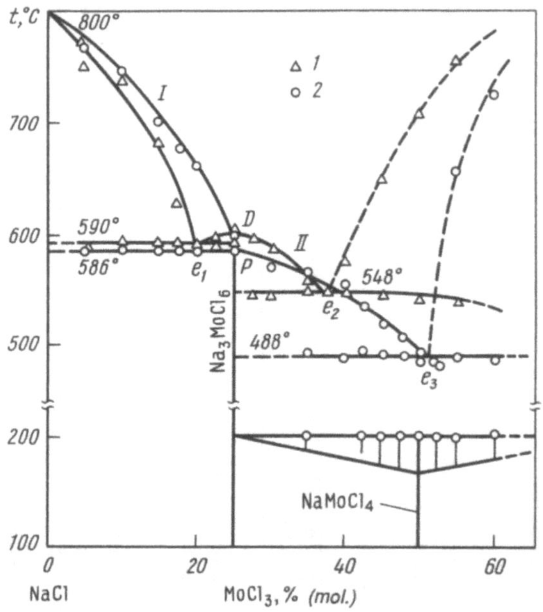

Fig. 217. $NaCl-MoCl_3$ system.

2nd Component	T°C	Conc. mol %	Phases involved	Def.	Method	Ref.
$SnCl_4$	46	92	$MoCl_3$ $SnCl_4$	Eut.	2a	1186
	-32	100	$SnCl_4$	M.P.		

$MoCl_5$

2nd Component	T°C	Conc. mol %	Phases involved	Def.	Method	Ref.
NaCl			Almost complete mutual insolubility. Eutectic close to pure $MoCl_5$		2,6c	1188
		0-67	Complete mutual insolubility		2b,2c, 5	1189
		>67	$2NaCl.MoCl_4$ appears			
	0-67		Complete mutual insolubility		2a,6c 5c	1190
	200-400	>67	$2NaCl.MoCl_4$ (vacuum)	Comp.		

$MoCl_5$

2nd Component	T°C	Conc. mol %	Phases involved	Def.	Method	Ref.
$NbCl_5$			Complete range of sol.sln.		2a	1191
RbCl		0-67	Complete mutual insolubility		2b,2c	1192
		>67	$2RbCl.MoCl_4$ appears			
$ReCl_5$			Complete range of sol.sln.		2a	1193
$SbCl_3$			Chlorinating reaction of $SbCl_3$		2d	1194
$SbCl_5$	194	0.0	$MoCl_5$	M.P.	2d,1	1195
	-	-	Indeterminate	Eut.		
$SeCl_4$	194	0.0	$MoCl_5$	M.P.	2d	1196
	174	36	$MoCl_5$,$MoCl_5.SeCl_4$	Eut.		
	192	50	$MoCl_5.SeCl_4$	Comp.		
	190	55	$MoCl_5.SeCl_4$,$SeCl_4$	Eut.		
	305	100	$SeCl_4$	M.P.		
	170	~42	52% $MoCl_5$ sol. sln. in $SeCl_4$ 20% $SeCl_4$ sol. sln. in $MoCl_5$	Eut.	1,2, 3b,1e	1197
$SnCl_4$	194	0.0	$MoCl_5$ Eutectic indeterminate	M.P.	2d,1	1198
$TaCl_5$			Complete range of sol. slns.		2a	1199
$TeCl_4$	132	58	$MoCl_5$, $TeCl_4$	Eut.	2a	1200
$TiCl_4$	194	0.0	$MoCl_5$ Eutectic indeterminate	M.P.	2d,1	1198

$MoCl_5$

2nd Component	T°C	Conc. mol %	Phases involved	Def.	Method	Ref.
$ZnCl_2$	194	0.0	$MoCl_5$	M.P.	2d, 1	1198
	~194	~0.0	$MoCl_5$, $ZnCl_2$	Eut.		
	240	25-75	Dissociation	P.T.		

MoF_6

2nd Component	T°C	Conc. mol %	Phases involved	Def.	Method	Ref.
SbF_3	-6		α-MoF_6, β-MoF_6	Per.	2a	1201
	-21	64	β-MoF_6, SbF_5	Eut.		
		28-62	Decomposition			
XeF_2			Fig. 218			1202
	9.6	0	MoF_6	M.P.	2d	1202
			~XeF_2.4MoF_6 (incongruent melting)	Comp.		
	124 ±1		XeF_2.MoF_6	Comp.		
	140	100	XeF_2	M.P.		

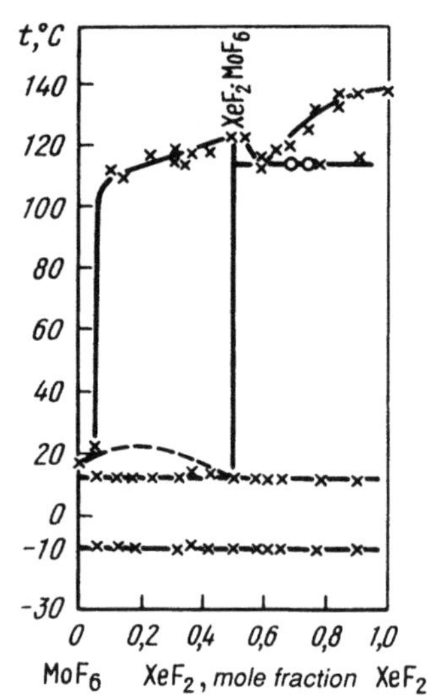

Fig. 218. MoF_6-XeF_2 system.

NH_4Br

2nd Component	T°C	Conc. mol %	Phases involved	Def.	Method	Ref.
NH_4Cl	~105.4	40	Solid sln.	Min.	2,3b	1203
	137.2	0	NH_4Br simple cubic ⇌ b.c.c.	P.T.	3b	1204
	183.1	100	NH_4Cl simple cubic ⇌ b.c.c.	P.T.		
NH_4I			Solid sln.		6m,3b	1205

NH_4Cl

2nd Component	T°C	Conc. mol %	Phases involved	Def.	Method	Ref.
$NbCl_5$	295	17	NH_4NbCl_6 decomposes	Per.	2a,6c	1206
	272	21	I - NH_4NbCl_6, II - NH_4NbCl_6			
	201	79	$NbCl_5$, NH_4NbCl_6	Eut.		
		57-86.5	Decomposition range			
$SbCl_5$	355	25	β-NH_4Cl, $3NH_4Cl.2SbCl_5$	Eut.	2d	1207
	347	43	$3NH_4Cl.2SbCl_5$, β-$NH_4Cl.SbCl_5$	Eut.		
	375	50	β-$NH_4Cl.SbCl_5$	Comp.		
	325	-	β-$NH_4Cl.SbCl_5$, β-$NH_4Cl.SbCl_5$	Per.		
	1	~100	α-$NH_4Cl.SbCl_5$, $SbCl_5$	Eut.		
	3	100	$SbCl_5$	M.P.		
	310	70-97	Decomposition			
$SnCl_2$			Strong volatility of NH_4Cl Limited investigation up to 50% NH_4Cl Table XII		1,2,2b	1208

NH_4Cl

TABLE XII

$SnCl_2$ Mol %	$SnCl_2$ Mass %	t°C Crystallization begins: Visually	Crystallization begins: Thermographically, From powder	Crystallization begins: Thermographically, From molten salt	Crystallization ends	Eutectic	P.T. α ⇌ β NH_4Cl	Phases involved	Def.
100	100	245	245	245	-	-	-	$SnCl_2$	
98	99.46	-	-	-	-	-	-	$SnCl_2$ based sol.sln.	
95	98.53	-	-	-	204	183	-		
90	97.0	214	223	219	-	179	-		
85	95.4	197	200	202	-	183	-		
82	94.5	188	-	-	-	182	-		
80	93.5	182	183	183	-	180	-	10% NH_4Cl sol.sln. in $SnCl_2$, $2SnCl_2.NH_4Cl$	Eut.
78	92.6	190	-	-	-	-	-	$2SnCl_2.NH_4Cl$	
75	91.5	196	194	196	-	186	-		
72	90.1	202	-	-	-	187	-		
70	89.4	204	204	210	-	-	-		
66.7	87.7	216	212	211	-	-	-		
65	86.9	210	206	204	-	-	-		
62	85.3	192	-	-	-	-	-		
60	84.0	189	187	188	-	-	-	$2SnCl_2.NH_4Cl$, NH_4Cl based sol. sln.	Eut.
57	82.5	192	-	-	-	-	-	NH_4Cl or NH_4Cl based sol. sln.	
55	81.5	204	204	200	-	-	-		
50	78.0	206	204	-	182	-	182		
45	74.4	209	210	-	188	-	182		

NH_4Cl

2nd Component	T°C	Conc. mol %	Phases involved	Def.	Method	Ref.
$SnCl_2$			Fig. 219			1209
	295	33	δ-$SnCl_2.3NH_4Cl$, NH_4Cl	Per.	2,6c	1209
	246	40	γ-$SnCl_2.3NH_4Cl$, δ-$SnCl_2.3NH_4Cl$	Per.		
	200	44	$SnCl_2.NH_4Cl$, γ-$SnCl_2.3NH_4Cl$	Eut.		
	206	50	$SnCl_2.NH_4Cl$	Comp.		
	173	58	$2SnCl_2.NH_4Cl$, $SnCl_2.NH_4Cl$	Eut.		
	188	67	$2SnCl_2NH_4Cl$	Comp.		
	172	81	$SnCl_2$,$2SnCl_2.NH_4Cl$	Eut.		
	154	-	α ⇌ β $SnCl_2.3NH_4Cl$	P.T.		
	200	-	β ⇌ γ $SnCl_2.3NH_4Cl$	P.T.		

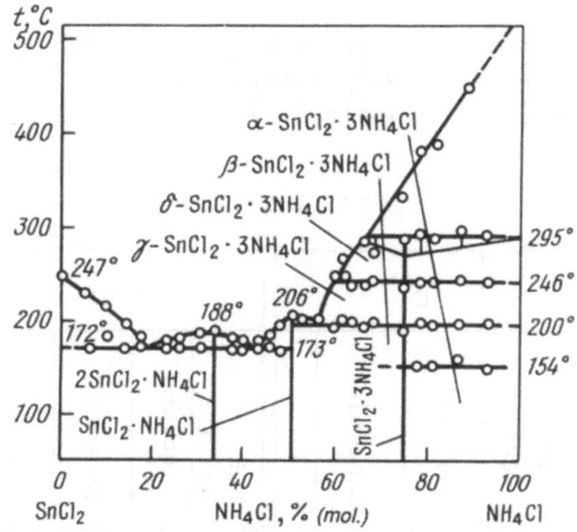

Fig. 219. $SnCl_2$-NH_4Cl system.

2nd Component	T°C	Conc. mol %	Phases involved	Def.	Method	Ref.
$TaCl_5$	243	30	II - NH_4TaCl_6, NH_4Cl	Eut.	2a,6c	1210
	268	-	I - NH_4TaCl_6, II - NH_4TaCl_6	Per.		
	304	50	I - NH_4TaCl_6	Comp.		
	214	96	$TaCl_5$, I - NH_4TaCl_6	Eut.		
		61.6-85.4	Decomposition			
$TeCl_4$	213	92	$2NH_4Cl.TeCl_4$,$TeCl_4$	Eut.	2d	1211
	224	100	$TeCl_4$	M.P.		

NH_4F

2nd Component	T°C	Conc. mol %	Phases involved	Def.	Method	Ref.
TlF	220.5	50	NH_4, TlF Increase of pressure shifts eutectic towards TlF and lower temp.	Eut.	2	1212

$NaBF_4$

2nd Component	T°C	Conc. mol %	Phases involved	Def.	Method	Ref.
NaF	368	0	$NaBF_4$	M.P.	2c,1,2	1213
	304	61	NaF, $NaBF_4$	Eut.		
	1026	100	NaF	M.P.		
		0-6	$NaBF_4$ crystallizes		2d,6d	1214
	384	7	NaBF, NaF	Eut.		
		7.5-77.7	NaF crystallizes Table XIII			

TABLE XIII

NaF mol %	t°C* Liquidus	t°C* Eutectic	t°C* α ⇌ β $NaBF_4$	t°C** Liquidus	t°C** Eutectic	NaF mol %	t°C* Liquidus	t°C* Eutectic	t°C* α ⇌ β $NaBF_4$	t°C** Liquidus	t°C** Eutectic
0.0	408	-	244	-	-	17.9	538	-	-	-	-
3.9	395	384	-	402±3	383±1	20.0	-	380	240	555±2	378±2
5.0	-	-	-	404±1	-	30.0	676	384	-	-	-
6.0	385	381	-	392±2	-	37.0	741	-	-	-	383±1
7.5	-	-	-	391±2	380±2	50.0	813	-	-	753±4	380±2
8.0	-	381	-	383±1	379±1	56.7	858	-	-	-	-
9.5	391	385	-	408±2	383±1	60.0	-	382	246	-	-
10.0	-	384	244	-	-	70.0	306	384	246	-	-
12,0	469	381	-	>420	381±1	77.7	937	384	247	-	-

*From thermographic data.
**From data on gradient guenching.

NaBr

2nd Component	T°C	Conc. mol%	Phases involved	Def.	Method	Ref.
$PbBr_2$	376	0.0	NaBr	M.P.	1,2	1215
	323	94.9	NaBr, $PbBr_2$	Eut.		1216
	380	90.7	$\alpha \rightleftharpoons \beta$ NaBr			
RbBr	754	0.0	NaBr	M.P.	2a	1217
	495	55	NaBr, RbBr	Eut.		
	680	100	RbBr	M.P.		
	747	0.0	NaBr	M.P.	1,2,2a	1218
	494	50	NaBr, RbBr	Eut.		
	690	100	RbBr	M.P.		
	747	0.0	NaBr	M.P.	1,2	1219
	507	50	NaBr, RbBr	Eut.		1220
	690	100	RbBr	M.P.		1221
$ScBr_3$	515	30.5	NaBr, Na_3ScBr_6	Per.	2d,3b	1222
	482	35.4	Na_3ScBr_6, $ScBr_3$	Eut.		
$SnBr_2$	750	0.0	NaBr	M.P.	1,2,2a	1223
	193	80	NaBr, $SnBr_2$	Eut.		
	226	100	$SnBr_2$	M.P.		
$SrBr_2$	456	64	NaBr, $SrBr_2$	Eut.	1,2	1224
	476	58	NaBr, $SrBr_2$	Eut.	1,2, 2d,3b	1225
$TeBr_4$	748	0.0	NaBr	M.P.	2,3b,1e	1226
	350	80.0	NaBr, $TeBr_4$	Eut.		
	386	100.0	$TeBr_4$	M.P.		
TlBr	452	25	NaBr, TlBr	Eut.	1,2	1227
	460	100	TlBr	M.P.		1228
NaCl	734	-	Continuous sol.sln.	Min.	1,2	1229

NaBr

2nd Component	T°C	Conc. mol %	Phases involved	Def.	Method	Ref.
NaCl	744	30	Continuous sol.sln.	Min.	1,2	1230
	747	0.0	NaBr	M.P.	1,2	1231
	740	50	Continuous sol.sln.	Min.		1232

NaCl

2nd Component	T°C	Conc. mol %	Phases involved	Def.	Method	Ref.
$NbCl_2$			$NbCl_2$ dissociates when heated, and the system is not binary. X-rays reveal $NbCl_4$ and metallic Nb.		2a,3b	1233
Nb_3Cl_8			Incongruently melting		2	1234
	718		$Na_7Nb_3Cl_{15}$ and	M.P.		
	704		$Na_4Nb_3Cl_{12}$	M.P.		
	520		$\sigma \rightleftharpoons \beta$ $Na_7Nb_3Cl_{15}$	P.T.		
	367		$\beta \rightleftharpoons \gamma$ $Na_7Nb_3Cl_{15}$	P.T.		
$NbCl_3$			Fig. 220			1235
	450	42	β-Na_2NbCl_5,NaCl	Per.	2a	1235
	362		Na_2NbCl_5	P.T.		

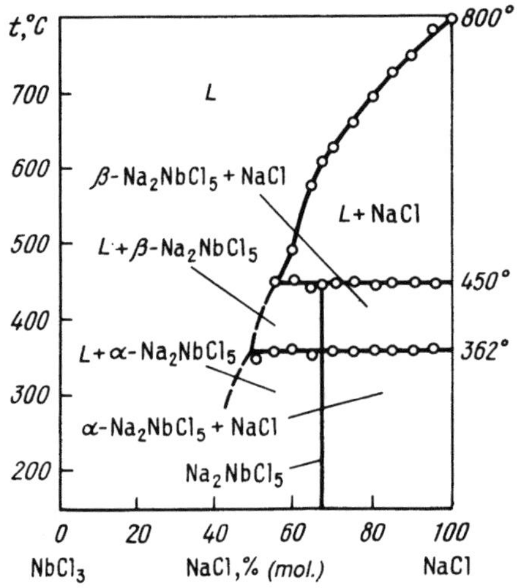

Fig. 220. $NbCl_3$-NaCl system.

NaCl

2nd Component	T°C	Conc. mol %	Phases involved	Def.	Method	Ref.
$NbCl_4$	248	67	$NbCl_4$, Na_2NbCl_6	Eut.	2c	1236
	582	33	Na_2NbCl_6	Comp.		
	530	30	NaCl, Na_2NbCl_6	Eut.		
	365		Na_2NbCl_6	P.T.		
$NbCl_5$			Fig. 221			1237
	430	53	β-NaCl.$NbCl_5$	Per.	2	1237
	256	-	α ⇌ β NaCl.$NbCl_5$	Per.		
	200	88	$NbCl_5$, α-NaCl.$NbCl_5$	Eut.		
	207	100	$NbCl_5$	M.P.		
	206	54-80	Dissociation			

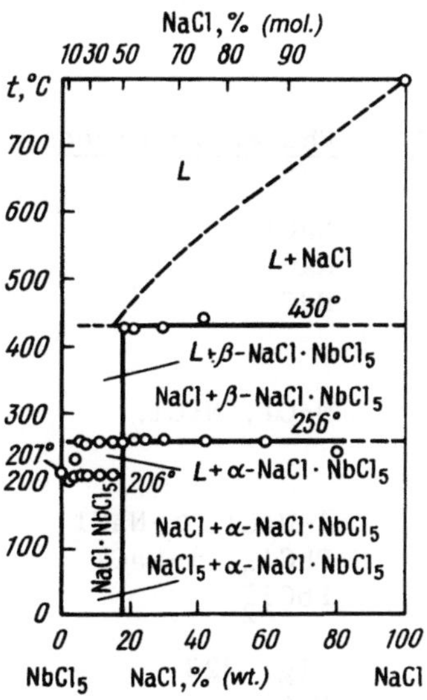

Fig. 221. $NbCl_5$-NaCl system.

			Fig. 222			1238
	450	-		Per.	2	1238
	~206	~100	$NaNbCl_6$, $NaCl_5$	Eut.		
		68.5-92.5	Dissociation region			
	259		$NaNbCl_6$	P.T.		
			Close to $NbCl_5$			
	378	73-91	Dissociation region	Eut.	1,2	1239

NaCl

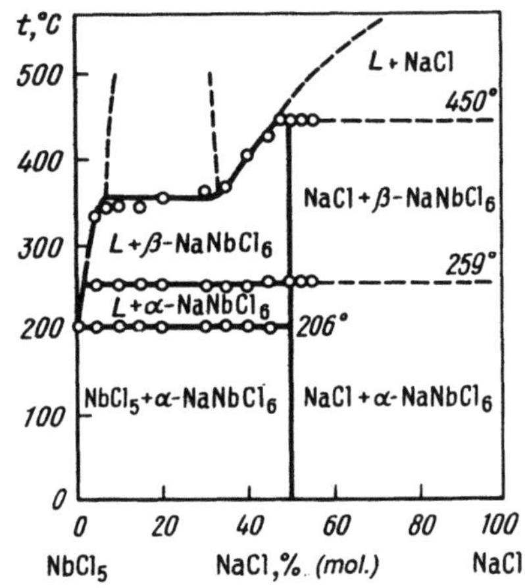

Fig. 222. $NbCl_5$-NaCl system.

2nd Component	T°C	Conc. mol %	Phases involved	Def.	Method	Ref.
$NdCl_3$	800	0.0	NaCl	M.P.	2	1240
	430	14	NaCl, $NdCl_3$	Eut.		
	763	100	$NdCl_3$	M.P.		
$NiCl_2$	563	32	NaCl, $NiCl_2$	Eut.	2, le	1241
$PbCl_2$	473	71.4	β-NaCl, α-NaCl	Per.	1,2	1242
	410	75.2	$PbCl_2$, β-NaCl	Eut.		1243
	496	100.0	$PbCl_2$	M.P.		
			Fig. 223			1244
	800	0.0	NaCl	M.P.	2	1244
	411	70	β-$PbCl_2$, NaCl (E_1)	Eut.		
	410	75	α-$PbCl_2$, β-$PbCl_2$ (C)	Per.		
	501	100	α-$PbCl_2$	M.P.		
			Diagram represents stable and metastable states			
	390	72	NaCl, α-$PbCl_2$ (E_2 - Metastable)	Eut.		
	422		Possibility of NaCl.2$PbCl_2$	Comp.		

NaCl

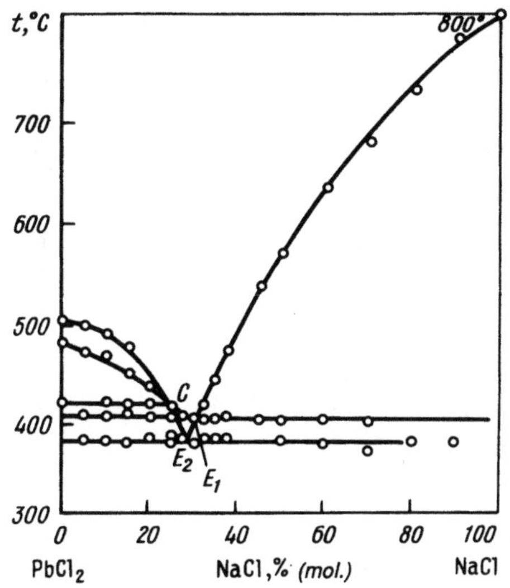

Fig. 223. $PbCl_2$-NaCl system.

2nd Component	T°C	Conc. mol %	Phases involved	Def.	Method	Ref.
$PbCl_2$	404	70	NaCl, $PbCl_2$	Eut.	2c	1245
	408	72	NaCl, $PbCl_2$	Eut.	2, 1e	1246
	800	0.0	NaCl	M.P.	2a, 3b,	1247
	406±2	68.9±0.7	NaCl, $PbCl_2$	Eut.	6k	
	497	100	$PbCl_2$	M.P.		
$PrCl_3$	480	41	NaCl, $PrCl_3$	Eut.	2a	1248
	468±2	37	NaCl, $PrCl_2$	Eut.	2c	1249
$PuCl_3$	453	36	NaCl, $PuCl_3$	Eut.	2c	1250
RbCl	542	59	NaCl, RbCl	Eut.	1,2	1251
	558	55	NaCl, RbCl	Eut.	2a	1252
			Solid slns. persist		6m, 3b	1253

NaCl

2nd Component	T°C	Conc. mol %	Phases involved	Def.	Method	Ref.
$RbCl$	800	0.0	$NaCl$	M.P.	1,2,2a	1254
	558	50	$NaCl$, $RbCl$	Eut.	3b	1255
	726	100	$RbCl$	M.P.		1256
						1257
	800	0.0	$NaCl$	M.P.	1,2	1258
	563	54	$NaCl$, $RbCl$	Eut.		
	726	100	$RbCl$	M.P.		
	800	0.0	$NaCl$	M.P.	1,2,2a	1259
	545	56.5	$NaCl$, $RbCl$	Eut.		
	716	100	$RbCl$	M.P.		
$SbCl_5$	800	0.0	$NaCl$	M.P.	2d	1260
	370	27	$NaCl$,$NaSbCl_6$	Per.		
	235	-	α-$NaSbCl_6$, β-$NaSbCl_6$			
	~3	~100	α-$NaSbCl_6$,$SbCl_5$	Eut.		
	3	100	$SbCl_5$	M.P.		
	310	80-98	Dissociation	-		
$ScCl_3$	566	28	$NaCl$,Na_3ScCl_6	Per	2,4a	1261
	480	38	Na_3ScCl_6, $ScCl_3$	Eut		1262
$SmCl_3$	425	40	$NaCl$, Na_2SmCl_5	Per.	12a	1263
	390	51	Na_2SmCl_5	Eut.		1264
$SnCl_2$	186	72	$NaCl$, $SnCl_2$	Eut.	2c	1265
$SrCl_2$	547	53	$NaCl$, $SrCl_2$	Eut.	1,2	1266
	556	52	$NaCl$, $SrCl_2$	Eut.	1,2	1267
	544	53	$NaCl$, $SrCl_2$	Eut.	1,2,2d	1268
			Limited solid slns.		3b	1269
$TaCl_3$	220	45	$NaCl$, $TaCl_3$	Eut.	2a	1270

NaCl

2nd Component	T°C	Conc. mol %	Phases involved	Def.	Method	Ref.
$TaCl_4$			Fig. 224			1271
	270	45		Eut.	2a	1271

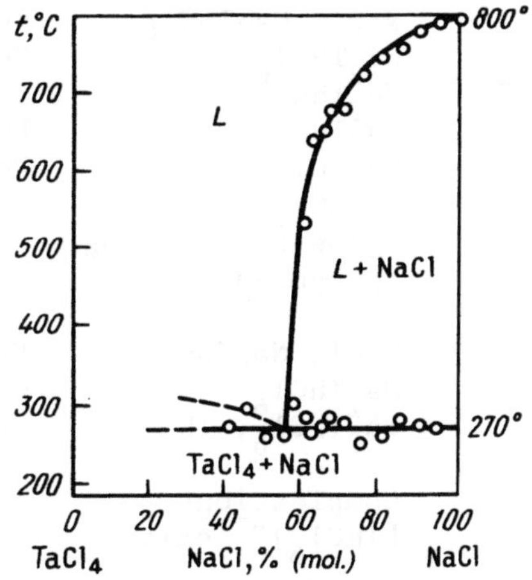

Fig. 224. $TaCl_4$-NaCl system.

2nd Component	T°C	Conc. mol %	Phases involved	Def.	Method	Ref.
$TaCl_5$			Fig. 225			1272
	470		NaCl, β-$NaTaCl_6$	Per.	2,6c	1272
	232		β-$NaTaCl_6$,α-$NaTaCl_6$	Per.		
	216	~100	α-$NaTaCl_6$,$TaCl_5$	Eut.		

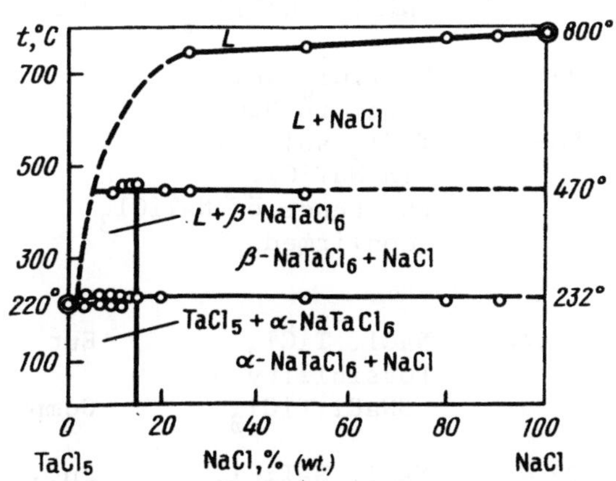

Fig. 225. $TaCl_5$-NaCl system.

NaCl

2nd Component	T°C	Conc. mol %	Phases involved	Def.	Method	Ref.
$TeCl_4$	214	95.5	NaCl, $TeCl_4$	Eut.	2d	1273
$ThCl_4$	346	29.7	NaCl, Na_2ThCl_6	Eut.	2	1274
	360	33	Na_2ThCl_6	Comp.		
	349	35.7	Na_2ThCl_6, $NaThCl_5$	Eut.		
	370	50	$NaThCl_5$, $ThCl_4$	Per.		
			$NeThCl_7$ - incongruently melting proposed			
	366	28	NaCl, Na_2ThCl_6	Eut.	2	1275
	435	-	Na_2ThCl_6	Comp.		
	397	44	Na_2ThCl_6, $ThCl_4$	Eut.		
			Complex ions $[ThCl_6]^{2-}$ exist		6f,6g	1276
$TiCl_2$	605	33.3	NaCl, $NaCl.TiCl_2$	Eut.	2	1277
	628	35	$NaCl.TiCl_2$, $TiCl_2$	Per.		
	548		$2NaCl.TiCl_2$	Comp.		
	625	34	10% $TiCl_2$ sol. sln. in NaCl, $NaTiCl_3$	Eut.	2	1278
	650	40	$NaTiCl_3$, $TiCl_2$	Per.		
	588		Na_2TiCl_4 ⇌ NaCl + $NaTiCl_3$	P.T.		
		0-4	Na_2TiCl_4 sol. sln. in NaCl		3b	1279
		50-60	$TiCl_2$ sol. sln. in $NaTiCl_3$			
			Existence of $NaTiCl_3$ confirmed			
$TiCl_3$	540	22.44	NaCl, $TiCl_3$	Eut.	2c	1280
			Possibility of $3NaCl.TiCl_3$	Comp.		1281
	549	22	NaCl, $NaTiCl_4$ (probably)	Eut.	2	1282

NaCl

2nd Component	T°C	Conc. mol %	Phases involved	Def.	Method	Ref.
$TiCl_3$	553	25	NaCl, Na_3TiCl_6	Per.	2d,3b,	1283
	462	40	Na_3TiCl_6, $TiCl_3$	Eut.	1c	
	550	23	NaCl,$TiCl_3$	Eut.	2c	1284
TlCl	412	94	NaCl, TlCl	Eut.	1,2	1285
	412	18	-	Eut.		1286
UCl_3	520	32	NaCl, UCl_3	Eut.		1286
	508±2	32	NaCl, UCl_3	2d,3b		1287
UCl_4	429	35	Na_2UCl_6, NaCl	Per.	2c	1288
	367	50	Na_2UCl_6, UCl_4	Eut.		
	435.4	30	NaCl, Na_2UCl_6	Eut.	2c,3b	1289
	445.5	33	Na_2UCl_6	Comp.		
	377.9	48	Na_2UCl_6, UCl_4	Eut.		
	-	-	NaCl, 2NaCl.UCl_4	Per.	2d	1290
	358±2	50	2NaCl.UCl_4, UCl_4	Eut.		
VCl_2	666	29.6	NaCl, VCl_2	Eut.	2d,3b	1291
	-	-	Max. solubility of VCl_2 in NaCl is 5-7%.		3b	1292
VCl_3	535±3	25	NaCl, 3NaCl.VCl_3	Per.	2c	1293
	425±5	29±1	3NaCl.VCl_3, Phase x	Eut.		
	555	30	Na_3VCl_6, NaCl	Per.	2c	1294
	508	37	Na_3VCl_6, phase of unknown composition	Eut.		
			Solid slns below eutectic.			
	800	0.0	NaCl	M.P.	2b	1295
	550	30.0	Na_3VCl_6, NaCl	Per.		
	360	55	VCl_3, Na_3VCl_6	Eut.		
	700	100	VCl_3	M.P.		

NaCl

2nd Component	T°C	Conc. mol %	Phases involved	Def.	Method	Ref.
VCl_3	552	25	Na_3VCl_6, NaCl	Per.	2	1296
	425	45	VCl_3, Na_3VCl_6	Eut.		
	700	100	VCl_3	M.P.		
WCl_5			Fig. 226			1297
	502 ±2	52	$NaWCl_6$, NaCl	Per.	2a	1297
	240	98	WCl_5, $NaWCl_6$	Eut.		
			Decomposition in the region 58-98% WCl_5			
			Line GH - P.T. of $NaWCl_6$			
			Lines AB and MN - sol.slns.			

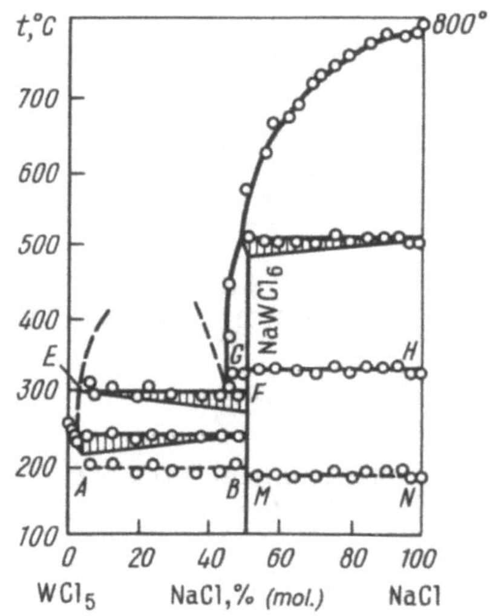

Fig. 226. WCl_5-NaCl system.

2nd Component	T°C	Conc. mol %	Phases involved	Def.	Method	Ref.
WCl_6			Fig. 227			1298
			Due to high vapor pressure of WCl_6 liquidus not obtained		2,6c	1298
	174	100	$\alpha \rightleftharpoons \beta$ WCl_6	P.T.		
	222	100	$\beta \rightleftharpoons \gamma$ WCl_6	P.T.		
			Complete insolubility of components. Two lines on the diagram correspond to melting and P.T. of WCl_6		2a,6c	1299 1300

NaCl

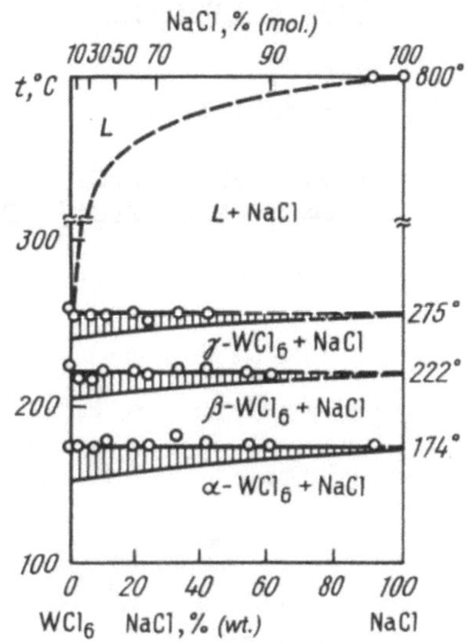

Fig. 227. WCl_6–NaCl system.

2nd Component	T°C	Conc. mol %	Phases involved	Def.	Method	Ref.
$YbCl_3$	550	30.0	Na_3YbCl_6, NaCl	Per.	2a,3b	1301
	454	60.0	$YbCl_3$, Na_3YbCl_6	Eut.		
$ZnCl_2$	400	38	$ZnCl_2.2NaCl$, NaCl	Per.	2a,3b,	1302
	262	66	$ZnCl_2$, $ZnCl_2.2NaCl$	Eut.	1c	
	410	38	$ZnCl_2.2NaCl$, NaCl	Per.	1,2	1303
	260	58	$ZnCl_2$, $ZnCl_2.2NaCl$	Eut.		
$ZrCl_2$	695±1	69	NaCl, $ZrCl_2$	Eut.	2c,2a, 3b	1304
$ZrCl_4$			Fig. 228			1305
	539	21	γ-$2NaCl.ZrCl_4$, NaCl	Eut.	2	1305
	695	34	γ-$2NaCl.ZrCl_4$	Comp.		
	377	42	β-$2NaCl.ZrCl_4$, γ-$2NaCl.ZrCl_4$	Per.		
	341	53	α-$2NaCl.ZrCl_4$, β-$2NaCl.ZrCl_4$	Per.		

NaCl

2nd Component	T°C	Conc. mol %	Phases involved	Def.	Method	Ref.
$ZrCl_4$	311	65	$ZrCl_4$, α-$2NaCl.ZrCl_4$	Eut.		1305
	437	100	$ZrCl_4$	M.P.		
	370		γ ⇌ β $2NaCl.ZrCl_4$	P.T.		
	340		β ⇌ α $2NaCl.ZrCl_4$	P.T.		

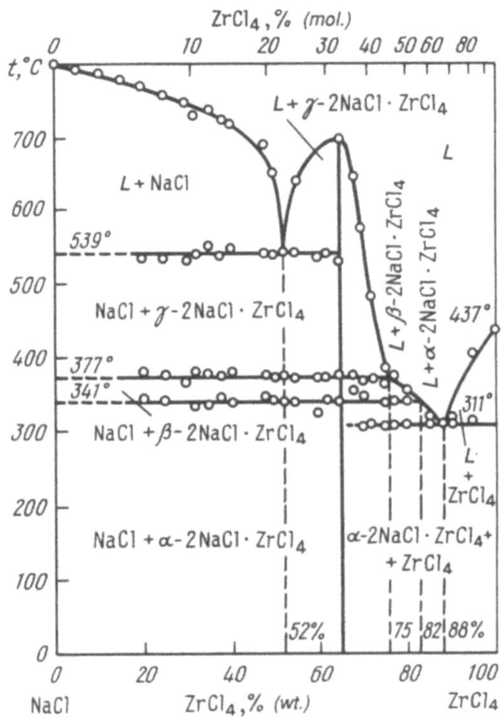

Fig. 228. $NaCl-ZrCl_4$ system.

2nd Component	T°C	Conc. mol %	Phases involved	Def.	Method	Ref.
$ZrCl_4$			Fig. 229			1306
	548	28.0	$2NaCl.ZrCl_4$, Phase X (B)	Eut.	1,2	1306
	648	33.3	$2NaCl.ZrCl_4$ (C)	Comp.	2c	
	381	52.0	Phase Y, $2NaCl.ZrCl_4$ (D)	Per.		
	314	62.4	$NaCl.2ZrCl_4$, Phase Y (E)	Eut.		
	317	65.0	$ZrCl_4$, $NaCl.2ZrCl_4$ (F)	Per.		

NaCl

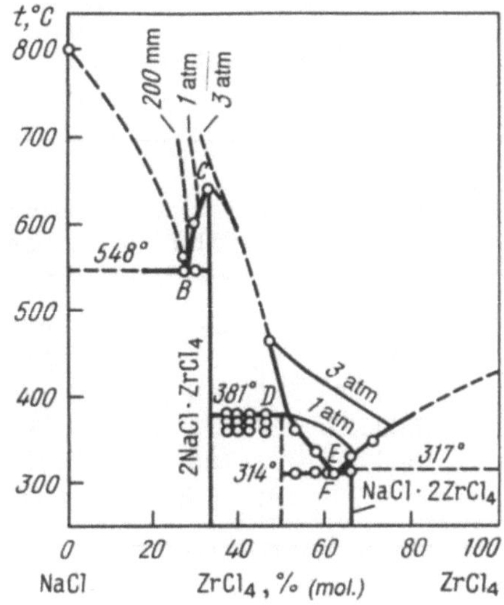

Fig. 229. $NaCl\text{-}ZrCl_4$ system.

2nd Component	T°C	Conc. mol %	Phases involved	Def.	Method	Ref.
$ZrCl_4$	548	20.5	NaCl, Na_2ZrCl_6	Eut.	3b,2c,2i	1307
NaF	681	34	NaF, NaCl	Eut.	2	1308
	800	0.0	NaCl	M.P.	2,1e	1309
	682	35	NaF, NaCl	Eut.		
	996	100	NaF			
	673	34.5	NaCl, NaF	Eut.	1,2	1310
	680	33.5	NaCl, NaF	Eut.	2d	1311
	680	33.3		Eut.	2a	1312
			Table XIV			
	675	-	-	Eut.	2	1313
NaI	587	62.5	NaCl, NaI	Eut.	1,1,2	1314
			Table XV		2a	1315
	570			Eut.	2	1316

NaCl

TABLE XIV

%NaF	t°C Liquidus	t°C Solidus	%NaF	t°C Liquidus	t°C Solidus
0.0	801	-	34.2	694.3	680.7
15	751.6	680.1	35	683.2	680.6
25	712.3	680.2	40	720	680.5
31.8	685.7	681.1	55	804	681.5
33	680.5	681.4	65	858.5	679.1
33.3	680.4	-	100	990	-

TABLE XV

%NaI	t°C Liquidus	t°C Solidus	%NaI	t°C Liquidus	t°C Solidus
0	801.0	-	60	576.2	574.0
15	755.1	572.4	61.4	572.9	-
30	706.6	573.2	63.1	576.6	573.0
40	671.1	573.2	65	581.1	572.9
50	634.0	574.4	69.9	594.7	573.1
52.5	615.1	572.5	75	605.1	572.5
55	602.2	573.1	80	519.0	572.6
57.5	589.0	572.9	90	638.3	572.3
			100	659.3	-

NaF

2nd Component	T°C	Conc. mol %	Phases involved	Def.	Method	Ref.
NdF_3			Fig. 230			1317
	732	26	NaF, $NaF.NdF_3$	Eut.	2,1f,3b	1318
	850	42	$NaF.NdF_3$, $5NaF.9NdF_3$	Per.		
	1056	58	$5NaF.9NdF_3$, NdF_3	Per.		
			Two types of $NaF.NdF_3$: (a) Fluorite cubic (high temp. exists with $5NaF.9NdF_3$ in sol.slns. dissociating at lower temp.) (b) Hexagonal			
	785	40	$5NaF.9NdF_3$ (sol.sln.) $\rightleftharpoons$ $NaF.NdF_3$ + NdF_3	P.T.		
	800		$\alpha \rightleftharpoons \beta$ NdF_3	P.T.		

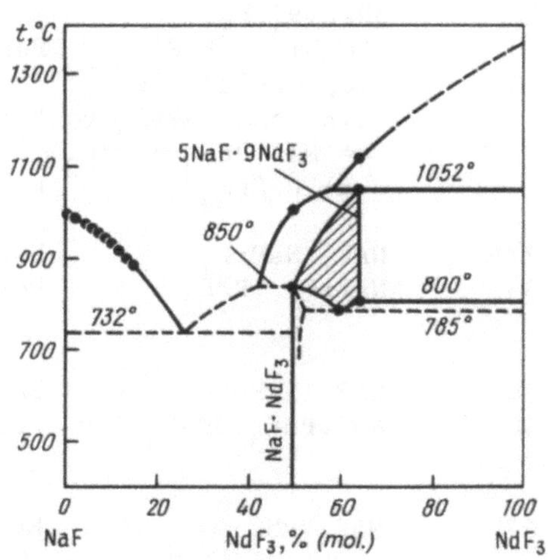

Fig. 230. $NaF-NdF_3$ system.

2nd Component	T°C	Conc. mol %	Phases involved	Def.	Method	Ref.
PbF_2	520	66	NaF, PbF_2	Eut.	2,3b	1319
PrF_3			Fig. 231			1320
	733	27	NaF,$NaF.PrF_3$ (hexagonal)	Eut.	2,3b,5 le	1321

NaF

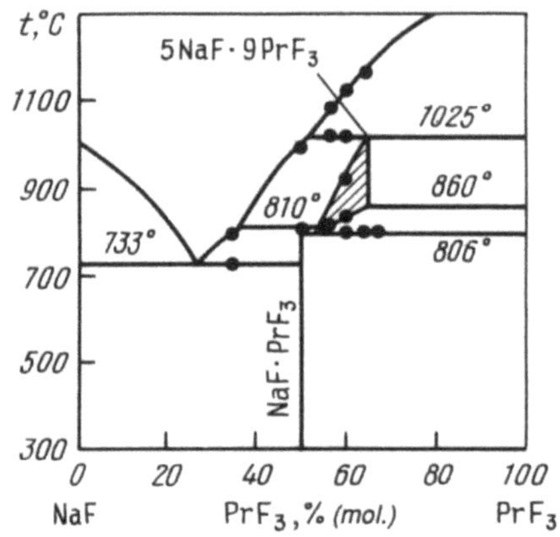

Fig. 231. $NaF-PrF_3$ system.

2nd Component	T°C	Conc. mol %	Phases involved	Def.	Method	Ref.
PrF_3	810	37	$NaF.PrF_3$ (hexagonal), $5NaF.9PrF_3$ (cubic)	Per.	2,3b,5 1e	1320 1321
	1025	52	$5NaF.9PrF_3$ (cubic), PrF_3	Per.		
	806	40	(Eutectoid) - sol. sln. $5NaF.9PrF_3$ ⇌ ⇌ $NaF.PrF_3$ + PrF_3	P.T.		
	860		α ⇌ β PrF_3	P.T.		
	730	27.5	NaF, $NaPrF_4$	Eut.	1,2,2d	1322
	800	35	$NaPrF_4$, PrF_3	Per.		
PuF_3	727±3	24	NaF, $NaF.PuF_3$	Eut.	2,1d	1323
	824	~40	$NaF.PuF_3$, PuF_3	Per.		
RbF	644	67	NaF, RbF	Eut.	1,2	1324
	680	50	NaF, RbF	Eut.	1,2	1325
	662	67	NaF, RbF	Eut.	1,2	1326
	667	67.2	NaF, RbF	Eut.	2,2d	1327

NaF

2nd Component	T°C	Conc. mol %	Phases involved	Def.	Method	Ref.
ScF_3	790	17	NaF, Na_3ScF_6	Eut.	1,2	1328
	890	25	Na_3ScF_6	Comp.		
	796	33	Na_3ScF_6, ScF_3	Eut.		
	825	26	NaF, Na_3ScF_6	Per.	2d	1329
	794	28	Na_3ScF_6, Na_2ScF_5	Eut.		
	888	33	Na_2ScF_5	Comp.		
	674	42	Na_2ScF_5, $NaScF_4$ (or supposedly ScF_3)	Eut.		
	660		Na_3ScF_6	P.T.		
	820	18	NaF, Na_3ScF_6	Eut.	1,2	1330
	894	25	Na_3ScF_6	Comp.		
	700	37.5	Na_3ScF_6, ScF_3	Eut.		
	820	18.0	NaF, Na_3ScF_6	Eut.	1,2,2a 3b	1331
	890	25.0	Na_3ScF_6	Comp.		
	680	37.5	Na_3ScF_6, ScF_3	Eut.		
	640	-	$NaScF_4$	P.T.		
	800	17	NaF, α-3NaF.ScF_3	Eut.	2,3b	1332
	885	25	α-3NaF.ScF_3	Comp.		
	680	35	α-3NaF.ScF_3, β-3NaF.ScF_3	Per.		
	650	38	β-3NaF.ScF_3, NaF.ScF_3	Eut.		
	660	42	NaF.ScF_3, ScF_3	Per.		
	1350	-	ScF_3	P.T.		
SmF_3			Fig. 232			1333
	725	25	NaF, NaF.SmF_3 (hexagonal)	Eut.	2,3b,5 le	1333
	834	38	NaF.SmF_3 (hexagonal), 5NaF.9SmF_3 (cubic)	Per.		
	1060	62	5NaF.9SmF_3 (cubic), SmF_3	Per.		
	765	40	(Eutectoid) Sol.sln. 5NaF.9SmF_3 ⇌ ⇌ NaF.SmF_3 + SmF_3	P.T.		
	733		α ⇌ β SmF_3			

NaF

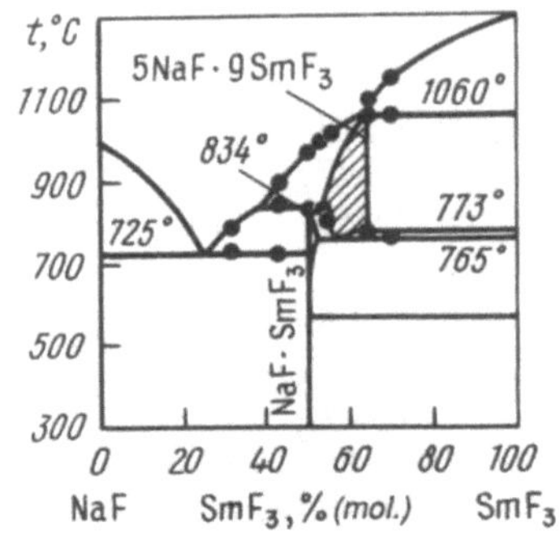

Fig. 232. NaF-SmF_3 system.

2nd Component	T°C	Conc. mol %	Phases involved	Def.	Method	Ref.
SnF_2	279		$6NaF.SnF_2$	Comp.	2c,3b	1334
	273		$2NaF.SnF_2$	Comp.		
	265		$NaF.SnF_2$	Comp.		
	299		$NaF.2SnF_2$	Comp.		
SrF_2	859	32	NaF, SrF_2	Eut.	2	1335
TbF_3	695	28	NaF, $NaF.TbF_3$	Eut.	2,1f,3b	1336
	790	39	$NaF.TbF_3$, $5NaF.9TbF_3$	Per.		
	1050	66	$5NaF.9TbF_3$	Comp.		
	960	71	$5NaF.9TbF_3$, TbF_3	Eut.		
			$NaF.Tb_3$ two forms: cubic (fluorite) at high temp. and hexagonal. Cubic forms sol.sln. with $5NaF.9TbF_3$			
			Fig. 233			1337
	696	27	NaF, $NaF.TbF_3$	Eut.	2,3b,5 le	1337
	789	39	$NaF.TbF_3$, $5NaF.9TbF_3$	Per.		
	1031	64.3	$5NaF.9TbF_3$	Comp.		
	994	68	$5NaF.9TbF_3$, TbF_3	Eut.		

NaF

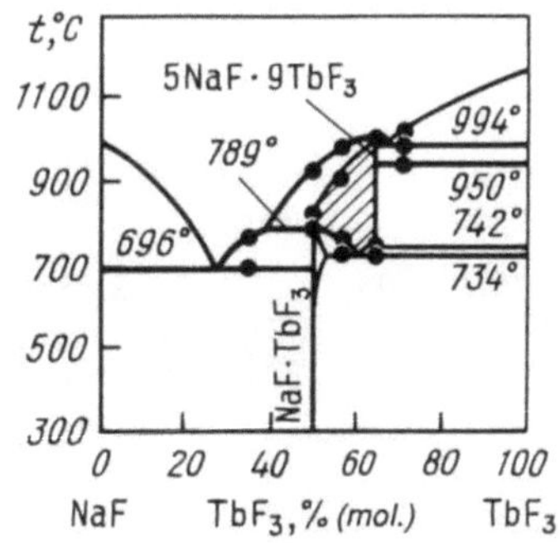

Fig. 233. NaF–TbF_3 system.

2nd Component	T°C	Conc. mol %	Phases involved	Def.	Method	Ref.
TbF_3	734	40	(Eutectoid) Sol. sln. $5NaF.9TbF_3 \rightleftharpoons$ $\rightleftharpoons NaF.TbF_3 + TbF_3$	P.T.	2,3b,5 le	1337
	742		$\alpha \rightleftharpoons \beta$ TbF_3	P.T.		
	450		$\beta \rightleftharpoons \gamma$ TbF_3	P.T.		
ThF_4			Fig. 234			1338
	648	22	NaF, Na_4ThF_8	Per.	2a,3b	1338
	620	24	Na_4ThF_8, α-Na_2ThF_6	Eut.		
	708	33.3	α-Na_2ThF_6	Comp.		
	700	36-37	α-Na_2ThF_6, $NaThF_5$	Eut.		
	760	45	$NaThF_5$	Comp.		
	750	53.5	$NaThF_5$,$NaTh_2F_9$	Eut.		
	830	62	$NaTh_2F_9$, ThF_4	Per.		
	606		$Na_4ThF_8 \rightleftharpoons$ Na_2ThF_6 + 2NaF	P.T.		
	568		$\alpha \rightleftharpoons \beta$ Na_2ThF_6	P.T.		
	645	21.5	NaF, α-$4NaF.ThF_4$	Per.	2b,2c 3b	1339
	618	22.5	α-$4NaF.ThF_4$, $2NaF.ThF_4$	Eut.		
	705	33.3	$2NaF.ThF_4$	Comp.		
	690	37.0	$2NaF.ThF_4$, $3NaF.2ThF_4$	Eut.		
	712	40.0	$3NaF.2ThF_4$	Comp.		
	705	41.0	$3NaF.2ThF_4$, α-$NaF.ThF_4$	Eut.		
	730	45.5	α-$NaF.ThF_4$, $NaF.2ThF_4$	Per.		

NaF

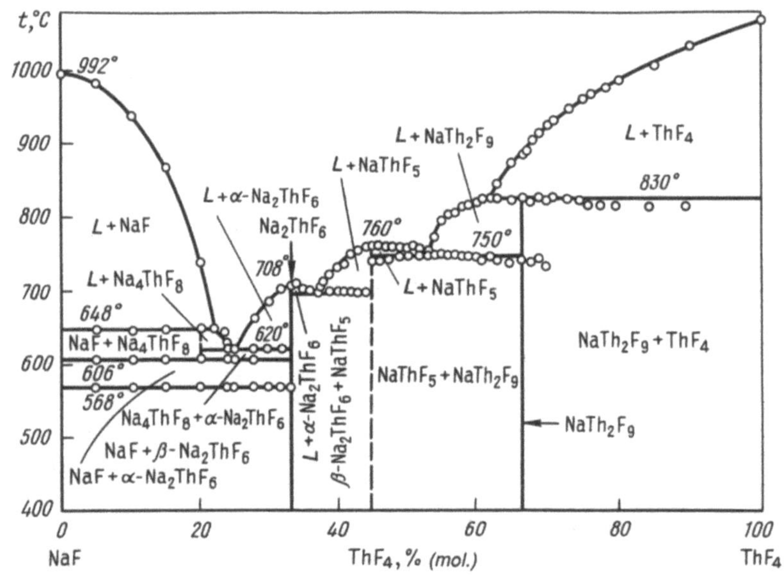

Fig. 234. NaF-ThF_4 system.

2nd Component	T°C	Conc. mol %	Phases involved	Def.	Method	Ref.
ThF_4	831	58.0	NaF.2ThF_4, 12% NaF sol.sln. in ThF_4	Per.	2b,2c, 3b	1339
	604		α ⇌ β NaF.ThF_4	P.T.		
	558		β-4NaF.ThF_4 ⇌ NaF + 2 NaF.ThF_4	P.T.		
	683		3NaF.2ThF_4 ⇌ 2NaF.ThF_4 + α-NaF.ThF_4	P.T.		
			Fig. 235			1339
	604-627		4NaF.ThF_4	Comp.	2,1f,3b	1340
	610		7NaF.2ThF_4	Comp.		1341
TiF_4			Fig. 236			1342
	990	0	NaF	M.P.	2a,2d,1, 2	
	618	28	NaF, Na_2TiF_4	Eut.		
	804	35	Na_2TiF_4	Comp.		
	320	51	Na_2TiF_4, TiF_4	Eut.		
	430	100	TiF_4	M.P.		
	560-600		Na_2TiF_6	P.T.		

NaF

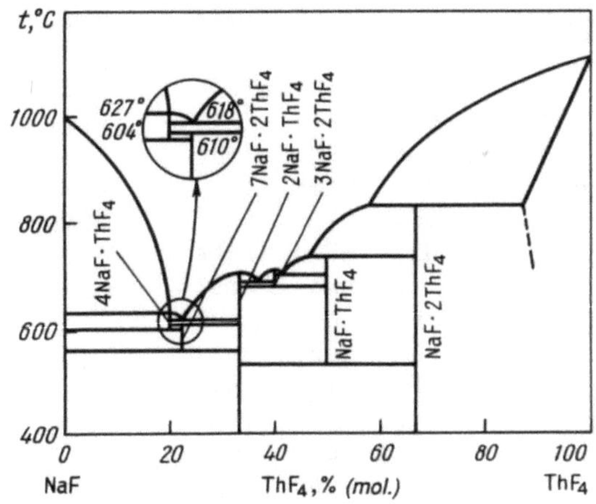

Fig. 235. NaF-ThF_4 system.

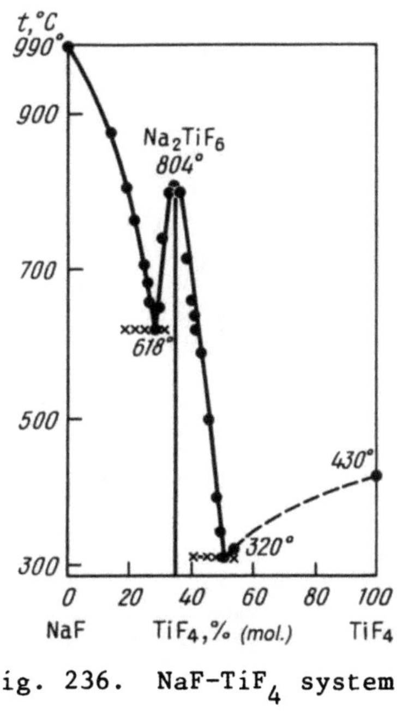

Fig. 236. NaF-TiF_4 system.

NaF

2nd Component	T°C	Conc. mol %	Phases involved	Def.	Method	Ref.
TmF_3	620	29	NaF, $NaF.TmF_3$	Eut.	2,1f,3b	1343
	650	31	$NaF.TmF_3$, $5NaF.9TmF_3$	Per.		
	950	64	$5NaF.9TmF_3$	Comp.		
	900	70	$5NaF.9TmF_3$	Eut.		
			$NaF.TmF_3$ exists in two forms: high temp. cubic (fluorite) and hexagonal. $5NaF.9TmF_3$ exists in two forms: high temp. cubic (fluorite) and low temp. orthorhombic. The cubic phases form mutual sol.sln., dissociating at lower temp.			
			Fig. 237			1344
	592	27	NaF, $NaF.TmF_3$	Eut.	2,3b,5, 1e	1344
	632	29	$NaF.TmF_3$, $5NaF.9TmF_3$	Per.		
	910	64.3	$5NaF.9TmF_3$	Comp.		
	905	~70	$5NaF.9TmF_3$, TmF_3	Eut.		
			$NaF.TmF_3$ and $5NaF.9TmF_3$ sol. sln. dissociate below 595.			

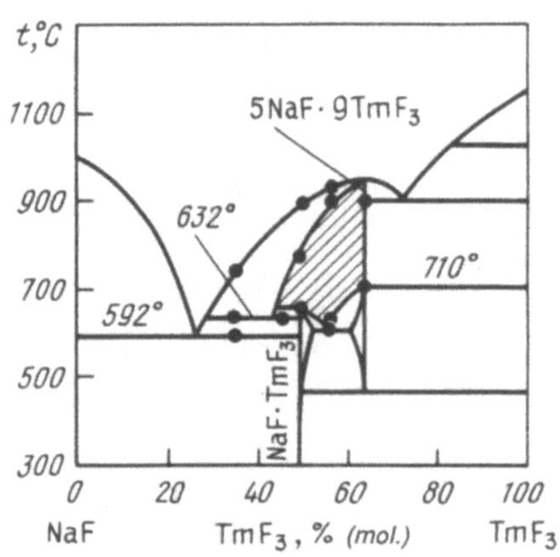

Fig. 237. NaF-TmF_3 system.

NaF

2nd Component	T°C	Conc. mol %	Phases involved	Def.	Method	Ref.
UF_3		0-33.3	$NaF + Na_2UF_5$		3b	1345
		33-100	$UF_3 + Na_2UF_5$			
			Fig. 238			1346
	715	27	$NaF, NaF.UF_3$	Eut.		
	770	37	$NaF.UF_3, UF_3$	Per.		

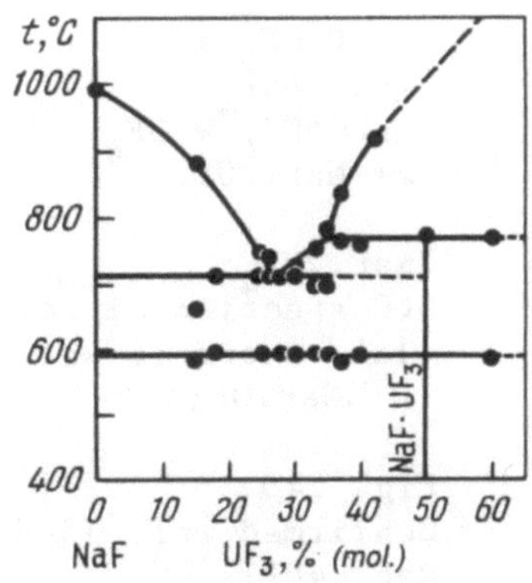

Fig. 238. $NaF-UF_3$ system.

2nd Component	T°C	Conc. mol %	Phases involved	Def.	Method	Ref.
UF_4			Fig. 239			1347
	618	21.5	$NaF, 3NaF.UF_4$	Eut.	2,2d,3b	1347
	629	25	$3NaF.UF_4$	Comp.	1	
	623	28	$3NaF.UF_4, 2NaF.UF_4$	Eut.		
	648	32.5	$2NaF.UF_4, 5NaF.3UF_4$	Per.		

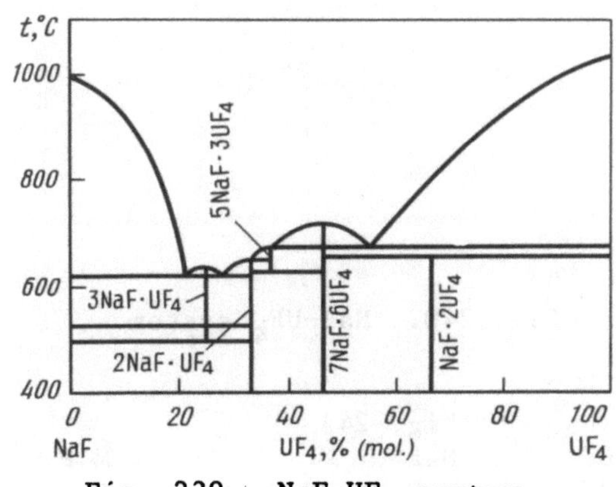

Fig. 239. $NaF-UF_4$ system.

NaF

2nd Component	T°C	Conc. mol %	Phases involved	Def.	Method	Ref.
UF_4	673	37	$5NaF.3UF_4$, $7NaF.6UF_4$	Per.		1347
	718	46.2	$7NaF.6UF_4$	Comp.		
	680	56	$7NaF.6UF_4$, UF_4	Eut.		
	497		$3NaF.UF_4 \rightleftharpoons NaF + 2NaF.UF_4$	P.T.		
	528		$\alpha \rightleftharpoons \beta$ $3NaF.UF_4$	P.T.		
	630		$5NaF.3UF_4 \rightleftharpoons 2NaF.UF_4 + 7NaF.6UF_4$	P.T.		
	660		$7NaF.6UF_4 + UF_4 \rightleftharpoons NaF.2UF_4$	P.T.		
		15-25	$7NaF.2UF_4$ (Correction of ref. 1347 where it was β-$3NaF.UF_4$)	P.T.	2,1f,3b	1348
			Fig. 240 Confirmed ref. 1348 $NaF_4.4UF_4$ - Additional solid phase.			1349

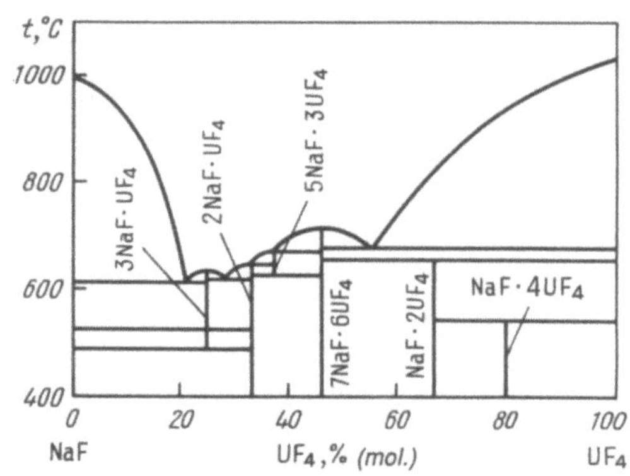

Fig. 240. NaF-UF_4 system.

2nd Component	T°C	Conc. mol %	Phases involved	Def.	Method	Ref.
VF_3			Fig. 241			1350
	995	0	NaF	M.P.	2d,3b	1350
	865±5	15	NaF, α-$3NaF.VF_3$	Eut.		
	1045±10	25	α-$3NaF.VF_3$	Comp.		

NaF

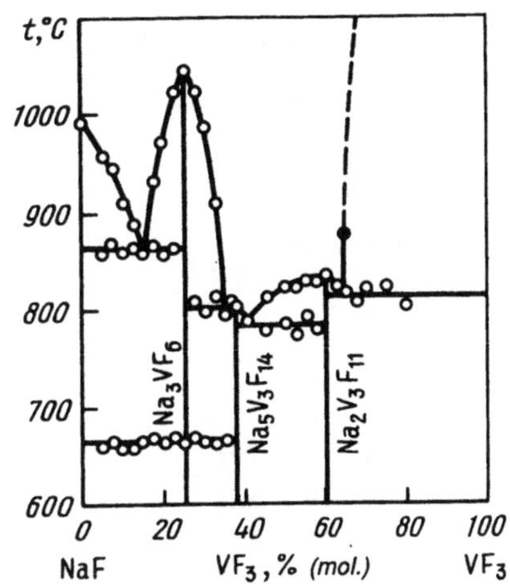

Fig. 241. NaF-VF_3 system.

2nd Component	T°C	Conc. mol %	Phases involved	Def.	Method	Ref.
VF_3	805±10	35	α-3NaF.VF_3, 5NaF.3VF_3	Eut.	2d,3b	1350
	810 ±10	37.5	5NaF.3VF_3	Comp.		
	785±10	40	5NaF.3VF_3, 2NaF.3VF_3	Eut.		
	830±10	60	2NaF.3VF_3	Comp.		
	815±10	64	2NaF.3VF_3, VF_3	Eut.		
	520		VF_3	P.T.		
	665		3NaF.VF_3	P.T.		
			Mass spectrometry revealed 3NaF.VF_3 base sol.sln.			
YF_3	638	28	NaF, NaF.YF_3	Eut.	2,1f,3b	1351
	710	32	NaF.YF_3,5NaF.9YF_3	Per.		
	980	65	5NaF.9YF_3	Comp.		
	940	75	5NaF.9YF_3	Eut.		
			NaF.YF_3 exists in two forms: fluorite cubic (high temp.) and hexagonal. The cubic and 5NaF.9YF_3 form sol.slns. dissociating at lower temp.			

NaF

2nd Component	T°C	Conc. mol %	Phases involved	Def.	Method	Ref.
YF_3	610	30	NaF, YF_3	Eut.	2a	1352
	554	50	NaF + YF_3 ⇌ $NaYF_4$	P.T.		
			Fig. 242			1353
	638	28	NaF, $NaF.YF_3$	Eut.	2,3b,5 1e	1353
	691	34	$NaF.YF_3$, $5NaF.9YF_3$	Per.		
	950	40	$5NaF.9YF_3$	Comp.		
	947	68	$5NaF.9YF_3$,α-YF_3	Eut.		
	1052	95	α-YF_3, β-YF_3	Per.		
	537		$NaF.YF_3$ and $5NaF.9YF_3$ sol.slns.dissociate.	P.T.		

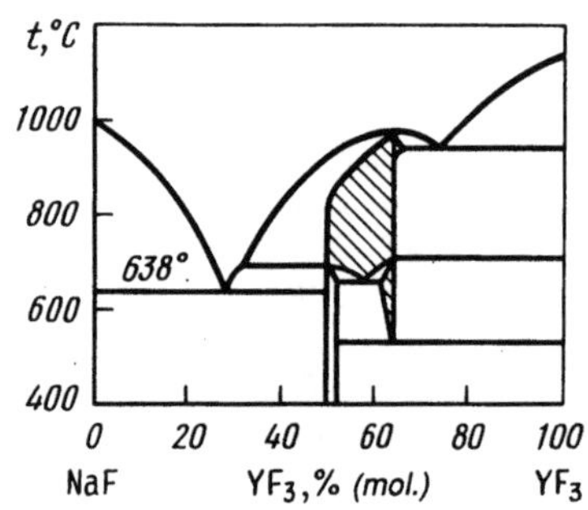

Fig. 242. NaF-YF_3 system.

2nd Component	T°C	Conc. mol %	Phases involved	Def.	Method	Ref.
YbF_3			Fig. 243			1354
	595	25	NaF, $NaF.YBF_3$	Eut.	2,1f,3b	1355
	-	-	$NaF.YbF_3$, $5NaF.9YbF_3$	Per.		
	928	63	$5NaF.9YbF_3$	Comp.		
	930	69	$5NaF.9YBF_3$,YbF_3	Eut.		
	950	75	α-YbF_3, β-YbF_3	Per.		
			$NaF.YbF_3$ exists in two forms: cubic (fluorite) - high temp. and hexagonal. $5NaF.9YbF_3$ exists in two forms: cubic (fluorite) - high temp. and orthorhombic - low temp. The cubic types form mutual sol.slns.			

NaF

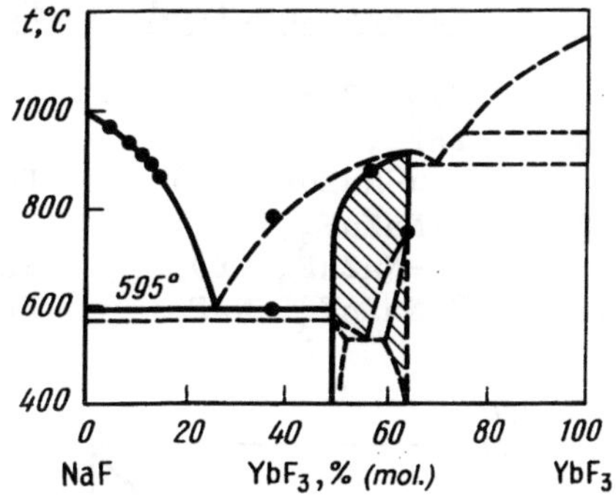

Fig. 243. NaF-YbF_3 system.

2nd Component	T°C	Conc. mol %	Phases involved	Def.	Method	Ref.
YbF_3			Fig. 244			1356
	600	27.5	NaF, $NaYbF_4$	Eut.	2a	1356
	930	42.5	$NaYbF_4$, YbF_3	Per.		
	574	25	3NaF.+ YbF_3 ⇌ Na_3YbF_6	P.T.		

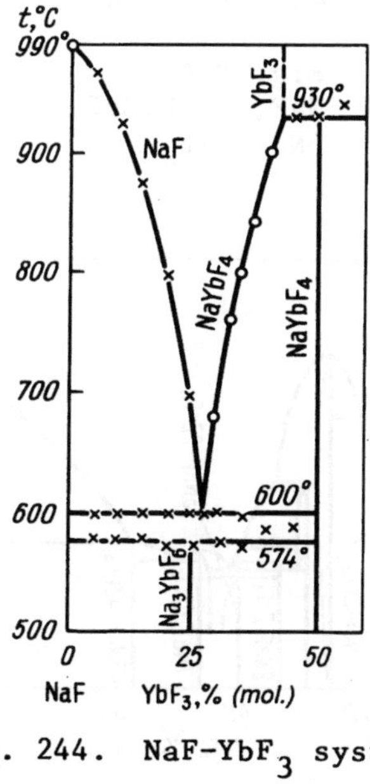

Fig. 244. NaF-YbF_3 system.

NaF

2nd Component	T°C	Conc. mol %	Phases involved	Def.	Method	Ref.
ZnF_2			Fig. 245			1357
	744	50.0	ZnF_2, α-$NaZnF_3$ (P)	Per.	2,3b	1357
	683	60.0	α-$NaZnF_3$, β-$NaZnF_3$ (B)	Per.		
	640	62.5	β-$NaZnF_3$, NaF	Eut.		

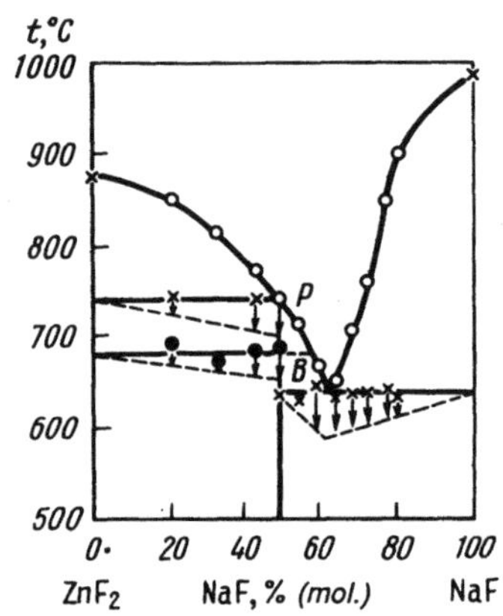

Fig. 245. ZnF_2-NaF system.

2nd Component	T°C	Conc. mol %	Phases involved	Def.	Method	Ref.
ZrF_4			Fig. 246			1358
	747	20	NaF, $3NaF.ZrF_4$	Eut.	2d,1,2	1358
	850	25	$3NaF.ZrF_4$	Comp.	3b	

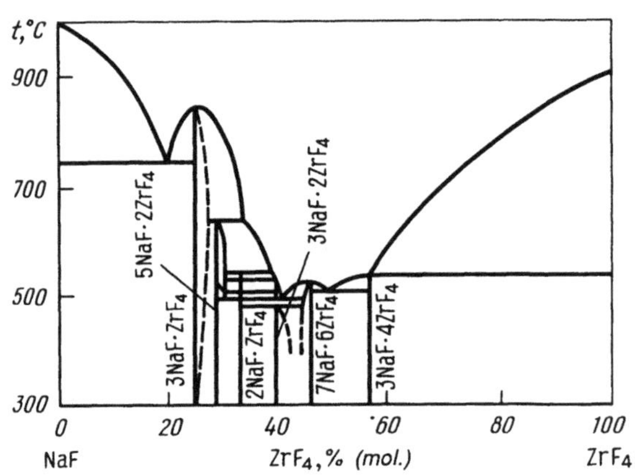

Fig. 246. $NaF-ZrF_4$ system.

NaF

2nd Component	T°C	Conc. mol %	Phases involved	Def.	Method	Ref.
ZrF_4	640	34	$3NaF.ZrF_4$, α-$5NaF.2ZrF_4$	Per.		
	544	39.5	α-$5NaF.2ZrF_4$, α-$2NaF.ZrF_4$	Per.		1358
	505	40.5	β-$2NaF.ZrF_4$, γ-$2NaF.ZrF_4$	Per.		
	500	40.5	γ-$2NaF.ZrF_4$, $7NaF.6ZrF_4$	Eut.		
	525	46.2	$7NaF.6ZrF_4$	Comp.		
	512	49.5	$7NaF.6ZrF_4$, $3NaF.4ZrF_4$	Eut.		
	537	56.5	$3NaF.4ZrF_4$, ZrF_4	Per.		
			$3NaF.ZrF_4$, $5NaF.2ZrF_4$ } Form sol.sln.			
			$7NaF.6ZrF_4$			
	523		$5NaF.2ZrF_4$	P.T.		
	533		$2NaF.ZrF_4$	P.T.		
	500		Eutectoid: α-$5NaF.2ZrF_4$ ⇌ β-$5NaF.2ZrF_4$ + γ-$2NaF.ZrF_4$	P.T.		
	487		γ-$2NaF.ZrF_4$ + $7NaF.6ZrF_4$ → $3NaF.2ZrF_4$	P.T.		
NaI			Table XVI		2a	1359
	603	-	-	Eut.	2	1360

TABLE XVI

%NaI	t°C Liquidus	t°C Solidus	Phases	%NaI	t°C Liquidus	t°C Solidus	Phases
0.0	990.0	-	NaF	81.5	597.1	-	NaI
45.0	855.8	597.4		85.0	609.5	597.3	
55.0	808.6	597.4		95.0	643.4	596.6	
65.0	742.8	597.3		100.0	659.3	-	
75.0	661.1	597.0					

NaI

2nd Component	T°C	Conc. mol %	Phases involved	Def.	Method	Ref.
NdI_3	654	0.0	NaI	M.P.	2d	1361
	433	28.0	$7NaI.NdI_3$	Per.		
	429	34	β-NdI_3, $7NaI.NdI_3$	Eut.		
	565	50	αNdI_3, β-NdI_3	Per.		
	780	100	αNdI_3	M.P.		
	385	-	-	P.T.		
PbI_2	378	71	NaI, PbI_2	Eut.	1,2	1362 1363
PrI_3	654	0.0	NaI	M.P.	2d	1364
	430	30	PrI_3, NaI	Eut.		
	730	100	PrI_3	M.P.		
RbI	475	50	Table XVII	Eut.	2a	1365

TABLE XVII

RbI%	Temperature arrests (t°C) 1	2	3	Phases
0	-	-	660	NaI
5	427	572	625	NaI base sol. sln.
10	450	-	610	NaI base sol. sln.
15	475	-	587	NaI base sol. sln.
30	480	-	532	NaI base sol. sln.
40	475	-	495	NaI base sol. sln.
50(E)	475	-	-	15% RbI sol.sln. in NaI, 20% NaI sol.sln. in RbI
60	475	-	580	RbI base sol. sln.
70	478	-	525	RbI base sol. sln.
80	475	-	570	RbI base sol. sln.
85	460	525	585	RbI base sol. sln.
90	440	548	590	RbI base sol. sln.
95	400	-	625	RbI base sol. sln.
100	-	-	655	RbI

NaI

2nd Component	T°C	Conc. mol %	Phases involved	Def.	Method	Ref.
RbI	512	52.5	Continuous sol.sln.	Min.	1,2	1366
			Solid sln.		3b,6h	1367
SmI_3	654	0	NaI	M.P.	2d	1368
	453	26	$3NaI.SmI_3$, NaI	Per.		
	443	34	SmI_3, $3NaI.SmI_3$	Eut.		
	670	52	α-SmI_3, β-SmI_3	Per.		
	335		$3NaI.SmI_3$	P.T.		
SnI_2	661	0.0	NaI	M.P.	2a,3b, 6c	1369
	290	78	NaI, SnI_2	Eut.		
	320	100	SnI_2	M.P.		
TeI_4	667	0.0	NaI	M.P.	2	1370
	264	80	NaI, TeI_4	Eut.		
	280	100	TeI_4	M.P.		
TlI	428	87.5	NaI, TlI	Eut.	1,2	1371

$NbCl_2$

2nd Component	T°C	Conc. mol %	Phases involved	Def.	Method	Ref.
RbCl	718	67	Rb_2NbCl_4	Comp.	2a,3b	1372
	648	82	Rb_2NbCl_4,$RbCl_2$	Eut.		

$NbCl_3$

2nd Component	T°C	Conc. mol %	Phases involved	Def.	Method	Ref.
RbCl	753	66	Rb_2NbCl_5	Comp.	2a,3b	1373
	605	85	Rb_2NbCl_5,RbCl	Eut.		
	718	100	RbCl	M.P.		

$NbCl_4$

2nd Component	T°C	Conc. mol %	Phases involved	Def.	Method	Ref.
RbCl		<50	Dissociation of $NbCl_4$		2b,2d	1374
	802		$RbNbCl_6$	Comp.		
	630	83	$RbNbCl_6$, $RbCl_4$	Eut.		

$NbCl_5$

2nd Component	T°C	Conc. mol %	Phases involved	Def.	Method	Ref.
$NiCl_2$			Temp.& composition of eutectic close to pure $NbCl_5$	Eut.	2d	1375
$ReCl_5$	204	10	13% $NbCl_5$ sol.sln. $ReCl_5$, 5% $ReCl_5$ sol. sln. in $NbCl_5$	Eut.	2a	1376
$SbCl_3$	73	~100	$SbCl_3$, $NbCl_5$	Eut.	1,2,2a	1377
$SeCl_4$			Fig. 247			1378
	182	68	$NbCl_3.SeCl_4$, $NbCl_5$	Eut.	2a	1378
	204	50	$SeCl_4$, $NbCl_5.SeCl_4$	Per.		

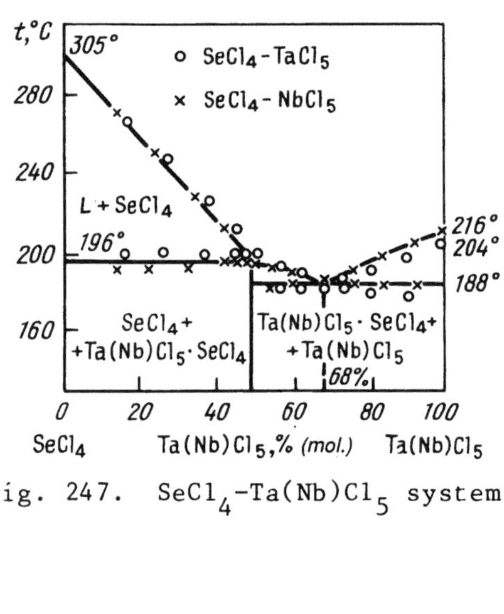

Fig. 247. $SeCl_4$-$Ta(Nb)Cl_5$ system.

$NbCl_5$

2nd Component	T°C	Conc. mol %	Phases involved	Def.	Method	Ref.
$SnCl_4$	-33	98.2	$SnCl_4$, $NbCl_5$	Eut.	2c	1379
$SrCl_2$	204	~100	$NbCl_5$,$SrNbCl_7$	Eut.	2a	1380
	208	~100	$SrNbCl_7$,$SrCl_2$	Per.		
$TaCl_5$			Continuous sol.sln.		22,6c	1381
$TeCl_4$			Fig. 248			1382
	168	50.0	$NbCl_5$, $NbCl_5.TeCl_4$	Per.	2a	1382
	152	62.0	$NbCl_5.TeCl_4$,$TeCl_4$	Eut.		

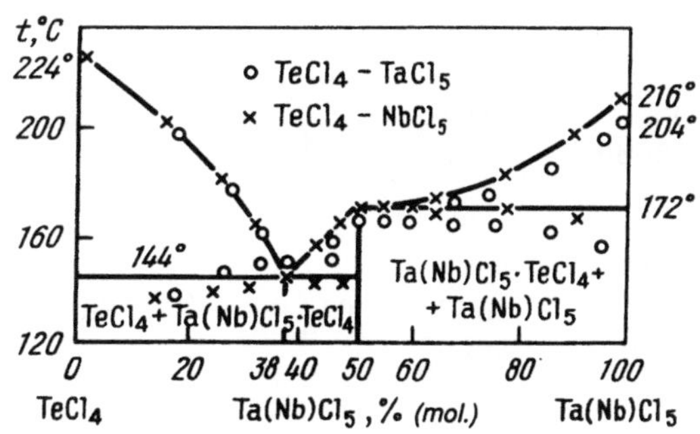

Fig. 248. $TeCl_4$-$Ta(Nb)Cl_5$ system.

2nd Component	T°C	Conc. mol %	Phases involved	Def.	Method	Ref.
	166	33.3	$NbCl_5$ $NbCl_5.TeCl_4$	Eut.	2a,3b	1383
	189	50.0	$NbCl_5.TeCl_4$	Comp.		
	144	68.0	$NbCl_5.TeCl_4$,$TeCl_4$	Eut.		
$ThCl_4$			Almost complete mutual insolubility		2,6c	1384
	203	0	$NbCl_5$,$ThCl_4$	Eut.		
$TiCl_4$	202	0	$NbCl_5$	M.P.	2,6n	1385
	-24	>99.5	$NbCl_5$,$ThCl_4$	Eut.		
	-24	100	$ThCl_4$	M.P.		
	-24.2	99.84	$NbCl_5$,$TiCl_4$	Eut.	1,2,2a	1386

$NbCl_5$

2nd Component	T°C	Conc. mol %	Phases involved	Def.	Method	Ref.
TlCl	204	~0	$NbCl_5$,$TlNbCl_6$	Eut.	1,2d	1387
	366	50	$TlNbCl_6$	Comp.		
	353	~26	TlCl,3TlCl.$NbCl_5$	Per.		
			$TlNbCl_6$ and 3TlCl.$NbCl_5$ complete mutual solubility			
	318	0-40	Dissociation			
	276		$TlNbCl_6$	P.T.		
VCl_4		~100	$NbCl_5$,VCl_4	Eut.	1,2,2a	1388
WCl_6	157	43.3	$NbCl_5$, α-WCl_6	Eut.	2a	1389
	174	47	α-WCl_6, β-WCl_6	Per.		
	222	70	β-WCl_6, γ-WCl_6	Per.		
$ZnCl_2$	204	~0	$NbCl_5$,$ZnCl_2$	Eut.	2d,1	1390
	266	18-88	Dissociation			
$ZrCl_4$	207	0.0	$NbCl_5$	M.P.	2	1391
	186	34	$NbCl_5$,$ZrCl_4$	Eut.		
	437	100	$ZrCl_4$			
NbF_5			Fig. 249			1392
	204	0	$NbCl_5$	M.P.	1,2,2i	1392
	81.6	100	NbF_5	M.P.		
			Beween the two extremes compounds of the type NbF_nCl_{5-n} - the most obvious $NbCl_4F$ and $NbClF_4$. Possibility of $NbCl_3F_2$ and $NbCl_2F_3$ suggested by other authors (1393, 1394, 1395). In the figure: 1 - liquid-vapour equilibrium at 760 mm Hg 2 - liquid-crystalline solid equilibrium.			

$NbCl_5$

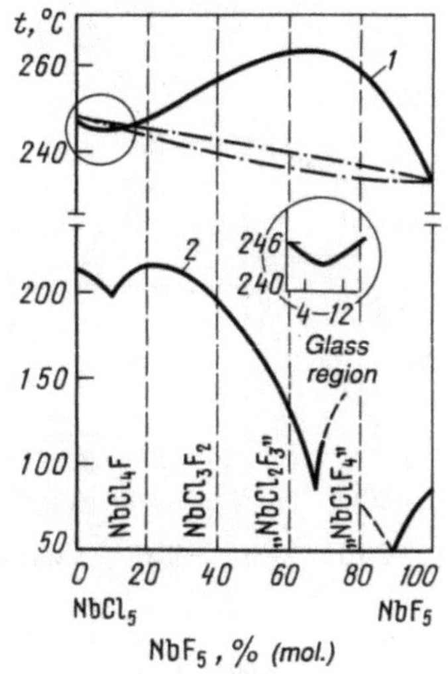

Fig. 249. $NbCl_5$-NbF_5 system.

2nd Component	T°C	conc. mol %	Phases involved	Def.	Method	Ref.
		60-80	Glass			
NbF_5	194.5	9.8	Significant points on on the liquidus line.			1393, 1394, 1395
	84.5	67.4				
	52.5	86.7				
	445.1	7	(~10 kPa) - Azeotrope on the liquid-vapor equil. line which significantly deviates from reality.			

NbF_5

TaF_5	80.2±0.2	0.0	NbF_5	M.P.	1,2	1396
			Continuous sol. sln.			
	97±0.2	100.0	TaF_5	M.P.		
XeF_2			Fig. 250			1397
	80	0.0	NbF_5	M.P.	2d	1397
	88	-	$XeF_2.2NbF_5$	Comp.		
	123-128		$2XeF_2.NbF_5$	Comp.		

NbF_5

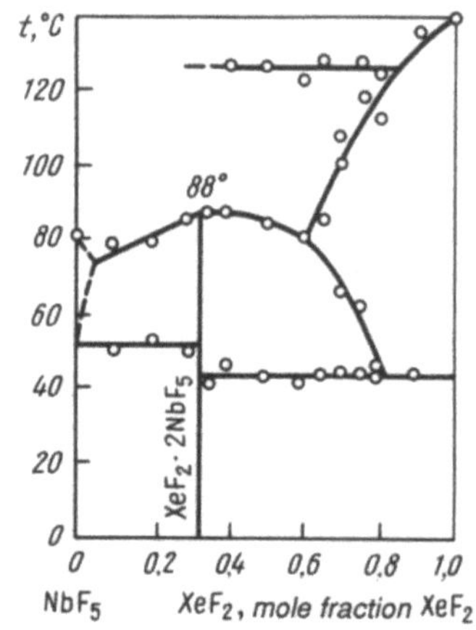

Fig. 250. NbF_5-XeF_2 system.

Nb_3I_8

2nd Component	T°C	Conc. mol %	Phases involved	Def.	Method	Ref.
SnI_4		>50	Complete insolubility in liquid state, some interaction in solid		2d	1398
	140	100	SnI_4	M.P.		

$NdCl_3$

2nd Component	T°C	Conc. mol %	Phases involved	Def.	Method	Ref.
RbCl			Fig. 251			1399
	584	39	$NdCl_3$,$RbNd_2Cl_7$	Per.	2	1399
	500	119.0	$RbNd_2Cl_7$,Rb_2NdCl_5	Eut.		
	570	51	Rb_2NdCl_5, β-Rb_3NdCl_6	Per.		
	773	74.5	β-Rb_3NdCl_6	Comp.		
	621	90.0	β-Rb_3NdCl_6,RbCl	Eut.		
	404		Rb_3NdCl_6	P.T.		
$SbCl_5$	~3	~100	$NdCl_3$	Eut.	2d	1400
	3	100	$SbCl_5$	M.P.		
$SrCl_2$	745	0	$NdCl_3$	M.P.	2a,3b	1401
	600	39	$NdCl_3$,Sr_3NdCl_9	Eut.		
	618	40	Sr_3NdCl_9,$SrCl_2$	Per.		
	868	100	$SrCl_2$	M.P.		

$NdCl_3$

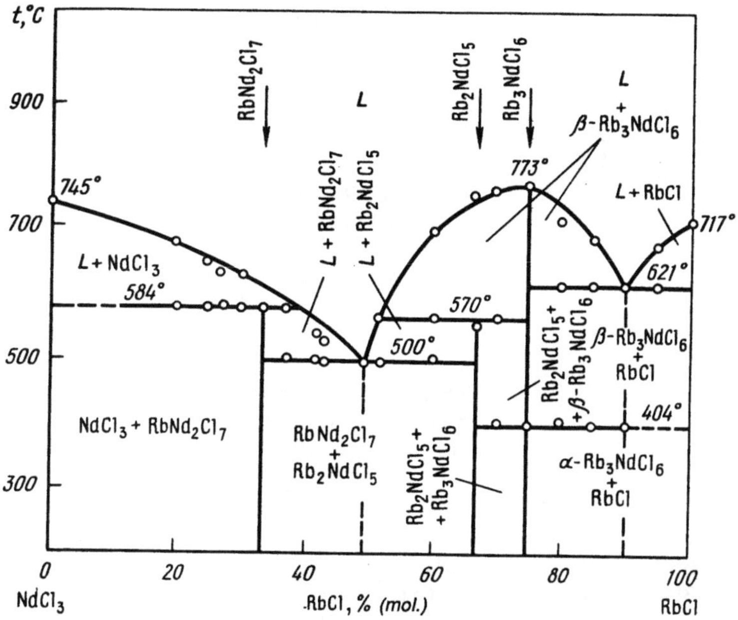

Fig. 251. $NdCl_3$-RbCl system.

2nd Component	T°C	Conc. mol %	Phases involved	Def.	Method	Ref.
YCl_3	586	69	10% YCl_3 sol.sln. $NdCl_3$, 5% $NdCl_3$ sol.sln. in YCl_3	Eut.	2	1402

NdF_3

2nd Component	T°C	Conc. mol %	Phases involved	Def.	Method	Ref.
SrF_2			Solid solutions of fluorite and hexagonal structure		3b	1403

$NiCl_2$

2nd Component	T°C	Conc. mol %	Phases involved	Def.	Method	Ref.
RbCl	505	25	$RbNiCl_3$, RbCl	Eut.	2a, 2b,	1404
	712	50	$NiCl_2$, $RbNiCl_3$	Per.	2c	

$NiCl_2$

2nd Component	T°C	Conc. mol %	Phases involved	Def.	Method	Ref.
RbCl	710	45	$RbNiCl_3$, $NiCl_2$	Eut.	1,2,2a	1405
	718	50	$RbNiCl_3$	Comp.	3a	
	502	64.5	RbCl, $RbNiCl_3$	Eut.		
	716	100	RbCl			
			$RbNiCl_3$ is of the same structure as $CsNiCl_3$ and $BaTiO_3$. Parameters a = 6.858Å c = 5.716Å			
$TaCl_5$			Temp. and composition of eutectic close to that of pure $TaCl_5$		2d,1	1406
TlCl	610	27	$TlNiCl_3$, $NiCl_2$	Eut.	1,2,2c	1407
	632	50	$TlNiCl_3$	Comp.	3a	
	382	74	6% $NiCl_2$ sol.sln. in TlCl, $TlNiCl_3$	Eut.		
	430	100	TlCl	M.P.		
			$TlNiCl_3$ is of hexagonal structure similar to $CsNiCl_3$ and $BaTO_3$, Parameters: a = 6.710Å c = 5.592Å			

NiF_2

2nd Component	T°C	Conc. mol %	Phases involved	Def.	Method	Ref.
SnF_2			Six congruently melting phases with molar ratios: 1:2, 2:3, 1:4, 1:8, 1:14. The compounds contain anions SnF_5^- and SnF_3^-.		2c	1408

$PbBr_2$

2nd Component	T°C	Conc. mol %	Phases involved	Def.	Method	Ref.
TlBr	352	10.4	$PbBr_2$, $TlBr.2PbBr_2$	Eut.	1,2	1409
	395	33.3	$TlBr.2PbBr_2$	Comp.		

$PbBr_2$

2nd Component	T°C	Conc. mol %	Phases involved	Def.	Method	Ref.
TlBr	356	54.0	$TlBr.2PbBr_2$, $3TlBr.PbBr_2$	Eut.		1409
	392	75.0	$3TlBr.PbBr_2$	Comp.		
	384	80.2	$3TlBr.PbBr_2$,TlBr	Eut.		
			Components form compounds $2PbBr_2 3TlBr$ and $PbBr_2.TlBr$		2a	1410
$PbCl_2$	376	0.0	$PbBr_2$	M.P.	1,2	1411
			Continuous sol.sln.			
	492	100	$PbCl_2$	M.P.		
			Fig. 252			1412
			Figure represents superposition of two states: stable $A_\alpha CB_\alpha$, and metastable $A_\beta C_1 B_\beta$. At low temp. there is a transition $\alpha \rightleftharpoons \beta$ in solid solutions AC_2B Assumption of compound PbClBr forming sol.slns. with $PbCl_2$ and $PbBr_2$. Curve 1 according to ref. 1413 Curve 2 " " " 1414 Curve 3 " " present authors		2a	1412
			Continuous solid sln.		2a	1415
PbI_2			Fig. 253			1416
			Figure represents susperposition of stable state $A_\alpha E_1 B_\alpha$ and metastable state $A_\beta E_2 B_\beta$(stable E_1)	Eut.	2b,2c	1416
	282	47	30% PbI sol.sln. in β-$PbBr_2$, 10% $PbBr_2$ sol.sln. in β-PbI_2			
	238	53	(Metastable E_2)35% PbI_2 sol.sln.in α-$PbBr_2$,	Eut.		
	344	-	$\alpha \rightleftharpoons \beta$ $PbBr_2$ (C)	P.T.		
	372	-	$\alpha \rightleftharpoons \beta$ PbI_2 (D)	P.T.		

$PbBr_2$

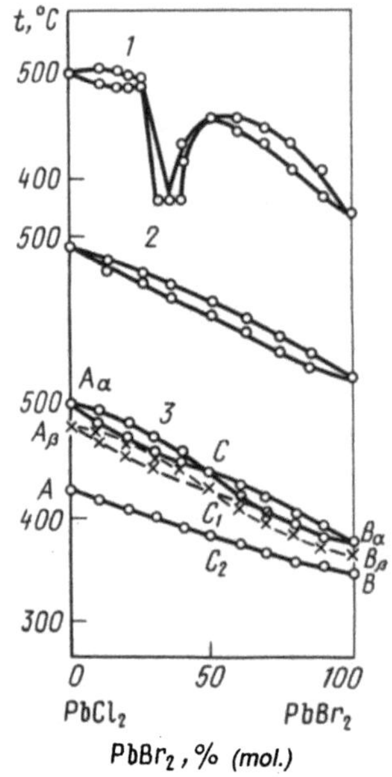

Fig. 252. $PbCl_2$-$PbBr_2$ system.

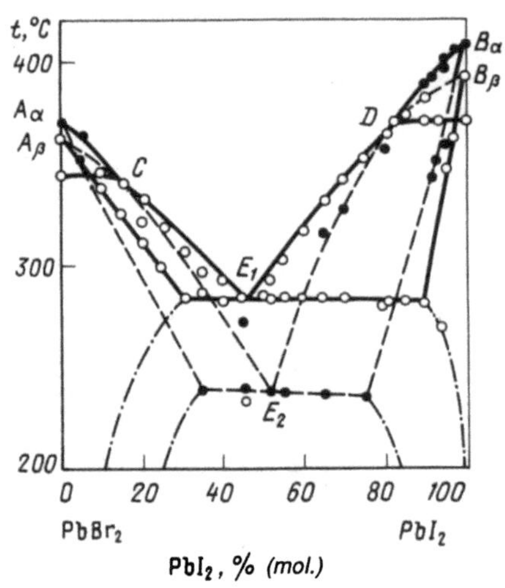

Fig. 253. PbI_2-$PbBr_2$ system.

$PbBr_2$

2nd Component	T°C	Conc. mol %	Phases involved	Def.	Method	Ref.
PbI_2	260	47	Continuous sol.sln.	Min.	1,2	1417
	265	48	Sol. slns. containing 38% and 95% PbI	Eut.	2a	1418
	320	85	Reactions liq + β ⇌ α-PbI_2	Per.		

$PbCl_2$

2nd Component	T°C	Conc. mol %	Phases involved	Def.	Method	Ref.
$PuCl_3$	476	15	$PbCl_2$, Pb_2PuCl_7	Eut.	2a	1419
	515	33	Pb_2PuCl_7	Comp.		
	502	40	Pb_2PuCl_7,$PuCl_3$	Eut.		
	755±3	100	$PuCl_3$	M.P.		
RbCl	496	0	$PbCl_2$	M.P.	1,2	1420
	414	25	$PbCl_2$, 2RbCl.3$PbCl_2$	Eut.		
	417	-	2RbCl.3$PbCl_2$	Comp.		
	417	43	2RbCl.3$PbCl_2$, $PbCl_2$.RbCl	Eut.		
	440	-	$PbCl_2$.RbCl	Comp.		
	426	62	$PbCl_2$.RbCl, $PbCl_2$.2RbCl	Eut.		
	450	67	$PbCl_2$.2RbCl,RbCl	Per.		
$SbCl_3$	72	100	$SbCl_3$ With slight addition of $PbCl_2$ to $SbCl_3$ the crystallization temp. rises rapidly	M.P.	2d	1421
$SbCl_5$	501	0.0	$PbCl_2$	M.P.	2d	1422
	~3	~100	$PbCl_2$, $SbCl_5$	Eut.		
	3	100	$SbCl_5$	M.P.		

$PbCl_2$

2nd Component	T°C	Conc. mol %	Phases involved	Def.	Method	Ref.
$SrCl_2$			$PbCl_2$ virtually insoluble in $SrCl_2$		3b, 1f	1423
	516	4	$PbCl_2$ based sol. sln., $2PbCl_2.SrCl_2$	Eut.	1,2	1424
	604	27	$2PbCl_2.SrCl_2$, $SrCl_2$	Per.		
$TeCl_4$	501	0.0	$PbCl_2$	M.P.	2d, 1	1425
	212	99.5	$TeCl_4$, $PbCl_2$	Eut.		
	498	10-96	Dissociation	P.T.		
$ThCl_4$	382	12.5	$PbCl_2$, $4PbCl_2.ThCl_4$	Eut.	2a	1426
	422	20	$4PbCl_2.ThCl_4$	Comp.		
	400	25.5	$4PbCl_2.ThCl_4$, $2PbCl_2.ThCl_4$	Eut.		
	544	33.3	$2PbCl_2.ThCl_4$	Comp.		
	376	48.5	$2PbCl_2.ThCl_4$, $ThCl_4$	Eut.		
TlCl	418	26	$PbCl_2$, $2PbCl_2$.TlCl	Eut.	1,2	1427
	427	33.3	$2PbCl_2$.TlCl	Comp.		
	371	58	$2PbCl_2$.TlCl, $PbCl_2$.3TlCl	Eut.		
	379	75	$PbCl_3$.3TlCl	Comp.		
	377	86	$PbCl_2$.3TlCl, TlCl	Eut.		
	262	86	-	Eut.	-	1428
			Following compounds established: $PbCl_2$.2TlCl, $2PbCl_2$.TlCl $PbCl_2$.3TlCl		3b	1429
			Following compounds established: $2PbCl_2$.TlCl, $PbCl_2$.3TlCl		3b	1430

$PbCl_2$

2nd Component	T°C	Conc. mol %	Phases involved	Def.	Method	Ref.
TlCl			Following compounds established: $2PbCl_2.TlCl$, $PbCl_2.TlCl$, $PbCl_2.3TlCl$		2a,2d	1431
	498	0	$PbCl_2$	M.P.	1,2	1432
	414	26	$PbCl_2$, $2PbCl_2.TlCl$	Eut.		
	420	33.3	$2PbCl_2.TlCl$	Comp.		
	390	52	$2PbCl_2.TlCl$, $PbCl_2.TlCl$	Per.		
	368	57	$PbCl_2.TlCl$, $PbCl_2.3TlCl$	Eut.		
	395	76	$PbCl_2.3TlCl$	Comp.		
	373	86	$PbCl_2.3TlCl$,TlCl	Eut.		
	430	100	TlCl	M.P.		
	418	31	$PbCl_2$, $2TlCl.3PbCl_2$	Eut.	1,2	1433
	430	-	$2TlCl.3PbCl_2$	Comp.		
	386	64	$2TlCl.3PbCl_2$, $PbCl_2.4TlCl$	Eut.		
	408	-	$PbCl_2.4TlCl$	Comp.		
	388	90	$PbCl_2.4TlCl$,TlCl	Eut.		
UCl_3	473	11	$PbCl_2$, UCl_3	Eut.	2c	1434
	471	11	-	Eut.	2d,3b	1435
	340	-	Noticeable thermal effect			
			Region of limited UCl_3 base sol.sln.		2d,5c	1436
	512	-	$4PbCl_2.UCl_3$	Comp.		
			Components react according to: $PbCl_2 + 2UCl_3 \rightarrow 2UCl_4 + Pb$ Max. degree of Pb generation is at equimolar ratio of $PbCl_2 + UCl_3$, whereby the reaction transforms into ternary $PbCl_2$-UCl_3-UCl_4			

$PbCl_2$

2nd Component	T°C	Conc. mol %	Phases involved	Def.	Method	Ref.
UCl_4	396	18.2	$PbCl_2$, $3PbCl_2.UCl_4$	Eut.	2a	1437
	482	25	$3PbCl_2.UCl_4$	Comp.		
	365	32.1	$3PbCl_2.UCl_4$, UCl_4	Eut.		
	347	0-25	-	P.T.		
	330	32.8	$UCl_4.3PbCl_2$, $PbCl_2$	Eut.	2a,3b, 1f,5	1438
	380	71.7	UCl_4, $UCl_4.3PbCl_2$	Per.		
	387-347		Incongruent melting	Comp.	2d,3b	1439
	498	0	$PbCl_2$	M.P.	2d,3b,	1440
	385	27	$PbCl_2$, $4PbCl_2.UCl_4$	Per.	1f,5	1441
	345	36	$4PbCl_2.UCl_4$, UCl_4	Eut.		
	590	100	UCl_4	M.P.		
		0-20	Melts tend to supercool			
$ZnCl_2$			Fig. 254			1442

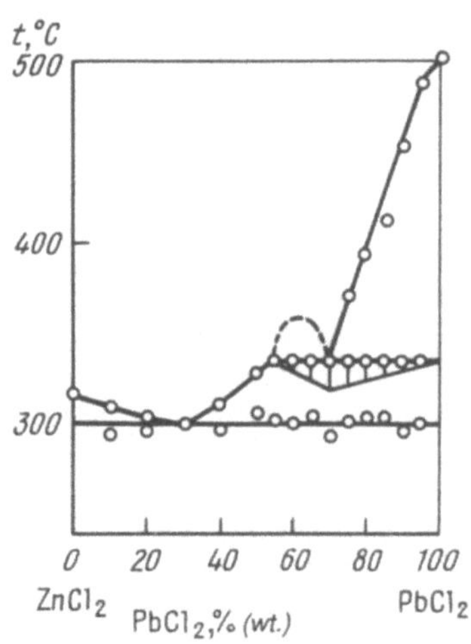

Fig. 254. $ZnCl_2$-$PbCl_2$ system.

$PbCl_2$

2nd Component	T°C	Conc. mol %	Phases involved	Def.	Method	Ref.
$ZnCl_2$	503	0	$PbCl_2$	M.P.	2d,4a,	1442
	302	83	$PbCl_2$, $ZnCl_2$	Eut.	1	
	316	100	$ZnCl_2$	M.P.		
	336	10-63	Dissociation	-		
	320	65	10% $ZnCl_2$ sol. sln. in $PbCl_2$, 50% $PbCl_2$ sol.sln. in $ZnCl_2$ Solubility of $ZnCl_2$ in $PbCl_2$ drops rapidly with lowering of temp.	Per.	3b	1443 1444
		10-11	Minimum of the liquidus line of the sol.sln. range	Min.		
	330	0.0, 75, 80, 85, 90	Dissociation of groups: $(ZnCl_2)_n$ with appearance of $ZnCl^-$		1i	1445
PbI_2			Fig. 255			1446

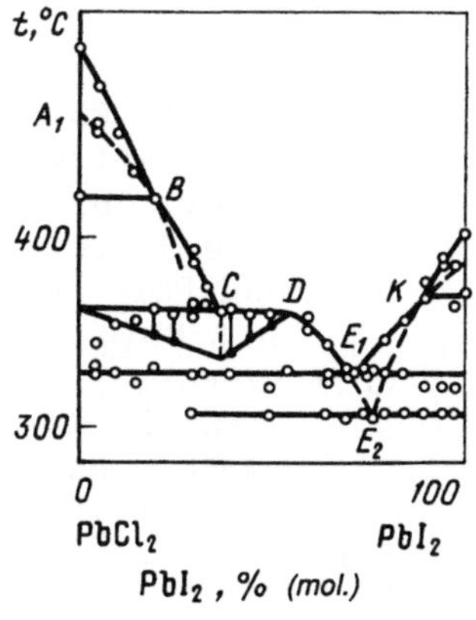

Fig. 255. $PbCl_2$-PbI_2 system.

$PbCl_2$

2nd Component	T°C	Conc. mol %	Phases involved	Def.	Method	Ref.
PbI_2	422	-	(B), β-$PbCl_2$, α-$PbCl_2$	Per.	2b,2c	1446
	328	72	(E), β-PbI_2, β-$PbCl_2$	Eut.		
	308	77	(E_2), $PbCl_2$, α-PbI_2 (metastable)	Eut.		
	372	-	(K), α-PbI_2, β-PbI_2	Per.		
		37-55	Dissociation (CD) BA, - Crystallization line of β-$PbCl_2$ in the metastable equilib.			
		90-98	Transformation between stable and metastable equilibrium	P.T.		
	496	0	$PbCl_2$	M.P.	1,2,3b	1447
	360	42	$PbCl_2$, PbI_2.$PbCl_2$	Eut.		1448
	334	73.5	PbI_2.$PbCl_2$, PbI_2	Eut.		
			Confirmation of existence of compound $PbCl_2$.PbI_2		2a	1449
			Fig. 256			1450
	498	0	$PbCl_2$	M.P.	2a,3a	1450
	365	36.5	(P_1), $PbCl_2$, $2PbCl_2$.PbI_2 base sol.sln.	Per.		
	335	72.5	(E), $2PbCl_2$.PbI_2 base sol.sln., 95% PbI_2 sol.sln. in β-PbI_2	Eut.		
	373	87.5	(P_2), β-PbI_2 base sol.sln., α-PbI_2 base sol.sln.	Per.		
	410	100	PbI_2	M.P.		
	372	-	PbI_2	P.T.		
		30-50	PbI_2 sol.sln. in $2PbCl_2$.PbI_2			
	361	38-60	Dissociation (ab)			

$PbCl_2$

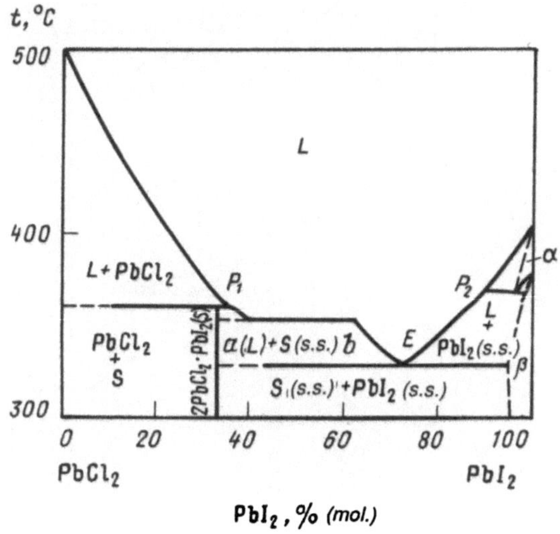

Fig. 256. $PbCl_2$-PbI_2 system.

PbF_2

2nd Component	T°C	Conc. mol %	Phases involved	Def.	Method	Ref.
RbF			Fig. 257			1451
	600	30	8% RbF sol.sln.in (11_H), 111_H (comp.of unknown compos.) sol.sln.	Per.	2,3b	1451
	550	45	111_H sol.sln., $RbPbF_3$	Eut.		
	560	50	$RbPbF_3$	Comp.		
	485	68	$RbPbF_3$, RbF	Eut.		
	320-350		$111_H \rightleftharpoons 111_T$	P.T.		
	420		(Eutectoid transf.) PbF_2 base sol. sln. (11_H) $\rightleftharpoons$ PbF_2(11_T) + 111_T	P.T.		
SmF_3			Limited solubility of SmF_3 in solid PbF_2 (~45%)		2,3b	1452
	1050	-	-	P.T.		

PbF_2

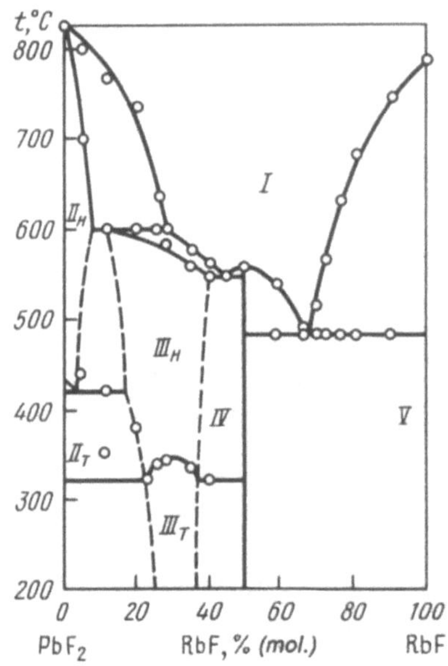

Fig. 257. PbF_2-RbF system.

2nd Component	T°C	Conc. mol %	Phases involved	Def.	Method	Ref.
TiF_3	-	0.0	α and β-PbF_2		3b, le,	1453
	575±15		$Pb_3(TiF_6)_2$	Comp.	2d	
			Solid sln. of orthorhombic type			
VF_3	566±7	-	α and β-PbF_2	Comp.	3b, le,	1454
			$PbVF_3$		2d	
			Solid sln. of orthorhombic type			
YF_3			Continuous sol.sln.		2d,3b	1455
	930	40	Beginning of crystallization	Max.		
	890	-	End of crystn.	Min.		

PbI_2

2nd Component	T°C	Conc. mol %	Phases involved	Def.	Method	Ref.
RbI	408	0	PbI_2	M.P.	2a	1456
	351	25	PbI_2, PbI_2.RbI	Eut.		
	416	50	PbI_2.RbI	Comp.		
	398	57	PbI_2.RbI, 9RbI.PbI_2	Eut.		
	432	65	9RbI.PbI_2, RbI	Per.		
	643	100	RbI	M.P.		

PbI_2

2nd Component	T°C	Conc. mol %	Phases involved	Def.	Method	Ref.
SnI_2			Continuous sol.sln.		2a,4c, 6c	1457
TeI_4	412	0.0	PbI_2	M.P.	2a	1458
	272	90	TeI_4, PbI_2	Eut.		
			A range of PbI base sol.sln.			
	280	100	TeI_4	M.P.		
TlI	325	31	PbI_2.TlI, PbI_2	Eut.	1,2	1459
	347	50	PbI_2.TlI	Comp.		
	336	54	PbI_2.3TlI, PbI_2.TlI	Eut.		
	353	75	PbI_2.3TlI	Comp.		
	346	77	PbI_2.3TlI, TlI	Eut.		
			PbI_2.3TlI, PbI_2.TlI } Confirmed	Comp.	3b	1460

$PrCl_3$

2nd Component	T°C	Conc. mol %	Phases involved	Def.	Method	Ref.
$SrCl_2$	795	0	$PrCl_3$	M.P.	2a,3b	1461
	620	45	$PrCl_3$, Sr_3PrCl_9	Eut		
	630	47	Sr_3PrCl_9, $SrCl_2$	Per.		
	868	100	$SrCl_2$	M.P.		

PrF_3

2nd Component	T°C	Conc. mol %	Phases involved	Def.	Method	Ref.
RbF	676	64	3RbF.PrF_3, PrF_3	Eut.	1,2,2d	1462
	778	75	3RbF.PrF_3	Comp.		
	660	85	RbF, 3RbF.PrF_3	Eut.		
	536		RbF.PrF_3	P.T.		

$PuCl_3$

2nd Component	T°C	Conc. mol %	Phases involved	Def.	Method	Ref.
RbCl			Fig. 258			1463
	584	36	$PuCl_3$, $RbPu_2Cl_7$	Per.	2, 1d	1463
	504	50	$RbPu_2Cl_7$, Rb_2PuCl_5	Eut.		
	560	58	Rb_2PuCl_5, Rb_3PuCl_6	Per.		
	744±3	73	Rb_3PuCl_6	Comp.		
	630	86	Rb_3PuCl_6, RbCl	Eut.		
	374	-	$RbPu_2Cl_7$	P.T.		
	398	-	Rb_3PuCl_2	P.T.		

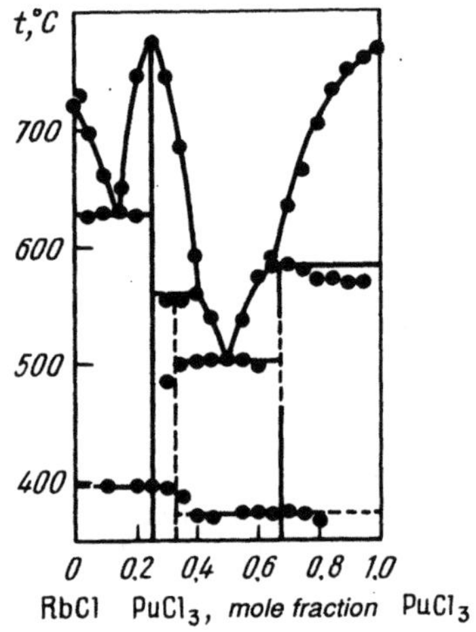

Fig. 258. RbCl-$PuCl_3$ system.

2nd Component	T°C	Conc. mol %	Phases involved	Def.	Method	Ref.
$SrCl_2$			Fig. 259			1464
	616	48	$PuCl_3$, Sr_3PuCl_9	Eut.	2, 1e	1464
	628	54	Sr_3PuCl_9, 10% $PuCl_3$ sol. sln. in $SrCl_2$	Per.		
$ThCl_4$	755±3	0.0	$PuCl_3$	M.P.	2a	1465
	620	67-68	14.5% $PuCl_3$ sol. sln. in $ThCl_4$ $PuCl_3$	Eut.		
	760	100	$ThCl_4$	M.P.		

$PuCl_3$

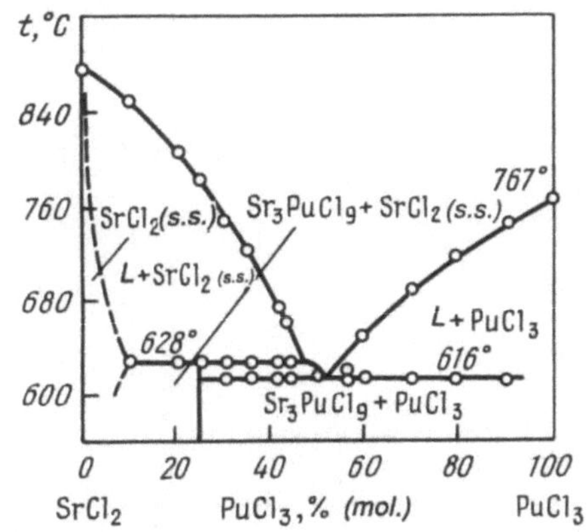

Fig. 259. $SrCl_2$-$PuCl_3$ system.

2nd Component	T°C	Conc. mol %	Phases involved	Def.	Method	Ref.
UCl_3	496	44	$PuCl_3$, $3UCl_3.PuCl_3$	Eut.	2d	1466
	700	70	$3UCl_3.PuCl_3$,UCl_3	Per.		
	762	0.0	$PuCl_3$	M.P.	2d	1467
	723	20	Continuous sol.sln.	Min.		
	839	100	UCl_3	M.P.		

$PuCl_4$

2nd Component	T°C	Conc. mol %	Phases involved	Def.	Method	Ref.
RbCl	556	88	-	Eut.	2d,3b	1468
			Numerous compounds		1g	1469
	502	-	Rb_2PuCl_6	P.T.		

PuF_6

2nd Component	T°C	Conc. mol %	Phases involved	Def.	Method	Ref.
UF_6			Continuous sol.sln.		2	1470

RbBr

2nd Component	T°C	Conc. mol %	Phases involved	Def.	Method	Ref.
$SbBr_3$			Fig. 260			1471
	490	~20	$4RbBr.SbBr_3$, RbBr	Per.	1,2	1471
	96	~100	$SbBr_3$. $4RbBr.SbBr_3$	Eut.		

RbBr

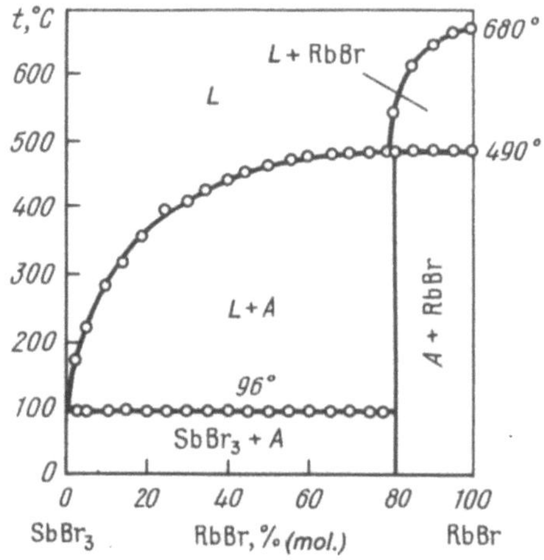

Fig. 260. $SbBr_3$-RbBr system.

2nd Component	T°C	Conc. mol %	Phases involved	Def.	Method	Ref.
$ScBr_3$	614	11.6	RbBr, Rb_3ScBr_6	Eut.	2d,3b	1472
	732	25	Rb_3ScBr_6	Comp.		
	634	37.8	Rb_3ScBr_6, $Rb_3Sc_2Br_9$	Eut.		
	640	40	$Rb_3Sc_2Br_9$	Comp.		
	585	49.5	$Rb_3Sc_2Br_9$, $ScBr_3$	Eut.		
	440	-	Rb_3ScBr_6	P.T.		
$SnBr_2$	682	0.0	RbBr	M.P.	1,2a	1473
	296	42.3	$2RbBr.SnBr_2$, RbBr	Per.		
	237	56	$RbBr.2SnBr_2$, $2RbBr.SnBr_2$	Eut.		
	292	66.7	$RbBr.2SnBr_2$	Comp.		
	202	91	6% $RbBr_2$ sol.sln. in $SnBr_2$, $RbBr.2SnBr_2$	Eut.		
	226	100	$SnBr_2$	M.P.		
		0-15	Due to dissociation of RbBr region not investigated			
$SrBr_2$	484	32.5	$RbBr.SrBr_2$, RbBr	Eut.	1,2	1474
	558	-	$RbBr.SrBr_2$	Comp.		

RbBr

2nd Component	T°C	Conc. mol %	Phases involved	Def.	Method	Ref.
$SrBr_2$	550	55	$RbBr.2SrBr_2$, $RbBr.SrBr_2$	Eut.	1,2	1474
	570	-	$RbBr.2SrBr_2$	Comp.		
	542	87.5	$SrBr_2$, $RbBr.2SrBr_2$	Eut.		
	507	28	$SrBr_2.RbBr$, RbBr	Eut.	2d,2,3b	1475
	596	50	$SrBr_2.RbBr$	Comp.		
	578	58	$2SrBr_2.RbBr$, $SrBr_2.RbBr$	Eut.		
	596	-	$2SrBr_2.RbBr$	Comp.		
	560	16	$SrBr_2$,$2SrBr_2.RbBr$	Eut.		
$TeBr_4$	680	0.0	RbBr	M.P.	2,3b,1c	1476
	480	15.0	PbBr,Rb_2TeBr_6	Eut.		
	653	33.3	Rb_2TeBr_6	Comp.		
	328	70.0	Rb_2TeBr_6,$TeBr_4$	Eut.		
	386	100.0	$TeBr_4$	M.P.		
$TiBr_2$	578	20.0	RbBr, $RbTiBr_3$	Eut.	2a	1477
	804	50.0	$RbTiBr_3$	Comp.		
	758	65.0	$RbTiBr_3$, $TiBr_2$	Eut.		
$TiBr_3$	604	10.0	RbBr,Rb_3TiBr_6	Eut.	2a	1477
	664	25.0	Rb_3TiBr_6	Comp.		
	617	35.0	Rb_3TiBr_6, $Rb_3Ti_2Br_9$	Eut.		
	647	40.0	$Rb_3Ti_2Br_9$	Comp.		
	576	50.0	$Rb_3Ti_2Br_9$,$TiBr_3$	Eut.		
$TiBr_4$	636	<3	RbBr,Rb_2TiBr_6	Eut.	2a	1477
	672	-	Rb_2TiBr_6	Comp.		
			System investigated to 35-37% $TiBr_4$			
RbCl	690	0.0	RbBr	M.P.	1,2	1483
	690		Continuous sol.sln.	Min.		
	726	100	RbCl	M.P.		
	687	0.0	RbBr	M.P.	1,2a,3b	1484
	673	30	Continuous sol.sln.	Min.		
	717	100	RbCl	M.P.		

RbBr

2nd Component	T°C	Conc. mol %	Phases involved	Def.	Method	Ref.
RbCl	20-470		Solid solutions		3b	1485
	669	23	Continuous sol.sln.	Min.	1,2	1486
			Continuous sol.sln. with minimum		2d,3b, 1,2	1487
RbF	530	50	RbBr, RbF	Eut.	1,2	1488
RbI			Fig. 261			1489
			Dissociation limits of sol.sln. determined		3f	1489
	73±3		(K) Critical temp.			
	618	50	Continuous sol.sln.	Min.	1,2	1490

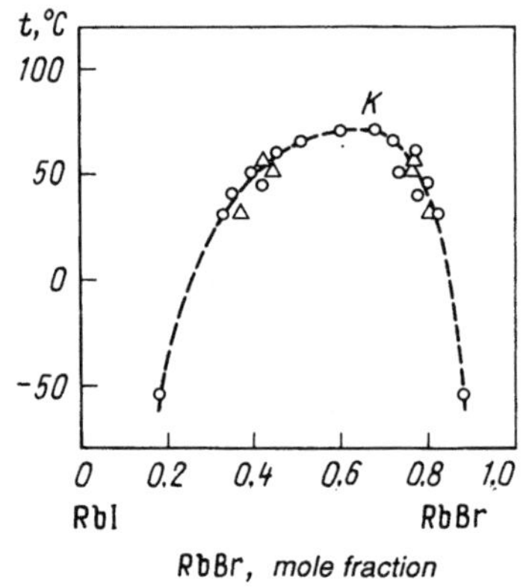

Fig. 261. RbI-RbBr system.

RbCl

2nd Component	T°C	Conc. mol %	Phases involved	Def.	Method	Ref.
$ReCl_3$			Fig. 262			1478
			Components affect each other according to the following reaction: $6RbCl + 4ReCl_3 \rightarrow 3Rb_2ReCl_6 + Re$ (the reaction proceeds to the right. Liquidus line corresponds to crystallization of pure Re		2a,3b,5	1478
	970	-	Rb_2ReCl_6	Comp.		

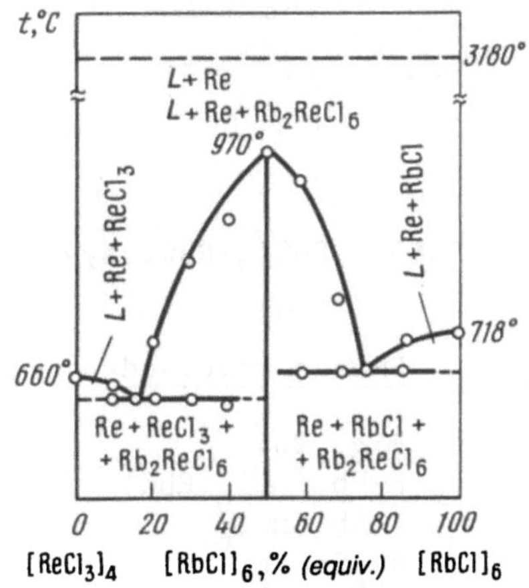

Fig. 262. $ReCl_3$-RbCl system.

2nd Component	T°C	Conc. mol %	Phases involved	Def.	Method	Ref.
$SbCl_3$			Fig. 263			1479
	450	28	$3RbCl.SbCl_3$, RbCl	Per.	1,2	1479
	128	70	$RbCl.2SbCl_3$, $3RbCl.SbCl_3$	Per.		
	44	85	$RbCl.2SbCl_3$, $SbCl_3$			
$ScCl_3$	655	9.4	RbCl, $3RbCl.ScCl_3$	Eut.	2d,4a	1480 1481
	820	25	$3RbCl.ScCl_3$	Comp.		
	605	41.6	$3RbCl.ScCl_3$, $3RbCl.2ScCl_3$	Per.		
	514	53.6	$3RbCl.2ScCl_3$, $ScCl_3$	Eut.		
	380		$\alpha \rightleftharpoons \beta$ $3RbCl.ScCl_3$	P.T.		
	400		$\beta \rightleftharpoons \gamma$ $3RbCl.ScCl_3$	P.T.		

RbCl

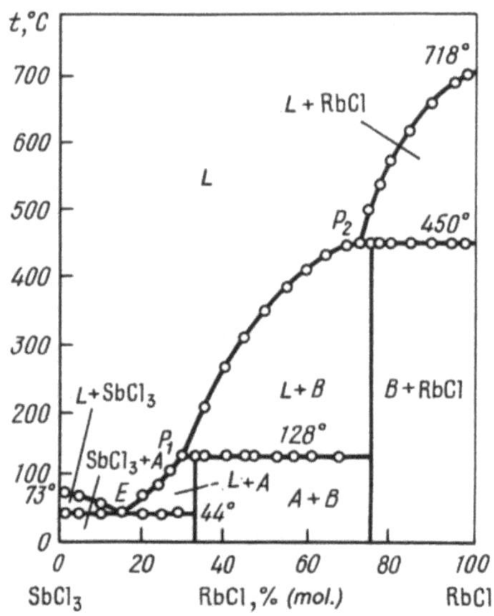

Fig. 263. $SbCl_3$-RbCl system.

2nd Component	T°C	Conc. mol %	Phases involved	Def.	Method	Ref.
$SmCl_3$			Fig. 264			1482
	630	10	β-Rb_3SmCl_6,RbCl	Eut.	2a	1482
	832	26	β-Rb_3SmCl_6	Comp.		
	480	37	Rb_2SmCl_5, Rb_3SmCl_6	Per.		
	464	44	$RbSm_2Cl_7$, Rb_2SmCl_5	Eut.		
	585	66	$RbSm_2Cl_7$	Comp.		
	560	78	$RbSm_2Cl_7$,$SmCl_3$	Eut.		
	440	-	Rb_3SmCl_6	P.T.		
$SnCl_2$	262	40.3	$2RbCl.SnCl_2$,RbCl	Per.	1,2	1491
	223	48.5	$RbCl.SnCl_2$, $2RbCl.SnCl_2$	Eut.		
	228	50	$RbCl.SnCl_2$	Comp.		
	203	55.5	$RbCl.2SnCl_2$, $RbCl.SnCl_2$	Eut.		
	231	66.7	$RbCl.2SnCl_2$	Comp.		
	192	83	$SnCl_2$,$RbCl.2SnCl_2$	Eut.		

RbCl

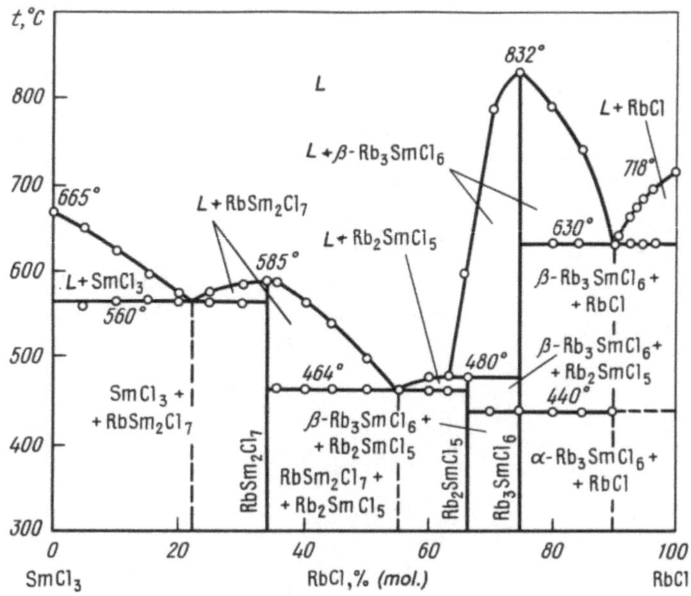

Fig. 264. $SmCl_3$-RbCl system.

2nd Component	T°C	Conc. mol %	Phases involved	Def.	Method	Ref.
$SnCl_2$			Fig. 265			1492

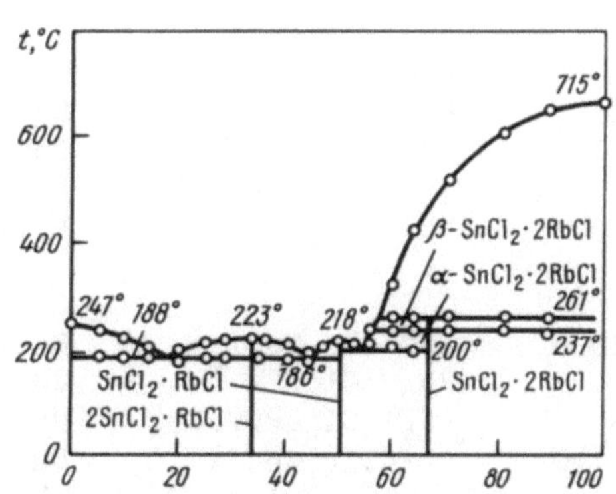

Fig. 265. $SnCl_2$-RbCl system.

RbCl

2nd Component	T°C	Conc. mol %	Phases involved	Def.	Method	Ref.
$SnCl_2$	261	42	β-$SnCl_2.2RbCl$, RbCl	Per.	2,6c	1492
	237	43	α-$SnCl_2.2RbCl$, β-$SnCl_2.2RbCl$	Per.		
	200	44	$SnCl_2.RbCl$, β-$SnCl_2.2RbCl$	Eut.		
	218	50	$SnCl_2.RbCl$	Comp.		
	186	56	$2SnCl_2.RbCl$, $SnCl_2.RbCl$	Eut.		
	223	66	$2SnCl_2.RbCl$	Comp.		
	188	82	$SnCl_2$, $2SnCl_2.RbCl$	Eut.		
$SrCl_2$	712	0	RbCl	M.P.	2a	1493
	528	26.5	RbCl, $RbCl.SrCl_2$	Eut.		1494
	660	50	$RbCl.SrCl_2$	Comp.		1495
	626	62.5	$RbCl.SrCl_2$, $RbCl.2SrCl_2$	Eut.		
	634	67	$RbCl.2SrCl_2$	Comp.		
	630	68.5	$RbCl.2SrCl_2$, $SrCl_2$	Eut.		
	74.5	Solid solutions		Eut.	8	1496
$TaCl_3$	592	20	RbCl, Rb_2TaCl_5	Eut.	2a	1497
	642	33	Rb_2TaCl_5	Comp.		
	300	45	$TaCl_3$, Rb_2TaCl_5	Eut.		
	360	-	Rb_2TaCl_5	P.T.		
$TaCl_4$	718	0	RbCl	M.P.	3b,2b, 6a	1498
	617	18	RbCl, β-Rb_2TaCl_6	Eut.		
	756	33.3	β-Rb_2TaCl_6	Comp.		
	284	47	β-Rb_2TaCl_6, $TaCl_4$	Eut.		
	323	-	Rb_2TaCl_6	P.T.		
$TeCl_4$	596	20.0	RbCl, Rb_2TeCl_6	Eut.	2d, le, 3b,6j	1499
	670	34	Rb_2TeCl_6	Comp.		
	518	63	Rb_2TeCl_6, $RbTeCl_5$	Per.		
	202	87.0	$RbTeCl_5.TeCl_5$	Eut.		
			Density of Rb_2TeCl_6 = 3.16 g/cm^{-3}, optically isotropic			

RbCl

2nd Component	T°C	Conc. mol %	Phases involved	Def.	Method	Ref.
$TiCl_2$	630	15	$RbCl, Rb_2TiCl_4$	Eut.	2d,3b	1500
	732	33	$Rb_2TiCl_4, RbTiCl_3$	Per.		
	852	50	$RbTiCl_3$	Comp.		
	763	65	$RbTiCl_3, TiCl_2$	Eut.		
$TiCl_3$	612	17	$Rb, TiCl_3.3RbCl$	Eut.	2	1501
	794	25	$TiCl_3.3RbCl$	Comp.		
	671	36	$TiCl_3.3RbCl$, $TiCl_3.RbCl$	Eut.		
	800	50	$TiCl_3.RbCl$	Comp.		
	662	10.0	$RbCl, Rb_3TiCl_6$	Eut.	2a	1502
	779	25.0	$RbTi_3Cl_6$	Comp.		
	658	38.0	Rb_3TiCl_6, $Rb_3Ti_2Cl_9$	Eut.		
	671	40.0	$Rb_3Ti_2Cl_9$	Comp.		
	610	60.0	$Rb_3Ti_2Cl_9$, $RbTiCl_4$	Per.		
	533	70.0	$RbTiCl_4, TiCl_3$	Eut.		
TlCl			Solid solutions form		6h,3b	1503
UCl_3			Fig. 266			1504
	605	15	$RbCl, 3RbCl.UCl_3$	Eut.		1504
	750	24	$3RbCl.UCl_3$	Comp.		

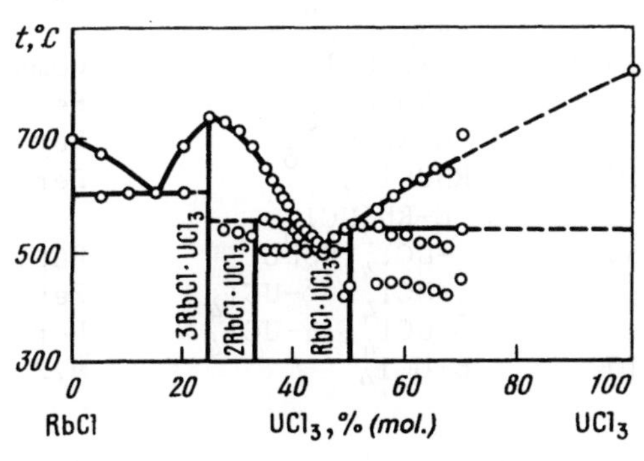

Fig. 266. $RbCl-UCl_3$ system.

RbCl

2nd Component	T°C	Conc. mol %	Phases involved	Def.	Method	Ref.
UCl_3	550	40	$3RbCl.UCl_3$, $2RbCl.UCl_3$	Per.		1504
	510	45	$2RbCl.UCl_2$, $RbCl.UCl_3$	Eut.		
	530	50	$RbCl.UCl_3$, UCl_3	Per.		
	757	15.5	$RbCl, 3RbCl.UCl_3$	Eut.	2d,3b	1505
	-	25	$3RbCl.UCl_3$	Comp.		
	535	-	$3RbCl.UCl_3$, $2RbCl.UCl_3$	Per.		
	500	45.0	$2RbCl.UCl_3$, UCl_3	Eut.		
UCl_4	546	29	$RbCl, Rb_2UCl_6$	Eut.	3b,2c	1506
	676	33	Rb_2UCl_6	Comp.		
	448	40	$Rb_2UCl_6, Rb_3U_2Cl_{11}$	Per.		
	367.8	44	$Rb_3U_2Cl_{11}, RbUCl_5$	Eut.		
	376	50	$RbUCl_5$	Comp.		
	370	58	$RbUCl_5, RbU_3Cl_{13}$	Eut.		
	417.5	62	Rb_3UCl_{13}, UCl_4	Per.		
	-	-	$RbCl, 3RbCl.2UCl_4$	Per.	2d	1507
	535±2	22	$3RbCl.2UCl_4$	Eut.		
	-	-	$2RbCl.UCl_4$	Comp.		
	345 ± 37		$2RbCl.UCl_4$, $RbCl.UCl_4$	Eut.		
	-	-	$RbCl.UCl_4$, UCl_4	Per.		
	720	0.0	RbCl	M.P.	2a,3b	1508
	560	22	Rb_4UCl_8, RbCl	Per.		
	530	24	α-Rb_2UCl_6, Rb_4UCl_8	Eut.		
	674	33.3	α-Rb_2UCl_6	Comp.		
	445	-	β-Rb_2UCl_6, α-Rb_2UCl_6	Per.		
				Per.		
	360	50	$RbUCl_5$, β-Rb_2UCl_6	Per.		
	350	55	δ-UCl_4, $RbUCl_5$	Eut.		
	-	-	γ-UCl_4, δ-UCl_4	Per.		
	485	-	β-UCl_4, γ-UCl_4	Per.		
	589	100	β-UCl_4	M.P.		
VCl_3	655	11	RbCl, Rb_3VCl_6	Eut.	2c	1509
	749	25	Rb_3VCl_6	Comp.		

RbCl

2nd Component	T°C	Conc. mol %	Phases involved	Def.	Method	Ref.
VCl_3	682	38	Rb_3VCl_6, $Rb_3V_2Cl_9$	Eut.	2c	1509
	688	40	$Rb_3V_2Cl_9$	Comp.		
	592	52	$Rb_3V_2Cl_9$,$RbVCl_4$	Per.		
	560	54	$RbVCl_4$, Undetermined phase	Eut.		
			3RbCl.VCl_3 - congruent melting; 3RbCl.2VCl_3, RbCl.VCl_3 } incongruent melting		2a,6c, 2h	1510
WCl_5			Fig. 267			1511

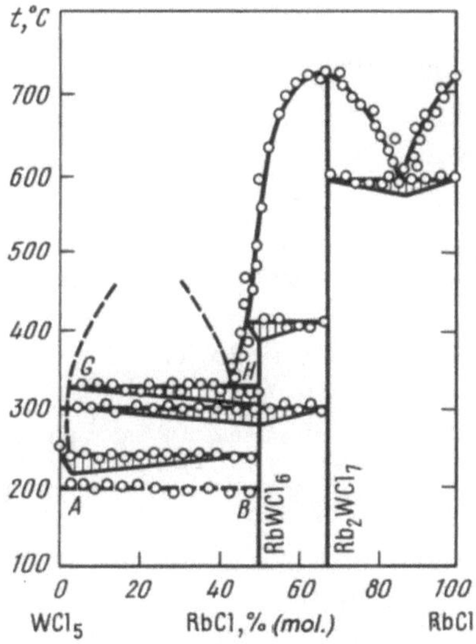

Fig. 267. WCl_5-RbCl system.

2nd Component	T°C	Conc. mol %	Phases involved	Def.	Method	Ref.
WCl_5	590	12	Rb_2WCl_7, RbCl	Eut.	2b,2c	1511
	726	33.3	Rb_2WCl_7	Comp.		
	410	54	β-$RbWCl_6$,Rb_2WCl_7	Per.		
	300	-	α-$RbWCl_6$,β-$RbWCl_6$	Per.		
	244	98-98.5	WCl_5, α-$RbWCl_6$	Eut.		
		58-98	(Line GH) - Dissociation			
			Line AB - sol.sln.			

RbCl

2nd Component	T°C	Conc. mol %	Phases involved	Def.	Method	Ref.
WCl_6			Fig. 268			1511
			System is not binary but multiphase - Cl_2 is evolved WCl_5 with RbCl form incongruently melting $RbWCl_6$		2b,2c	1511
	282	100	(Line II) - WCl_6	M.P.		
	226	10-0	(Line I) - WCl_6	P.T.		
			(Line IV) - $RbWCl_6$	Per.		
			(line III) - $RbWCl_6$	P.T.		

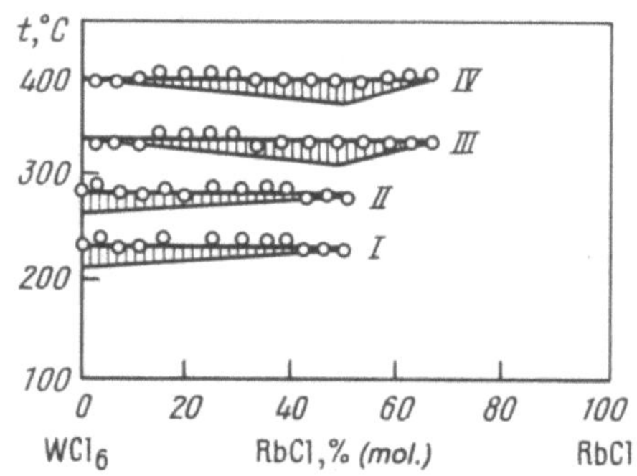

Fig. 268. WCl_6-RbCl system.

2nd Component	T°C	Conc. mol %	Phases involved	Def.	Method	Ref.
YCl_3	669	6.0	RbCl,β-Rb_3YCl_6	Eut.	2	1512
	858	26	β-Rb_3YCl_6	Comp.		
	483	45.0	β-Rb_3YCl_6,RbY_3Cl_{10}	Eut.		
	569	61	RbY_3Cl_{10},YCl_3	Per.		
	441	-	Rb_3YCl_6	P.T.		
$ZnCl_2$	718	0	RbCl	M.P.	1,2	1513
	510	25	$ZnCl_2$.2RbCl,RbCl	Eut.		
	528	33.3	$ZnCl_2$.2RbCl	Comp.		
	249	52.5	RbCl.2$ZnCl_2$, $ZnCl_2$.2RbCl	Eut.		
	306	66.7	RbCl.2$ZnCl_2$	Eut.		
	277	83	$ZnCl_2$, RbCl.2$ZnCl_2$	Eut.		
	314	100	$ZnCl_2$	M.P.		
RbF	546	46.5	RbF, RbCl	Eut.	1,2	1514

RbCl

2nd Component	T°C	Conc. mol %	Phases involved	Def.	Method	Ref.
RbI	720	0.0	RbCl	M.P.	1,2	1515
	570	56	RbCl, RbI	Eut.		
	642	100	RbI	M.P.		
	720	0.0	RbCl	M.P.	1,2,2a	1516
	563	55	45% RbCl sol.sln. in RbI, 12.5% RbI sol.sln. in RbCl	Eut.		
	644	100	RbI	M.P.		
	560	55	Continuous sol.sln.	Min.	1,2	1517

RbF

2nd Component	T°C	Conc. mol %	Phases involved	Def.	Method	Ref.
ScF_3	760 (746)	5.0	RbF, Rb_3ScF_6	Eut.	1,2,2a, 3b	1518
	1120	25.0	Rb_3ScF_6	Comp.		
	1070	30.0	Rb_3ScF_6, ScF_3	Eut.		
			$3RbF.ScF_3$, $RbF.ScF_3$ } Exist. confirmed		5b	1519
SrF_2	678	9	RbF, SrF_2	Eut.		1520
TiF_4			Fig. 269			1521
	772	0	RbF	M.P.	2d,3b, 6j	1521
	680	4	RbF based sol.sln., Rb_3TiF_7	Eut.		
	806	12.5	Rb_3TiF_7	Comp.		
	733	15.2	Rb_3TiF_7,Rb_2TiF_6	Eut.		
	826	25	Rb_2TiF_6	Comp.		
			Densities: Rb_3TiF_7 → 3.63 g/cm^3; Rb_2TiF_6 → 3.60 g/cm^3			
UF_4			Fig. 270			1522
	710	10	Rb,$3RbF.UF_4$	Eut.	1,2,1f, 3b	1522
	995	25	$3RbF.UF_4$	Comp.		
	818	38	$3RbF.UF_4$, $2RbF.UF_4$	Per.		

RbF

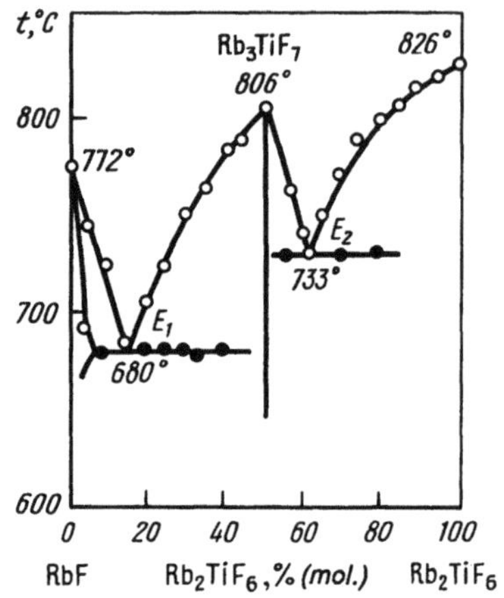

Fig. 269. RbF-Rb_2TiF_6 system.

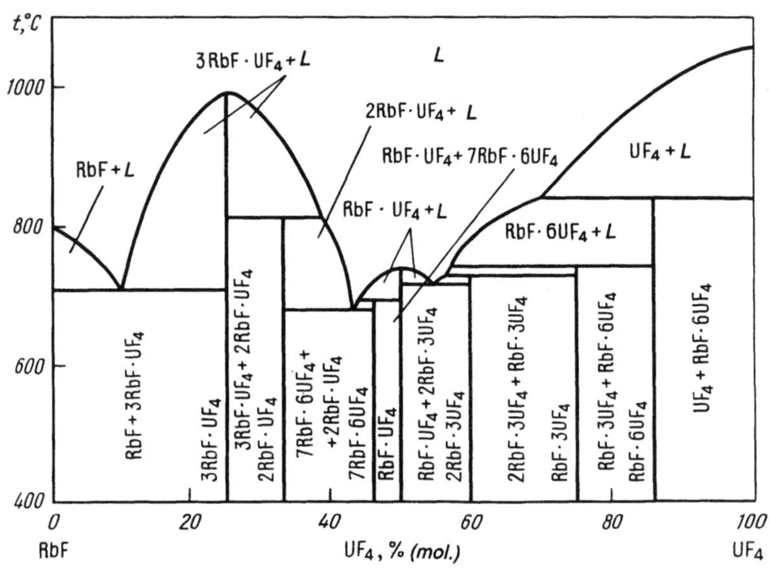

Fig. 270. RbF-UF_4 system.

RbF

2nd Component	T°C	Conc. mol %	Phases involved	Def.	Method	Ref.
UF_4	675	43.5	$2RbF.UF_4$, $7RbF.6UF_4$	Eut.	1,2,1f, 3b	1522
	693	44	$7RbF.6UF_4$,$RbF.UF_4$	Per.		
	735	50	$RbF.UF_4$	Comp.		
	714	55	$RbF.UF_4$,$2RbF.3UF_4$	Eut.		
	722	56.5	$2RbF.3UF_4$,$RbF.3UF_4$	Per.		
	730	57	$RbF.3UF_4$,$RbF.6UF_4$	Per.		
	832	70.5	$RbF.6UF_4$,UF_4	Per.		
VF_3	755	3.5	Rb_3VF_6, RbF	Eut.	2d,3b	1523
	1128	-	Rb_3VF_6	Comp.		
	828	39.5	$RbVF_4$,Rb_3VF_6	Eut.		
	928	48	$Rb_2V_5F_{17}$,$RbVF_4$	Per.		
	996	52	VF_3,$Rb_2V_5F_{17}$	Per.		
	690		$2Rb_2VF_5 \rightleftharpoons Rb_3VF_6 + RbVF_4$	P.T.		
	115		$\alpha \rightleftharpoons \beta$ $Rb_2V_5F_{17}$	P.T.		
	130		$\alpha \rightleftharpoons \beta$ } Rb_3VF_6			
	258		$\beta \rightleftharpoons \gamma$ } Rb_3VF_6	P.T.		
	345		$\gamma \rightleftharpoons \delta$ } Rb_3VF_6			
	755	3.5	RbF, Rb_3VF_6	Eut.	2d,3b	1524
	1128	-	Rb_3VF_6	Comp.		
	828	39.5	Rb_3VF_6, $RbVF_4$	Eut.		
	928	48	$RbVF_4$, $Rb_2V_5F_{17}$	Per.		
	996	52	$Rb_2V_5F_{17}$, VF_3	Per.		
	345		$\alpha \rightleftharpoons \beta$ Rb_2VF_6	P.T.		
	115		$\alpha \rightleftharpoons \beta$ $RbVF_4$	P.T.		
	690		Dissociation of Rb_2VF_5	P.T.		
ZnF_2	596	23	Rb_2ZnF_4,RbF	Eut.	2,3b	1525
	648	34	$RbZnF_3$, Rb_2ZnF_4	Per.		
	740	50	$RbZnF_3$	Comp.		
	667	65	ZnF_2, $RbZnF_3$	Eut.		
ZrF_4			Fig. 271			1526
	703	10	RbF,$3RbF.ZrF_4$	Eut.		1526
	900	25	$3RbF.ZrF_4$	Comp.		
	615	34	28% ZrF_4 sol.sln. in $3RbF.ZrF_4$, $2RbF.ZrF_4$	Per.		

RbF

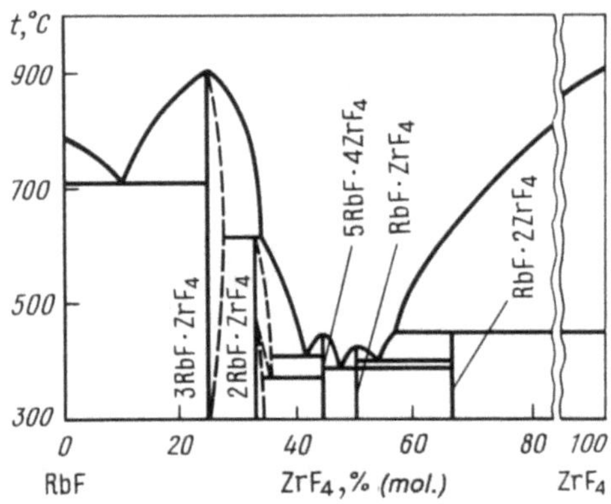

Fig. 271. RbF-ZrF_4 system.

2nd Component	T°C	Conc. mol %	Phases involved	Def.	Method	Ref.
ZrF_4	405	43	~35% ZrF_4 sol.sln. in $2RbF.ZrF_4$, $5RbF.4ZrF_4$	Eut.		1526
	430	45	$5RbF.4ZrF_4$	Comp.		
	380	48	$5RbF.4ZrF_4$, $RbF.ZrF_4$	Eut.		
	410	50	$RbF.ZrF_4$	Comp.		
	400	54	$RbF.ZrF_4$,$RbF.2ZrF_4$	Eut.		
	440	57	$Rb.2ZrF_4$, ZrF_4	Per.		
	375		α-$2RbF.ZrF_4$ ⇌ β-$2RbF.ZrF_4$ + $5RbF.4ZrF_4$	P.T.		
	395		$RbF.ZrF_4$ + $RbF.2ZrF_4$ ⇌ phase x	P.T.		

RbI

2nd Component	T°C	Conc. mol %	Phases involved	Def.	Method	Ref.
SbI_3	682	0.0	RbI	M.P.	1,2,2d, 3b	1527
	482	27	$3RbI.2SbI_3$,RbI	Eut.		
	496	40	$3RbI.2SbI_3$	Comp.		
	165	82	SbI_3,$3RbI.2SbI_3$	Eut.		
	170	100	SbI_3	M.P.		
	490	28	$3RbI.2SbI_3$,RbI	Eut.	6c	1528
	528	40	$3RbI.2SbI_3$	Comp.		
	-	-	SbI_3,$3RbI.2SbI_3$	Eut.		

RbI

2nd Component	T°C	Conc. mol %	Phases involved	Def.	Method	Ref.
SnI_2	289	49.62	$RbI, RbSnI_3$	Eut.	2a,3b	1529
	310	50.0	$RbSnI_3$	Comp.		
	279	62.0	$RbSnI_3, RbSn_2I_5$	Eut.		
	284	67.0	$RbSn_2I_5$	Comp.		
	271	81.0	$RbSn_2I_5, SnI_2$	Eut.		
	643	0.0	RbI	M.P.	2a	1530
	350	50	$3RbI.2SnI_2, RbI$	Per.		
	300	80	$SnI_2, 3RbI.2SnI_2$	Eut.		
	330	100	SnI_2	M.P.		
TeI_4	642	0.0	RbI	M.P.	2,3b,1e	1531
	458	21.5	RbI, Rb_2TeI_6	Eut.		
	484	33	Rb_2TeI_6	Comp.		
	268	73	Rb_2TeI_6, TeI_4	Eut.		
	280	100	TeI_4	M.P.		
TlI	430	~85	~17% TlI sol.sln. in RbI, 4% RbI sol. sln. in TlI	Eut.	2d,1g	1532
		~100	α-TlI ⇌ β-TlI based sol.sln. + RbI	P.T.		
		100	α ⇌ β TlI - with moderate cooling this transformation occurs	P.T.		
			2-3% RbI sol.sln. in TlI of limited extent		2,6d,1e	1533

$ReCl_3$

2nd Component	T°C	Conc. mol %	Phases involved	Def.	Method	Ref.
$ReCl_5$	350	84	$ReCl_3, β\text{-}ReCl_4$	Per.	2,3b,5c	1534
	260	85	$β\text{-}ReCl_4, ReCl_5$	Eut.		
	273	100	$ReCl_5$	M.P.		

$ReCl_5$

2nd Component	T°C	Conc. mol %	Phases involved	Def.	Method	Ref.
$SbCl_3$	60	97	$ReCl_5$, $SbCl_3$	Eut.	2	1535
$SbCl_5$			Fig. 272			1535
	-2	~100	$ReCl_5$, $SbCl_5$	Eut.	2	1535
$SnCl_4$			Fig. 273			1535
	-34	~100	$ReCl_5$, $SnCl_4$	Eut.	2	1535

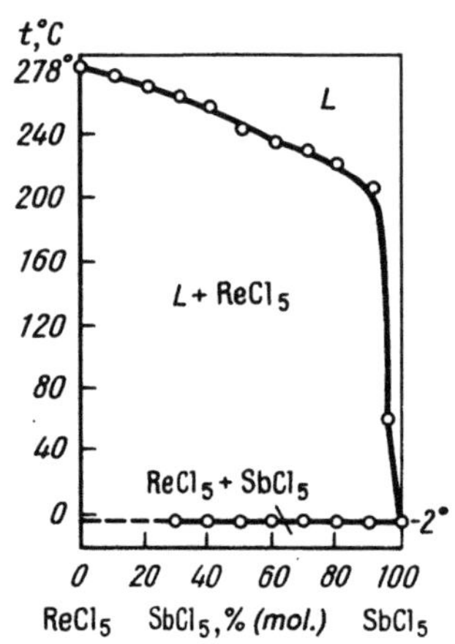

Fig. 272. $ReCl_5$-$SbCl_5$ system.

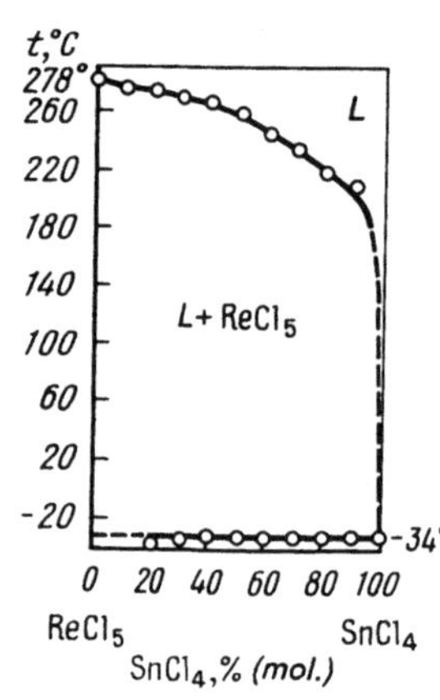

Fig. 273. $ReCl_5$-$SnCl_4$ system.

$ReCl_5$

2nd Component	T°C	Conc. mol %	Phases involved	Def.	Method	Ref.
$TaCl_5$			Fig. 274			1536
	196	80	20% $TaCl_5$ sol. sln. in $ReCl_5$, ~5% $ReCl_5$ sol.sln. in $TaCl_5$	Eut.	2a	1536
WCl_6			Fig. 275			1536
	222	20	β-WCl_6, γ-WCl_6	Per.	2a	1536
	200	75	10% WCl_6 sol.sln. in $ReCl_5$ β-WCl_6	Eut.		
	174	-	WCl_6	P.T.		

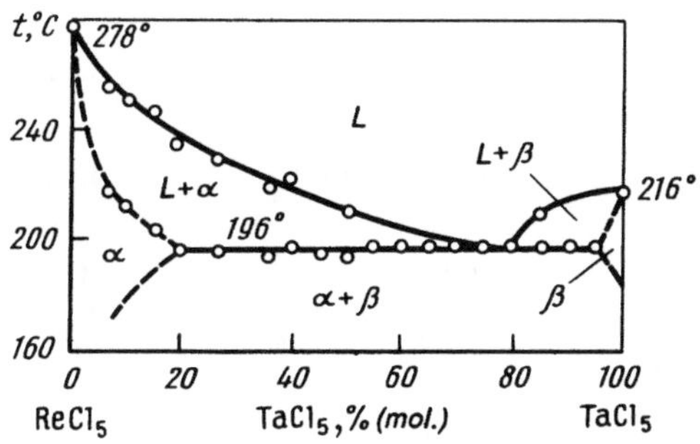

Fig. 274. $ReCl_5$-$TaCl_5$ system.

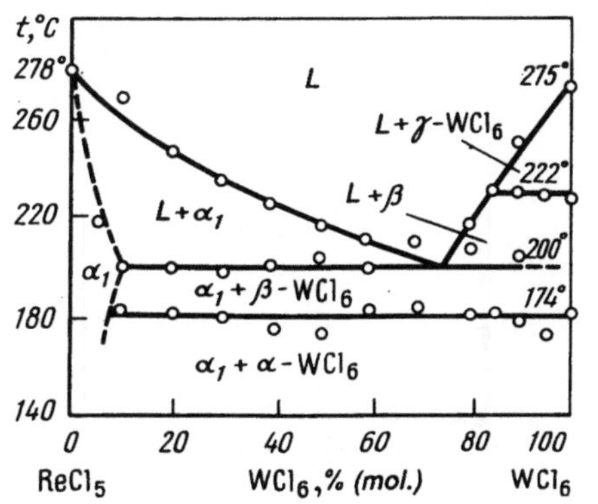

Fig. 275. $ReCl_5$-WCl_6 system.

$SbBr_3$

2nd Component	T°C	Conc. mol %	Phases involved	Def.	Method	Ref.
$TeBr_4$	97	0.0	$SbBr_3$	M.P.	2d	1537
	84	3-4	$SbBr_3$, $TeBr_4$	Eut.		
	386	100	$TeBr_4$	M.P.		

$SbCl_3$

2nd Component	T°C	Conc. mol %	Phases involved	Def.	Method	Ref.
$SbCl_5$	75	0.0	$SbCl_3$	M.P.	2d	1538
	1	95	$SbCl_3$, $SbCl_5$	Eut.		
	3	100	$SbCl_5$	M.P.		
$SeCl_4$	66	~6	~32% $SbCl_3$ sol. in $SeCl_4$, $SbCl_3$	Eut.	1,2,3b, le	1539
$TaCl_5$	73	~0	$SbCl_3$, $TaCl_5$	Eut.	1,2,2a	1540
$TeCl_4$	72	~0	$SbCl_3$, $TeCl_4$	Eut.	2a	1541
$TiCl_4$	73.3	0	$SbCl_3$	M.P.	1,2,4a	1542
	-25	97.25	$SbCl_3$, $TiCl_4$	Eut.		
	-23.1	100	$TiCl_4$			
			Fig. 276			1543

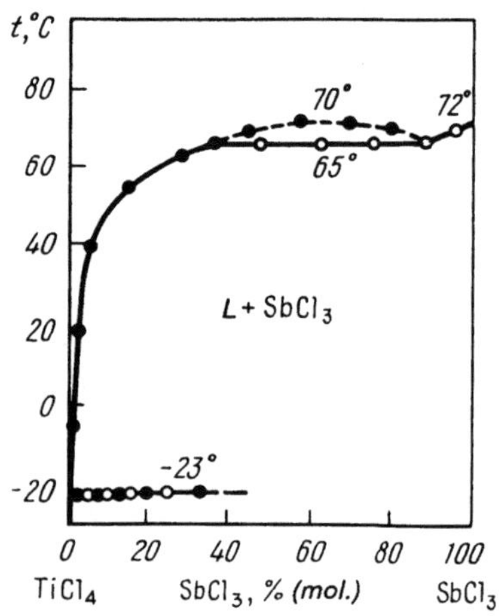

Fig. 276. $TiCl_4$-$SbCl_3$ system.

$SbCl_3$

2nd Component	T°C	Conc. mol %	Phases involved	Def.	Method	Ref.
$TiCl_4$	72	0.0	$SbCl_3$	M.P.	2d	1543
		~100	$TiCl_4$, $SbCl_3$	Eut.		
	65	12-64	Dissociation	P.T.		
TlCl			Fig. 277			1544
	72	0.0	$SbCl_3$	M.P.	2d	1544
	70	6	$2TlCl.3SbCl_3$,$SbCl_3$	Eut.		
	135	40	$2TlCl.SbCl_3$, $2TlCl.3SbCl_3$	Per.		
	235	57	$3TlCl.SbCl_3$, $2TlCl.SbCl_3$	Per.		
	282	74	TlCl, $3TlCl.SbCl_3$	Per.		

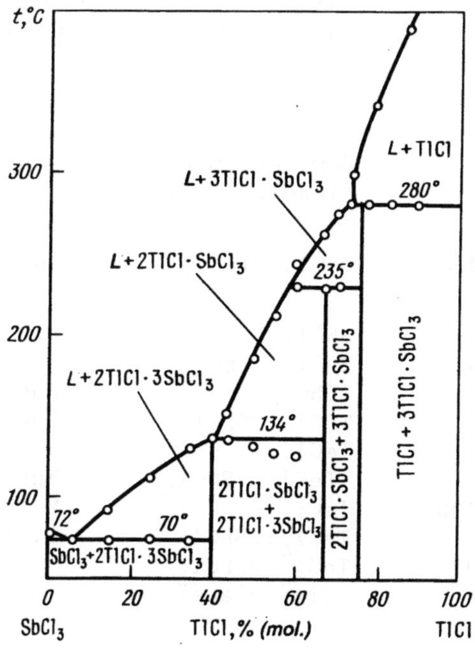

Fig. 277. $SbCl_3$-TlCl system.

WCl_6	75	0	$SbCl_3$	M.P.	2	1545
	73±2	~0	$SbCl_3$, α-WCl_6	Eut.		
	174	38	α-WCl_6, β-WCl_6	Per.		
	222	62	β-WCl_6, γ-WCl_6	Per.		
	275	100	γ-WCl_6	M.P.		

$SbCl_3$

2nd Component	T°C	Conc. mol %	Phases involved	Def.	Method	Ref.
$ZnCl_2$			Components insoluble either in liquid or in solid state		2d	1544
$ZrCl_4$	72	0.0	$SbCl_3$ Addition of $ZrCl_4$ rapidly raises solidification pt.	M.P.	2d	1543

$SbCl_5$

2nd Component	T°C	Conc. mol %	Phases involved	Def.	Method	Ref.
$SeCl_4$	3	0.0	$SbCl_5$	M.P.	2d	1546
	1	~0.0	$SeCl_4.2SbCl_5$, $SbCl_5$	Eut.		
	-	66.7	$SeCl_4.2SbCl_5$	Comp.		
	218	44	$SeCl_4$, $SeCl_4.2SbCl_5$	Eut.		
$SnCl_4$			Complete mutual insolubility		2d	1547
$TaCl_5$		0.0	$SbCl_5$, $TaCl_5$	Eut.	1,2,2a	1540
$TeCl_4$	3	0.0	$SbCl_5$	M.P.	2a,3b	1548
	1	0.0	$TeCl_4.2SbCl_5$, $SbCl_5$	Eut.		
	144	33.3	$TeCl_4.2SbCl_5$	Comp.		
	119	60.0	$TeCl_4$, $TeCl_4.2SbCl_5$	Eut.		
	224	100	$TeCl_4$	M.P.		
	-4	-	$TeCl_4.2SbCl_5$	P.T.		
	144	97.57	$TeCl_4.2SbCl_5$ dissociates			
$TiCl_4$	-47.5	65	-	Eut.	2	1548a
TlCl	3	~100	$SbCl_5$	M.P.	2d	1546
	3	10.0	$2TlCl.3SbCl_5$,$SbCl_5$	Eut.		
	238	65	$2TlCl.3SbCl_5$, $3TlCl.SbCl_5$ base sol.sln.	Eut.		

$SbCl_5$

2nd Component	T°C	Conc. mol %	Phases involved	Def.	Method	Ref.
TlCl	255	75	TlCl,3TlCl.$SbCl_5$ base sol.sln.	Per.	2d	1546
	430	~100	TlCl	M.P.		
	295	3-40	Dissociation			
WCl_6			Fig. 278			1549
	2	~0.0	-	Eut.	2,1	1549
	174	-	α ⇌ β WCl_6	P.T.		
	222		β ⇌ γ WCl_6	P.T.		

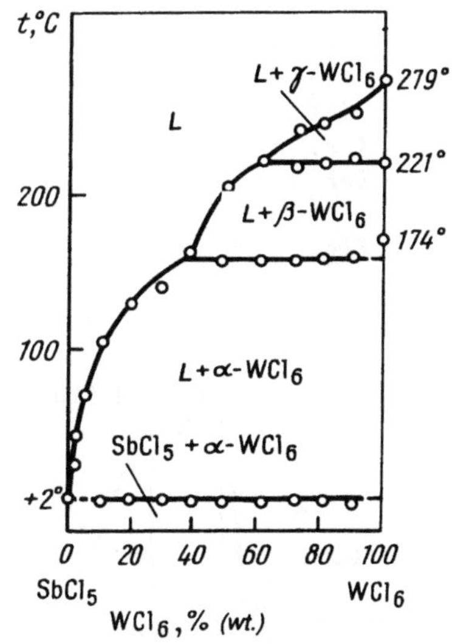

Fig. 278. $SbCl_5$-WCl_6 system.

2nd Component	T°C	Conc. mol %	Phases involved	Def.	Method	Ref.
$ZnCl_2$	3	0.0	$SbCl_5$	M.P.	2d	1546
	~3	~0.0	$SbCl_5$, $ZnCl_2$	Eut.		
	312	100	$ZnCl_2$	M.P.		
	2-94		Dissociation			

SbF_5

2nd Component	T°C	Conc. mol %	Phases involved	Def.	Method	Ref.
UF_6	-5.5	19.5	UF_6, SbF_5	Eut.	2	1550
		23.3-39	Dissociation (max. temp. 73.5)			
XeF_2	2	1.5	SbF_5, $XeF_2.6SbF_5$	Eut.	2d,2b	1551
	34	14	$XeF_2.6SbF_5$	Comp.		
	25	19	$XeF_2.6SbF_5$, $XeF_2.2SbF_5$	Eut.		
	63	33	$XeF_2.2SbF_5$	Comp.		
	50	38	$XeF_2.2SbF_5$, $XeF_2.1.5SbF_5$	Eut.		
	86.5	40	$XeF_2.1.5SbF_5$	Comp.		
	73.5	43	$XeF_2.1.5SbF_5$, $XeF_2.SbF_5$	Eut.		
	161	50	$XeF_2.SbF_5$	Comp.		
	186	86	$XeF_2.SbF_5$, XeF_2	Eut.		
			Fig. 279			1552

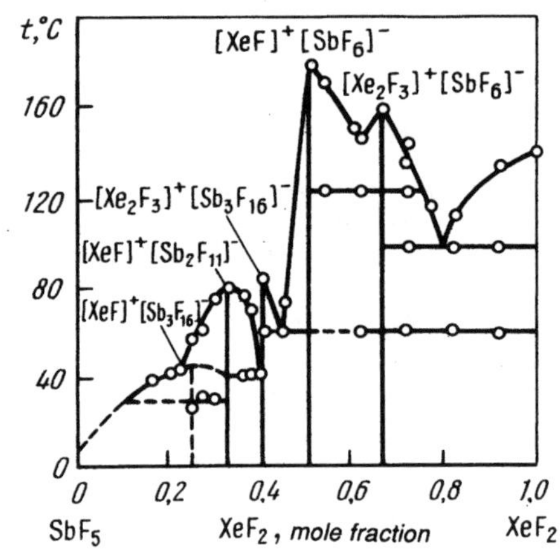

Fig. 279. SbF_5-XeF_2 system.

	~44	-	$[XeF]^+[Sb_3F_{16}]^-$ (incongruent melting)	Comp.	2a	1552

SbF_5

2nd Component	T°C	Conc. mol %	Phases involved	Def.	Method	Ref.
XeF_2	79	-	$[XeF]^+[Sb_2F_{11}]^-$	Comp.	2a	1552
	44	-	$[Xe_2F_3]^+[Sb_3F_{16}]^-$	Comp.		
	178	-	$[XeF]^+[SbF_6]^-$	Comp.		
	159	-	$[Xe_2F_3]^+[SbF_6]^-$	Comp.		

SbI_3

2nd Component	T°C	Conc. mol %	Phases involved	Def.	Method	Ref.
TeI_4	176	0.0	SbI_3	M.P.	2,3b	1553
	160	15	SbI_3, TeI_4	Eut.		
	280	100	TeI_4	M.P.		

$SeCl_4$

2nd Component	T°C	Conc. mol %	Phases involved	Def.	Method	Ref.
$TaCl_5$			Fig. 91			
	196	50	$SeCl_4$, $TaCl_5.SeCl_4$	Per.	2a	1554
	188	68	$TaCl_5.SeCl_4$,$TaCl_5$	Eut.		
$TeCl_4$			Continuous sol.sln.		2a	1555
	>270	$SeCl_4$	Dissociation			
WCl_6	213	60	WCl_6 base sol. sln., $SeCl_4$	Eut.	2d	1556
	158	-	WCl_6	P.T.		
	217	-	WCl_6	P.T.		

$SmCl_3$

2nd Component	T°C	Conc. mol %	Phases involved	Def.	Method	Ref.
$SrCl_2$	660	0.0	$SmCl_3$	M.P.	2a,3b	1557
	540	40	$SmCl_3$,Sr_3SmCl_9	Eut.		
	580	49	Sr_3SmCl_9,$SrCl_2$	Per.		
	868	100	$SrCl_2$	M.P.		
YCl_3	662	0.0	$SmCl_3$	M.P.	2d,3b	1558
	564	50.5	9% YCl_3 sol.sln. in $SmCl_3$, 9%$SmCl_3$ sol.sln.in YCl_3	Eut.		
	716	100	YCl_3	M.P.		

SmF_3

2nd Component	T°C	Conc. mol %	Phases involved	Def.	Method	Ref.
SrF_2			Limited sol.sln. of hexagonal form		3b	1559

$SnBr_2$

2nd Component	T°C	Conc. mol %	Phases involved	Def.	Method	Ref.
SnI_2			Fig. 280			
	240	-	15% SnBr sol.sln.in SnI (β) [monoclinic structure]		2d,3b	1560
			$SnBr_2$ base sol.sln. (α) [rhombic structure]			
		0-63	Range of α sol. sln.			

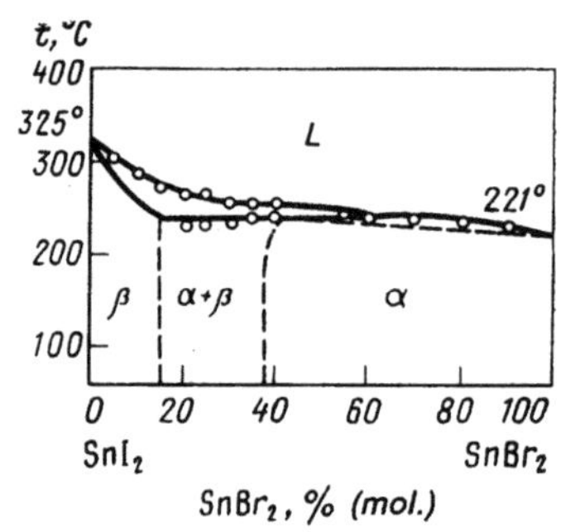

Fig. 280. SnI_2-$SnBr_2$ system.

$SnBr_4$

2nd Component	T°C	Conc. mol %	Phases involved	Def.	Method	Ref.
$TiBr_4$			Continuous sol.sln.		2,1c	1565

$SnCl_2$

2nd Component	T°C	Conc. mol %	Phases involved	Def.	Method	Ref.
$SnCl_4$			Complete mutual insolubility		2d	1547

$SnCl_2$

2nd Component	T°C	Conc. mol %	Phases involved	Def.	Method	Ref.
$TaCl_5$	154	30.5	$SnCl_2$, $2TaCl_5.SnCl_2$	Eut.	2c	1561
	239	66.6	$2TaCl_5.SnCl_2$	Comp.		
	208	88.9	$2TaCl_5.SnCl_2$, $TaCl_5$	Eut.		
TlCl	178	28	-	Eut.	-	1562
	191	21	$TlCl.2SnCl_2$, $SnCl_2$	Eut.	2d	1563
	207	33	$TlCl.2SnCl_2$	Comp.		
	204	34	$TlCl.SnCl_2$, $TlCl.2SnCl_2$	Eut.		
	256	50	$TlCl.SnCl_2$	Comp.		
	251	56	$3TlCl.SnCl_2$, $TlCl.SnCl_2$	Eut.		
	318	75	$3TlCl.SnCl_2$	Comp.		
	315	76	TlCl, $3TlCl.SnCl_2$	Eut.		
YCl_3	247	0.0	$SnCl_2$	M.P.	2	1564
	244	10.4	YCl_3 base sol.sln.	Eut.		
	700	100	YCl_3	M.P.		
	230	-	$SnCl_2.YCl_3$	P.T.		
$ZnCl_2$	200	26	$ZnCl_2$, $SnCl_2$	Eut.	1,2	1565
$ZrCl_4$	210	13	$SnCl_2$, $2ZrCl_4.SnCl_2$ (?)	Eut.	1,2	1565
	315	52	$2ZrCl_4.SnCl_2$ (?) $ZrCl_4$	Per.		

$SnCl_4$

2nd Component	T°C	Conc. mol %	Phases involved	Def.	Method	Ref.
$TaCl_5$	-33	<1.5	$SnCl_4$, $TaCl_5$	Eut.	2c	1561
WCl_6	~-32	~0	-	Eut.	1,2	1549
	174		$\alpha \rightleftharpoons \beta$ WCl_6	P.T.		
	222		$\beta \rightleftharpoons \gamma$ WCl_6	P.T.		

$SrBr_2$

2nd Component	T°C	Conc. mol %	Phases involved	Def.	Method	Ref.
$SrCl_2$	643	0.0	$SrBr_2$	M.P.	1,2	1566
	641	7.5	Continuous sol.sln.	Min.		
	867	100	$SrCl_2$	M.P.		
		0-18	SrBr type sol.sln. (cubic)		3b	1567
		18-30	Solid sln.of cubic and orthorhombic structure			
		30-85	Solid sln. of orthorhombic ($PbCl_2$) structure			
		85-93	Orthorhombic and cubic sol.slns.			
		93-100	Sol.slns. of cubic ($SrCl_2$) structure			
SrI_2	450	0-10	SrBr type str. (1)	sol. slns.	2a,3b, 1e	1567
	450	10-32	mixture of (1) & (11)	sol. slns.		
	450	32-56	Orthorhombic, $PbCl_2$ type (11)	sol. slns.		
	450	56-82	Mixture of (11) * (111)	sol. slns.		
	450	82-100	Orthorhombic, SrI_2 type (111)	sol. slns.		

$SrCl_2$

2nd Component	T°C	Conc. mol %	Phases involved	Def.	Method	Ref.
$TaCl_5$	296	-	$SrCl_2$, $SrTaCl_7$	Per.	2a	1568
	216	~0.0	$TaCl_5$, $SrTaCl_7$	Eut.		
$ThCl_4$	874±2	0.0	$SrCl_2$	M.P.	2d	1569
	496±2	58	$SrCl_2$, $ThCl_4$	Eut.		
	768±2	100	$ThCl_4$	M.P.		
UCl_4	398±2	59	$SrCl_2$, UCl_4	Eut.	2d,3b	1570

SrF_2

2nd Component	T°C	Conc. mol %	Phases involved	Def.	Method	Ref.
TbF_3			Limited sol.sln. of hexagonal structure		3b	1559

SrF_2

2nd Component	T°C	Conc. mol %	Phases involved	Def.	Method	Ref.
TiF_3	719±10		$SrTiF_5$ (monclinic str.)	Comp.	2d,3b, le	1571
	735±15		$Sr_3(TiF_6)_2$ (cubic str.)	Comp.		
TuF_3			SrF_2 base sol. sln. (fluorite str.), TuF_3 base sol.sln (tisonite str.), Mixture of above between the two limits		3b	1559
UF_3		0-33	Solid solutions		3b	1572
VF_3	820		$Sr_3(VF_6)_2$, SrF_2	Per.	2d,3b	1573
	822		$SrVF_5$,$Sr_3(VF_6)_2$	Eut.		
	860		VF_3,$SrVF_5$	Per.		
YF_3			$SrF_2.2YF_3$ (partially soluble in components)	Comp.	3b	1574
		0-38	SrF_2 base sol.sln (fluorite str.)		3b	1559
		38-65	Fluorite phase + tisonite phase			
		65-80	Tisonite phase			
		80-100	Tisonite phase + YF_3			
YbF_3			Limited solid slns: SrF_2 base - fluorite str. YbF_3 base - tisonite str. Mixture of the above two between the limits		3b	1559

$TaCl_5$

2nd Component	T°C	Conc. mol %	Phases involved	Def.	Method	Ref.
$TeCl_4$	178	32.5	$TaCl_5$, $TaCl_5.TeCl_4$	Eut.	2a	1575
	182	50	$TaCl_5.TeCl_4$	Comp.		
	152	65.5	$TaCl_5.TeCl_4$, $TeCl_4$	Eut.		
			Fig. 248			
	172	50.0	$TaCl_5.TeCl_4$, $TaCl_5$	Per.	2a	1554
	144	62.0	$TeCl_4$,$TaCl_5.TeCl_4$	Eut.		
$ThCl_4$			Almost complete mutual insolubility		2,6c	1576
	220	~0.0	$TaCl_5$,$ThCl_4$	Eut.		
$TiCl_4$	219	0.0	$TaCl_5$	M.P.	2,6m	1577
	-24	~100	-	Eut.		
	-24	100	$TiCl_4$	M.P.		
	-24.2	99.86	$TaCl_5$,$TiCl_4$	Eut.	1,2,2a	1578
TlCl	215	~0.0	$TlTaCl_6$,$TaCl_5$	Eut.	2d,1	1579
	370	50	$TlTaCl_6$	Comp.		
	276	64	Tl_3TaCl_8,$TlTaCl_6$	Eut.		
	350	~72	TlCl,Tl_3TaCl_8 (unstable)	Per.		
	333	0-60	Dissociation			
	300		$TlTaCl_6$	P.T.		
UCl_4	198	3	$TaCl_5$,UCl_4	Eut.	2	1580
VCl_4		~100	$TaCl_5$,VCl_4	Eut.	1,2,2a	1581
WCl_6	158	50.1	$TaCl_5$,α-WCl_6	Eut.	2a	1582
	174	52	α-WCl_6,β-WCl_6	Per.		
	222	73	β-WCl_6,γ-WCl_6			
$ZnCl_2$	215	~0.0	$TaCl_5$,$ZnCl_2$	Eut.	2d,1	1583
	250	18-88	Dissociation			

$TaCl_5$

2nd Component	T°C	Conc. mol %	Phases involved	Def.	Method	Ref.
TaF_5			Fig. 281			1584
	216	0.0	$TaCl_5$	M.P.	1,2,2i	1584
	198.6	10.1	$TaCl_5,TaCl_4F$			
	179.8	38.6	$TaCl_4,TaCl_3F_2$			
	101.5	54.6	$TaCl_3F_2,TaCl_2F_3$(?)			
	72	88.5	$TaClF_4$(?) TaF_5			
	96.8	100.0	TaF_5	M.P.		
			The upper curve → liq.-vap.equil.			
			The lower curve → liq.-solid equil.			
		60-80	Glassy str.			
	231	3.5	760mm Hg Azeotropic			

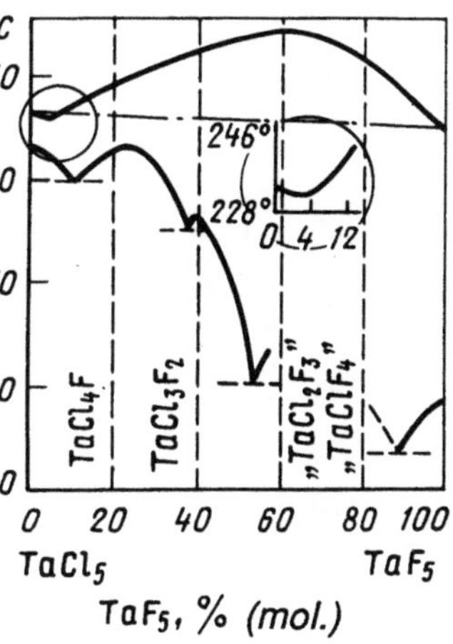

Fig. 281. $TaCl_5$-TaF_5 system.

TaF_5

2nd Component	T°C	Conc. mol %	Phases involved	Def.	Method	Ref.
UF_6	52	83	TaF_5,UF_6	Eut.	2	1585
		26-76	Dissociation			
XeF_2	96	0.0	TaF_5	M.P.		1586
	60	25	$TaF_5,XeF_2.2TaF_5$	Eut.		
	84	33	$XeF_2.2TaF_5$	Comp.		
	65	40	$XeF_2.2TaF_5$, $XeF_2.TaF_5$	Eut.		
	92	50	$XeF_2.TaF_5$	Comp.		
	70	68	$XeF_2.TaF_5,XeF_2$	Eut.		
	140	100	XeF_2	M.P.		

$TeBr_4$

2nd Component	T°C	Conc. mol %	Phases involved	Def.	Method	Ref.
TlBr	388	0.0	$TeBr_4$	M.P.	2a,2e, 3b	1587
	356	18.0	$TeBr_4$, Tl_2TeBr_6	Eut.		
	536	66	Tl_2TeBr_6	Comp.		
	424	90	Tl_2TeBr_6,TlBr	Eut.		
	456	100	TlBr	M.P.		
$TeCl_4$	386	0.0	$TeBr_4$	M.P.	2a	1588
			Continuous sol.sln.			
	224	100	$TeCl_4$	M.P.		
TeI_4	386	0.0	$TeBr_4$	M.P.	2a	1588
	267	60	Continuous sol.sln.	Min.		
	280	100	TeI_4	M.P.		

$TeCl_4$

2nd Component	T°C	Conc. mol %	Phases involved	Def.	Method	Ref.
$TiCl_4$	224	0.0	$TeCl_4$	M.P.	2d	1589
	-23	~100	-	Eut.		
	-23	100	$TiCl_4$	M.P.		
TlCl	210	14	$TeCl_4$, $2TlCl.TeCl_4$	Eut.	1,2a	1590
	488	66.7	$2TlCl.TeCl_4$	Comp.		
	388	89	$2TlCl.TeCl_4$,TlCl	Eut.		
	224	0	$TeCl_4$	M.P.		1591
	201	11	$TeCl_4$,Tl_2TeCl_6	Eut.		
	457	66.7	Tl_2TeCl_6	Comp.		
	375	88	Tl_2TeCl_6,TlCl	Eut.		
	429	100	TlCl	M.P.		
WCl_6	182	45	β-WCl_6,$TeCl_4$	Eut.	2a	1592
	224	-	$\gamma \rightleftharpoons \beta$ WCl_6	P.T.		
	174	-	$\beta \rightleftharpoons \alpha$ WCl_6	P.T.		
TeI_4	224	0.0	$TeCl_4$	M.P.	2a	1588
	202	65	Continuous sol.sln.	Min.		
	280	100	TeI_4	M.P.		

TeI_4

2nd Component	T°C	Conc. mol %	Phases involved	Def.	Method	Ref.
TlI	280	0.0	TeI_4	M.P.	2a,2e,	1593
	268	20.0	TeI_4,Tl_2TeI_6	Eut.	6c,3b	
	426	66.7	Tl_2TeI_6	Comp.		
	358	74.0	Tl_2TeI_6, α-TlI	Eut.		
	440	100	TlI	M.P.		
	170	-	TlI	P.T.		
ZnI_2	280	0.0	TeI_4	M.P.	2a	1594
	260	8.0	TeI_4,ZnI_2	Eut.		
	446	100	ZnI_2	M.P.		
	380	40-70	Dissociation			

$ThCl_4$

2nd Component	T°C	Conc. mol %	Phases involved	Def.	Method	Ref.
UCl_3	632	30	$ThCl_4$, $3UCl_3.ThCl_4$	Eut.	2d	1595
	750	65	$3UCl_3.ThCl_4$,UCl_3	Per.		

ThF_4

2nd Component	T°C	Conc. mol %	Phases involved	Def.	Method	Ref.
UF_4			Continuous sol.sln.		2,1f,3b	1596

$TiBr_4$

2nd Component	T°C	Conc. mol %	Phases involved	Def.	Method	Ref.
$TiCl_4$	13.4		$TiClBr_3$	Comp.		1597
	-5.7		$TiCl_2Br_2$	Comp.		
	-19.4		$TiCl_3Br$	Comp.		
	202		$TiClBr_3$ } Boiling points			
			$TiCl_2Br_2$ } with significant			
	155		$TiCl_3Br$ } decomposition			
			Continuous solid slns. between the compounds			
	38.4	0.0	$TiBr_4$	M.P.	2d,4a	1598
		100	Continuous sol.sln.	Min.		

$TiCl_4$

2nd Component	T°C	Conc. mol %	Phases involved	Def.	Method	Ref.
VCl_4			Continuous sol.sln.		2d	1599
WCl_6			Fig. 282			1600
	-23	~0.0	-	Eut.	1,2	1600
	174	-	$\alpha \rightleftharpoons \beta$ WCl_6	P.T.		
	222		$\beta \rightleftharpoons \gamma$ WCl_6	P.T.		

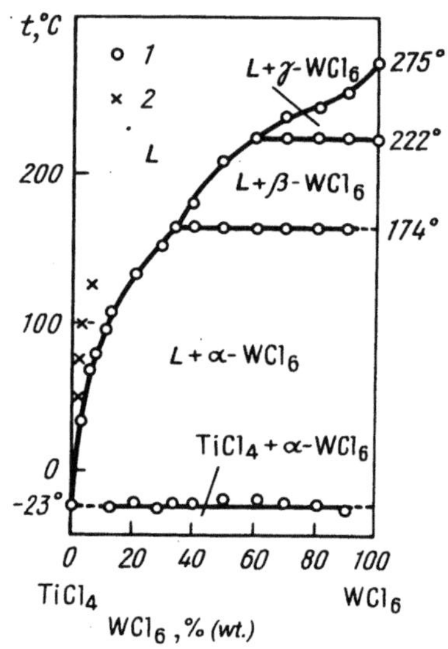

Fig. 282. $TiCl_4$-WCl_6 system.

TlBr

2nd Component	T°C	Conc. mol %	Phases involved	Def.	Method	Ref.
TlCl	423	60	Sol.sln.	Min.	1,2	1601
	460	0.0	TlBr	M.P.	1,2	1602
	425	65	Continuous sol.sln.	Min.		
	429	100	TlCl	M.P.		
	418	~70	Continuous sol.sln.	Min.	2d	1603
	-	70	Continuous sol.sln.	Min.	1e	1604

TlBr

2nd Component	T°C	Conc. mol %	Phases involved	Def.	Method	Ref.
TlI	405	50	Continuous sol.sln. Phase transf. in solid state	Min.	1,2	1605
	411.7	50	Continuous sol.sln.	Min.	2d	1603
		58	Continuous sol.sln.	Min.	3a,3b,	1604
		80-100	Sol.sln.in two phases β - cubic (high temp.) α - rhombic (low temp.) α ⇌ β temp. increases with TlI content α not obtained below 20°C		2a,5	1606

$TlBr_2$

2nd Component	T°C	Conc. mol %	Phases involved	Def.	Method	Ref.
$TlCl_2$			Sol.sln. of the form $TlCl_{2-x}Br_x$ le ($0 \leq x < 2$) having the following structures: $x \geqslant 0.28$ - tetragonal ($CaWO_4$ type) $x \leqslant 1.83$ - orthorhombic $0.28 \leqslant x \leqslant 1.83$ - different orthorhombic (TlClBr type) - similar to NiAs with Tl^+ in position of As and $TlCl_4$ in position of Ni		5c,3b,3	1607

TlCl

2nd Component	T°C	Conc. mol %	Phases involved	Def.	Method	Ref.
$TlCl_3$			Fig. 283			1608
	365	25.6	β-Tl_2Cl_3, $TlCl_2$	Eut.	2	1608
	388	40.0	β-Tl_2Cl_3	Comp.		
	369	50.0	α-Tl_2Cl_3,β-Tl_2Cl_3	Per.		
	232	67.2	TlCl,α-Tl_2Cl_3	Per.		
$ZnCl_2$			Fig. 284			1609
	334	22	TlCl, $3TlCl.ZnCl_2$	Eut.	2d,1	1610

TlCl

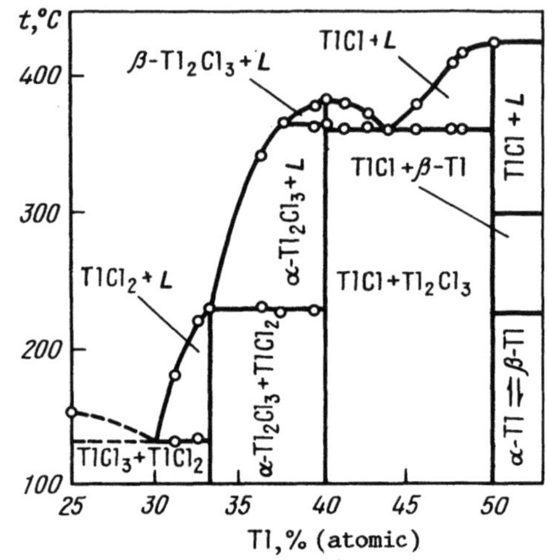

Fig. 283. $TlCl_3$-TlCl system.

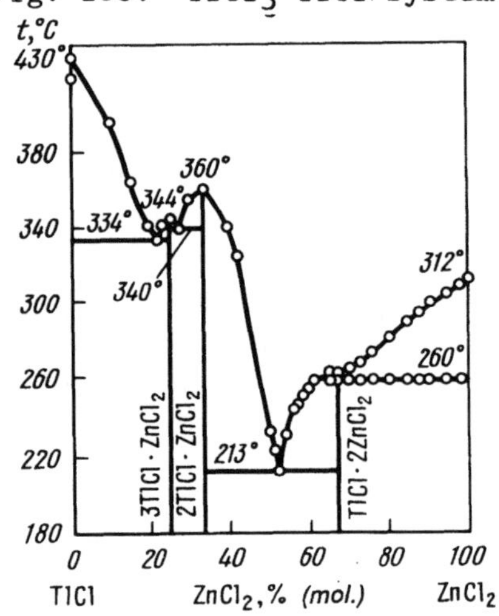

Fig. 284. TlCl-$ZnCl_2$ system.

2nd Component	T°C	Conc. mol %	Phases involved	Def.	Method	Ref.
$ZnCl_2$	344	25	$3TlCl.ZnCl_2$	Comp.	2d, l	1609
	340	27	$3TlCl.ZnCl_2$, $2TlCl.ZnCl_2$	Eut.		1610
	360	33.3	$2TlCl.ZnCl_2$	Comp.		
	213	52	$2TlCl.ZnCl_2$, $TlCl.2ZnCl_2$	Eut.		
	260	61	$TlCl.2ZnCl_2$, $ZnCl_2$	Per.		

TlCl

2nd Component	T°C	Conc. mol %	Phases involved	Def.	Method	Ref.
$ZnCl_2$			Congruently melting compound $ZnCl_2.2TlCl$ exists, whose stability decreases with increase of temp.		2a	1611
	450		Increasing the content of TlCl in the melt causes dissociation of $(ZnCl_2)_n$ and complex ions $ZnCl_3^-$ and $ZnCl_4^{--}$ appear		1i	1612
TlI	429	0.0	TlCl	M.P.	1,2	1613
	316	47.5	TlCl, TlI	Eut.		

TlF

2nd Component	T°C	Conc. mol %	Phases involved	Def.	Method	Ref.
VF_3	310	3	TlF, Tl_3VF_6	Eut.	2d,3b	1614
	736	-	Tl_3VF_6	Comp.		
	678	33	$Tl_3VF_6, TlVF_4$	Eut.		
	885	47	$TlVF_4, VF_3$	Per.		
	115		Tl_3VF_6	P.T.		

UCl_3

2nd Component	T°C	Conc. mol %	Phases involved	Def.	Method	Ref.
UCl_4			Fig. 285			1615

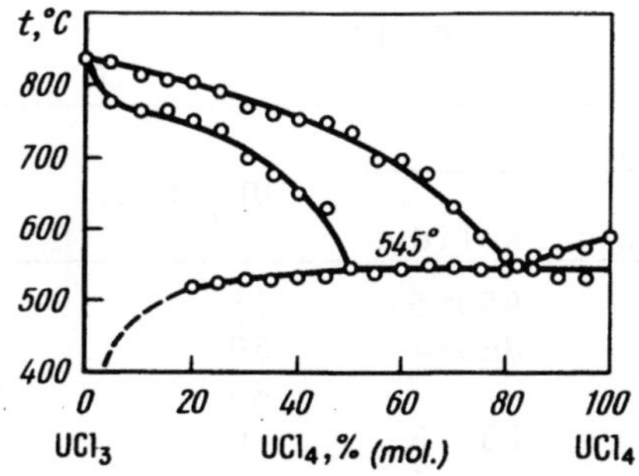

Fig. 285. UCl_3-UCl_4 system.

UCl_3

2nd Component	T°C	Conc. mol %	Phases involved	Def.	Method	Ref.
UCl_4	545	83	UCl_4, 50% UCl_4 sol.sln. in UCl_3	Eut.	2d	1615
UF_4	836±2	0.0	UCl_3	M.P.	2d	1616
	625±2	-	UCl_3,$2UCl_3.UF_4$	Per.		
	590±2	58	$2UCl_3.UF_4$, $UCl_3.7UF_4$	Eut.		
	760±2	-	$UCl_3.7UF_4$, II-UF_4	Per.		
	863	-	II-UF_4, I-UF_4	Per.		
	1036±2	100	I-UF_4	M.P.		

UCl_4

2nd Component	T°C	Conc. mol %	Phases involved	Def.	Method	Ref.
UF_4	543	3.5	β-UCl_4, α-UCl_4	Per.	2d,3b	1617
	444±2	38.0	α-UCl_4, UCl_3F	Per.		
	426	33.0	UCl_3F,UCl_2F_2	Eut.		
	460±3	38.5	UCl_2F_2,$UClF_3$	Per.		
	530±6	44.0	$UClF_3$,α-UF_4	Per.		
	833	80.0	α-UF_4, β-UF_4	Per.		

UF_3

2nd Component	T°C	Conc. mol %	Phases involved	Def.	Method	Ref.
UF_4			UF_4 insoluble in UF_3; UF_3 slightly soluble in UF_4		3b	1618
	879	67	UF_3, UF_4	Eut.	2a,3b	1619
	841		UF_4	P.T.		
	865	68	33% UF_4 sol.sln. in UF_3; No UF_4 base sol.sln.	Eut.		1620

TABLE XVIII

UF_4%(mol)	Temp °C		UF_4%(mol)	Temp °C	
	Liquidus	solidus		Liquidus	solidus
32±4	-	950±5	75	921±5	-
50	-	865±4	80	951±5	868±17
57±2	970±5	-	85	975±5	868±5
60	-	837±5	90	1066±5	865±5
70	890±5	876±5			

UF_4

2nd Component	T°C	Conc. mol %	Phases involved	Def.	Method	Ref.
ZrF_4			Fig. 286			1621
	765	77	Continuous sol.sln.	Min.	2d,1,2, 3b	

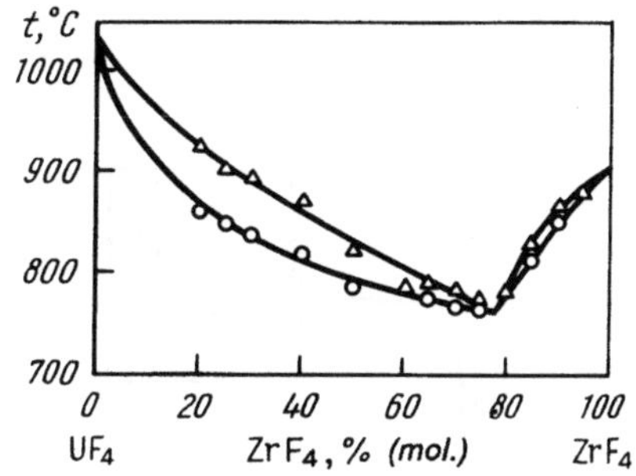

Fig. 286. UF_4-ZrF_4 system.

UF_6

2nd Component	T°C	Conc. mol %	Phases involved	Def.	Method	Ref.
VF_5	12	64	UF_6, VF_5	Eut.	2a	1622

VF_5

2nd Component	T°C	Conc. mol %	Phases involved	Def.	Method	Ref.
XeF_2	19.0	0	VF_5	M.P.	2d	1623
	12.0	30	VF_5,$XeF_2.VF_5$	Eut.		
	38 ±1	50	$XeF_2.VF_5$	Comp.		
	28.0	57	$XeF_2.VF_5$,$2XeF_2.VF_5$	Eut.		
	44.0	61	$2XeF_2.VF_5$,XeF_2	Per.		
	140.0	100	XeF_2	M.P.		

WCl_6

2nd Component	T°C	Conc. mol %	Phases involved	Def.	Method	Ref.
$ZnCl_2$	256	0.0	WCl_6	M.P.	2d	1624
			Complete insolubility			
	310	100	$ZnCl_2$	M.P.		
	224	-	$\alpha \rightleftharpoons \beta$ WCl_6	P.T.		
	174	-	$\beta \rightleftharpoons \gamma$ WCl_6	P.T.		

WF_6

2nd Component	T°C	Conc. mol %	Phases involved	Def.	Method	Ref.
XeF_2			Fig. 287			1625
	-8.5	0.0	WF_6	M.P.	2d	1625
	-		$XeF_2.4WF_6$ (incongruent melting pt)	Comp.		
	125±1		$XeF_2.WF_6$	Comp.		
	140	100	XeF_2	M.P.		

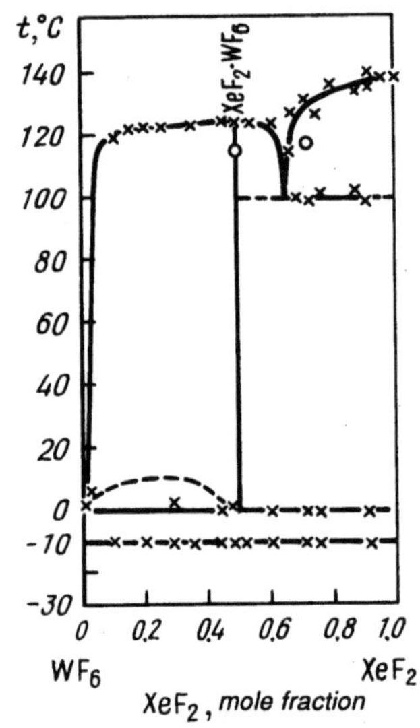

Fig. 287. WF_6-XeF_2 system.

XeF_2

2nd Component	T°C	Conc. mol %	Phases involved	Def.	Method	Ref.
XeF_6	140±1	0.0	XeF_2	M.P.	2d, 1h	1626
	30±1	50	XeF_2, α-XeF_6	Eut.		
	49±1	100	XeF_6	M.P.		
	15	-	α ⇌ β XeF_6			

$ZnCl_2$

2nd Component	T°C	Conc. mol %	Phases involved	Def.	Method	Ref.
$ZrCl_4$	308.5	7	$ZnCl_2$,$ZrCl_4$	Eut.	1,2	1627

$ZrBr_4$

2nd Component	T°C	Conc. mol %	Phases involved	Def.	Method	Ref.
$ZrCl_4$			Continuous sol.sln. with minimum		1,2	1628

$ZrCl_4$

2nd Component	T°C	Conc. mol %	Phases involved	Def.	Method	Ref.
ZrI_4	368	42	Continuous sol.sln.	Min.	1,2	1628

REFERENCES

1. I.K. Tovmas'yan, V.I. Monastyrskaya, The Physical Chemistry and Electrochemistry of Salts and Slags, Naukova Dumka, Kiev (1969), Part 1, p. 342
2. J.N. Bradley, P.D. Greene, Trans. Farad. Soc., **63**:424 (1967).
3. S.D. Dionis'ev, A.V. Buryakov, Zh. Neorg. Khim., **17**:1747 (1972).
4. W. Jost, S. Frhr. von Salmuth, Z. Phys. Chem. Neue Folge, **16**:277 (1958).
5. J. Teltow, Z. Phys. Chem. Neue Folge, **18**:134 (1958).
6. A.G. Dudareva, Yu.É. Bogatov, Akkelidu Konstanidiya, B.N. Ivanov-Émin, P.I. Fedorov, Izv. Vyssh. Uchebn. Zaved., Khim. Khim. Tekhnol., **14**:1299 (1971).
7. A.G. Bergman, A.S. Arabadzhan, Zh. Neorg. Khim., **8**:2148 (1963).
8. T. Tsugi, S. Murakoshi, K. Fucki, T. Mukaibo, Bull. Chem. Soc. Jpn., **43**:1343 (1970).
9. M.A. Zakharchenko, A.G. Bergman, Nauchn. Tr. Novocherk. Politekh. Inst., **27**(41):3 (1956).
   C. Sinistri, G. Flor, P. Ferloni, Z. Naturforsch., **23A**:2073 (1968).
10. C. Sinistri, R. Riccardi, C. Margheritis, P. Titarelli, Z. Naturforsch., **27A**:149 (1972).
11. V.V. Safonov, V.A. Grin'ko, M.B. Varfolomeev, E.S. Malysheva, V.I. Ksenzenko, Zh. Neorg. Khim., **18**:2827 (1973).
12. N.D. Chikanov, V.A. Ugai, Zh. Neorg. Khim., **18**:819 (1973).
13. Yu.I. Vesnin, Izv. Sib. Otd. Akad. Nauk SSSR, Ser. Khim. Nauk, Issue 2, No. 4, 62 (1867).
14. G. Flor, C. Sinistri, Ric. Sci., **38**(3):227 (1968).
15. M.S. Golubeva, R.D. Sidakova, The Physicochemical Analysis of Salt Systems, Izd. Rostov. Gos. Univ., Rostov (1962), p. 189.
16. C.R. Tipton, Reactor Handbook (Materials), 2nd ed., New York (1960), p. 442.
17. P.I. Fedorov, V.M. Yakunina, Zh. Neorg. Khim., **8**:2103 (1963).
18. P.I. Fedorov, N.I. Il'ina, Zh. Neorg. Khim., **9**:1207 (1964).
19. I.G. Murgulescu, S. Sternberg, Rev. Chim. Acad. Rep. Popul. Roum., **3**:55 (1958).
20. M.A. Zakharchenko, A.G. Bergman, Nauchn. Tr. Novocherk. Politekh. Inst., 27(41):3 (1956).
21. R.J. Stokes, C.H. Li, Acta Metall., **10**:535 (1962).
22. N.D. Chikanov, Zh. Neorg. Khim., **14**:1120 (1969).
23. G.G. Urazov, A.S. Karnaukhov, Zh. Neorg. Khim., **1**:733 (1956).
24. I.G. Murgulescu, S. Sternberg, Rev. Chim. Acad. Rep. Popul. Roum., **3**:55 (1958).
25. R.T. Trent, B.J. Welch, Austral. J. Chem., **19**:1085 (1966).

26. S.I. Dionis'ev, A.G. Bergman, Zh. Neorg. Khim., **15**:2579 (1970).
27. C. Sinistri, G. Flor, Ric. Sci., **38**:967 (1968).
28. N.D. Chikanov, Zh. Neorg. Khim., **17**:2558 (1972).
29. N.D. Chikanov, Zh. Neorg. Khim., **14**:1120 (1969).
30. V.V. Safonov, A.V. Konov, B.G. Korshunov, Zh. Neorg. Khim., **15**:1150 (1970).
31. V.I. Monastyrskaya, The Physical Chemistry and Electrochemistry of Molten Salts and Slags, Naukova Dumka, Kiev (1969), Part 1, p. 324.
32. S.I. Dionis'ev, A.G. Bergman, Zh. Neorg. Khim., **15**:2579 (1970).
33. O. Joshio, J. Yasuhisa, N. Akira, K. Motohisa, N. Atsuhiko, J. Chem. Soc. Jpn., Ind. Chem. Sec., **69**:1716 (1966).
34. V.I. Monastyrskaya, I.K. Tovmas'yan, Research in Thermography and Corrosion, Rostov. Knizhn. Izd., Rostov (1970), p. 156.
35. V.P. Kornienko, G.M. Shapovalova, V.N. Kolesnikov, Zh. Neorg. Khim., **15**:262 (1970).
36. P.I. Fedorov, G.A. Kot, Zh. Neorg. Khim., **12**:2313 (1967).
37. P.I. Fedorov, A.G. Dudareva, M.S. Nosova, Zh. Neorg. Khim., **12**:274 (1967).
38. E.A. Caimi, Sci. Invest., **25**:512 (1969).
39. V.V. Safonov, O.V. Lemeshko, Zh. Neorg. Khim., **18**:1957 (1973).
40. V.A. Robin, Teploenergetika (Moscow), No. 7, 27 (1956).
41. S.M. Arkhipov, T.V. Revzina, V.I. Mikheeva, Zh. Neorg. Khim., **15**:3383 (1970).
42. C.T.H.M. Cronenberg, J.W. van Spronsen, Z. Anorg. Allg. Chem., **354**:103 (1967).
43. V.I. Mikheeva, S.M. Arkhipov, T.V. Revzina, Zh. Neorg. Khim., **13**:1697 (1968).
44. G.M. Kurdyumov, V.A. Molochko, L.K. Agureeva, Zh. Neorg. Khim., **17**:2309 (1972).
45. N.G. Chovnyk, V.V. Vashchenko, M.S. Polyantseva, Ukr. Khim. Zh., **32**:454 (1966).
46. M.A. Kuvakin, L.I. Talanova, A.I. Kulikova, Zh. Neorg. Khim., **18**:1137 (1973).
47. K.N. Semenenko, T.N. Naumova, L.N. Gorokhov, Dokl. Akad. Nauk SSSR, **154**:648 (1964).
48. B.G. Korshunov, N.I. Kaloev, L.A. Nisel'son, O.R. Gavrilov, Izv. Vyssh. Uchebn. Zaved., Tsvetn. Metall., **1**:6, 66, 73 (1968); Zh. Neorg. Khim., **13**:1956 (1968).
49. S.M. Arkhipov, T.V. Revzina, V.I. Mikheeva, Zh. Neorg. Khim., **15**:3383 (1970).
50. I.S. Morozov, A.T. Simonin, Zh. Neorg. Khim., **2**:1907 (1957).
51. G. Boef, S.H. Bruins, R.A.W. van Leouwen, H. Wessels, J.W. van Spronsen, Z. Anorg. Allg. Chem., **353**:93 (1967).
52. L.K. van der Kamp, J.W. van Spronsen, Z. Anorg. Allg. Chem., **361**:328 (1968).

53. N.V. Ostrikova, Zh. Neorg. Khim., **15**:1142 (1970).
54. B.G. Korshunov, G.A. Lovetskaya, A.A. Palant, Zh. Neorg. Khim., **12**:203 (1967).
55. I.S. Morozov, Zh. Neorg. Khim., **1**:2792 (1956).
56. K.N. Semenenko, T.N. Naumova, L.N. Gorokhov, G.A. Semenova, A.V. Novoselova, Dokl. Akad. Nauk SSSR, **154**:169 (1964).
57. P.I. Fedorov, G.A. Lovetskaya, Zh. Neorg. Khim., **13**:3357 (1968).
58. A.P. Palkin, N.V. Ostrikova, Zh. Neorg. Khim., **7**:2635 (1962).
59. P.I. Fedorov, S.K. Nedev, Zh. Neorg. Khim., **10**:2717 (1965).
60. G.M. Kurdyumov, V.A. Molochko, T.B. Taganova, Zh. Neorg. Khim., **17**:1469 (1972).
61. I.S. Morozov, V.A. Tverskov, G.I. Kurapova, Zh. Neorg. Khim., **9**:2196 (1964).
62. O.R. Gavrilov, A.S. Krivoshein, L.A. Nisel'son, Zh. Neorg. Khim., **11**:2392 (1966).
63. N.D. Denisova, A.P. Baskova, Zh. Fiz. Khim., **47**:2450 (1973).
64. R.J. Clark, E. Griswold, J. Kleinberg, J. Am. Chem. Soc., **80**:4764 (1958).
65. T.N. Larionova, E.P. Vognikova, Zh. Neorg. Khim., **12**:2164 (1967).
66. T.N. Larionova, N.V. Ostrikova, Zh. Neorg. Khim., **13**:888 (1968).
67. T.N. Larionova, T.M. Pustyl'nik, Zh. Neorg. Khim., **14**:3151 (1969).
68. P.I. Fedorov, N.S. Malova, G.V. Vorob'eva, Zh. Neorg. Khim., **15**:2284 (1970).
69. T.N. Larionova, L.A. Lozgacheva, Zh. Neorg. Khim., **13**:2883 (1968).
70. W. Fischer, A.-L. Simon, Z. Anorg. Allg. Chem., **306**:1 (1960).
71. D.V. Drobot, B.G. Korshunov, L.M. Novikova, Izv. Vyssh. Uchebn. Zaved., Khim. Khim. Tekhnol., **12**:6 (1969).
72. A.I. Morozov, V.G. Kuznetsov, S.T. Maksimova, Zh. Neorg. Khim., **16**:3353 (1971).
73. B.G. Korshunov, V.I. Gol'din, Zh. Neorg. Khim., **6**:1642 (1961).
74. Lyu Yu-Sob, Zh. Neorg. Khim., **5**:2804 (1960).
75. M.A. Kuvakin, L.I. Talanova, A.I. Kulikova, Zh. Neorg. Khim., **18**:1137 (1973).
76. I.S. Morozov, B.G. Korshunov, A.T. Simonich, Zh. Neorg. Khim., **1**:1646 (1956).
77. G. Boef, S.H. Bruins, R.A.W. van Leouwen, H. Wessels, J.W. van Spronsen, Z. Anorg. Allg. Chem., **353**:93 (1967).
78. L.K. van der Kamp, J.W. van Spronsen, Z. Anorg. Allg. Chem., **361**:328 (1968).
79. N.V. Ostrikova, Z.A. Miroshnik, V.D. Kuleshova, N.P. Dudkina, A.I. Ilyushchenko, Zh. Neorg. Khim., **14**:2229 (1969).

80. D.V. Drobot, B.G. Korshunov, V.A. Aleksandrova, Zh. Neorg. Khim., 13:3354 (1968).
81. L.A. Nisel'son, Z.N. Orshanskaya, K.V. Tret'yakova, Zh. Neorg. Khim., 19:1060 (1974).
82. N.D. Chikanov, Zh. Neorg. Khim., 17:2558 (1972).
83. H. Houtgraaf, H.J. Rang, L. Vollbracht, Recl. Trav. Chim. Pays-Bas, 72:978 (1953).
84. D.V. Drobot, B.G. Korshunov, L.M. Novikova, Izv. Vyssh. Uchebn. Zaved., Khim. Khim. Tekhnol., 12:6 (1969).
85. I.S. Morozov, Li Chi-Fa, Zh. Neorg. Khim., 8:2733 (1963).
86. B.A. Voitovich, A.S. Barabanova, Ukr. Khim. Zh., 33:1019 (1967).
87. H. Houtgraaf, H.J. Rang, L. Vollbracht, Recl. Trav. Chim. Pays-Bas, 72:978 (1953).
88. B.G. Korshunov, V.I. Ionov, T.A. Baklashova, V.V. Kokorev, Izv. Vyssh. Uchebn. Zaved., Tsvetn. Metall., Issue 6, 114 (1960).
89. Z.W. Wilchinsky, R.W. Looney, Adv. X-Ray Anal. (Denver, Colorado), 7:203 (1964).
90. E.G.M. Tornqvist, J.T. Richardson, Z.W. Wilchinsky, R.W. Looney, J. Catal., 8:189 (1967).
91. L.N. Éingorn, Ukr. Khim. Zh., 16:404 (1950).
92. I.S. Morozov, D.Ya. Toptygin, Zh. Neorg. Khim., 2:1915 (1957).
93. B.A. Voitovich, A.S. Barabanova, Ukr. Khim. Zh., 29:1264 (1963).
94. I.P. Palyura, A.P. Palkin, Zh. Neorg. Khim., 9:2668 (1964).
95. D.V. Drobot, A.V. Nikolaev, A.V. Belousov, Zh. Neorg. Khim., 18:2243 (1973).
96. B.G. Korshunov, V.I. Gol'din, Zh. Neorg. Khim., 6:1642 (1961).
97. B.G. Korshunov, A.M. Reznik, I.S. Morozov, The Chemistry and Technology of Rare Elements, Izd. GNTI Tekst., Legk. i Poligr. Promst., Moscow (1958), p. 127.
98. A.J. Shor, W.T. Smith, M.A. Breding, J. Phys. Chem., 70:1151(1966).
99. O.R. Gavrilov, A.S. Krivoshein, L.A. Nisel'son, Zh. Neorg. Khim., 11:2392 (1966).
100. N.D. Denisova, E.K. Safronov, O.N. Bystrova, Zh. Neorg. Khim., 11:2412 (1966).
101. J. Ravez, Bull. Soc. Chim. France, No. 5, 1583 (1969).
102. M.A. Kuvakin, A.A. Sunin, L.I. Talanova, Zh. Neorg. Khim., 17:2785 (1972).
103. M. Malinovský, J. Vrbenská, I. Čakajdová, Chem. Zvesti, 35:35 (1969).
104. M.A. Kuvakin, E.N. Evstifeev, Zh. Neorg. Khim., 18:276 (1973).
105. J. Ravez, D. Dumora, C. R. Acad. Sci., C269:331 (1969).
106. V.T. Mal'tsev, in: The Phase Diagrams of Some Salt Systems, Izd. Rostov. Gos. Univ., Rostov (1964), p. 24.

107. G.A. Bukhalova, V.T. Mal'tsev, Zh. Neorg. Khim., **10**:189 (1965).
108. B. Phillips, C.H. Warshaw, I. Mockrin, J. Am. Ceram. Soc., **49**:631 (1966).
109. M. Rolin, R. Muhlethaler, Bull. Soc. Chim. France, No. 10, 2593 (1964).
110. M. Malinovský, I. Čakajdová, K. Matiašovský, Chem. Zvesti, **21**:794 (1967).
111. T. Matsushima, Denki Kagaku, **37**:778 (1969).
112. M.A. Kuvakin, T.D. Vol'khina, Uch. Zap. Mordovsk. Gos. Univ., Ser. Khim. Nauk, No. 81, 11 (1971).
113. A.A. Kostyukov, A.B. Karpov, The Electrometallurgy of Nonferrous Metals, ONTI, Leningrad (1957), p. 58.
114. J.-C. Cousseins, W. Freundlich, A. Erb, C. R. Acad. Sci., **C268**:717 (1969).
115. Yamaguchi Takashi, Takei Takeshi, Sci. Papers Inst. Phys. Chem. Res., **53**:284 (1959).
116. M.A. Kuvakin, P.S. Kusakin, Problems of Control and Complex Use of Raw Materials in Metallurgy, Izd. Ural. Fil., Akad. Nauk SSSR, Sverdlovsk (1960), p. 149.
117. V. Gottardi, M. Tornati, B. Locardi, Vetro Silicati, **4**: 14 (1960).
118. M. Rolin, Bull. Soc. Chim. France, 7:1351 (1961).
119. V.T. Mal'tsev, in: The Phase Diagrams of Some Salt Systems, Izd. Rostov. Gos. Univ., Rostov (1964), p. 29.
120. G.A. Bukhalova, V.T. Mal'tsev, Zh. Neorg. Khim., **10**:189 (1965).
121. G. Boef, S.H. Bruins, R.A.W. von Leouwen, H. Wessels, J.W. van Spronsen, Z. Anorg. Allg. Chem., **353**:93 (1967).
122. V.I. Mikheeva, S.M. Arkhipov, T.V. Revzina, Zh. Neorg. Khim., **13**:1946 (1968).
123. P.I. Fedorov, G.A. Kot, L.L. Nikol'skaya, Zh. Neorg. Khim., **12**:3172 (1967).
124. P.I. Fedorov, N.S. Malova, Yu. N. Denisov, Zh. Neorg. Khim., **16**:3347 (1971).
125. V.V. Safonov, A.G. Ivanchenko, B.G. Korshunov, Zh. Neorg. Khim., **19**:852 (1974).
126. V.V. Safonov, I.V. Vasilishcheva, V.M. Khazan, B.G. Korshunov, P.I. Fedorov, Izv. Vyssh. Uchebn. Zaved., Tsvetn. Metall., No. 4, 82 (1971).
127. L.A. Nil'son, V.V. Mogucheva, Zh. Neorg. Khim., **11**:144 (1966).
128. A.V. Konov, V.V. Safonov, Zh. Neorg. Khim., **19**:1134 (1974).
129. L.N. Éingorn, Ukr. Khim. Zh., **16**:404 (1950).
130. P.I. Fedorov, G.V. Sukhov, V.M. Khazan, L.G. Nosova, N.S. Sysoeva, Izv. Akad. Nauk SSSR, Neorg. Mater., **8**:178 (1972).
131. P.I. Fedorov, V.M. Khazan, L.G. Nosova, Zh. Neorg. Khim., **15**:2298 (1970).
132. A.V. Konov, V.V. Safonov, Zh. Neorg. Khim., **19**:1134 (1974).

133. P.I. Fedorov, V.A. Molochko, G.M. Kurdyumov, M.B. Braginskaya, Zh. Neorg. Khim., **15**:1398 (1970).
134. J. Ravez, P. Hagenmuller, Bull. Soc. Chim. France, **8**:1811 (1964).
135. E.A. Shyrginov, in: The Proceedings of the Ninth Conference of Candidates - Series of Topical Sciences, Izd. Rostov. Gos. Univ., Rostov (1969), p. 37
136. R. Riccardi, C. Sinistri, V. Campari, A. Magistris, Z. Naturforsch., **25a**:781 (1970).
137. I.N. Belyaev, A.K. Doroshenko, in: The Physicochemical Analysis of Salt Systems, Izd. Rostov. Gos. Univ., Rostov (1962), p. 12
138. M.A. Zakharchenko, S.M. Mkrtchyan, Zh. Neorg. Khim., **8**:1450 (1963).
139. N.D. Chikanov, V.A. Ugai, Zh. Neorg. Khim., **18**:849 (1973).
140. G.I. Nagornyi, T.D. Zimina, N.A. Finkel'shtein, Izv. Fiz.-Khim. Nauchn.-Issled. Inst. Irkutsk. Univ., **85**:15 (1961).
141. G.A. Bukhalova, V.T. Berezhnaya, Zh. Neorg. Khim., **7**:1399 (1962).
142. N.A. Finkel'shtein, A.G. Bergman, G.I. Nagornyi, Zh. Neorg. Khim., **10**:1895 (1965).
143. Fam Ngok Tien, I.S. Morozov, Zh. Neorg. Khim., **16**:2279 (1971).
144. A.V. Storonkin, I.V. Vasil'kova, O.D. Grebennikova, I.I. Kozhina, Vestn. Leningrad. Univ., Ser. Fiz., Khim., Issue 4, 80 (1973).
145. Z.A. Mateiko, E.S. Yagub'yan, G.A. Bukhalova, Zh. Neorg. Khim., **11**:2405 (1966).
146. I.N. Belyaev, D.S. Lesnykh, I.G. Éikhenbaum, Zh. Neorg. Khim., **15**:846 (1970).
147. N.I. Lisov, N.P. Burmistrova, A.S. Trunin, G.E. Shter, Ukr. Khim. Zh., **40**:603 (1974).
148. P.I. Fedorov, T.A. Lovetskaya, E.E. Shariadi, Izv. Vyssh. Uchebn. Zaved., Tsvetn. Metall., **6**:63 (1969).
149. P.I. Fedorov, S.K. Nedev, Liu Go-yuan, Zh. Neorg. Khim., **11**:406 (1966).
150. P.I. Fedorov, N.I. Il'ina, Zh. Neorg. Khim., **11**:1693 (1966).
151. E.I. Banashek, A.G. Bergman, Izv. Sekt. Fiz.-Khim. Anal., Akad. Nauk SSSR, **20**:94 (1950).
152. G.I. Nagornyi, N.A. Finkel'shtein, Izv. Fiz.-Khim. Nauchn.-Issled. Inst. Irkutsk. Gos. Univ., **4**:94 (1959).
153. G.I. Nagornyi, T.L. Zimina, Izv. Fiz.-Khim. Nauchn.-Issled. Inst. Irkutsk. Gos. Univ., **5**:30 (1961).
154. G.I. Nagornyi, N.A. Finkel'shtein, N.A. Samoilova, Izv. Fiz.-Khim. Nauchn.-Issled. Inst. Irkutsk. Gos. Univ., **5**:227 (1961).
155. C. Krohn, Acta Chem. Scand., **20**:255 (1966).
156. M.N. Zakhvalinskii, V.G. Romanovskaya, V.F. Tkachenko, N.A. Finkel'shtein, Zh. Neorg. Khim., **16**:3315 (1971).

157. N.V. Bondarenko, Zh. Neorg. Khim., **7**:1387 (1962).
158. N.V. Bondarenko, Kh.L. Strelets, in: The Physical Chemistry of Molten Salts and Slags, Naukova Dumka, Kiev (1970), p. 216.
159. M.A. Kuvakin, Uch. Zap. Mordovsk. Gos. Univ., Ser. Khim. Nauk, No. 81, 20 (1971).
160. M.A. Kuvakin, L.I. Talanova, A.I. Kulikova, Zh. Neorg. Khim., **18**:1137 (1973).
161. N.V. Maksina, L.K. Sosnovskaya, N.A. Finkel'shtein, N.M. Ryutina, Zh. Neorg. Khim., **18**:1098 (1973).
162. Fam Ngok Tien, I.S. Morozov, Zh. Neorg. Khim., **16**:1667 (1970).
163. K.W.R. Johnson, M. Kahn, J.A. Leary, J. Phys. Chem., **65**:2226 (1961).
164. N.D. Chikanov, Zh. Neorg. Khim., **13**:2616 (1968).
165. L.V. Gorbunov, V.N. Desyatnik, S.P. Raspopin, K.I. Trifonov, Zh. Neorg. Khim., **19**:3093 (1974).
166. Kuroda Tadashi, Suzuki Tadashi, J. Electrochem. Soc. Jpn., **26**:140 (1958).
167. V.N. Desyatnik, N.M. Emel'yanov, Yu. T. Mel'nikov, S.P. Raspopin, A.M. Tsybizov, Zh. Fiz. Khim., **46**:1886 (1972).
168. Fam Ngok Tien, I.S. Morozov, Zh. Neorg. Khim., **14**:1361 (1969).
169. M.A. Kuvakin, A.N. Sunin, L.I. Talanova, Uch. Zap. Mordovsk. Univ., Ser. Khim. Nauk, No. 81, 25 (1971).
170. D.F. Kirkina, A.V. Novoselova, Yu. P. Simanov, Zh. Neorg. Khim., **1**:125 (1956).
171. V.T. Berezhnaya, G.A. Bukhalova, Zh. Neorg. Khim., **2**:1408 (1957).
172. B.M. Zhigarnovskii, E.G. Ippolitov, Izv. Akad. Nauk. SSSR, Neorg. Mater., **5**:1558 (1969).
173. E.G. Ippolitov, L.S. Garashina, B.N. Zhigarnovskii, Izv. Sib. Otd. Akad. Nauk SSSR, Ser. Khim. Nauk, No. 4, 49 (1968).
174. J.-C. Cousseins, M. Samouël, C. R. Acad. Sci., **C266**:915 (1968).
175. V.T. Berezhnaya, G.A. Bukhalova, Zh. Neorg. Khim., **12**:2179 (1967).
176. M.Rolin, M. Clausier, Rev. Int. Hautes Temp. Refract., **4**:39 (1967).
177. J. Ravez, J. Viollet, R. de Pape, P. Hagenmuller, Bull. Soc. Chim. France, No. 4, 1325 (1967).
178. Ts.G. Tsiklauri, E.G. Ippolitov, S.V. Petrov, B.M. Zhigarnovskii, Zh. Neorg. Khim., **19**:2850 (1974).
179. B.M. Zhigarnovskii, E.G. Ippolitov, Izv. Akad. Nauk SSSR Neorg. Mater., **6**:1598 (1970).
180. J. Grannec, J. Ravez, C. R. Acad. Sci., **C270**:2059 (1970).
181. B.M. Zhigarnovskii, E.G. Ippolitov, Izv. Akad. Nauk SSSR, Neorg. Mater., **5**:1806 (1969).

182. N.A. Bychkova, A.G. Bergman, Zh. Obshch. Khim., **26**:639 (1956).
183. G.A. Bukhalova, V.T. Berezhnaya, Zh. Neorg. Khim., **4**:1141 (1959).
184. J.-C. Cousseins, M. Samouël, C. R. Acad. Sci., **C265**:1121 (1967).
185. S.V. Petrov, E.G. Ippolitov, Izv. Akad. Nauk SSSR, Neorg. Mater., **7**:876 (1971).
186. E.G. Ippolitov, A.G. Machlachkov, Izv. Akad. Nauk SSSR, Neorg. Mater., **6**:1423 (1970).
187. E.G. Chernevskaya, Opt.-Mekh. Promst., Issue 7, 51 (1966).
188. J. Ravez, M. Vassiliadis, P. Hagenmuller, C. R. Acad. Sci., **C268**:1876 (1969).
189. R.W.M. D'Eye, F.S. Martin, Repts. At. En. Res. Est., No. C/M292, 4 (1956).
190. B.P. Sobolev, E.G. Ippolitov, B.M. Zhigarnovskii, L.S. Garashina, Izv. Akad. Nauk SSSR, Neorg. Mater., **1**:362 (1965).
191. J.-C. Cretenet, C. R. Acad. Sci., **C268**:945 (1969).
192. E.G. Ippolitov, A.G. Maklachko, Izv. Akad. Nauk, SSSR, Neorg. Mater., **6**:146 (1970).
193. B.M. Zhigarnovskii, E.G. Ippolitov, Izv. Akad. Nauk SSSR, Neorg. Mater., **6**:1182 (1970).
194. M. Samouël, A. de Kozak, C. R. Acad. Sci., **C268**:2312 (1969).
195. K.A. Bol'shakov, P.I. Fedorov, S.B. Stenina, L.M. Akulkina, M.N. Shakhova, Zh. Neorg. Khim., **7**:605 (1962).
196. P.I. Fedorov, S.K. Nedev, Liu Go-yuan, Zh. Neorg. Khim., **11**:406 (1966).
197. A.E. Mordovin, I.F. Nichkov, V.A. Tomashov, Zh. Fiz. Khim., **48**:1469 (1974).
198. K.N. Semenenko, V.N. Surov, T.E. Alymova, Zh. Neorg. Khim., **13**:1669 (1968).
199. N.D. Chikanov, Zh. Neorg. Khim., **16**:1452 (1971).
200. E. Furby, K.L. Wilkinson, J. Inorg. Nucl. Chem., **14**:123 (1960).
201. C.R. Tipton, Reactor Handbook (Materials), 2nd ed., New York (1960), p. 442.
202. N.D. Chikanov, Zh. Neorg. Khim., **14**:1430 (1969).
203. V.N. Desyatnik, N.M. Emel'yanov, Yu. T. Mel'nikov, S.P. Raspopin, A.M. Tsybizov, Zh. Fiz. Khim., **46**:1886 (1972).
204. N.D. Chikanov, Izv. Vyssh. Uchebn. Zaved., Khim. Khim. Tekhnol., **13**:1407 (1970).
205. O.N. Kuvyrkin, O.N. Breusov, A.V. Novoselova, Nauchn. Dokl. Vyssh. Shkol. Khim. Khim. Tekhnol., No. 4, 660 (1958).
206. O.N. Breusov, A.V. Novoselova, Yu.P. Simanov, Dokl. Akad. Nauk SSSR, **118**:935 (1958).

207. M.P. Borzenkova, A.V. Novoselova, Yu.P. Simanov, V.I. Chernykh, E.I. Yarembash, Zh. Neorg. Khim., **1**:2071 (1956).
208. C.R. Tipton, Reactor Handbook (Materials), 2nd ed., New York (1960), p. 435.
209. Hu Chi-Tsu, A.V. Novoselova, Zh. Neorg. Khim., **6**:2148 (1961).
210. A.V. Novoselova, Yu.M. Korenev, M.P. Borzenkova, Zh. Neorg. Khim., **9**:2042 (1964).
211. T.B. Rhinehammer, C.R. Hudgens, J. Am. Ceram. Soc., **45**:79 (1962).
212. E.I. Evstyukhin, Metall. Metalloved. Chistykh Met., **1**:91 (1959).
213. A.V. Novoselova, M.E. Levina, M.P. Savel'eva, Zh. Neorg. Khim., **3**:2562 (1958).
214. R.G. Grebenshchikov, Dokl. Akad. Nauk. SSSR, **114**:316 (1957).
215. O.N. Breusov, G. Trank, A.V. Novoselova, Yu. P. Simanov, Zh. Neorg. Khim., **4**:671 (1959).
216. R.E. Thoma, H. Insley, H.A. Friedman, C.F. Weaver, J. Phys. Chem., **64**:865 (1960).
217. L.V. Jones, D.E. Etter, C.R. Hudgens, et al., J. Am. Ceram. Soc., **45**:79 (1962).
218. K.P. Borzenkova, A.V. Novoselova, G.N. Kuznetsova, Izv. Sib. Otd. Akad. Nauk SSSR, Ser. Khim. Nauk, Issue 1, 5 (1968).
219. M.P. Borzenkova, V.N. Galkina, A.V. Novoselova, Izv. Akad. Nauk SSSR, Neorg. Mater., **6**:25 (1970).
220. R.E. Thoma, H. Insley, H.A. Friedman, G.M. Hebert, J. Nucl. Mater., **27**:166 (1968).
221. V.E. Plyushchev, S.B. Stepina, I.V. Vlasova, I.N. Vasilenko, Izv. Vyssh. Uchebn. Zaved., Tsvetn. Metall., **1**:106 (1966).
222. V.A. Grin'ko, V.V. Safonov, V.I. Ksenzenko, Zh. Neorg. Khim., **19**:853 (1974).
223. A.I. Pustyl'nik, N.D. Denisova, L.G. Nekhamkin, L.A. Nisel'son, Zh. Neorg. Khim., **12**:1031 (1967).
224. N.I. Kaloev, O.I. Egorev, K.B. Dzeranova, L.K. Kulova, Zh. Neorg. Khim., **21**:290 (1976).
225. N.D. Chikanov, V.A. Ugai, Zh. Neorg. Khim., **18**:849 (1973).
226. C.C. Addison, W.D. Halstead, J. Chem. Soc., **9**:1236 (1966).
227. B.G. Korshunov, N.I. Kaloev, Izv. Vyssh. Uchebn. Zaved., Tsvetn. Metall., **6**:73 (1968).
228. J.-C. Couturier, C. R. Acad. Sci., **C624**:1378 (1967).
229. L.A. Nisel'son, V.P. Borisova, K.V. Tret'yakova, Zh. Neorg. Khim., **19**:3358 (1974).
230. L.A. Nisel'son, B.N. Ivanov-Émin, L.E. Larionova, Zh. Neorg. Khim., **6**:186 (1961).
231. P.I. Fedorov, N.I. Il'ina, Zh. Neorg Khim., **11**:205 (1966).
232. L.E. Topol, S.W. Mayer, L.D. Ransom, J. Phys. Chem., **64**:862 (1960).
233. N.I. Kaloev, A.K. Tebiev, Zh. Neorg. Khim., **18**:1349 (1973).

234. N.D. Chikanov, Zh. Neorg. Khim., 16:1452 (1971).
235. B.G. Korshunov, N.I. Kaloev, Izv. Vyssh. Uchebn. Zaved., Tsvetn. Metall., 6:73 (1968).
236. B.G. Korshunov, N.I. Kaloev, L.A. Nisel'son, O.R. Gavrilov, Zh. Neorg. Khim., 13:1956 (1968).
237. N.I. Glistenko, N.D. Chikanov, Zh. Neorg. Khim., 13:293 (1968).
238. K.A. Seleznev, Nauchn. Tr. Khim. Khim. Tekhnol. Gor'kov. Gos. Univ., Issue 3, 513 (1959).
239. N.D. Chikanov, Zh. Neorg. Khim., 17:2558 (1972).
240. V.V. Safonov, B.G. Korshunov, Zh. Neorg. Khim., 15:2300 (1970).
241. V.V. Safonov, B.G. Korshunov, Zh. Neorg. Khim., 13:2804 (1968).
242. B.G. Korshunov, V.N. Bezuevskaya, Zh. Neorg. Khim., 12:1304 (1967).
243. V.E. Plyushchev, S.B. Stepina, G.B. Zimina, O.P. Molchanova, L.V. Savel'eva, D.A. Yashkov, Izv. Vyssh. Uchebn. Zaved., Tsvetn. Metall., Issue 1, 65 (1970).
244. P.I. Fedorov, G.A. Kot, L.L. Nikol'skaya, Zh. Neorg. Khim., 12:3172 (1967).
245. P.I. Fedorov, A.G. Dudareva, M.Ts. Tsygankova, Zh. Neorg. Khim., 11:2409 (1966).
246. V.G. Viter, D.P. Belotskii, I.N. Antipov, Zh. Neorg. Khim., 13:3134 (1968).
247. T.N. Mel'nichenko, I.P. Mihal'ko, A.A. Kikineshi, D.G. Semak, I.D. Turyanitsa, Izv. Akad. Nauk SSSR, Neorg. Mater., 10:363 (1974).
248. V.V. Safonov, I.V. Visilishcheva, V.M. Khazan, B.G. Korshunov, P.I. Fedorov, Izv. Vyssh. Uchebn. Zaved., Tsvetn. Metall., No. 4, 82 (1971)
249. B.G. Korshunov, I.S. Morozov, V.I. Ionov, Izv. Vyssh. Uchebn. Zaved., Khim. Khim. Tekhnol., 3:402 (1960).
250. A.P. Lystsov, N.V. Galitskii, N.I. Shcherbina, Ukr. Khim. Zh., 34:1012 (1968).
251. N.V. Galitskii, A.I. Lystsov, N.I. Shcherbina, A.P. Sidorenko, E.V. Sergach, Obshch. Prikl. Khim., Issue 1, 46 (1969).
252. N.V. Galitskii, A.I. Lystsov, N.I. Shcherbina, A.P. Sidorenko, E.V. Sergach, The Metallurgy and Chemistry of Titanium, Metallurgiya, Moscow (1969), P. 103.
253. I.V. Shakhno, V.E. Plyushchev, Zh. Neorg. Khim., 5:1172 (1960).
254. V.M. Burlakova, G.A. Bukhalova, Zh. Neorg. Khim., 13:2539 (1968).
255. D.S. Coleman, P.D.A. Lacy, Trans. Instr. Mining Metall., C77:170 (1968).
256. P.I. Fedorov, S.K. Nedev, Liu Go-yuan, Zh. Neorg. Khim., 11:406 (1966).

257. P.I. Fedorov, N.I. Il'ina, Zh. Neorg. Khim., **14**:1432 (1969).
258. M.S. Golubeva, A.G. Bergman, Zh. Obshch. Khim., **26**:328 (1956).
259. V.E. Plyushchev, L.N. Komissarova, L.V. Meshchaninova, L.M. Akulkina, Zh. Neorg. Khim., **1**:820 (1956).
260. G.I. Nagornyi, N.A. Finkel'shtein, N.A. Samoilova, Izv. Fiz.-Khim. Nauchn.-Issled. Inst. Irkutsk. Gos. Univ., **5**:227 (1961).
261. K.B. Zaborenko, V.P. Polyakov, Yu.G. Shoroshev, Radiokhimiya, **7**:324 (1965).
262. V.M. Burlakova, G.A. Bukhalova, Izv. Vyssh. Uchebn. Zaved., Khim. Khim. Tekhnol., **10**:487 (1967).
263. Z.N. Shevtsova, I.S. Morozov, L.N. Prikhodkina, in: The Chemistry and Technology of Rare and Dispersed Elements, GNTI Tekst., Legkoi, i Poligr. Promst., Moscow (1958), p. 117.
264. G.A. Bukhalova, A.S. Arabadzhan, Zh. Neorg. Khim., **7**:2230 (1962).
265. Č. Barta, J. Žemlička, S. Stankov, Kristall Tech., **3**:35 (1968).
266. K. Matiašovsky, Chem. Zvesti, **13**:69 (1959).
267. Grjotheim Kai Holm Jan Lützow, Malmo Jardar, Acta Chem. Scand., **24**:77 (1970).
268. A. Seltveit, H. Flood, Acta Chem. Scand., **12**:1030 (1958).
269. G.A. Bukhalova, Zh. Neorg. Khim., **4**:117 (1959).
270. N.D. Chikanov, Zh. Neorg. Khim., **13**:2616 (1968).
271. I.S. Morozov, Z.N. Shevtsova, L.V. Klyukina, Zh. Neorg. Khim., **2**:1639 (1957).
272. L.Sh. Tsemekhman, M.A. Nemoitin, S.E. Vaisburd, Zh. Prikl. Khim., **42**:1402 (1969).
273. A.I. Efimov, V.F. Mikromanov, Zh. Prikl. Khim., **43**:1835 (1970).
274. K.W.R. Johnson, M. Kahn, J.A. Leary, J. Phys. Chem., **65**:2226 (1961).
275. I.V. Shakhno, V.E. Plyushchev, Zh. Neorg. Khim., **5**:1172 (1960).
276. B.G. Korshunov, V.I. Ionov, T.A. Baklashova, V.V. Kokorev, Izv. Vyssh. Uchebn. Zaved., Tsvetn. Metall., **6**:114 (1960).
277. L.V. Gorbunov, V.N. Desyatnik, S.P. Raspopin, K.I. Trifonov, Zh. Neorg. Khim., **19**:3093 (1974).
278. V.N. Desyatnik, Yu.T. Mel'nikov, I.I. Trifonov, The Physical Chemistry and Electrochemistry of Salt Melts, Metallurgiya, Sverdlovsk-Moscow (1973), p. 15.
279. M.P. Vorobei, V.N. Desyatnik, O.V. Skiba, N.M. Emel'yanov, Zh. Neorg. Khim., **18**:3060 (1973).
280. Kuroda Tadashi, Suzuki Tadashi, J. Electrochem. Soc. Jpn., **26**:140 (1958).

281. V.N. Desyatnik, N.M. Emel'yanov, Yu.T. Mel'nikov, S.P. Raspopin, A.M. Tsybizov, Zh. Fiz. Khim., **46**:1886 (1972).
282. I.S. Morozov, Z.I. Shevtsova, Li Chi-Fa, Zh. Neorg. Khim., **9**:2606 (1964).
283. I.V. Krivousova, N.N. Karavaeva, Nauchn. Tr. Tambovsk. Inst. Khim. Mashinostr., Issue 3, 129 (1969).
284. G.A. Reutova, É.E. Lukashenko, Zh. Prikl. Khim., **44**:426 (1971).
285. M.A. Kuvakin, E.N. Evstifeev, Zh. Neorg. Khim., **18**:276 (1974).
286. A.M. Vokhmyakov, V.N. Desyatnik, S.F. Katyshev, Yu.F. Chervinskii, The Physical Chemistry of Metallurgical Processes, Issue 2, Izd. UPI, Sverdlovsk (1974), p. 70.
287. R.A. Sharma, High Temp. Sci., No. 4, 423 (1969).
288. Z.A. Zagorskaya, N.A. Ivanova, B.I. Maksakov, Izv. Sib. Otd. Akad. Nauk SSSR, Ser. Khim. Nauk, No. 2, 27 (1968).
289. E.G. Ippolitov, L.S. Garashina, B.N. Zhigarnovskii, Izv. Sib. Otd. Akad. Nauk SSSR, Ser. Khim. Nauk, No. 4, 49 (1968).
290. Z.A. Mateiko, G.A. Bukhalova, D.V. Sementsova, Izv. Vyssh. Uchebn. Zaved., Khim. Khim. Tekhnol., **10**:856 (1967).
291. D.V. Sementsova, G.A. Bukhalova, Z.A. Mateiko, Zh. Neorg. Khim., **12**:1645 (1967).
292. D. Dumora, J. Ravez, P. Hagenmuller, Bull. Soc. Chim. France, No. 4, 1301 (1970).
293. N.G. Gogadze, E.G. Ippolitov, B.M. Zhigarnovskii, Zh. Neorg. Khim., **17**:576 (1972).
294. N.G. Gogadze, E.G. Ippolitov, B.M. Zhigarnovskii, Zh. Neorg. Khim., **17**:2588 (1972).
295. J. Ravez, J. Viollet, R. Pape, P. Hagenmuller, Bull. Soc. Chim. France, No. 4, 1325 (1967).
296. N.G. Gogadze, E.G. Ippolitov, B.M. Zhigarnovskii, Zh. Neorg. Khim., **17**:1152 (1972).
297. E.G. Ippolitov, N.G. Gogadze, B.M. Zhigarnovskii, Zh. Neorg. Khim., **15**:3318 (1970).
298. W.E. Roake, J. Electrochem. Soc., **104**:661 (1957).
299. G.A. Bukhalova, V.G. Berezhnaya, Zh. Neorg. Khim., **2**:1408 (1957).
300. D.L. Deadmore, J.S. Machin, J. Phys. Chem., **64**:824 (1960).
301. Maurice Rolin, Maurice Clausier, Rev. Int. Hautes. Temp. Refract., **4**:39 (1967).
302. J.E. Antill, P. Murray, Silicates Ind., **20**:293 (1955).
303. G.A. Bukhalova, K. Sulaimankulov, A.K. Bostandzhiyan, Zh. Neorg. Khim., **4**:1138 (1959).
304. G.A. Bukhalova, V.T. Berezhnaya, Z.A. Mateiko, Zh. Neorg. Khim., **7**:2233 (1962).
305. C.R. Tipton, Reactor Handbook (Materials), 2nd ed., New York (1960), p 440.
306. É.G. Chernevskaya, Opt.-Mekh. Promst., No. 7, 51 (1966).

307. J. Ravez, M. Vassiliadis, P. Hagenmuller, C. R. Acad. Sci., **C268**:1876 (1969).
308. N.P. Nekrasova, E.N. Oblomeev, V.N. Golovanova, A.V. Beznosikova, At. Energ., **22**:293 (1967).
309. J. Short, R. Roy, J. Phys. Chem., **67**:1860 (1963).
310. B.P. Sobolev, E.G. Ippolitov, B.M. Zhigarnovskii, L.S. Garashina, Izv. Akad. Nauk SSSR, Neorg. Mater., **1**:362 (1965).
311. J.-C. Cretenet, C. R. Acad. Sci., **C268**:945 (1969).
312. I.I. Il'yasov, Zh. Neorg. Khim., **9**:1411, 1416 (1964).
313. C. Sinistri, R. Riccardi, A. Magistris, Ber. Bunsenges. Phys. Chem., **71**:376 (1967).
314. I.I. Il'yasov, D.G. Barsegov, I.G. Berikashvili, L.P. Danilenko, Zh. Neorg. Khim., **14**:1484 (1969).
315. V.P. Radina, I.K. Tovmas'yan, in: Research in Thermography and Corrosion, Rostov. Knizhn. Izd., Rostov (1970), p. 188.
316. I.I. Il'yasov, V.N. Mirsoyanov, Yu.V. Korotkov, Zh. Neorg. Khim., **4**:909 (1959).
317. I.I. Il'yasov, S.D. Dionis'ev, A.G. Bergman, Zh. Neorg. Khim., **5**:664 (1960).
318. V.I. Monastyrskaya, in: The Physical Chemistry and Electrochemistry of Molten Salts and Slags, Part I, Naukova Dumka, Kiev (1969), p. 324.
319. I.N. Belyaev, A.K. Doroshenko, in: The Physicochemical Analysis of Salt Systems, Izd. Rostov. Gos. Univ., Rostov (1962), p. 26.
320. I.I. Il'yasov, L.V. Rozhkovskaya, A.G. Bergman, Zh. Neorg. Khim., **2**:2174 (1957).
321. I.I. Il'yasov, G.G. Shemeleva, A.G. Bergman, Zh. Neorg. Khim., **4**:906 (1959).
322. V.A. Grin'ko, V.V. Safonov, V.I. Ksenzenko, Zh. Neorg. Khim., **19**:853 (1974).
323. I.I. Il'yasov, S.D. Dionis'ev, A.G. Bergman, Zh. Neorg. Khim., **7**:625 (1962).
324. V.I. Monastyrskaya, I.K. Tovmas'yan, in: Research in Thermography and Corrosion, Rostov. Knizhn. Izd., Rostov (1970), p. 156.
325. I.I. Il'yasov, Zh. Neorg. Khim., **7**:2604 (1962).
326. H.-J. Seifert, F.W. Koknat, Z. Anorg. Allg. Chem., **357**:314 (1968).
327. I.I. Il'yasov, A.G. Bergman, Ukr. Khim. Zh., **35**:173 (1969).
328. I.N. Belyaev, D.S. Lesyknh, A.K. Doroshenko, I.G. Éikhenbaum, Zh. Prikl. Khim., **45**:665 (1972).
329. I.K. Tovmas'yan, V.P. Radina, in: Research in Thermography and Corrosion, Rostov. Knizhn. Izd., Rostov (1970), pp. 188, 198.
330. P.I. Fedorov, T.A. Lovetskaya, E.E. Shariadi, Izv. Vyssh. Uchebn. Zaved.,, Tsvetn. Matall., **6**:63 (1969).
331. P.I. Fedorov, V.V. Tsimbalist, Liu Go-yuan, Zh. Neorg. Khim., **9**:1681 (1964).

332. P.I. Fedorov, N.S. Malova, G.V. Vorob'eva, Zh. Neorg. Khim., **15**:2281 (1970).
333. P.I. Fedorov, N.I. Il'ina, Zh. Neorg. Khim., **11**:1693 (1966).
334. A.G. Bergman, E.L. Bakumskaya, Zh. Obshch. Khim., **26**:629 (1956).
335. I.I. Il'yasov, A.K. Bostandzhiyan, A.G. Bergman, Zh. Neorg. Khim., **1**:2543 (1956).
336. I.I. Il'yasov, A.K. Bostandzhiyan, A.G. Bergman, Zh. Neorg. Khim., **2**:172, 395 (1957).
337. I.N. Belyaev, A.K. Doroshenko, Zh. Prikl. Khim., **42**:2704 (1969).
338. T.I. Drobasheva, I.I. Il'yasov, I.A. Tokman, Zh. Neorg. Khim., **16**:2853 (1971).
339. A.G. Bergman, Zh.V. Misler, Zh. Neorg. Khim., **8**:403 (1963).
340. A.G. Bergman, A.P. Sadovskii, Zh.V. Misler, Zh. Neorg. Khim., **8**:954 (1963).
341. I.I. Il'yasov, A.G. Bergman, Zh. Obshch. Khim., **26**:1288 (1956).
342. I.I. Il'yasov, A.K. Bostandzhiyan, A.G. Bergman, Zh. Neorg. Khim., **2**:172 (1957).
343. I.I. Il'yasov, A.G. Bergman, Zh. Neorg. Khim., **7**:356 (1962).
344. M.M. Shul'ts, I.M. Bushueva, Vestn. Leningrad. Univ., **22**: 120 (1965).
345. N.D. Chikanov, Zh. Neorg. Khim., **14**:1430 (1969).
346. I.I. Il'yasov, A.G. Bergman, Zh. Neorg. Khim., **2**:2159 (1957).
347. H.-J. Seifert, F.W. Koknat, Z. Anorg. Allg. Chem., **357**:314 (1968).
348. I.I. Il'yasov, M. Davranov, Zh. Neorg. Khim., **18**:279 (1973).
349. I.V. Vasil'kova, I.I. Kozhina, S.V. Mendeleva, A.V. Storonkin, Zh. Fiz. Khim., **47**:1684 (1973).
350. Yu.P. Afinogenov, A.E. Shulyakovskii, Yu.A. Litvinov, Zh. Neorg. Khim., **15**:1436 (1970).
351. A.K. Sementsova, A.G. Bergman, D.S. Lesnykh, Zh. Neorg. Khim., **1**:163 (1956).
352. I.P. Palyura, A.P. Palkin, Zh. Neorg. Khim., **4**:238 (1959).
353. A.K. Bostandzhiyan, A.G. Bergman, Zh. Neorg. Khim., **4**:2564 (1959).
354. A.P. Palkin, G.G. Korotkikh, N.B. Vlasenko, Zh. Neorg. Khim., **5**:637 (1960).
355. N.I. Glistenko, T.I. Burlakova, Nauchn. Tr. Voronezh. Gos. Univ., **40**:17 (1956).
356. A. De Kozak, C. R. Acad. Sci., **C268**:2184 (1969).
357. J.-C. Cousseins, C. R. Acad. Sci., **C263**:585 (1966).
358. I.N. Belyaev, A.K. Doroshenko, Zh. Neorg. Khim., **13**:1655 (1968).

359. J.-C. Cousseins, C. Pina Perez, C. R. Acad. Sci., **C264**:2059 (1967).
360 I.I. Il'yasov, A.G. Bergman, Zh. Neorg. Khim., **9**:1416 (1964).
361. I.N. Belyaev, E.A. Shurginov, N.S. Kudryashov, Zh. Neorg. Khim., **17**:2812 (1972).
362. P.I. Fedorov, G.A. Kot, L.L. Nikol'skaya, Zh. Neorg. Khim., **12**:2818 (1967).
363. Yu.N. Denisov, N.S. Malova, P.I. Fedorov, Zh. Neorg. Khim., **19**:815 (1974).
364. Yu.N. Denisov, N.S. Malova, P.I. Fedorov, Zh. Neorg. Khim., **18**:1363 (1973).
365. P.I. Fedorov, A.G. Dudareva, Zh. Neorg. Khim., **6**:252 (1961).
366. A.G. Bergman, I.I. Il'yasov, Zh. Neorg. Khim., **2**:395 (1957).
367. I.N. Belyaev, E.A. Shurginov, N.S. Kudryashov, Zh. Neorg. Khim., **17**:2812 (1972).
368. I.I. Il'yasov, N.I. Chaurskii, A.G. Bergman, S.D. Dionis'ev, Zh. Neorg. Khim., **7**:618 (1962).
369. I.I. Il'yasov, A.K. Bostandzhiyan, Zh. Neorg. Khim., **2**:167 (1957).
370. D.P. Belotskii, T.N. Sushkevich, Ya.A. Lyuter, Izv. Akad. Nauk SSSR, Neorg. Mater., **10**:1899 (1974).
371. D.P. Belotskii, V.F. Nadtochii, L.F. Tkachuk, Ukr. Khim. Zh., **36**:30 (1970).
372. V.V. Safonov, I.V. Vasilishcheva, B.G. Korshunov, Zh. Neorg. Khim., **16**:2310 (1971).
373. I.I. Il'yasov, A.G. Bergman, Zh. Neorg. Khim., **4**:913 (1959).
374. I.I. Il'yasov, D.G. Barsegov, I.G. Berikashvili, L.P. Danilenko, Zh. Neorg. Khim., **14**:1484 (1969).
375. D.P. Belotskii, I.N. Antipov, V.F. Nadtochii, S.M. Dodik, Izv. Akad. Nauk, SSSR. Neorg. Mater., **5**:1852 (1969).
376. Sun In-chzhu, I.S. Morozov, Zh. Neorg. Khim., **3**:1914 (1958).
377. Li Chi-Fa, I.S. Morozov, Zh. Neorg. Khim., **7**:2765 (1962).
378. G.I. Novikov, A.K. Baev, O.G. Polyachenok, in: The Chemistry of Rare Elements, Izd. Leningrad. Gos. Univ., Leningrad (1964), p. 63.
379. G.I. Novikov, A.K. Baev, Vestn. Leningrad. Gos. Univ., Ser. Fiz., Khim., No. 22, 116 (1961).
380. B.G. Korshunov, I.S. Morozov, V.I. Ionov, Izv. Vyssh. Uchebn. Zaved., Khim. Khim. Tekhnol., **3**:402 (1960).
381. G.I. Novikov, A.K. Baev, Vestn. Leningrad. Gos. Univ., Ser. Fiz., Khim., No. 22, 116 (1961).
382. A.V. Storonkin, I.V. Vasil'kova, O.D. Grebennikova, I.I. Kozhina, Vestn. Leningrad. Gos. Univ., Ser. Fiz., Khim., No. 22, 84 (1973).

383. Fam Ngok Tien, I.S. Morozov, Zh. Neorg. Khim., 16:2279 (1971).
384. V.I. Ionov, B.G. Korshunov, V.V. Kokorev, I.S. Morozov, Izv. Vyssh. Uchebn. Zaved., Tsvetn. Metall., 3:102; 6:114 (1960).
385. W.J. Asker, A.W. Wylie, Austr. J. Chem., 18:969 (1965).
386. G.A. Bukhalova, É.P. Babaeva, Zh. Neorg. Khim., 11:624 (1966).
387. R.E. Thoma, Rare-Earth Halides, ORNL (1965), p. 3804.
388. L.N. Starkov, N.K. Starkova, A.R. Babenko, Chemistry and Chemical Technology, Permsk. Knizhn. Izd., Perm (1968), p. 45.
389. C.Y. Barton, Y.D. Redman, R.A. Strehlow, J. Inorg. Nucl. Chem., 20:45 (1961).
390. R.E. Thoma, H. Insley, G.M. Hebert, Inorg. Chem., 5:1222 (1966).
391. R.E. Thoma, Rare-Earth Research. III. Proceedings of IV Conf., New York (1964), p. 561.
392. V.A. Grin'ko, V.V. Safonov, V.I. Ksenzenko, Zh. Neorg. Khim., 19:853 (1974).
393. H.-J. Seifert, Z. Anorg. Allg. Chem. 307:137 (1961).
394. N.G. Korzhukov, Yu.G. Kobanin, Vestn. Mosk. Gos. Univ., Ser. II, Khim., Issue 2, 66 (1966).
395. P.I. Fedorov, S.K. Nedev, Zh. Neorg. Khim., 12:1301 (1967).
396. K.A. Bol'shakov, P.I. Fedorov, G.D. Agashkina, Zh. Neorg. Khim., 2:1115 (1957).
397. L.G. Moskalenko, Tsvetn. Met., No. 12, 39 (1957).
398. K.A. Bol'shakov, P.I. Fedorov, G.D. Agashkina, Zh. Neorg. Khim., 3:1891 (1958).
399. N.G. Korzhukov, K.G. Khomyakov, Vestn. Mosk. Gos. Univ., Ser. II, Khim., No. 5, 53 (1966).
400. V.I. Fadeeva, N.G. Korzhukov, K.G. Khomyakov, Vestn. Mosk. Gos. Univ., Ser. II, Khim., Issue 2, 66 (1969).
401. I.N. Belyaev, D.S. Lesnykh, I.G. Éikhenbaum, Zh. Neorg. Khim., 15:846 (1970).
402. I.N. Belyaev, S.A. Shilov, Zh. Neorg. Khim., 14:1994 (1969).
403. J.D. Donaldson, R. Oteng, J. Chem. Soc., A17:2696 (1969).
404. P.I. Fedorov, V.M. Khazan, N.I. Evdokimova, Zh. Neorg. Khim., 15:2568 (1970).
405. R. Gut, R. Gnehm, Chimia, 16:289 (1962).
406. H.-J. Seifert, K. Klatyk, Naturwissenschaften, 49:539 (1962).
406a. H.-J. Seifert, K. Klatyk, Z. Anorg. Allg. Chem., 334:113 (1964).
407. N.V. Galitskii, A.I. Lystsov, N.I. Shcherbina, A.P. Sidorenko, E.V. Sergach, Obshch. Prikl. Khim., Issue 1, 46 (1969).

407a. N.V. Galitskii, A.I. Lystsov, N.I. Shcherbina, A.P. Sidorenko, E.V. Sergach, The Metallurgy and Chemistry of Titanium, Metallurgiya, Moscow (1969), p. 103.
408. A.A. Solomenko, Ya.E. Vil'nyanskii, Zh. Neorg. Khim., **14**:1993 (1969).
409. S.N. Shkol'nikov, A.M. Volkov, Izv. Vyssh. Uchebn. Zaved., Tsvetn. Metall., **6**:82 (1964).
410. J.-C. Shiloff, J. Phys. Chem., **64**:1566 (1960).
411. S.N. Shkol'nikov, Electrolysis in Metallurgy, Metallurgiya, Moscow (1967), p. 30.
412. A.I. Efimov, I.I. Kozhina, E.K. Lupenko, Vestn. Leningrad. Gos. Univ., No. 4, 138 (1967).
413. A.N. Lystsov, N.V. Galitskii, N.I. Shcherbina, Ukr. Khim. Zh., **34**:1012 (1968).
414. V.G. Gopienko, S.N. Shkol'nikov, E.F. Klyuchnikova, Zh. Prikl. Khim., **46**:1123 (1973).
415. A.I. Efimov, B.Z. Pitirimov, Zh. Neorg. Khim., **8**:2001 (1963).
416. S.N. Shkol'nikov, A.M. Volkov, Izv. Vyssh. Uchebn. Zaved., Tsvetn. Metall., Issue 2, 65 (1962).
417. B.G. Korshunov, B.Ya. Raskin, Zh. Neorg. Khim., **7**:1137 (1962).
418. C.M. Cook, J. Inorg. Nucl. Chem., **25**:123 (1963).
419. I.V. Vasil'kova, A.I. Efimov, B.Z. Pitirimov, Zh. Neorg. Khim., **9**:900 (1964).
420. A.V. Storonkin, I.V. Vasil'kova, Yu.A. Fedorov, Zh. Fiz. Khim., **41**:695 (1967).
421. I.V. Vasil'kova, I.V. Krivousova, M.P. Susarev, S.S. Tolkachev, Vestn. Leningrad. Gos. Univ., Ser. Fiz., Khim., No. 16, 126 (1965).
422. T. Oki, Z. Ganikawa, Denki Kagaku, **36**:119 (1968).
423. I.V. Vasil'kova, I.V. Krivousova, M.P. Susarev, S.S. Tolkachev, Vestn. Leningrad. Gos. Univ., Ser. Fiz., Khim., No. 16, 126 (1968).
424. N.D. Chikanov, Zh. Neorg. Khim., **13**:2884 (1968).
425. P.S. Shapkin, I.V. Vasil'kova, M.P. Susarev, Zh. Neorg. Khim., **12**:2838 (1967).
426. N.D. Chikanov, Zh. Neorg. Khim., **17**:2558 (1972).
427. B.-J. Sturm, Inorg. Chem., **1**:665 (1962).
428. J.-C. Cousseins, S. De Kozak, C. R. Acad. Sci., **C265**:991 (1967).
429. A. De Kozak, C. R. Acad. Sci., **C268**:416 (1969).
430. D. Dumora, J. Ravez, P. Hagenmuller, Bull. Soc. Chim. France, **5**:175 (1970).
431. A. De Kozak, J.-C. Cousseins, C. R. Acad. Sci., **C267**:74 (1968).
432. A.G. Dudareva, T.B. Fedorova, Yu.É. Bogatov, P.I. Fedorov, Zh. Neorg. Khim., **19**:1607 (1974).
433. A.G. Dudareva, Yu.É. Bogatov, B.N. Ivanov-Émin, P.I. Fedorov, Zh. Neorg. Khim., **16**:2568 (1971).

434. V.E. Plyushchev, R.G. Samuseva, Zh. Neorg. Khim., **9**:2179 (1964).
435. I.K. Tovmas'yan, T.M. Tsushba, A.G. Bergman, Zh. Neorg. Khim., **15**:2268 (1970).
436. D.S. Lesnykh, E.P. Garmatina, Zh. Neorg. Khim., **12**:1307 (1967).
437. É.M. Moroz, S.V. Ketchik, S.S. Batsanov, Zh. Neorg. Khim., **17**:1755 (1972).
439. I.I. Il'yasov, Ukr. Khim. Zh., **31**:930 (1965).
440. D.S. Lesnykh, I.G. Éikhenbaum, S.A. Chernyakhovskaya, Zh. Neorg. Khim., **15**:824 (1970).
441. I.K. Tovmas'yan, T.M. Tsushba, A.G. Bergman, Ukr. Khim. Zh., **36**:796 (1970); Zh. Neorg. Khim., **17**:2796 (1972).
442. I.I. Il'yasov, N.I. Chaurskii, Zh. Neorg. Khim., **12**:560 (1967).
443. Ng.B. Chanh, B. de Pelisse, Bull. Soc. Franc. Mineral. Cristallogr., **91**:13 (1968).
444. I.K. Tovmas'yan, T.M. Tsushba, A.G. Bergman, Zh. Neorg. Khim., **15**:2268 (1970); Ukr. Khim. Zh., **36**:796 (1970).
445. V.E. Plyushchev, S.B. Stepina, G.V. Zimina, V.G. Zhilyakov, Izv. Vyssh. Uchebn. Zaved., Tsvetn. Metall., Issue 4, 112 (1964).
446. N.Ya. Fedorov, E.S. Petrov, Izv. Sib. Otd. Akad. Nauk, Ser. Khim. Nauk, No. 14, 55 (1969).
447. I.N. Belyaev, E.A. Shurginov, Zh. Neorg. Khim., **15**:883 (1970).
448. E.A. Shurginov, in: Proceedings of the IX Conf. of Aspirants, Izd. Rostov. Gos. Univ., Rostov (1969), p. 37.
449. R. Riccardi, C. Sinistri, V. Campari, A. Magistris, Z. Naturforsch., **25a**:781 (1970).
450. V.A. Grin'ko, V.V. Safonov, V.I. Ksenzenko, B.G. Korshunov, Zh. Neorg. Khim., **17**:1755 (1972).
451. S.A. Shchukarev, I.V. Vasil'kova, D.V. Korol'kov, Zh. Neorg. Khim., **8**:1933 (1963).
452. I.I. Il'yasov, D.G. Barsegov, Ukr. Khim. Zh., **34**:473 (1968).
453. V.I. Stepanov, M.V. Smirnov, Dep. VINITI No. 2427-70.
454. Nguyen-Ba-Chanh, I. Bouillaud, Bull. Soc. Chim. France, No. 7, 2513 (1971).
455. I.K. Tovmas'yan, T.M. Tsushba, A.G. Bergman, Ukr. Khim. Zh., **36**:1007 (1970).
456. I.K. Tovmas'yan, T.M. Tsushba, Zh. Neorg. Khim., **17**:2796 (1972).
457. T.E. Bykova, in: Biochemical and Technological Processes in the Food Industry, Buryatsk. Knizhn. Izd., Ulan-Udé (1974), p. 27.
458. G.A. Bukhalova, G.A. Shegurova, T.M. Khliyan, E.S. Yagubyan, Zh. Neorg. Khim., **18**:1106, 2578 (1973).
459. I.I. Il'yasov, A.G. Bergman, Zh. Neorg. Khim., **9**:1416 (1964).

460. M. Ahtee, Soumalais. Tiedeakat. Toimituks, Ser. IV, 3 (1969).
461. V.I. Ionov, I.S. Morozov, B.G. Korshunov, Zh. Neorg. Khim., 5:1248 (1960).
462. H.-J. Seifert, K. Klatyk, Z. Anorg. Allg. Chem., 342:1 (1966).
463. P.I. Fedorov, V.V. Tsimbalist, Zh. Neorg. Khim., 9:1676 (1964).
464. V.A. Sryvtsev, V.N. Arbekov, E.S. Petrov, in: Rare Alkaline Elements, Nauka, Novosibirsk (1967), p. 191.
465. N.V. Ostrikova, Zh. Neorg. Khim., 15:1142 (1970).
466. I.S. Morozov, Sun In-Chzhu, Zh. Neorg. Khim., 4:678 (1959).
467. I.I. Il'yasov, A.G. Bergman, Zh. Neorg. Khim., 7:695 (1962).
468. Zh.G. Moisseenko, E.K. Akopov, L.A. Panieva, Zh. Neorg. Khim., 17:3098 (1972).
469. Sun In-Chzhu, I.S. Morozov, Zh. Neorg. Khim., 3:1914 (1958).
470. G.A. Bukhalova, D.V. Sementsova, Zh. Neorg. Khim., 10:1886 (1965).
471. E.K. Akopov, Zh.G. Moiseenko, Zh. Neorg. Khim., 18:566 (1973).
472. L.M. Kukis, in: The Interaction of Some Alkali Metals in Melts, Irkutsk. Knizhn. Izd., Irkutsk (1973), p. 138.
473. B.F. Markov, R.V. Chernov, Ukr. Khim. Zh., 24:139 (1958).
474. A.I. Efimov, L.P. Belorukova, A.M. Ryndina, Zh. Neorg. Khim., 8:1168 (1963).
475. R.G. Samuseva, V.E. Plyushchev, Zh. Neorg. Khim., 6:2139 (1961).
476. É.M. Moroz, S.V. Ketchik, S.S. Batsanov, Zh. Neorg. Khim., 17:1775 (1972).
477. T.P. Bortinkova, E.K. Akopov, V.A. Ocheretnyi, Zh. Neorg. Khim., 19:1066 (1974).
478. N.M. Ryutina, N.V. Maksina, N.A. Finkel'shtein, Zh. Neorg. Khim., 18:1101 (1973).
479. V.V. Safonov, B.G. Korshunov, S.N. Stel'bovskaya, Zh. Neorg. Khim., 11:2143 (1966).
480. V.V. Safonov, B.G. Korshunov, T.N. Zimina, Zh. Neorg. Khim., 11:906 (1966).
481. V.V. Safonov, B.G. Korshunov, Z.N. Shevtsova, Zh. Neorg. Khim., 7:1979 (1962).
482. E. Iberson, R. Gut, D.M. Gruen, J. Phys. Chem., 66:65 (1962).
483. A.G. Bergman, G.I. Maslennikova, Zh. Neorg. Khim., 7:1382 (1962).
484. I.I. Il'yasov, Zh. Neorg. Khim., 8:1230 (1963).
485. I.I. Il'yasov, A.G. Bergman, N.I. Chaurskii, Zh. Neorg. Khim., 10:1256 (1965).
486. R. Benz, R.M. Douglass, J. Phys. Chem., 65:1461 (1961).

487. M.P. Vorobei, O.V. Skiba, A.S. Bevz, V.N. Desyatnik, The Physical Chemistry and Electrochemistry of Salt Melts, Metallurgiya, Moscow (1973), p. 3.
488. M.P. Vorobei, O.V. Skiba, A.S. Bevz, V.N. Desyatnik, Zh. Neorg. Khim., **18**:2506 (1973).
489. L.J. Wood, Ch. Sweeney, M.T. Derbes, Sr., J. Am. Chem. Soc., **81**:6148 (1959).
490. Ng.B. Chanh, B. de Pelissac, Bull. Soc. Chim. France, No. 2, 695 (1967).
491. Ng.B. Chanh, B. de Pelissac, Bull. Soc. Franc. Mineral. Cristallogr., **91**:13 (1968).
492. L.J. Wood, W. Secunda, H. Clifford, McBride, J. Am. Chem. Soc., **80**:307 (1958).
493. Krogh-Moc, J. Am. Chem. Soc., **82**:6196 (1960).
494. V.M. Burlakova, in: Proceedings of a Scientific-Theoretical Conference of Aspirants, Izd. Rostov. Gos. Ped. Inst., Rostov (1967), p. 435.
495. I.K. Tovmas'yan, T.M. Tsushba, A.G. Bergman, Ukr. Khim. Zh., **40**:376 (1974).
496. I.I. Il'yasov, M. Davranov, K.I. Iskandarov, Zh. Neorg. Khim., **19**:1919 (1974).
497. D.V. Drobot, B.G. Korshunov, Nad' B. Sharkadi, Zh. Neorg. Khim., **14**:3158 (1969).
498. R. Gut, D.M. Gruen, J. Inorg. Nucl. Chem., **21**:259 (1961).
499. N.Ya. Fedorov, E.S. Petrov, in: Rare Alkaline Elements, Nauka, Novosibirsk (1967), p. 185.
500. N.Ya. Fedorov, E.S. Petrov, Izv. Sib. Otd. Akad. Nauk SSSR, Ser. Khim. Nauk, **2**:57 (1967).
501. B.G. Korshunov, D.V. Drobot, V.V. Bukhtyarov, Z.N. Shevtsova, Zh. Neorg. Khim., **9**:1427 (1964).
502. D.V. Drobot, B.G. Korshunov, Z.N. Shevtsova, in: The Physical Chemistry of Molten Salts, Metallurgiya, Moscow (1965), p. 48.
503. B.F. Markov, R.V. Chernov, Ukr. Khim. Zh., **24**:143 (1958).
504. Li Chi-Fa, I.S. Morozov, Zh. Neorg. Khim., **8**:708 (1963).
505. A.G. Bergman, T.D. Zimina, Zh. Neorg. Khim., **10**:2326 (1965).
506. G.A. Bukhalova, V.M. Burlakova, Zh. Neorg. Khim., **11**:2812 (1966).
507. A.N. Kirgintsev, V.A. Isaenko, V.I. Kosyakov, Izv. Akad. Nauk SSSR, Ser. Khim., No. 2, 439 (1970).
508. V.V. Safonov, B.G. Korshunov, Z.N. Shevtsova, S.I. Bakum, Zh. Neorg. Khim., **9**:1687 (1964).
509. V.V. Safonov, B.G. Korshunov, Z.N. Shevtsova, L.G. Shadrova, Zh. Neorg. Khim., **9**:1406 (1964).
510. I.S. Morozov, A.T. Simonich, Zh. Neorg. Khim., **2**:1907 (1957).
511. V.V. Safonov, A.V. Konov, B.G. Korshunov, Zh. Neorg. Khim., **14**:2880 (1969).

512. P. Ehrlich, R. Schmitt, Z. Anorg. Allg. Chem., **308**:91 (1961).
513. B.F. Markov, R.V. Chernov, Ukr. Khim. Zh., **25**:279 (1959).
514. D.V. Korol'kov, G.N. Kudryashova, Zh. Neorg. Khim., **13**:1626 (1968).
515. V.P. Radina, I.K. Tovmas'yan, in: Research in Thermography and Corrosion, Rostov. Knizhn. Izd., Rostov (1970), pp. 188, 198.
516. V.N. Desyatnik, B.N. Dubinin, S.P. Raspopin, Zh. Fiz. Khim., **47**:2726 (1973).
517. A. Bogacz, W. Trzebiatowski, Rocz. Chem., **38**:729 (1964).
518. V.M. Vdovenko, V.A. Volkov, I.I. Kozhina, I.G. Suglobova, Radiokhimiya, **14**:482 (1972).
519. V.N. Desyatnik, N.N. Kurbatov, S.P. Raspopin, Yu.F. Chervinskii, Zh. Fiz. Khim., **46**:2159 (1972).
520. H.-I. Seifert, P. Ehrlich, Z. Anorg. Allg. Chem., **302**:284 (1959).
521. S.A. Shchukarev, I.L. Perfilova, L.N. Garin, Zh. Neorg. Khim., **11**:1451 (1966).
522. I.V. Vasil'kova, N.D. Zaitseva, V.A. Petrova, Zh. Neorg. Khim., **8**:2369 (1963).
523. A.I. Efimov, I.V. Vasil'kova, E.K. Smirnova, N.D. Zaitseva, T.S. Shemyakina, I.A. Perfilova, in: The Chemistry of Rare Elements, Izd. Leningrad. Gos. Univ., Leningrad (1964), p. 38.
524. A.V. Drobot, G.P. Anikina, L.V. Durinina, B.G. Korshunov, Zh. Neorg. Khim., **10**:562 (1965).
525. B.F. Markov, I.D. Panchenko, T.G. Kostenko, Ukr. Khim. Zh., **22**:287 (1956).
526. I.S. Morozov, Sun In-Chzhu, Zh. Neorg. Khim., **4**:678 (1959).
527. A. Chrétien, J. Chassaing, C. R. Acad. Sci., **C263**:1301 (1966).
528. J. Chassaing, Rev. Chim. Miner., **5**:1115 (1968).
529. A.A. Kosorukov, V.Ya. Frenkel', Yu.M. Korenev, A.V. Novoselova, Zh. Neorg. Khim., **18**:1938 (1973).
530. R.G. Samuseva, V.E. Plyushchev, Zh. Neorg. Khim., **10**:1270 (1965).
531. I.N. Belyaev, O.Ya. Revina, Zh. Neorg. Khim., **11**:1952 (1966).
532. R.E. Thoma, Oak Ridge Natl. Lab., ORNL-2548, 17 (1959).
533. D.L. Deadmore, J.S. Machin, J. Phys. Chem., **64**:824 (1960).
534. G.A. Bukhalova, D.V. Sementseva, Zh. Neorg. Khim., **10**:1880 (1965).
535. G.A. Bukhalova, É.P. Babaeva, Zh. Neorg. Khim., **10**:1883 (1965).
536. G.A. Bukhalova, É.P. Babaeva, Zh. Neorg. Khim., **11**:648 (1966).
537. V.T. Berezhnaya, G.A. Bukhalova, Zh. Neorg. Khim., **12**:2179 (1967).

538. I.N. Belyaev, O.Ya. Revina, in: The Physicochemical Analysis of Salt Systems, Izd. Rostov. Gos. Univ., Rostov (1962), p. 77.
539. I.N. Belyaev, O.Ya. Revina, Zh. Neorg. Khim., **11**:1446 (1966).
540. R.G. Samuseva, V.E. Plyushchev, Zh. Neorg. Khim., **6**:2139 (1961).
541. O. Schmitz-Dumont, G. Bergerhoff, Z. Anorg. Allg. Chem., **283**:314 (1956).
542. G.A. Bukhalova, N.A. Litovskaya, V.A. Lyutsedarskii, Izv. Akad. Nauk SSSR, Neorg. Mater., **5**:510 (1969).
543. É.P. Babaeva, G.A. Bukhalova, Zh. Neorg. Khim., **11**:648 (1966).
544. R.E. Thoma, T.S. Carlton, J. Inorg. Nucl. Chem., **17**:88 (1961).
545. R.V. Chernov, I.M. Ermolenko, Zh. Neorg. Khim., **19**:3096 (1974).
546. C.R. Tipton, Reactor Handbook (Materiáls), 2nd ed., New York (1960), p. 427.
547. J. Cretenet, C. R. Acad. Sci., **C267**:240 (1968).
548. G.A. Bukhalova, É.P. Babaeva, Zh. Neorg. Khim., **11**:402 (1966).
549. O. Schmitz-Dumont, H. Bornefeld, Z. Anorg. Allg. Chem., **287**:120 (1956).
550. G.D. Robbins, R.E. Thoma, H. Insley, J. Inorg. Nucl. Chem., **27**:559 (1965).
551. A.A. Kosorukov, V.Ya. Frenkel', Yu.M. Korenev, A.V. Novoselova, Zh. Neorg. Khim., **18**:1938 (1973).
552. G.A. Bukhalova, G.A. Shegurova, T.M. Khliyan, E.S. Yagub'yan, Zh. Neorg. Khim., **18**:1106 (1973).
553. G.A. Bukhalova, G.A. Shegurova, E.S. Yagub'yan, Zh. Neorg. Khim., **18**:2578 (1973).
554. I.I. Il'yasov, A.G. Bergman, Zh. Neorg. Khim., **9**:1416 (1964).
555. M. Ahtee, Soumalais. Tiedeakat. Toimituks, Ser. VI, No. 312, 3 (1969).
556. V.E. Plyushchev, S.B. Stepina, G.V. Zimina, O.P. Molchanova, L.V. Savel'eva, D.A. Yashkov, Izv. Vyssh. Uchebn. Zaved., Tsvetn. Metall., No. 1, 65 (1970).
557. J. Kutscher, A. Schneider, Z. Anorg. Allg. Chem., **386**:38 (1971).
558. I.N. Belyaev, E.A. Shurginov, N.S. Kudryashov, Zh. Neorg. Khim., **17**:2812 (1972).
559. P.I. Fedorov, N.S. Malova, V.A. Zhikharev, Zh. Neorg. Khim., **17**:548 (1972).
560. P.I. Fedorov, A.G. Dudareva, M.S. Nosova, Zh. Neorg. Khim., **12**:274 (1967).
561. R.G. Samuseva, V.E. Plyushchev, Zh. Neorg. Khim., **9**:2433 (1964).
562. I.I. Il'yasov, A.G. Bergman, Ukr. Khim. Zh., **31**:772 (1965).

563. I.I. Il'yasov, N.I. Chaurskii, D.G. Barsegov, Zh. Neorg. Khim., **11**:1985 (1966).
564. J. Kutscher, A. Schneider, Z. Anorg. Allg. Chem., **386**:38 (1971).
565. R.G. Samuseva, V.E. Plyushchev, Zh. Neorg. Khim., **6**:2139 (1961).
566. I.I. Il'yasov, A.G. Bergman, Zh. Neorg. Khim., **9**:1416 (1964).
567. I.I. Il'yasov, N.I. Chaurskii, D.G. Barsegov, A.G. Bergman, Zh. Neorg. Khim., **12**:2210 (1967).
568. I.I. Il'yasov, D.G. Barsegov, I.G. Barikashvili, L.P. Danilenko, Zh. Neorg. Khim., **14**:1484 (1969).
569. I.N. Belyaev, E.A. Shurginov, N.S. Kudryashov, Z. Neorg. Khim., **17**:2812 (1972).
570. J. Kutscher, A. Schneider, Z. Anorg. Allg. Chem., **386**:38 (1971).
571. R.G. Samuseva, V.E. Plyushchev, Zh. Neorg. Khim., **9**:2433 (1964).
572. I.I. Il'yasov, M. Davranov, K.I. Iskandarov, Zh. Neorg. Khim., **19**:1919 (1974).
573. V.E. Plyushchev, S.B. Stepina, G.V. Zimina, O.P. Molchanova, L. Savel'eva, D.A. Yashkov, Izv. Vyssh. Uchebn. Zaved., Tsvetn. Metall., No. 1, 65 (1970).
574. J. Kutscher, A. Schneider, Z. Anorg. Allg. Chem., **386**:38 (1971).
575. T.N. Sevast'yanova, N.V. Karpenko, Zh. Neorg. Khim., **14**:3130 (1969).
576. I.N. Belyaev, E.A. Shurginov, N.S. Kudryashov, Zh. Neorg. Khim., **17**:2812 (1972).
577. V.V. Safonov, O.V. Lemeshko, B.G. Korshunov, Zh. Neorg. Khim., **16**:2283 (1971).
578. T.I. Darvoid, P.I. Fedorov, M.A. Popova, L.A. Radushkevich, Izv. Akad. Nauk SSSR, Neorg. Mater., **3**:333 (1967).
579. I.I. Il'yasov, D.G. Barsegov, Zh. Neorg. Khim., **12**:2515 (1967).
580. M.A. Ol'skaya, T.I. Darvoid, L.I. Gosteva, in: The Processes of Formation of Salts and Single Crystals of Thalium Halides, Metallurgiya, Moscow-Leningrad (1970), p. 21.
581. V.V. Safonov, V.A. Grin'ko, M.B. Varfolomeev, E.S. Malysheva, V.I. Ksenzenko, Zh. Neorg. Khim., **18**:2827 (1973).
582. J. Teltow, J. Phys. Chem., **211**:241 (1959).
583. I.A. Kirovskaya, O.A. Belyavskaya, Izv. Akad. Nauk SSSR, Neorg. Mater., **9**:23 (1973).
584. P.I. Fedorov, V.M. Yakunina, Zh. Neorg. Khim., **8**:2103 (1963).
585. P.I. Fedorov, N.I. Il'ina, Zh. Neorg. Khim., **9**:1207 (1964).
586. C.R. Tipton, Reactor Handbook (Materials), 2nd ed., New York (1960), p. 442.

587. D.S. Coleman, R. Pollit, Trans. Inst. Mining Metall., **C78**:233 (1969).
588. D.S. Coleman, B. Brown, R. Pollit, Trans. Inst. Mining Metall., **C78**:148 (1969).
589. D.Ya. Toptygin, Zh. Neorg. Khim., **8**:1187 (1963).
590. N.D. Chikanov, Zh. Neorg. Khim., **14**:1120 (1969).
591. D.S. Coleman, P.D.A. Lacy, Trans. Inst. Mining Metall., **C77**:170 (1968).
592. A.P. Palkin, G.P. Chepurko, Zh. Neorg. Khim., **1**:1832 (1956).
593. M.P. Vorobei, O.V. Skiba, Zh. Neorg. Khim., **15**:1414 (1970).
594. N.D. Chikanov, Zh. Neorg. Khim., **14**:1430 (1969).
595. N.D. Chikanov, Zh. Neorg. Khim., **17**:2558 (1972).
596. M.S. Golubeva, R.D. Sidakova, in: The Physicochemical Analysis of Salt Systems, Izd. Rostov. Gos. Univ., Rostov (1962), p. 189.
597. P.I. Fedorov, G.A. Kot, Zh. Neorg. Khim., **12**:2313 (1967).
598. P.I. Fedorov, A.G. Dudareva, N.F. Drobot, Zh. Neorg. Khim., **8**:1286 (1963).
599. I.N. Bradley, P.D. Greene, Trans. Faraday Soc., **63**:424 (1967).
600. V.V. Safonov, A.G. Ivanchenko, B.G. Korshunov, Zh. Neorg. Khim., **19**:852 (1974).
601. V.V. Safonov, O.V. Lemeshko, Zh. Neorg. Khim., **18**:1957 (1973).
602. D. Dumora, J. Ravez, P. Hagenmuller, Bull. Soc. Chim. France, No. 4, 1301 (1970).
603. B.G. Korshunov, D.V. Drobot, Z.M. Shevtsova, Zh. Neorg. Khim., **10**:939 (1965).
604. D.V. Drobot, B.G. Korshunov, Z.M. Shevtsova, in: The Physical Chemistry of Molten Salts, Metallurgiya, Moscow (1965), p. 48.
605. E.I. Ardashnikova, E.I. Borzenkova, A.V. Novoselova, Izv. Akad. Nauk SSSR, Neorg. Mater., **10**:1632 (1974).
606. R.E. Thoma, Rare Earth Halides, ORNL-3804 (1965).
607. R.E. Thoma, Rare Earth Research. III. Proceedings of IV Conf. on Rare Earth Research, New York (1964) p. 565.
608. J. Kutscher, A. Schneider, Z. Anorg. Allg. Chem., **386**:38 (1971).
609. B.G. Korshunov, D.V. Drobot, I.E. Galchenko, Z.M. Shevtsova, Zh. Neorg. Khim., **11**:441 (1966).
610. F.G. Gavryuchenkova, G.I. Novikov, Zh. Neorg. Khim., **11**:2172 (1966).
611. B.G. Korshunov, D.V. Drobot, Z.M. Shevtsova, Zh. Neorg. Khim., **10**:2310 (1965).
612. R.E. Thoma, Rare Earth Research. III. Proceedings of IV Conf. on Rare Earth Research, New York (1964), p. 561.
613. R.E. Thoma, H. Insley, G.M. Hebert, Inorg. Chem., **5**:1222 (1966).

614. E.G. Ippolitov, L.S. Garashina, B.N. Zhigarnovskii, Izv. Sib. Otd. Akad. Nauk SSSR, Ser. Khim. Nauk, No. 4, 49 (1968).
615. G. Brauer, M. Knausenberger, Z. Anorg. Allg. Chem., **320**:54 (1963).
616. B.G. Korshunov, D.V. Drobot, Zh. Neorg. Khim., **10**:2310 (1965).
617. I.S. Morozov, V.L. Chatova, Zh. Neorg. Khim., **13**:2810 (1968).
618. P.I. Fedorov, N.I. Il'ina, Zh. Neorg. Khim., **11**:205 (1966).
619. H.L. Pinch, J.M. Hirshon, J. Am. Chem. Soc., **79**:6149 (1957).
620. V.I. Ionov, I.S. Morozov, B.G. Korshunov, Zh. Neorg. Khim., **5**:1248 (1960).
621. N.V. Galitskii, V.I. Borodin, A.N. Lystsov, Ukr. Khim. Zh., **32**:695 (1966).
622. Li Chi-Fa, I.S. Morozov, Zh. Neorg. Khim., **7**:2765 (1962).
623. G.I. Novikov, A.K. Baev, O.G. Polyachenok, in: The Chemistry of Rare Elements, Izd. Leningrad. Gos. Univ., Leningrad (1964), p. 63.
624. N.G. Korzhukov, M.P. Ozerova, K.G. Khomyakov, L.D. Onikienko, Zh. Neorg. Khim., **11**:202 (1966).
625. V.I. Fadeeva, N.G. Korzhukov, K.G. Khomyakov, Vestn. Mosk. Gos. Univ., Ser. II, Khim., No. 2, 66 (1969).
626. I.S. Morozov, V.L. Chatova, Zh. Neorg. Khim., **13**:2810 (1968).
627. B.G. Korshunov, G.A. Lovetskaya, A.A. Palant, Zh. Neorg. Khim., **12**:203 (1967).
628. N.D. Chikanov, Zh. Neorg. Khim., **14**:1430 (1969).
629. N.G. Korzhukov, K.G. Khomyakov, Vestn. Mosk. Gos. Univ., Ser. II, Khim., No. 4 68 (1966).
630. V.I. Fadeeva, N.G. Korzhukov, K.G. Khomyakov, Vestn. Mosk. Gos. Univ., Ser. II, Khim., No. 2, 66 (1969).
631. H.-J. Seifert, K. Klatyk, Z. Anorg. Allg. Chem., **342**:1 (1966).
632. I.S. Morozov, Z.N. Shevtsova, Li Chi-Fa, Zh. Neorg. Khim., **9**:2606 (1964).
633. P.I. Fedorov, Liu Go-Yuan, S.K. Nedev, Zh. Neorg. Khim., **10**:2720 (1965).
634. G.M. Kurdyumov, V.A. Molochko, T.B. Taganova, Zh. Neorg. Khim., **17**:1469 (1972).
635. N.D. Denisova, A.P. Baskova, Investigation of the Chemistry and Metallurgy of Zirconium and Hafnium and Their Compounds, Metallurgiya, Moscow-Leningrad (1972), p. 51.
636. P.I. Fedorov, N.I. Il'ina, Zh. Neorg. Khim., **11**:205 (1966).
637. I.S. Morozov, L. Tsegledi, Zh. Neorg. Khim., **6**:2766 (1961).
638. Ch.M. Cook, W.E. Dunn, J. Phys. Chem., **65**:1505 (1961).
639. N.V. Galitskii, V.I. Borodin, A.N. Lystsov, Ukr. Khim. Zh., **34**:223 (1968).
640. A.G. Krokhina, V.A. Krokhin, I.S. Morozov, Zh. Neorg. Khim., **18**:1359 (1973).

641. A.V. Storonkin, I.V. Vasil'kova, M.D. Pyatunin, Zh. Neorg. Khim., **46**:1605 (1972).
642. B.G. Korshunov, V.I. Gol'din, Zh. Neorg. Khim., **6**:1642 (1961).
643. A.I. Efimov, I.V. Vasil'kova, E.K. Smirnova, N.D. Zaitseva, T.S. Shemyakina, I.L. Perfilova, Vestn. Leningrad. Gos. Univ., Ser. Fiz.-Khim., No. 16, Issue 3, 98 (1962).
644. A.I. Efimov, I.V. Vasil'kova, E.K. Smirnova, N.D. Zaitseva, T.S. Shemyakina, I.L. Perfilova, in: The Chemistry of Rare Elements, Izd. Leningrad. Gos. Univ., Leningrad (1964), p. 38.
645. I.S. Morozov, D.Ya. Toptygin, Zh. Neorg. Khim., **2**:2129 (1957).
646. B.G. Korshunov, I.S. Morozov, V.I. Ionov, M.A. Zorina, Izv. Vyssh. Uchebn. Zaved., Tsvetn. Metall., **5**:67 (1960).
647. I.S. Morozov, Zh. Neorg. Khim., **1**:2792 (1956).
648. C.M. Cook, R.B. Hand, J. Phys. Chem., **65**:1467 (1961).
649. D.V. Drobot, B.G. Korshunov, V.A. Aleksandrova, Zh. Neorg. Khim., **13**:3354 (1968).
650. N.D. Chikanov, Zh. Neorg. Khim., **18**:277 (1973).
651. N.D. Chikanov, Zh. Neorg. Khim., **17**:2558 (1972).
652. I.S. Morozov, Li Chi-Fa, Zh. Neorg. Khim., **8**:2733 (1963).
653. I.S. Morozov, Zh. Neorg. Khim., **1**:2792 (1956).
654. V.N. Safonov, B.G. Korshunov, Izv. Vyssh. Uchebn. Zaved., Tsvetn. Metall., **6**:81 (1968).
655. B.G. Korshunov, S.I. Ionov, T.A. Baklashova, S.S. Kokorev, Izv. Vyssh. Uchebn. Zaved., Tsvetn. Metall., **6**:114 (1960).
656. A.G. Krokhina, V.A. Krokhin, I.S. Morozov, Zh. Neorg. Khim., **18**:1359 (1973).
657. D.V. Drobot, A.V. Nikolaev, A.V. Belousov, Zh. Neorg. Khim., **18**:2243 (1973).
658. I.S. Morozov, A.T. Simonich, Zh. Neorg. Khim., **6**:937 (1961).
659. A. Tressland, R. De Pape, J. Portier, P. Hagenmuller, C. R. Acad. Sci., **C266**:984 (1968).
660. J.D. Donaldson, R. Oteng, J. Chem. Soc., **A18**:2696 (1969).
661. J. Ravez, M. Duale, C. R. Acad. Sci., **C270**:56 (1970).
661a. J. Ravez, J. Viollet, R. De Pape, P. Hagenmuller, Bull. Soc. Chim. France, No. 4, 1325 (1967).
662. P.I. Fedorov, G.A. Kot, L.L. Nikol'skaya, Zh. Neorg. Khim., **12**:2818 (1967).
663. P.I. Fedorov, V.M. Khazan, N.I. Evdokimova, Zh. Neorg. Khim., **15**:2568 (1970).
664. V.V. Safonov, I.V. Vasilishcheva, V.M. Khazan, B.G. Korshunov, P.I. Fedorov, Izv. Vyssh. Uchebn. Zaved., Tsvetn. Metall., No. 4, 82 (1971).
665. P.I. Fedorov, G.V. Sukhov, V.M. Khazan, L.G. Nosova, N.S. Sysoeva, Izv. Akad. Nauk SSSR, Neorg. Mater., **8**:178 (1972).

666. J.D. Corbett, A. Hershaft, J. Am. Chem. Soc., **80**:1530 (1958).
667. P.I. Fedorov, G.A. Lovetskaya, Zh. Neorg. Khim., **15**:1973 (1970).
668. A.G. Dudareva, N.N. Mel'nikov, P.I. Fedorov, Zh. Neorg. Khim., **15**:2273 (1970).
669. A.G. Dudareva, T.V. Fedorova, Yu.É. Bogatov, P.I. Fedorov, Zh. Neorg. Khim., **19**:1607 (1974).
670. A.G. Dudareva, A.K. Molodkin, É. Uipe Nova, M. Rabanni, Zh. Neorg. Khim., **22**:2313 (1977).
671. P.I. Fedorov, G.A. Lovetskaya, Zh. Neorg. Khim., **14**:3175 (1969).
672. P.I. Fedorov, G.A. Lovetskaya, E.E. Shariadi, Izv. Vyssh. Uchebn. Zaved., Tsvetn. Metall., **6**:63 (1969).
673. P.I. Fedorov, G.A. Lovetskaya, Zh. Neorg. Khim., **13**:3392 (1968).
674. P.I. Fedorov, S.K. Nedev, Zh. Neorg. Khim., **11**:2413 (1966).
675. P.I. Fedorov, V.V. Tsimbalist, Liu Go-Yuan, Zh. Neorg. Khim., **9**:1681 (1964).
676. P.I. Fedorov, G.A. Lovetskaya, Zh. Neorg. Khim., **16**:1993 (1971).
677. A.P. Palkin, N.V. Ostrikova, Zh. Neorg. Khim., **9**:2045 (1964).
678. P.I. Fedorov, S.K. Nedev, Zh. Neorg. Khim., **10**:2717 (1965).
679. P.I. Fedorov, V.V. Tsimbalist, Zh. Neorg. Khim., **9**:1676 (1964).
680. V.N. Arbekov, E.S. Petrov, Izv. Sib. Otd. Akad. Nauk SSSR, Ser. Khim. Nauk, Issue 2, 56 (1967).
681. V.N. Arbekov, E.S. Petrov, Izv. Sib. Otd. Akad. Nauk SSSR, Ser. Khim. Nauk, No. 9, 130 (1968).
682. P.I. Fedorov, S.K. Nedev, Liu Go-Yuan, Zh. Neorg. Khim., **11**:406 (1966).
683. P.I. Fedorov, S.K. Nedev, Zh. Neorg. Khim., **12**:1301 (1967).
684. P.I. Fedorov, L.P. Khagleeva, Zh. Neorg. Khim., **12**:818 (1967).
685. P.I. Fedorov, V.M. Yakunina, Zh. Neorg. Khim., **8**:2103 (1963).
686. I.S. Morozov, V.L. Chatova, Zh. Neorg. Khim., **13**:2810 (1968).
687. N.D. Chikanov, Zh. Neorg. Khim., **13**:2616 (1968).
688. V.A. Sryvtsev, V.N. Arbekov, E.S. Petrov, in: Rare Alkaline Elements, Nauka, Novosibirsk (1967), p. 91.
689. N.V. Ostrikova, Z.A. Miroshnik, V.D. Kuleshova, N.P. Dudkina, A.I. Ilyushchenko, Zh. Neorg. Khim., **14**:2229 (1969).
690. L.A. Nisel'son, Z.N. Orshanskaya, K.V. Tret'yakova, Zh. Neorg. Khim., **19**:1060 (1974).
691. N.D. Chikanov, Zh. Neorg. Khim., **17**:2558 (1972).
692. P.I. Fedorov, L.P. Khagleeva, Zh. Neorg. Khim., **11**:2174 (1966).

693. P.I. Fedorov, G.A. Lovetskaya, E.E. Sharkadi, Zh. Neorg. Khim., **14**:2595 (1969).
694. N.D. Chikanov, Izv. Vyssh. Uchebn. Zaved., Khim. Khim. Tekhnol., **13**:1047 (1970).
695. J. Chassaing, Rev. Chim. Miner., **5**:1115 (1968).
696. A. Chrétien, J. Chassaing, C. R. Acad. Sci., **C263**:1301 (1966).
697. J. Chassaing, A. Erb, C. R. Acad. Sci., **C270**:949 (1970).
698. J. Grannec, J. Ravez, Bull. Soc. Chim. France, **5**:1753 (1970).
699. P.I. Fedorov, G.A. Kot, L.L. Nikol'skaya, Zh. Neorg. Khim., **12**:3172 (1967).
700. P.I. Fedorov, G.A. Kot, L.L. Nikol'skaya, Zh. Neorg. Khim., **12**:2818 (1967).
701. P.I. Fedorov, G.A. Kot, Zh. Neorg. Khim., **12**:2313 (1967).
702. B.G. Korshunov, D.V. Drobot, Z.M. Shevtsova, Zh. Neorg. Khim., **10**:939 (1965).
703. D.V. Drobot, B.G. Korshunov, Z.M. Shevtsova, in: The Physical Chemistry of Molten Salts, Metallurgiya, Moscow (1965), p. 48.
704. Fam Ngok Tien, I.S. Morozov, Zh. Neorg. Khim., **14**:1361 (1969).
705. R.E. Thoma, Rare Earth Halides, ORNL-3804 (1965).
706. R.E. Thoma, Rare Earth Research. III. Proceedings of IV Conf. on Rare Earth Research, New York (1964).
707. R.E. Thoma, H. Insley, G.M. Hebert, Inorg. Chem., **5**:1222 (1966).
708. E.G. Ippolitov, L.S. Garashina, B.N. Zhigarnovskii, Izv. Sib. Otd. Akad. Nauk SSSR, Ser. Khim. Nauk, No. 4, 49 (1968).
709. J. Kutscher, A. Schneider, Z. Anorg. Allg. Chem., **386**:38 (1971).
710. G.M. Kurdyumov, V.A. Molochko, L.K. Agureeva, Zh. Neorg. Khim., **17**:2309 (1972).
711. N.D. Chikanov, Zh. Neorg. Khim., **16**:1452 (1971).
712. P.I. Fedorov, V.A. Molochko, G.M. Kurdyumov, M.B. Braginskaya, Zh. Neorg. Khim., **15**:1398 (1970).
713. V.V. Safonov, A.V. Konov, Zh. Neorg. Khim., **17**:3363 (1972).
714. G.M. Kurdyumov, V.A. Molochko, T.B. Taganova, Zh. Neorg. Khim., **17**:1469 (1972).
715. N.D. Chikanov, Izv. Vyssh. Uchebn. Zaved., Khim. Khim. Tekhnol., **13**:1047 (1970).
716. P.I. Fedorov, V.M. Khazan, L.G. Nosova, Zh. Neorg. Khim., **15**:2298 (1970).
717. V.A. Molochko, G.M. Kurdyumov, T.N. Sokhanova, Zh. Neorg. Khim., **17**:2307 (1972).
718. L.A. Nisel'son, Zh. Neorg. Khim., **7**:693 (1962).
719. I.S. Morozov, Sun In-Chzhu, Zh. Neorg. Khim., **4**:678 (1959).
720. A.S. Roy, L.I. Howell, H.H. Kellogg, Trans. Metall. Soc. AIME, **212**:817 (1958-59).

721. L.A. Nisel'son, B.N. Ivanov-Émin, L.E. Larionova, Zh. Neorg. Khim., **6**:186 (1961).
722. I.N. Sheiko, V.T. Mal'tsev, G.A. Bukhalova, Ukr. Khim. Zh., **32**:1292 (1966); Dokl. Akad. Nauk UkrSSR, **6**:782 (1966).
723. I.N. Sheiko, V.T. Mal'tsev, G.A. Bukhalova, Ukr. Khim. Zh., **32**:1292 (1966); Dokl. Akad. Nauk SSSR, **5**:628 (1966).
724. J. Huart, Bull. Soc. Frans. Mineral. Cristallogr., **89**:23 (1966).
725. A.G. Dudareva, A.K. Molodkin, É. Uipe Nova, M. Rabbani, Zh. Neorg. Khim., **22**:2313 (1977).
726. I.D. Turyanitsa, V.V. Khiminiets, Kristallografiya, **18**:1092 (1973).
727. L.F. Grantham, J. Chem. Phys., **49**:3835 (1968).
728. N.D. Chikanov, Zh. Neorg. Khim., **14**:1120 (1969).
729. H. Mikler, E. Hayer, K.L. Komarek, Monatsh. Chem., **104**:1251 (1973).
730. P.I. Fedorov, N.I. Il'ina, Zh. Neorg. Khim., **11**:1693 (1966).
731. N.D. Chikanov, Zh. Neorg. Khim., **14**:1430 (1969).
732. I.N. Belyaev, D.S. Lesnykh, I.G. Éikhenbaum, Zh. Neorg. Khim., **15**:846 (1970).
733. N.D. Chikanov, Zh. Neorg. Khim., **17**:2558 (1972).
734. N.D. Chikanov, Izv. Vyssh. Uchebn. Zaved., Khim. Khim. Tekhnol., **13**:1047 (1970).
735. P.I. Fedorov, A.G. Dudareva, M.S. Tsygankova, Zh. Neorg. Khim., **11**:2410 (1966).
736. I.N. Belyaev, E.A. Shurginov, N.S. Kudryashov, Zh. Neorg. Khim., **17**:2812 (1972).
737. V.V. Safonov, A.G. Ivanchenko, B.G. Korshunov, Zh. Neorg. Khim., **19**:852 (1974).
738. A. Gäumann, Chimia, **20**:82 (1966).
739. V.V. Safonov, I.V. Vasilishcheva, B.G. Korshunov, Zh. Neorg. Khim., **16**:2310 (1971).
740. B.G. Korshunov, D.V. Drobot, I.E. Galchenko, Z.N. Shevtsova, Zh. Neorg. Khim., **11**:411 (1966).
741. B.G. Korshunov, D.V. Drobot, Zh. Neorg. Khim., **10**:2310 (1965).
742. I.B. Taimuradov, L.P. Reshetnikova, A.V. Novoselova, Izv. Akad. Nauk SSSR, Neorg. Mater., **10**:1468 (1974).
743. E.G. Ippolitov, L.S. Garashina, B.N. Zhigarnovskii, Izv. Sib. Otd. Akad. Nauk SSSR, Ser. Khim. Nauk, No. 4, 49 (1968).
744. A.G. Dudareva, Gladis Kiros, P.I. Fedorov, A.I. Ezhov, V.M. Akimov, Zh. Neorg. Khim., **18**:216 (1973).
745. A.G. Dudareva, Konstantidiya Akkelidu, B.N. Ivanov-Émin, P.I. Fedorov, Zh. Neorg. Khim., **16**:1742 (1971).
746. A.G. Dudareva, Yu.É. Bogatov, Konstantidiya Akkelidu, V.N. Ivanov-Émin, P.I. Fedorov, Izv. Vyssh. Uchebn. Zaved., Khim. Khim. Tekhnol., **14**:1299 (1971).

747. R.J. Clark, E. Griswold, J. Kleinberg, J. Am. Chem. Soc., 80:4764 (1958).
748. P.I. Fedorov, N.S. Malova, G.V. Vorob'eva, Zh. Neorg. Khim., 16:2598 (1971).
749. P.I. Fedorov, Zh. Neorg. Khim., 15:2853 (1970).
750. J. Josiak, J. Terpilowski, Rocz. Chem., 39:1777 (1965).
751. T.N. Larionova, T.M. Pustyl'nik, Zh. Neorg. Khim., 14:3151 (1969).
752. P.I. Fedorov, N.S. Malova, G.V. Vorob'eva, Zh. Neorg. Khim., 15:2853 (1970).
753. N.M. Demidov, A.N. Kokoev, A.S. Malyugin, A.D. Pogorelyi, V.I. Shereshkova, Izv. Vyssh. Uchebn. Zaved., Tsvetn. Metall., 3:29 (1969).
754. P.I. Fedorov, N.S. Malova, G.V. Vorob'eva, Zh. Neorg. Khim., 15:2281 (1970).
755. A.N. Kokoev, A.S. Malyugin, V.I. Shereshkova, Zh. Prikl. Khim., 43:683 (1970).
756. P.F. Fedorov, N.I. Il'ina, Zh. Neorg. Khim., 9:1207 (1964).
757. V.A. Sryvtsev, E.S. Petrov, Izv. Sib. Otd. Akad. Nauk SSSR, Ser. Khim. Nauk, Issue 2, 67 (1968).
758. D.V. Drobot, B.G. Korshunov, L.M. Novikova, Izv. Vyssh. Uchebn. Zaved., Khim. Khim. Tekhnol., 12:6 (1969).
759. P.I. Fedorov, N.I. Il'ina, Zh. Neorg. Khim., 11:1693 (1966).
760. P.I. Fedorov, N.I. Il'ina, Zh. Neorg. Khim., 11:1432 (1969).
761. V.A. Sryvtsev, E.S. Petrov, Izv. Sib. Otd. Akad. Nauk SSSR, Ser. Khim. Nauk, Issue 1, 107 (1964).
762. P.I. Fedorov, Tzhan Tsulian, Zh. Neorg. Khim., 9:231 (1964).
763. V.L. Chatova, I.S. Morozov, Zh. Neorg. Khim., 13:1645 (1968).
764. P.I. Fedorov, Tzhan Tsulian, Zh. Neorg. Khim., 6:2605 (1961).
765. V.A. Sryvtsev, V.N. Arbekov, E.S. Petrov, in: Rare Alkaline Elements, Nauka, Novosibirsk (1967), p. 191.
766. A.P. Palkin, T.N. Vigutova, L.I. Glotova, Zh. Neorg. Khim., 8:253 (1963).
767. T.N. Larionova, L.A. Lozgacheva, Zh. Neorg. Khim., 13:2883 (1968).
768. D.V. Drobot, B.G. Korshunov, L.M. Novikova, Izv. Vyssh. Uchebn. Zaved., Khim. Khim. Tekhnol., 12:6 (1969).
769. J. Granec, J.-C. Champarnaud-Mosjard, J. Portier, Bull. Soc. Chim. France, XXIII:3862 (1970).
770. J.E. Roberts, A.W. Laubengayer, J. Am. Chem. Soc., 79:5895 (1957).
771. J. Ravez, J. Solid State Chem., 3:234 (1971).
772. J. Granec, J.-C. Champarnaud-Mosjard, J.P. Costy, J.-C. Cousseins, B. Gadreau, Rev. Chim. Miner., 9:569 (1972).

773. J. Granec, J. Ravez, C. R. Acad. Sci., **C270**:2059 (1970).
774. J.-C. Champarnaud-Mosjard, J.-P. Laval, B. Gaudreau, Rev. Chim. Miner., **11**:735 (1974).
775. P.I. Fedorov, N.S. Malova, V.N. Zhikharev, Zh. Neorg. Khim., **17**:548 (1972).
776. P.I. Fedorov, N.S. Malova, Yu.N. Denisov, Zh. Neorg. Khim., **16**:3347 (1971).
777. Yu.N. Denisov, N.S. Malova, P.I. Fedorov, Zh. Neorg. Khim., **19**:815 (1974).
778. O.N. Postnikova, Yu.N. Denisov, P.I. Fedorov, N.S. Malova, L.A. Radushkevich, Zh. Neorg. Khim., **18**:1436 (1973).
779. Yu.N. Denisov, N.S. Malova, P.I. Fedorov, Zh. Neorg. Khim., **18**:1363 (1973).
780. P.I. Fedorov, A.G. Dudareva, M.S. Nosova, Zh. Neorg. Khim., **12**:274 (1967).
781. A.G. Dudareva, N.N. Mel'nikov, P.I. Fedorov, Zh. Neorg. Khim., **15**:2273 (1970).
782. P.I. Fedorov, V.M. Khazan, N.I. Evdokimova, Zh. Neorg. Khim., **15**:2568 (1970).
783. S.M. Arkhipov, T.V. Revzina, N.I. Kashina, G.E. Revzin, Zh. Neorg. Khim., **13**:1204 (1968).
784. P.I. Fedorov, A.G. Dudareva, É.M. Kuz'mina, Zh. Neorg. Khim., **6**:1378 (1961).
785. P.I. Fedorov, A.G. Dudareva, V.A. Nesterov, Zh. Neorg. Khim., **15**:2570 (1970).
786. G.V. Samsonov, V.A. Obolonchik, G.N. Kulichkina, Khim. Nauka Promst., **4**:804 (1959).
787. V.G. Selivanov, V.V. Stender, Zh. Neorg. Khim., **3**:447 (1958).
788. S. Pawlenko, Z. Anorg. Allg. Chem., **336**:172 (1965).
789. C.J. Barton, L.O. Gilpatrick, J.A. Bornmann, H.H. Stone, T.N. McVay, H. Insley, J. Inorg. Nucl. Chem., **33**:337 (1971).
790. V.K. Pyrina, M.E. Proetanov, Zh. Neorg. Khim., **20**:1140 (1975).
791. E. Aukrust, B. Björge, H. Flood, T. Forland, Ann. N.Y. Acad. Sci., **79**(11):830 (1960).
792. N.N. Volkov, T.V. Kuryleva, M.N. Zakhvalinskii, Izv. Fiz.-Khim. Nauchn.-Issled. Inst. Irkutsk. Gos. Univ., Issue 2, 152 (1961).
793. A.I. Agaev, A.G. Bergman, Uch. Zap. Azerb. Gos. Univ., Ser. Khim., Issue 1, 3 (1963).
794. A.S. Arabadzhan, A.G. Bergman, Zh. Neorg. Khim., **8**:720 (1963).
795. A.G. Bergman, Yu.I. Andryushchenko, Zh. Neorg. Khim., 12:1025 (1967).
796. N.N. Volkov, N.V. Popova, Nauch. Tr. Irkutsk. Med. Inst., Issue 95, 124 (1969).
797. I.I. Il'yasov, V.N. Mirsoyanov, Yu.V. Korotkov, Zh. Neorg. Khim., **4**:909 (1959).

798. I.I. Il'yasov, Zh. Neorg. Khim., **7**:169 (1962).
799. I.I. Il'yasov, A.G. Bergman, Zh. Obshch. Khim., **31**:368 (1961).
800. Nguyen Ba-Chahn, J. Chim. Phys. Phys.-Chim. Biol., **61**(10):1428 (1964).
801. T.M. Tsushba, in: Proc. of the Interuniversity Scientific-Theoretical Conference of Post Graduate Students, Izd. Rostovsk. Gos. Univ., Rostov (1970), p. 202.
802. I.K. Tovmas'yan, T.M. Tsushba, Ukr. Khim. Zh., **39**:952 (1973).
803. A.K. Bostandzhiyan, I.I. Il'yasov, A.G. Bergman, Zh. Neorg. Khim., **4**:2079 (1959).
804. I.I. Il'yasov, S.D. Dionis'ev, A.G. Bergman, Zh. Neorg. Khim., **5**:664, 1135 (1960).
805. I.I. Il'yasov, D.G. Barsegov, I.G. Berikashvili, L.P. Danilenko, Zh. Neorg. Khim., **14**:1484 (1969).
806. G.G. Diogenov, V.I. Ermachkov, Zh. Neorg. Khim., **12**:573 (1967).
807. I.K. Tovmas'yan, T.M. Tsushba, A.G. Bergman, Zh. Neorg. Khim., **15**:2268 (1970).
808. I.K. Tovmas'yan, T.M. Tsushba, V.P. Radina, Zh. Neorg. Khim., **15**:2535 (1970).
809. V.E. Plyushchev, S.B. Stepina, G.B. Zimina, B.G. Zhilyakov, Izv. Vyssh. Uchebn. Zaved., Tsvetn. Metall., No. 4, 112 (1964).
810. N.Ya. Fedorov, E.S. Petrov, Izv. Sib. Otd. Akad. Nauk SSSR, Ser. Khim. Nauk, No. 12, Issue 5, 47 (1969).
811. I.N. Belyaev, E.A. Shurginov, Zh. Neorg. Khim., **15**:883 (1970).
812. I.N. Belyaev, A.K. Doroshenko, in: The Physicochemical Analysis of Salt Systems, Izd. Rostovsk. Gos. Univ., Rostov (1962), p. 12.
813. R. Riccardi, C. Sinistri, V. Campari, A. Magistris, Z. Naturforsch., **25a**(5):781 (1970).
814. V.N. Prokhorov, I.V. Krivousova, I.I. Kozhina, A.N. Efimov, Vestn. Leningrad. Gos. Univ., Ser. Fiz.-Khim., No. 4, 89 (1974).
815. G.A. Bukhalov, A.M. Burlakov, Zh. Neorg. Khim., **11**:164 (1966).
816. V.A. Grin'ko, V.V. Safonov, V.I. Ksenzenko, B.G. Korshunov, Zh. Neorg. Khim., **17**:1755 (1972).
817. P. Ehrlich, G. Kaupa, K. Blankenstein, Z. Anorg. Allg. Chem., **299**:213 (1959).
818. I.I. Il'yasov, L.V. Rozhkovskaya, A.G. Bergman, Zh. Neorg. Khim., **2**:1883 (1957).
819. S.D. Dionis'ev, I.I. Il'yasov, A.G. Bergman, Zh. Neorg. Khim., **5**:1135 (1960).
820. S.D. Dionis'ev, I.I. Il'yasov, A.G. Bergman, Zh. Neorg. Khim., 7:169 (1962).

821. E. Vilhonen, Turun Yliopiston Julkaisuja, Ser. A1, No. 57, 42 (1962).
822. É.M. Moroz, S.V. Ketchik, S.S. Batsanov, Zh. Neorg. Khim., **17**:1775 (1972).
823. I.K. Tovmas'yan, T.M. Tsushba, Ukr. Khim. Zh., **39**:952 (1973).
824. L. Pertti, T. Olli, Suomen Kem., **39**(11):B220 (1966).
825. I.I. Il'yasov, D.G. Barsegov, Ukr. Khim. Zh., **35**:873 (1969).
826. Sun In-Chu, I.S. Morozov, Zh. Neorg. Khim., **3**:1914 (1958).
827. G.I. Novikov, A.K. Baev, Vestn. Leningrad. Gos. Univ., Ser. Fiz.-Khim., No. 22, Issue 4, 116 (1961).
828. N.N. Volkov, M.N. Zakhvalinskii, Izv. Fiz.-Khim. Nauchn.-Issled. Inst. Irkutsk. Gos. Univ., Issue 1, 107 (1959).
829. E. Aukrust, B. Björge, H. Flood, T. Forland, Ann. N.Y. Acad. Sci., **79**(11):830 (1960).
830. C.R. Tipton, Reactor Handbook (Materials), 2nd ed., New York (1960), p. 442.
831. A.I. Kislova, A.G. Bergman, Zh. Neorg. Khim., **5**:2499 (1960).
832. N.N. Volkov, T.V. Kuryleva, M.N. Zakhvalinskii, Izv. Fiz.-Khim. Nauchn.-Issled. Inst. Irkutsk. Gos. Univ., Issue 2, 152 (1961).
833. S.I. Berezina, A.G. Bergman, E.L. Bakumskaya, Zh. Neorg. Khim., **8**:2140 (1963).
834. C. Liteanu, E. Cordos, Rev. Roum. Chim., **12**(8):983 (1967).
835. R.P. Clark, R.L. Blucher, H.J. Goldsmith, J. Chem. Eng. Data, **14**(4):465 (1969).
836. Zh.G. Moiseenko, E.K. Akopov, L.A. Panieva, Zh. Neorg. Khim., **17**:3098 (1972).
837. G.N. Litvinova, V.T. Berezhnaya, in: Research in General and Natural Science, Izd. Rostovsk. Inzh.-Stroit. Inst., Rostov (1973), p. 90.
838. N.D. Chikanov, V.T. Il'ginova, Zh. Neorg. Khim., **11**:1455 (1966).
839. H.-J. Seifert, F.W. Koknat, Z. Anorg. Allg. Chem., **341**:269 (1965).
840. V.V. Safonov, B.G. Korshunov, L.N. Taranyuk, Zh. Neorg. Khim., **13**:1950 (1968).
841. R.N. Yakovleva, I.I. Kozhina, I.V. Krivousova, I.V. Vasil'kova, Vestn. Leningrad. Gos. Univ., Ser. Fiz.-Khim., Issue 3, No. 16, 87 (1970).
842. A.S. Kuz'menko, É.N. Ryabov, R.A. Sandler, Zh. Neorg. Khim., **19**:2550 (1974).
845. É.N. Ryabov, R.A. Sandler, E.F. Klyuchnikova, L.P. Dorfman, Izv. Vyssh. Uchebn. Zaved., Tsvetn. Metall., No. 2, 65 (1972).
846. D.V. Drobot, É.A. Sapranova, Zh. Neorg. Khim., **18**:2008 (1973).

847. B.G. Korshunov, V.I. Gol'din, L.A. Averkieva, Izv. Vyssh. Uchebn. Zaved., Tsvetn. Metall., 1:101 (1962).
848. A.I. Efimov, L.P. Belorukova, A.M. Ryndina, Zh. Neorg. Khim., 8:1168 (1963).
849. A.I. Soklakov, N.A. Vasilenko, A.M. Pavlidis, N.A. Portnova, Zh. Neorg. Khim., 13:1193 (1968).
850. V.V. Rubleva, A.G. Bergman, Zh. Obshch. Khim., 26:561 (1956).
851. M.A. Zakharchenko, Zh. Neorg. Khim., 2:2178 (1957).
852. G.I. Nagornyi, T.D. Zimina, Izv. Fiz.-Khim. Nauchn.-Issled., Irkutsk. Gos. Univ., Issue 2, 30 (1961).
853. Nguyen Ba-Chahn, J. Chim. Phys. Phys.-Chim. Biol., 61(10):1428 (1964).
854. Pertti Luova, Olli Tanilla, Suomen Kem., 39(11):B220 (1966).
855. K. Heide, U. Brückner, Chem. Erde, 26:235 (1967).
856. D.S. Coleman, P.D.A. Lacy, Mater. Res. Bull., 2(10):935 (1967).
857. A.G. Morachevskii, A.F. Alabyshev, N.I. Berdichevskii, in: The Physical Chemistry and Electrochemistry of Molten Salts and Slags, Khimiya, Leningrad (1968), p. 110.
858. Nguyen Ba-Chahn, J.-P. Bastide, J. Chim. Phys. Phys.-Chim. Biol., 65:1425 (1968).
859. R. Cohen-Adad, A. Sebaoun, Bull. Soc. Chim. France, No. 8, 2597 (1969).
860. I.K. Tovmas'yan, T.M. Tsushba, A.G. Bergman, Ukr. Khim. Zh., 36:796 (1970).
861. I.K. Tovmas'yan, T.M. Tsushba, Ukr. Khim. Zh., 38:952 (1973).
862. V.V. Safonov, B.G. Korshunov, S.N. Stel'bovskaya, Zh. Neorg. Khim., 11:2143 (1966).
863. Y. Sacki, T. Suzuki, S. Yamaki, J. Electrochem. Soc. Jpn., 33:199 (1965).
864. V.V. Safonov, B.G. Korshunov, T.N. Zimina, Z.N. Shevtsova, Zh. Neorg. Khim., 11:2139 (1966).
865. B.G. Korshunov, V.V. Safonov, Zh. Neorg. Khim., 6:753 (1961).
866. A.P. Palkin, N.D. Chikanov, Zh. Neorg. Khim., 7:370 (1962).
867. O.R. Gavrilov, L.A. Nisel'son, Zh. Neorg. Khim., 11:210 (1966).
868. I.I. Kozhina, V.P. Kustovskaya, I.V. Vasil'kova, M.P. Susarev, Vestn. Leningrad. Gos. Univ., Ser. Fiz.-Khim., No. 16, Issue 3, 113 (1968).
869. Sun In-Chzhu, I.S. Morozov, Zh. Neorg. Khim., 3:1914 (1958).
870. G.I. Novikov, A.K. Baev, Vestn. Leningrad. Gos. Univ., Ser. Fiz.-Khim., No. 22, Issue 4, 116 (1961).
871. I.N. Belyaev, D.S. Lesnykh, I.G. Éikhenbaum, Zh. Neorg. Khim., 15:846 (1970).

872. I.I. Il'yasov, A.G. Bergman, Zh. Obshch. Khim., **26**:981 (1956).
873. I.I. Il'yasov, A.K. Bostandzhiyan, A.G. Bergman, Zh. Neorg. Khim., **1**:2543 (1956).
874. I.I. Il'yasov, A.K. Bostandzhiyan, A.G. Bergman, Zh. Neorg. Khim., **4**:2079 (1959).
875. C.R. Tipton, Reactor Handbook (Materials), 2nd ed., New York (1960), p. 442.
876. T.N. Sumarokova, T.P. Modestova, Zh. Neorg. Khim., **5**:2477 (1960).
877. I.I. Il'yasov, D.G. Barsegov, I.G. Berikashvili, L.P. Danilenko, Zh. Neorg. Khim., **14**:1484 (1969).
878. T.I. Drobasheva, A.G. Bergman, Zh. Neorg. Khim., **14**:1348 (1969).
879. P.V. Polyakov, É.M. Gil'debrant, G.A. Kotel'nikova, V.N. Mozhaeva, Zh. Fiz. Khim., **47**:2155 (1973).
880. Z.N. Shevtsova, I.S. Morozov, O.A. Efremova, Izv. Vyssh. Uchebn. Zaved., Tsvetn. Metall. No. 3, 109 (1960).
881. G.I. Novikov, A.K. Baev, Vestn. Leningrad. Gos. Univ., Ser. Fiz.-Khim., No. 22, Issue 4, 116 (1961).
882. R. Benz, M. Kahn, J.A. Leary, J. Phys. Chem., **63**(11):1983 (1959).
883. M.P. Vorobei, O.V. Skiba, A.S. Bevz, V.N. Desyatnik, The Physical Chemistry and Electrochemistry of Salt Melts, Metallurgiya, Moscow (1973), p. 3.
884. M.P. Vorobei, O.V. Skiba, A.S. Bevz, Zh. Neorg. Khim., **18**:2506 (1973).
885. G.M. Unzhakov, Physicochemical Methods of Investigation of Biological Objects and Salt Systems, Irkutsk. Knizhn. Izd., Irkutsk (1969), p. 163.
886. I.K. Tovmas'yan, T.M. Tsushba, A.G. Bergman, Ukr. Khim. Zh., **3**:796 (1970).
887. I.K. Tovmas'yan, T.M. Tsushba, Ukr. Khim. Zh., **39**:952 (1973).
888. N.D. Chikanov, Zh. Neorg. Khim., **17**:2558 (1972).
889. N.Ya. Fedorov, B.S. Petrov, Izv. Sib. Otd. Akad. Nauk SSSR, Ser. Khim. Nauk, No. 11, Issue 3, 154 (1964); No. 2, Issue 1, 57 (1967).
890. G.I. Novikov, O.G. Polyachenok, S.A. Frid, Zh. Neorg. Khim., **9**:472 (1964).
891. Sun Yui-lin, G.I. Novikov, Zh. Neorg. Khim., **8**:700 (1963).
892. G.I. Novikov, O.G. Polyachenok, S.A. Frid, Zh. Neorg. Khim., **9**:472 (1964).
893. B.G. Korshunov, D.V. Drobot, V.V. Bukhtiyarov, Z.N. Shevtsova, Zh. Neorg. Khim., **9**:1427 (1964).
894. B.G. Korskhunov, D.V. Drobot, V.V. Bukhtiyarov, Z.N. Shevtsova, in: The Physical Chemistry of Molten Salts, Metallurgiya, Moscow (1965), p. 48.
895. Li Chi-fa, I.S. Morozov, Zh. Neorg. Khim., **8**:708 (1963).

896. M.V. Tokareva, A.G. Bergman, Zh. Neorg. Khim., 2:1895 (1957).
897. I.N. Belyaev, A.K. Doroshenko, in: The Physicochemical Analysis of Salt Systems, Izd. Rostovsk. Gos. Univ., Rostov (1962), p. 12.
898. V.M. Burlakova, G.A. Bukhalova, Izv. Vyssh. Uchebn. Zaved., Khim. Khim. Tekhnol., **10**:487 (1967).
899. V.M. Burlakova, G.A. Bukhalova, Zh. Neorg. Khim., **11**:164 (1966).
900. T.I. Drobasheva, A.G. Bergman, Zh. Neorg. Khim., **14**:1348 (1969).
901. A.N. Kirgintsev, V.A. Isaenko, V.I. Kisyakov, Izv. Akad. Nauk SSSR, Ser. Khim., No. 2, 439 (1970).
902. V.N. Prokhorov, I.V. Krivousova, N.N. Kozhina, A.N. Efimov, Vestn. Leningrad. Gos. Univ., Ser. Fiz.-Khim., No. 4, Issue 1, 89 (1974).
903. V.V. Safonov, B.G. Korshunov, Z.N. Shevtsova, S.I. Bakum, Zh. Neorg. Khim., **9**:1687 (1964).
904. V.V. Safonov, B.G. Korshunov, Z.N. Shevtsova, L.G. Shadrova, Zh. Neorg. Khim., **10**:669 (1965).
905. A.P. Palkin, N.D. Chikanov, Zh. Neorg. Khim., **7**:2388 (1962).
906. I.I. Kozhina, V.P. Kustovskaya, I.V. Vasil'kova, M.P. Susarev, Vestn. Leningrad. Gos. Univ., Ser. Fiz.-Khim., No. 16, Issue 3, 113 (1968).
907. V.V. Safonov, A.V. Konov, B.G. Korshunov, Zh. Neorg. Khim., **14**:2880 (1969).
908. V.I. Ionov, B.G. Korshunov, V.V. Kokorev, I.S. Morozov, Izv. Vyssh. Uchebn. Zaved., Tsvetn. Metall., **3**:102 (1960).
909. S. Tanii, Denki Kagaku, **32**:167 (1964).
910. V.N. Desyatnik, V.A. Dubinin, I.F. Nichkov, S.P. Raspopin, Izv. Vyssh. Uchebn. Zaved., Tsvetn. Metall., No. 4, 87 (1966).
911. P. Ehrlich, H. Kühnl, Z. Anorg. Allg. Chem., **292**:146 (1957).
912. Li Si-Chan, D.M. Chizhikov, Izv. Akad. Nauk SSSR, Otd. Tekh. Nauk Metall. Topl., No. 5, 22 (1961).
913. Lu Cu-Chan, Acta Metallurg. Sinica, **8**(2):187 (1965).
914. Z.L. Artem'eva, I.V. Vasil'kova, M.P. Susarev, I.I. Kozhina, Vestn. Leningrad. Gos. Univ., Ser. Fiz.-Khim., No. 22, Issue 4, 107 (1968); No. 4, Issue 1, 121 (1969).
915. A.V. Storonkin, I.V. Vasil'kova, I.I. Kozhina, É.N. Ryabov, Z.L. Artem'eva, Zh. Fiz. Khim., **43**:1010 (1969).
916. M.V. Kamenetskii, Nauch. Tekh. Inform. Byull. LPI, No. 3, 3 (1957).
917. M.V. Kamenetskii, Tsvetn. Met., No. 2, 39 (1958).
918. B.F. Markov, R.V. Chernov, Ukr. Khim. Zh., **25**:279 (1959).
919. M.V. Kamenetskii, A.A. Kostyukov, Siao De-Chuan, in: The Physical Chemistry of Molten Salts and Slags, Metallurgizdat, Moscow (1962), p. 54.

920. P. Ehrlich, G. Kaupa, K. Blankenstein, Z. Anorg. Allg. Chem., **299**:213 (1959).
921. I.V. Vasil'kova, I.I. Kozhina, M.P. Susarev, Z.L. Artem'eva, Vestn. Leningrad. Gos. Univ., Ser. Fiz.-Khim., No. 4, 121 (1969).
922. A.V. Storonkin, I.V. Vasil'kova, I.I. Kozhina, É.I. Ryabov, Zh. Fiz. Khim., **43**(4):1008 (1969).
923. C.R. Tipton, Reactor Handbook (Materials), 2nd ed., New York (1960), p. 443.
924. V.N. Desyatnik, B.N. Dubinin, S.P. Raspopin, Zh. Fiz. Khim., **47**:2726 (1973).
925. T. Kuroda, T. Suzuki, J. Electrochem. Soc. Jpn, **26**(7):140 (1958).
926. A. Bogacz, W. Trzebiatowski, Rocz. Chem., **38**(5):729 (1964).
927. A.S. Martynova, I.V. Vasil'kova, M.P. Susarev, At. Energ., **18**:616 (1965).
928. A.G. Krokhina, I.S. Morozov, Zh. Neorg. Khim., **14**:1402 (1969).
929. V.N. Desyatnik, N.N. Kurbatov, S.P. Raspopin, Yu.F. Chervinskii, Zh. Fiz. Khim., **46**:2159 (1972).
930. H.-J. Seifert, P. Ehrlich, Z. Anorg. Allg. Chem., **302**:284 (1959).
931. E.K. Lupenko, I.V. Vasil'kova, I.I. Kozhina, Vestn. Leningrad. Gos. Univ., Ser. Fiz.-Khim., No. 16, Issue 3, 101 (1969).
932. B.Ya. Raskin, B.G. Korshunov, Yu.U. Samson, V.A. Kotelevskii, Izv. Vyssh. Uchebn. Zaved., Tsvetn. Metall., No. 5, 95 (1971).
933. C. Grena, Bull. Soc. Chim. France, No. 4, 655 (1960).
934. S.A. Shchukarev, I.L. Perfilova, Zh. Neorg. Khim., **8**:2106 (1963).
935. I.S. Morozov, Z.I. Shevtsova, Li Chi-Fa, Zh. Neorg. Khim., **9**:2605 (1964).
936. B.G. Korshunov, D.V. Drobot, Zh. Neorg. Khim., **9**:222 (1964).
937. A.I. Efimov, I.V. Vasil'kova, E.K. Smirnova, N.D. Zaitseva, T.S. Shemyakina, I.L. Perfilova, in: The Chemistry of Rare Elements, Izd. Leningrad. Gos. Univ., Leningrad (1964), p. 38.
938. I.V. Vasil'kova, I.V. Krivousova, M.P. Susarev, S.S. Tolkachev, Vestn. Leningrad. Gos. Univ., Ser. Fiz.-Khim., No. 16, Issue 3, 126 (1965).
939. A.V. Storonkin, I.V. Vasil'kova, Yu.A. Fedorov, Zh. Fiz. Khim., **41**:695 (1967).
940. I.V. Vasil'kova, N.D. Zaitseva, P.S. Shapkin, Zh. Neorg. Khim., **8**:2360 (1963).
941. B.G. Korshunov, V.I. Gol'din, L.A. Averkieva, Izv. Vyssh. Uchebn. Zaved., Tsvetn. Metall., **1**:161 (1962).
942. I.V. Vasil'kova, N.D. Zaitseva, P.S. Shapkin, Zh. Neorg. Khim., **8**:2360 (1963).

943. A.I. Efimov, I.V. Vasil'kova, E.K. Smirnova, N.D. Zaitseva, T.S. Shemyakina, I.A. Perfilova, in: The Chemistry of Rare Elements, Izd. Leningrad. Gos. Univ., Leningrad (1964), p. 38.
944. D.V. Drobot, B.G. Korshunov, Z.M. Shevtsova, in: The Physical Chemistry of Molten Salts, Metallurgiya, Moscow (1965), p. 48.
945. I.F. Nichkov, V.A. Tomashov, A.E. Mordovin, Zh. Neorg. Khim., **19**:823 (1974).
946. G.I. Novikov, O.G. Polyachenok, S.A. Frid, Zh. Neorg. Khim., **9**:472 (1964).
947. F.R. Duke, R.A. Fleming, J. Electrochem. Soc., **104**(4):251 (1957).
948. B. Swaroop, S.N. Flengas, Can. J. Chem., **44**(2):199 (1966).
949. V.T. Barchuk, I.N. Sheiko, T.N. Grechina, in: Electrochemical Refining of Heavy, Low-Melting Metals from Molten Salts, Naukova Dumka, Kiev (1971), p. 116.
950. L.J. Howell, R.C. Sommer, H.H. Kellog, J. Met., **9**(1):193 (1957).
951. I.S. Morozov, Sun In-Chzhu, Zh. Neorg. Khim., **4**:678 (1959).
952. K.L. Lister, S.N. Flengas, Can. J. Chem., **43**(11):2947 (1965).
953. G.A. Bukhalova, E.S. Yagub'yan, Yu.V. Allakhverdova, in: Rare Earth Elements, Nauka, Moscow-Novosibirsk (1969), p. 158.
954. A.J. Darnell, W.S. McCollum, High Temp. Sci., **3**(1):73 (1971).
955. M.A. Kuvakin, Z.M. Novikova, Zh. Neorg. Khim., **18**:1356 (1973).
956. I.I. Il'yasov, A.G. Bergman, Zh. Obshch. Khim., **26**:981 (1956); Zh. Neorg. Khim., **2**:395 (1957).
957. V. Sholokhova, in: Proceedings of the Interuniversity Scientific-Theoretical Conference of Aspirants, Izd. Rostovsk. Gos. Ped. Inst., Rostov (1972), p. 434.
958. V.T. Berezhnaya, G.A. Bukhalova, Zh. Neorg. Khim., **12**:2179 (1967).
959. Hu Chi-Tszu, A.V. Novoselova, Zh. Neorg. Khim., **6**:2149 (1961).
960. G.A. Bukhalova, É.P. Babaeva, T.M. Khliyan, Zh. Neorg. Khim., **10**:2127 (1965).
961. A.G. Bergman, N.A. Bychkova-Shul'ga, Zh. Neorg. Khim., **2**:179 (1957).
962. N.P. Sigida, I.N. Belyaev, Zh. Neorg. Khim., **2**:1128 (1957).
963. E. Aukrust, B. Björge, H. Flood, T. Forland, Ann. N.Y. Acad. Sci., **79**(11):830 (1960).
964. S.I. Berezina, A.G. Bergman, E.L. Bakumskaya, Zh. Neorg. Khim., **8**:2140 (1963).
965. G.A. Bukhalova, Z.N. Topshinoeva, V.G. Akhtyrskii, Zh. Neorg. Khim., **19**:235 (1974).

966. G.G. Diogenov, V.I. Ermachkov, Nauchn. Tr. Irkutsk. Politekh. Inst., Ser. Khim., Issue 44, 116 (1971).
967. H. Remy, F. Hansen, Z. Anorg. Allg. Chem., **283**(1-6):277 (1956).
968. M.A. Kuvakin, Z.M. Novikova, Zh. Neorg. Khim., **18**:1356 (1973).
969. I.N. Belyaev, O.Ya. Revina, in: The Physicochemical Analysis of Salt Systems, Izd. Rostovsk. Gos. Univ., Rostov (1962), p. 77.
970. I.N. Belyaev, O.Ya. Revina, Zh. Neorg. Khim., **11**:1446 (1966).
971. A.G. Bergman, V.V. Rubleva, Zh. Neorg. Khim., **4**:138 (1959).
972. Nguyen Ba-Chahn, J. Chim. Phys. Phys.-Chim. Biol., **61**(10):1428 (1964).
973. Jan Lützow Holm, Acta Chem. Scand., **19**(3):638 (1965).
974. G.A. Bukhalova, É.P. Babaeva, Zh. Neorg. Khim., **11**:644 (1966).
975. A. Trunin, V. Goryacheva, in: Proceedings of the Interuniversity Scientific-Theoretical Conference of Aspirants, Izd. Rostovsk. Gos. Ped. Inst., Rostov (1967), p. 438.
976. A. Mukhtar, R. Winant, C. R. Acad. Sci., **260**(13):3674 (1965).
977. L.A. Kamenskaya, V.I. Konstantinov, Zh. Neorg. Khim., **16**:2003 (1971).
978. B.S. Zakharova, L.P. Reshetnikova, A.V. Novoselova, Vestn. Mosk. Gov. Univ., Ser. II, Khim., No. 3, 102 (1967).
979. V.A. Gladushchenko, A.G. Bergman, Zh. Obshch. Khim., **26**:339 (1956).
980. O. Schmitz-Dumont, G. Bergerhoff, Z. Anorg. Allg. Chem., **283**(1-6):314 (1956).
981. G.A. Bukhalova, N.A. Litovskaya, V.A. Lyutsedarskii, Izv. Akad. Nauk SSSR, Neorg. Mater., 5(3):510 (1969).
982. N.N. Volkov, Izv. Fiz.-Khim. Nauchn.-Issled. Inst. Irkutsk. Gos. Univ., 5(2):158 (1961).
983. L.N. Sidorov, A.S. Alikhanyan, L.V. Novikova, V.B. Shol'ts, Zh. Fiz. Khim., **47**:2934 (1973).
984. J.D. Donaldson, J.D. O'Donogue, R. Oteng, J. Chem. Soc., No. 6, 3876 (1965).
985. V.T. Berezhnaya, G.A. Bukhalova, Zh. Neorg. Khim., **5**:925, 2061 (1960).
986. Tsui Bin-Sin, N.P. Luzhnaya, V.I. Konstantinov, Zh. Neorg. Khim., **8**:389 (1963).
987. I.D. Fedorov, M.D. Lantratov, Zh. Prikl. Khim., **37**:2521 (1964).
988. F.V. Kovalev, V.E. Kartsev, V.M. Ioffe, M.E. Leonov, Zh. Neorg. Khim., **18**:1352 (1973).
989. L.P. Batsanova, V.P. Doronina, Izv. Akad. Nauk SSSR, Neorg. Mater., **8**:2142 (1972).
990. V.S. Emel'yanov, A.I. Evstyukhin, At. Energ., **4**:80 (1956).

991. L.N. Antipin, V.V. Nerubashchenko, A.V. Kuleshova, Yu.I. Rys'eva, V.N. Vovk, The Metallurgy and Chemistry of Titanium, Metallurgiya, Moscow (1968), p. 204.
992. L.N. Antipin, V.V. Nerubashchenko, S.F. Vazhenin, Research in the Metallurgy of Titanium Chloride, Metallurgiya, Moscow (1969), p. 230.
993. V.V. Nerubashchenko, L.N. Antipin, V.N. Vovk, É.V. Sergach, S.I. Bubyr', A.P. Sidorenko, in: The Interuniversity Conference on the Theory of Processes in Nonferrous Metallurgy, Izd. Kaz. Politekh. Inst., Alma-Ata (1971), p. 275.
994. R.V. Chernov, I.M. Ermolenko, Zh. Neorg. Khim., **18**:2238 (1973).
995. R.E. Thoma, H. Insley, B.S. Landau, H.A. Friedman, W.R. Grimes, J. Am. Ceram. Soc., **41**(12):538 (1958).
996. J.-C. Cretenet, J.-C. Cousseins, C. R. Acad. Sci., **C267**(3):240 (1968).
997. G.A. Bukhalova, É.P. Babaeva, Zh. Neorg. Khim., **11**:644 (1966).
998. M.P. Borzenkova, A.V. Novoselova, G.N. Kuznetsova, Izv. Sib. Otd. Akad. Nauk, Ser. Khim. Nauk, Issue 1, No. 2, 5 (1968).
999. M.P. Borzenkova, G.P. Kuznetsova, A.V. Novoselova, Izv. Akad. Nauk SSSR, Neorg. Mater., **7**:242 (1971).
1000. O. Schmitz-Du Mont, H. Bornfeld, Z. Anorg. Allg. Chem., **287**:120 (1956).
1001. A.I. Evstyukhin, V.S. Emel'yanov, Yu.G. Godin, in: The Physical Chemistry of Molten Salts and Slags, Metallurgizdat, Moscow (1962), p. 63.
1002. I.N. Sheiko, R.V. Chernov, V.S. Kikhno, in: The Physical Chemistry of Molten Salts and Slags, Metallurgizdat, Moscow (1962), p. 72.
1003. A.V. Novoselova, Yu.M. Korenev, Yu.P. Simanov, Dokl. Akad. Nauk SSSR, **139**:892 (1961).
1004. I.N. Sheiko, R.V. Chernov, V.S. Kikhno, Ukr. Khim. Zh. **27**:469 (1961).
1005. J. Kutscher, A. Schneider, Z. Anorg. Allg. Chem., **386**(1):38 (1971).
1006. D.B. Leiser, O.J. Whittemore, J. Am. Ceram. Soc., **50**(1):60 (1967).
1007. Yu.S. Surkov, Izv. Vyssh. Uchebn. Zaved., Khim. Khim. Tekhnol., **11**(10):1190 (1968).
1008. C.H. Liu, L.R. Lieto, J. Chem. Eng. Data, **14**(1):83 (1969).
1009. R. Sridhar, C.E. Johnson, E.J. Cairns, J. Chem. Eng. Data, **15**(2):244-245 (1970).
1010. R.N. Nyankovskaya, Zh. Neorg. Khim., **1**:783 (1956).
1011. I.I. Il'yasov, G.G. Shchemeleva, R.G. Bergman, Zh. Neorg. Khim., **6**:699 (1961).
1012. Nguyen Ba-Chahn, J. Chim. Phys. Phys.-Chim. Biol., **61**(10):1428 (1964).

1013. J. Kutscher, A. Schneider, Z. Anorg. Allg. Chem., **386**(1):38 (1971).
1014. I.I. Il'yasov, A.G. Bergman, Zh. Obshch. Khim., **26**:981 (1956).
1015. I.I. Il'yasov, S.D. Dionis'ev, A.G. Bergman, Zh. Neorg. Khim., **9**:422 (1964).
1016. I.I. Il'yasov, D.G. Barsegov, I.G. Berikashvili, L.P. Danilenko, Zh. Neorg. Khim., **14**:1484 (1969).
1017. I.N. Belyaev, E.A. Shurginov, N.S. Kudryashov, Zh. Neorg. Khim., **17**:2812 (1972).
1018. J. Kutscher, A. Schneider, Z. Anorg. Allg. Chem., **386**(1):38 (1971).
1019. R.G. Samuseva, V.E. Plyushchev, Zh. Neorg. Khim., **9**:2433 (1964).
1020. I.I. Il'yasov, Zh. Neorg. Khim., **11**:211 (1966).
1021. V.E. Plyushchev, S.B. Stepina, G.V. Zimina, O.P. Molchanova, L.V. Savel'eva, D.A. Yashkov, Izv. Vyssh. Uchebn. Zaved., Tsvetn. Metall., No. 1, 65 (1970).
1022. J. Kustcher, A. Schneider, Z. Anorg. Allg. Chem., **386**(1):38 (1971).
1023. T.N. Sevast'yanova, N.V. Karpenko, Zh. Neorg. Khim., **14**:3130 (1969)
1024. I.N. Belyaev, E.A. Shurginov, N.S. Kudryashov, Zh. Neorg. Khim., **17**:2812 (1972).
1025. V.V. Safonov, O.V. Lemeshko, B.G. Korshunov, Zh. Neorg. Khim., **16**:2283 (1971)
1026. I.I. Il'yasov, G.G. Shemeleva, A.G. Bergman, Zh. Neorg. Khim., **6**:699 (1961).
1027. Z.N. Shevtsova, I.S. Morozov, L.I. Prikhodkina, The Chemistry and Technology of Rare and Dispersed Elements, GNTI Tekst., Legk., Poligr. Promst., Moscow (1958), p. 117.
1028. N.D. Chikanov, Zh. Neorg. Khim., **18**:277 (1973).
1029. N.D. Chikanov, Zh. Neorg. Khim., **17**:2558 (1972).
1030. B.G. Korshunov, D.V. Drobot, L.V. Durinina, Zh. Neorg. Khim., **10**:2120 (1965).
1031. Li Chi-fa, I.S. Morozov, Zh. Neorg. Khim., **7**:2765 (1962).
1032. B.G. Korshunov, D.V. Drobot, L.V. Durinina, Zh. Neorg. Khim., **10**:2120 (1965).
1033. L.A. Khripin, Izv. Sib. Otd. Akad. Nauk SSSR, Ser. Khim. Nauk, No. 7, Issue 2, 107 (1963).
1034. G.A. Bukhalova, É.P. Babaeva, Zh. Neorg. Khim., **10**:1883 (1965).
1035. R.E. Thoma, Rare Earth Halides, ORNL-3804 (1965).
1036. L.N. Starkov, N.I. Starkova, A.R. Babenko, in: Chemistry and Chemical Technology, Vol. 29, Permsk. Knizhn. Izd., Perm (1968), p. 45.
1037. F. Matthes, S. Holz, Z. Chem., **2**:22 (1962).
1038. R.E. Thoma, Rare Earth Research. III. Proceedings of IV Conf. on Rare Earth Research New York (1964), p. 561.

1039. G.A. Bukhalova, É.P. Babaeva, T. M. Khliyan, Zh. Neorg. Khim., **10**:2127 (1965).
1040. R.E. Thoma, H. Insley, G.M. Hebert, Inorg. Chem., **5**(7):1222 (1966).
1041. E.G. Ippolitov, L.S. Garashina, B.N. Zhigarnovskii, Izv. Sib. Otd. Akad. Nauk SSSR, Ser. Khim. Nauk, No. 4, Issue 2, 49 (1968).
1042. J. Kutscher, A. Schneider, Z. Anorg. Allg. Chem., **386**(1):38 (1971).
1043. K.N. Semenenko, A.P. Chavgun, V.N. Surov, Zh. Neorg. Khim., **16**:513 (1971).
1044. N.I. Vilutis, N.A. Reshetnikov, Izv. Sib. Otd. Akad. Nauk SSSR, Ser. Khim. Nauk, No. 9, 88 (1960).
1045. N.N. Volkov, T.F. Kuryleva, M.N. Zakhvalinskii, Izv. Fiz.-Khim. Nauchn.-Issled. Inst. Irkutsk. Gos. Univ., 5(2):152 (1961).
1046. A.S. Arabadzhan, A.G. Bergman, Zh. Neorg. Khim., **7**:2226 (1962).
1047. A.S. Arabadzhan, A.G. Bergman, Zh. Neorg. Khim., **8**:720 (1963).
1048. G.G. Diogenov, V.I. Ermachkov, Zh. Neorg. Khim., **12**:2517 (1967).
1049. A.G. Bergman, Yu.I. Andryushchenko, Zh. Neorg. Khim., **12**:1025 (1967).
1050. N.N. Volkov, M.N. Zakhvalinskii, in: Medicinal Raw Material Resources in Irkutsk Region, Vol. 3, Irkutsk. Obl. Gos. Izd., Irkutsk (1961), p. 216.
1051. M.N. Zakhvalinskii, G.G. Diogenov, G.M. Belozerova, in: Interactions of Some Alkali-Metal Salts in Melts, Irkutsk. Knizhn. Izd., Irkutsk (1973), p. 18.
1052. N.Ya. Fedorov, E.S. Petrov, V.E. Bogatyr', Izv. Sib. Otd. Akad. Nauk SSSR, Ser. Khim. Nauk, No. 7, Issue 3, 109 (1970).
1053. I.N. Belyaev, E.A. Shurginov, Zh. Neorg. Khim., **15**:883 (1970).
1054. I.N. Belyaev, A.K. Doroshenko, in: The Physicochemical Analysis of Salt Systems, Izd. Rostovsk. Gos. Univ., Rostov (1962), p. 12.
1055. R. Riccardi, C. Sinistri, V. Campari, A. Magistris, Z. Naturforsch., **25a**(5):781 (1970).
1056. A.G. Bergman, A.S. Arabadzhan, Zh. Neorg. Khim., **8**:1453 (1963).
1057. N.N. Volkov, T.F. Kuryleva, M.N. Zakhvalinskii, Izv. Fiz.-Khim. Nauchn.-Issled. Inst. Irkutsk. Gos. Univ., 5(2):152 (1961).
1058. I.N. Belyaev, A.K. Doroshenko, in: The Physicochemical Analysis of Salt Systems, Izd. Rostovsk. Gos. Univ., Rostov (1962), p. 12.
1059. A.G. Bergman, A.S. Arabadzhan, Zh. Neorg. Khim., **8**:1228 (1963).

1060. L.K. Kuznetsova, Kh.L. Strelets, in: The Physical Chemistry and Electrochemistry of Molten Salts and Slags, Khimiya, Leningrad (1968), p. 85.
1061. R.S. Umyarova, V.S. Urusov, K.I. Tobelko, A.B. Slutskii, Zh. Fiz. Khim., **48**:2141 (1974).
1062. D.S. Lesnykh, Z.M. Karmanova, Uch. Zap. Rostovsk. Gos. Univ., **25**(7):19 (1955).
1063. E.K. Akopov, Zh. Neorg. Khim., **1**:1019 (1956).
1064. T.F. Kuryleva, Izv. Fiz-Khim. Nauchn.-Issled. Inst. Irkutsk. Gos. Univ., **4**(1):126 (1959).
1065. N.N. Volkov, T.F. Kuryleva, M.N. Zakhvalinskii, Izv. Fiz.-Khim. Nauchn. Issled. Inst. Irkutsk. Gos. Univ., **5**(2):152 (1961).
1066. A.S. Arabadzhan, A.G. Bergman, Zh. Neorg. Khim., **7**:2226 (1962).
1067. A.G. Bergman, E.L. Kozachenko, S.I. Berezina, Zh. Neorg. Khim., **9**:1214 (1964).
1068. E.K. Akopov, E.I. Korobka, Ukr. Khim. Zh., **35**:588 (1969).
1069. D.S. Lesnykh, I.G. Éikhenbaum, S.A. Cheryakhovskaya, Zh. Neorg. Khim., **15**:824 (1970).
1070. V.M. Burlakova, in: Research in General and Natural Science, Izd. Rostovsk. Inzh.-Stroit. Inst., Rostov (1973), p. 86.
1071. A.V. Storonkin, O.D. Grebennikova, I.V. Krivousova, I.I. Kozhina, I.V. Vasil'kova, Vestn. Leningrad. Gos. Univ., Ser. Fiz.-Khim., Issue 2, No. 10, 70 (1973).
1072. D.Ya. Toptygin, Zh. Neorg. Khim., **8**:1187 (1963).
1073. M.S. Golubeva, B.S. Medvedev, Zh. Neorg. Khim., **7**:2600 (1962).
1074. D.S. Lesnykh, A.G. Bergman, Uch. Zap. Rostovsk. Gos. Univ., **20**:19 (1954).
1075. A.G. Bergman, Yu.I. Andryushchenko, R.K. Bineeva, Zh. Neorg. Khim., **8**:1693 (1963).
1076. V.I. Monastyrskaya, in: The Physical Chemistry and Electrochemistry of Molten Salts and Slags, Naukova Dumka, Kiev (1969), p. 324.
1077. C.W. Bjorklund, J.C. Reans, J.A. Leary, K.A. Walsh, J. Phys. Chem., **63**(10):1774 (1959).
1078. G.G. Diogenov, L.M. Kukis, Nauchn. Tr. Angarsk Fil. Irkutsk. Politekh. Inst., Issue 4, 205 (1970).
1079. L.M. Kukis, in: Interactions of Some Alkali-Metal Salts in Melts, Irkutsk. Obl. Gos. Izd., Irkutsk (1973), p. 138.
1080. N.D. Chikanov, Zh. Neorg. Khim., **17**:2558 (1972).
1081. N.Ya. Fedorov, E.S. Petrov, in: Rare Alkaline Elements, Nauka, Novosibirsk (1967), p. 182.
1082. N.Ya. Fedorov, E.S. Petrov, Izv. Sib. Otd. Akad. Nauk SSSR, Ser. Khim. Nauk, No. 2, Issue 1, 48 (1967).
1083. C.R. Tipton, Reactor Handbook (Materials), 2nd ed., New York (1960), p. 442.

1084. I.N. Belyaev, A.K. Doroshenko, in: The Physicochemical Analysis of Salt Systems, Izd. Rostovsk. Gos. Univ., Rostov (1962), p. 12.
1085. G.A. Bukhalova, V.M. Burlakova, Izv. Vyssh. Uchebn. Zaved., Khim. Khim. Tekhnol., **11**:955 (1968).
1086. V.M. Burlakova, Research in General and Natural Science, Izd. Rostovsk. Inzh.-Stroit. Inst., Rostov (1973), p. 86.
1087. D.Ya. Toptygin, Zh. Neorg. Khim., **8**:1187 (1963).
1088. V.V. Safonov, A.V. Konov, B.G. Korshunov, Zh. Neorg. Khim., **14**:2880 (1969).
1089. S. Tanii, Denki Kagaku, **32**:167 (1964).
1090. P. Ehrlich, R. Schmitt, Z. Anorg. Allg. Chem., **308**:91 (1961).
1091. E.F. Klyuchnikova, V.G. Gopienko, É.N. Ryabov, I.V. Goldun, Zh. Prikl. Khim., **46**:1947 (1973).
1092. C.R. Tipton, Reactor Handbook (Materials), 2nd ed., New York (1960), p. 443.
1093. V.N. Desyatnik, N.N. Kurbatov, S.P. Raspopin, I.I. Trifonov, At. Energ., **32**:479 (1972).
1094. V.N. Desyatnik, B.N. Dubinin, S.P. Raspopin, Zh. Fiz. Khim., **47**:2726 (1973).
1095. A. Bogacz, W. Trzebiatowski, Rocz. Chem., **38**(5):729 (1964).
1096. V.N. Desyatnik, N.N. Kurbatov, S.P. Raspopin, Yu.F. Chervinskii, Zh. Fiz. Khim., **46**:2159 (1972).
1097. V.N. Desyatnik, N.N. Kurbatov, S.P. Raspopin, I.I. Trifonov, At. Energ., **32**:479 (1972).
1098. B.Ya. Raskin, B.G. Korshunov, Yu.U. Samson, V.A. Kotelevskii, Izv. Vyssh. Uchebn. Zaved., Tsvetn. Metall., No. 5, 95 (1971).
1099. V.I. Monastyrskaya, in: The Physical Chemistry and Electrochemistry of Salts and Slags, Naukova Dumka, Kiev (1969), p. 324.
1100. H.M. Haendler, P.S. Sennett, C.M. Wheeler, J. Electrochem. Soc., **106**(3):264 (1959).
1101. I.N. Belyaev, A.K. Doroshenko, in: The Physicochemical Analysis of Salt Systems, Izd. Rostovsk. Gos. Univ., Rostov (1962), p. 26.
1102. A.G. Bergman, A.S. Arabadzhan, Zh. Neorg. Khim., **8**:1228 (1963).
1103. S.I. Berezina, A.G. Bergman, E.L. Bakumskaya, Zh. Neorg. Khim., **8**:2140 (1963).
1104. A.G. Bergman, E.A. Kozachenko, S.I. Berezina, Zh. Neorg. Khim., **9**:1214 (1964).
1105. C.E. Johnson, E.J. Hathaway, J. Electrochem. Soc., **118**(4):631 (1971).
1106. C.E. Johnson, S.E. Wood, C.E. Crouthamel, Inorg. Chem., **3**(11):1487 (1964).
1107. C.E. Johnson, E.J. Hathaway, J. Chem. Eng. Data, **14**(2):174 (1970).

1108. C.E. Johnson, M.S. Foster, J. Electrochem. Soc., **116**(11):1612 (1969).
1109. C.E. Johnson, E.J. Hathaway, J. Chem. Eng. Data, **14**(2):174 (1969); J. Electrochem. Soc., **118**(4):631 (1971).
1110. R.E. Thoma, Rare-Earth Halides, ORNL-3804 (1965).
1111. I.N. Belyaev, O.Ya. Revina, in: The Physicochemical Analysis of Salt Systems, Izd. Rostovsk. Gos. Univ., Rostov (1962), p. 77.
1112. I.N. Belyaev, O.Ya. Revina, Zh. Neorg. Khim., **11**:1446 (1966).
1113. A.J. Singh, R.G. Ross, R.E. Thoma, Oak Ridge Natl. Lab., ORNL-3658 (1965).
1114. I.N. Belyaev, N.P. Sigida, Zh. Neorg. Khim., **2**:1119 (1957).
1115. V.P. Goryacheva, A.G. Bergman, A.I. Kislova, Zh. Neorg. Khim., **4**:2744 (1959).
1116. C.R. Tipton, Reactor Handbook (Materials), 2nd ed., New York (1960), p. 439.
1117. J.L. Holm, Acta Chem. Scand., **19**(3):638 (1965).
1118. K. Matiašovský, I. Čakajdova, M. Malinovský, Chem. Zvesti, **19**:513 (1965).
1119. M.A. Kuvakin, T.D. Vol'khina, Uch. Zap. Mordovsk. Gos. Univ., Ser. Khim. Nauk, No. 81, 11 (1971).
1120. R.E. Thoma, Rare-Earth Halides, ORNL-3804 (1965).
1121. V.A. Gladushchenko, A.G. Bergman, Nauchn. Tr. Novocherkask. Politekh. Inst., 27(41):49 (1956).
1122. R.E. Thoma, Rare-Earth Halides, ORNL-3804 (1965).
1123. G.A. Bukhalova, N.A. Litovskaya, V.A. Lyutsedarskii, Izv. Akad. Nauk SSSR, Neorg. Mater., **5**(3):510 (1969).
1124. C.I. Barton, R.A. Strehlow, J. Inorg. Nucl. Chem., **18**:143 (1961).
1125. C.R. Tipton, Reactor Handbook (Materials), 2nd ed., New York (1960), p. 439.
1126. G.G. Diogenov, V.I. Ermachkov, Nauchn. Tr. Irkutsk. Politekh. Inst., Ser. Khim., Issue 44, 116 (1971).
1127. É.P. Babaeva, in: Phase Diagrams of Some Salt Systems, Izd. Rostovsk. Gos. Univ., Rostov (1964), p. 42.
1128. É.P. Babaeva, G.A. Bukhalova, Zh. Neorg. Khim., **11**:648 (1959).
1129. Z.A. Mateiko, G.A. Bukhalova, Zh. Neorg. Khim., **7**:165 (1962).
1130. R.E. Thoma, H. Insley, B.S. Landau, H.A. Friedman, W.P. Grimes, J. Phys. Chem., **63**(8):1267 (1959).
1131. R.V. Chernov, I.M. Ermolenko, Zh. Neorg. Khim., **17**:2563 (1972).
1132. C.R. Tipton, Reactor Handbook (Materials), 2nd ed., New York (1960), p. 427.
1133. C.J. Barton, H.A. Friedman, W.R. Grimes, H. Insley, K.E. Moore, R.E. Thoma, J. Am. Ceram. Soc., **41**(2):63 (1958).
1134. R.E. Thoma, C.F. Weaver, H.A. Friedman, H. Insley, L.A. Harris, H.A. Yakel, J. Phys. Chem., **65**(7):1096 (1961).

1135. G.A. Bukhalova, É.P. Babaeva, Zh. Neorg. Khim., **11**:402 (1966).
1136. G.A. Bukhalova, É.P. Babaeva, Zh. Neorg. Khim., **11**:624 (1966).
1137. O. Schmitz-Du Mont, H. Bornefeld, Z. Anorg. Allg. Chem., **287**:120 (1956).
1138. Yu.M. Korenev, A.V. Novoselova, K.K. Glinskii, V.V. Gornikov, Izv. Akad. Nauk SSSR, Neorg. Mater., **1**(2):201 (1965).
1139. R.E. Thoma, H. Insley, H.A. Friedman, G.M. Hebert, J. Chem. Eng. Data, **10**(3):219 (1965).
1140. R.E. Thoma, H. Insley, H.A. Friedman, G.M. Hebert, J. Nucl. Mater., **27**(2):166 (1968).
1141. Ch.E. Messer, J. Mellor, J. Phys. Chem., **64**(4):503 (1960).
1142. C.E. Johnson, E. Hathaway, J. Electrochem. Soc., **118**(4):631 (1971).
1143. C.E. Johnson, E. Hathaway, J. Chem. Eng. Data, **14**(2):174 (1970).
1144. V.V. Safonov, O.V. Lemeshko, Zh. Neorg. Khim., **18**:1957 (1973).
1145. D.V. Drobot, B.G. Korshunov, G.P. Borodulenko, Zh. Neorg. Khim., **13**:1635 (1968).
1146. R.E. Thoma, Rare Earth Research. III. Proceedings of IV Conference on Rare Earth Research, New York (1964), p. 561.
1147. R.E. Thoma, H. Insley, G.M. Hebert, Inorg. Chem., 5(7):1222 (1966).
1148. E.G. Ippolitov, L.S. Garashina, B.N. Zhigarnovskii, Izv. Sib. Otd. Akad. Nauk SSSR, Ser. Khim. Nauk, No. 4, Issue 2, 49 (1968).
1149. K. Matiašovský, Chem. Zvesti, **13**(2):69 (1959).
1150. N.V. Bondarenko, Kh.L. Strelets, Zh. Neorg. Khim., **7**:1387 (1962).
1151. N.V. Bondarenko, Kh.L. Strelets, in: The Physical Chemistry of Molten Salts and Slags, Naukova Dumka, Kiev (1970), p. 216.
1152. B.G. Korshunov, N.A. Sil'vestrova, The Chemistry and Technology of Rare and Dispersed Elements, GNTI Tekstil., Legk. i Poligraf. Promst., Moscow (1956), p. 21.
1153. I.S. Morozov, V.I. Ionov, B.G. Korshunov, Zh. Neorg. Khim., **4**:1457 (1959).
1154. C.R. Tipton, Reactor Handbook (Materials), 2nd ed., New York (1960), p. 442.
1155. D.S. Coleman, R. Pollitt, Trans. Instr. Mining Met., **678**: 233 (1969).
1156. Z.N. Shevtsova, I.S. Morozov, O.A. Efremova, Izv. Vyssh. Uchebn. Zaved., Tsvetn. Metall., No. 3, 109 (1960).
1157. K.W.R. Johnson, M. Kahn, J.A. Leary, J. Phys. Chem., **65**(12):2226 (1961).
1158. N.D. Chikanov, Zh. Neorg. Khim., **17**:2558 (1972).

1159. B.G. Korshunov, V.I. Ionov, T.A. Baklashova, V.V. Kokorev, Izv. Vyssh. Uchebn. Zaved., Tsvetn. Metall., **6**:114 (1960).
1160. L.V. Gorbunov, V.N. Desyatnik, S.P. Raspopin, K.I. Trifonov, Zh. Neorg. Khim., **19**:3093 (1974).
1161. K. Komarek, P. Herasymenko, J. Electrochem. Soc., **105**(4): 210 (1958).
1162. V.N. Desyatnik, Yu.A. Izmodenov, Yu.T. Mel'nikov, I.F. Nichkov, S.P. Raspopin, At. Energ., **26**:549 (1969).
1163. Ya.M. Sterlin, V.V. Artamonov, At. Energ., **22**:473 (1967).
1164. V.N. Desyatnik, Yu.A. Izmodenov, Yu.T. Mel'nikov, I.F. Nichkov, S.P. Raspopin, At. Energ., **26**:549 (1969).
1165. V.N. Desyatnik, N.M. Emelyanov, Yu.T. Mel'nikov, S.P. Raspopin, A.M. Tsybizov, Zh. Fiz. Khim., **46**:1886 (1972).
1166. I.S. Morozov, Z.N. Shevtsova, Li Chi-Fa, Zh. Neorg. Khim., **9**:2606 (1964).
1167. B.G. Korshunov, N.A. Sil'vestrova, in: The Chemistry and Technology of Rare and Dispersed Elements, GNTI Tekstil., Legk. i Poligraf. Promst., Moscow (1956), p. 21.
1168. Ram A. Sharma, Irving Johnson, J. Am. Ceram. Soc., **52**(11): 612 (1969).
1169. M.A. Kuvakin, Zh. Neorg. Khim., **16**:3290 (1971).
1170. A.A. Kostyukov, A.B. Karpov, The Electrometallurgy of Nonferrous Metals, ONTI, Leningrad (1957), p. 58.
1171. L.N. Komissarova, B.I. Pokrovskii, Dokl. Akad. Nauk SSSR, **149**:599 (1963).
1172. Z.A. Mateiko, G.A. Bukhalova, Zh. Neorg. Khim., **7**:165 (1962).
1173. C.R. Tipton, Reactor Handbook (Materials), 2nd ed., New York (1960), p. 438.
1174. H.J. Seifert, F.W. Koknat, Z. Anorg. Allg. Chem., **341**(6):269 (1965).
1175. V.V. Safonov, B.G. Korshunov, L.N. Taranyuk, Zh. Neorg. Khim., **13**:1950 (1968).
1176. R.N. Yakovleva, I.I. Kozhina, I.V. Krivousova, I.V. Vasil'kova, Vestn. Leningrad. Gos. Univ., Ser. Fiz.-Khim., No. 16, Issue 3, 88 (1970).
1177. N.D. Chikanov, Zh. Neorg. Khim., **14**:1430 (1969).
1178. N.G. Korzhukhov, K.G. Khomyakov, Zh. Neorg. Khim., **12**:789 (1967).
1179. B.F. Markov, R.V. Chernov, Ukr. Khim. Zh., **24**:139 (1958).
1180. N.D. Chikanov, Zh. Neorg. Khim., **17**:2558 (1972).
1181. N.D. Chikanov, Zh. Neorg. Khim., **14**:1430 (1969).
1182. A.S. Kuz'menko, É.N. Ryabov, R.A. Sandler, Zh. Neorg. Khim., **19**:2550 (1974).
1183. I.N. Belyaev, O.Ya. Revina, Zh. Neorg. Khim., **13**:2542 (1968).
1184. I.N. Belyaev, O.Ya. Revina, in: The Physicochemical Analysis of Salt Systems, Izd. Rostovsk. Gos. Univ., Rostov (1962), p. 77.

1185. I.N. Belyaev, O.Ya. Revina, Zh. Neorg. Khim., **11**:1446 (1966).
1186. D.V. Drobot, É.A. Sapranova, Zh. Neorg. Khim., **19**:228 (1974).
1187. Nguyen Nyuk Kuan, R.A. Sandler, É.N. Ryabov, I.V. Vasil'kova, I.I. Kozhina, E.F. Klyuchnikova, Zh. Neorg. Khim., **17**:2253 (1972).
1188. B.G. Korshunov, V.I. Gol'din, L.A. Averkieva, Izv. Vyssh. Uchebn. Zaved., Tsvetn. Metall., **1**:101 (1962).
1189. A.I. Efimov, L.P. Belorukova, A.M. Ryndina, Zh. Neorg. Khim., **8**:1168 (1963).
1190. A.I. Efimov, I.V. Vasil'kova, E.K. Smirnova, N.D. Zaitseva, T.S. Shemyakina, I.L. Perfilova, in: The Chemistry of Rare Elements, Izd. Leningrad. Gos. Univ., Leningrad (1964), p. 38.
1191. N.D. Chikanov, Zh. Neorg. Khim., **13**:2884 (1968).
1192. A.I. Efimov, L.P. Belorukova, A.M. Ryndina, Zh. Neorg. Khim., **8**:1168 (1963).
1193. D.V. Drobot, B.G. Korshunov, V.A. Aleksandrova, Zh. Neorg. Khim., **13**:3354 (1968).
1194. N.D. Chikanov, Zh. Neorg. Khim., **18**:277 (1973).
1195. N.D. Chikanov, Zh. Neorg. Khim., **16**:1452 (1971).
1196. N.D. Chikanov, Izv. Vyssh. Uchebn. Zaved., Khim. Khim. Tekhnol., **13**:1378 (1970).
1197. V.V. Safonov, B.G. Korshunov, V. Arshanskaya, Zh. Neorg. Khim., **15**:2300 (1970).
1198. N.D. Chikanov, Zh. Neorg. Khim., **16**:1452 (1971).
1199. N.D. Chikanov, Zh. Neorg. Khim., **13**:2884 (1968).
1200. V.V. Safonov, B.G. Korshunov, Zh. Neorg. Khim., **13**:2804 (1968).
1201. V.K. Ezhov, Zh. Neorg. Khim., **17**:2043 (1972).
1202. V.A. Legasov, A.S. Marinin, Zh. Fiz. Khim., **46**:2649 (1972).
1203. P.S. Costich, G.J. Maas, N.O. Smith, J. Chem. Eng. Data, **8**: 26 (1963).
1204. J.E. Callanan, N.O. Smith, J. Chem. Thermodyn., **3**(4):531 (1971).
1205. S.S. Batsanov, L.I. Kopanova, E.V. Lazareva, in: Proceedings of 1st All-Union Symposium on Impulsive Pressures, VNIIFTRI, Moscow (1974), p. 78.
1206. I.S. Morozov, D.Ya. Toptygin, Zh. Neorg. Khim., **3**:1637 (1958).
1207. N.D. Chikanov, Zh. Neorg. Khim., **17**:2558 (1972).
1208. S.I. Berul', M.V. Pevnitskaya, Nauchn. Tr. NIIChermet, Issue 28, 178 (1962).
1209. Li Chi-fa, I.S. Morozov, Zh. Neorg. Khim., **8**:708 (1963).
1210. I.S. Morozov, D.Ya. Toptygin, Zh. Neorg. Khim., **3**:1635 (1958).
1211. V.V. Safonov, A.V. Konov, B.G. Korshunov, Zh. Neorg. Khim., **14**:2880 (1969).

1212. A. Tranquard, M. Riviere, G. Coffy, K. Cohen-Adad, C. R. Acad. Sci., **C270**(25):2056 (1970).
1213. V.G. Selivanov, V.V. Stender, Zh. Neorg. Khim., **3**:447 (1958).
1214. C.J. Barton, L.O. Gilpatrick, J.A. Bormann, H.H. Stone, T.N. McVay, H. Insley, J. Inorg. Nucl. Chem., **33**(2):337 (1971).
1215. I.I. Il'yasov, G.G. Shemeleva, A.G. Bergman, Zh. Neorg. Khim., **2**:2168 (1957).
1216. I.I. Il'yasov, G.G. Shemeleva, A.G. Bergman, Zh. Neorg. Khim., **4**:906 (1959).
1217. R.G. Samuseva, V.E. Plyushchev, Zh. Neorg. Khim., **9**:2436 (1964).
1218. I.K. Tovmas'yan, T.M. Tsushba, Zh. Neorg. Khim., **17**:2796 (1972).
1219. I.K. Tovmas'yan, T.M. Tsushba, A.G. Bergman, Ukr. Khim. Zh., **36**:96 (1970).
1220. I.K. Tovmas'yan, T.M. Tsushba, V.P. Radina, Zh. Neorg. Khim., **15**:2535 (1970).
1221. I.K. Tovmas'yan, T.M. Tsushba, A.G. Bergman, Ukr. Khim. Zh., **39**:952 (1973).
1222. N.Ya. Fedorov, E.S. Petrov, Izv. Sib. Otd. Akad. Nauk SSSR, Ser. Khim., No. 12, Issue 5, 47 (1969).
1223. I.N. Belyaev, E.A. Shurginov, Zh. Neorg. Khim., **15**:883 (1970).
1224. I.N. Belyaev, A.K. Doroshenko, in: The Physicochemical Analysis of Salt Systems, Izd. Rostovsk. Gos. Univ., Rostov (1962), p. 12.
1225. R. Riccardi, C. Sinistri, V. Campari, A. Magistris, Z. Naturforsch., **25a**(5):781 (1970).
1226. V.A. Grin'ko, V.V. Safonov, V.I. Ksenzenko, B.G. Korshunov, Zh. Neorg. Khim., **17**:1755 (1972).
1227. I.I. Il'yasov, V.M. Fonardzhian, A.G. Bergman, Zh. Neorg. Khim., **2**:2154 (1957).
1228. I.I. Il'yasov, Zh. Neorg. Khim., **7**:169 (1962).
1229. N.N. Volkov, T.F. Kuryleva, M.N. Zakhvalinskii, Izv. Fiz.-Khim. Nauchn.-Issled. Inst. Irkutsk. Gos. Univ., 5(2):152 (1961).
1230. I.N. Belyaev, A.K. Doroshenko, in: The Physicochemical Analysis of Salt Systems, Izd. Rostovsk. Gos. Univ., Rostov (1962), p. 12.
1231. I.K. Tovmas'yan, T.M. Tsushba, V.P. Radina, Zh. Neorg. Khim., **15**:2535 (1970).
1232. I.K. Tovmas'yan, T.M. Tsushba, Zh. Neorg. Khim., **17**:2796 (1972).
1233. V.V. Safonov, B.G. Korshunov, S.N. Stel'bovskaya, Zh. Neorg. Khim., **11**:2143 (1966).
1234. S. Yuzo, S. Tadashi, Y. Shinsuke, Denki Kagaku, **33**:656 (1965).
1235. V.V. Safonov, B.G. Korshunov, T.N. Zimina, Z.N. Shevtsova, Zh. Neorg. Khim., **11**:2139 (1966).

1236. B.G. Korshunov, V.V. Safonov, Zh. Neorg. Khim., **6**:753 (1961).
1237. I.S. Morozov, B.G. Korshunov, Zh. Neorg. Khim., **1**:145 (1956).
1238. A.P. Palkin, N.D. Chikanov, Zh. Neorg. Khim., **4**:898 (1959).
1239. O.R. Gavrilov, L.A. Nisel'son, Zh. Neorg. Khim., **11**:209 (1966).
1240. I.S. Morozov, Z.N. Shevtsova, L.V. Klyukina, Zh. Neorg. Khim., **2**:1639 (1957).
1241. K.A. Bol'shakov, P.I. Fedorov, G.D. Agashkina, Zh. Neorg. Khim., **2**:1115 (1957).
1242. I.I. Il'yasov, G.G. Shemeleva, A.G. Bergman, Zh. Neorg. Khim., **2**:2168 (1957).
1243. I.I. Il'yasov, A.G. Bergman, Zh. Neorg. Khim., **7**:356 (1962).
1244. T.N. Sumarokova, T.P. Modestova, Zh. Neorg. Khim., **6**:679 (1961).
1245. S. Sternberg, J. Adorian, Studii si cercetari Chim. Acad. Rep. Popul. Roum., **11**(3-4):315 (1963).
1246. D.S. Coleman, B. Bown, R. Pollitt, Trans. Inst. Mining Metall., **C78**:148 (1969).
1247. A.I. Efimov, V.F. Miromanov, Zh. Prikl. Khim., **43**:1835 (1970).
1248. G.I. Novikov, A.K. Baev, Vestn. Leningrad. Gos. Univ., Ser. Fiz.-Khim., No. 22, Issue 4, 116 (1961).
1249. Z.N. Shevtsova, E.N. Korzina, B.G. Korshunov, Zh. Neorg. Khim., **7**:2596 (1962).
1250. C.W. Bjorklund, J.G. Reavis, J.A. Leary, K.A. Walsh, J. Phys. Chem., **63**(10):1774 (1959).
1251. I.N. Belyaev, A.K. Doroshenko, in: The Physicochemical Analysis of Salt Systems, Izd. Rostovsk. Gos. Univ., Rostov (1962), p. 21.
1252. P.S. Shapkin, I.V. Vasil'kov, M.P. Susarev, Zh. Neorg. Khim., **12**:2838 (1967).
1253. É.M. Moroz, S.V. Ketchik, S.S. Batsanov, Zh. Neorg. Khim., **17**:1775 (1972).
1254. I.K. Tovmas'yan, T.M. Tsushba, A.G. Bergman, Zh. Neorg. Khim., **15**:2268 (1970).
1255. I.K. Tovmas'yan, T.M. Tsushba, V.P. Radina, Zh. Neorg. Khim., **15**:2535 (1970).
1256. I.K. Tovmas'yan, T.M. Tsushba, Zh. Neorg. Khim., **17**:2796 (1972).
1257. I.K. Tovmas'yan, T.M. Tsushba, Ukr. Khim. Zh., **39**:952 (1973).
1258. I.I. Il'yasov, M. Davranov, Zh. Neorg. Khim., **18**:279 (1973).
1259. N.V. Maksina, L.K. Sosnovskaya, N.A. Finkel'shtein, N.M. Ryutina, Zh. Neorg. Khim., **18**:1098 (1973).
1260. N.D. Chikanov, Zh. Neorg. Khim., **17**:2558 (1972).

1261. N.Ya. Fedorov, S.I. Sklyarenko, E.S. Petrov, Izv. Sib. Otd. Akad. Nauk SSSR, Ser. Khim. Nauk, No. 11, Issue 3, 120 (1963).
1262. N.Ya. Fedorov, E.S. Petrov, Izv. Sib. Otd. Akad. Nauk SSSR, Ser. Khim. Nauk, No. 2, Issue 1, 48 (1967).
1263. B.G. Korshunov, D.V. Drobot, V.V. Bukhtiyarov, Z.N. Shevtsova, Zh. Neorg. Khim., **9**:1427 (1964).
1264. D.V. Drobot, B.G. Korshunov, Z.N. Shevtsova, in: The Physical Chemistry of Molten Salts, Metallurgiya, Moscow (1965), p. 48.
1265. Li Chi-Fa, I.S. Morozov, Zh. Neorg. Khim., **7**:2765 (1962).
1266. M.V. Tokarev, Nauchn. Zap. Voroshilovgradsk. Ped. Inst., Issue 6, 3 (1956).
1267. I.N. Belyaev, A.K. Doroshenko, in: The Physicochemical Analysis of Salt Systems, Izd. Rostovsk. Gos. Univ., Rostov (1962), p. 12.
1268. G.A. Bukhalova, V.M. Burlakova, Izv. Vyssh. Uchebn. Zaved., Khim. Khim. Tekhnol., **11**:955 (1968).
1269. I.V. Vasil'kova, P.I. Kozhina, S.V. Mendeleva, A.V. Stronkin, Zh. Fiz. Khim., **47**:1684 (1973).
1270. V.V. Safonov, B.G. Korshunov, Z.N. Shevtsova, S.I. Bakum, Zh. Neorg. Khim., **9**:1687 (1964).
1271. V.V. Safonov, B.G. Korshunov, Z.N. Shevtsova, L.G. Shadrova, Zh. Neorg. Khim., **10**:669 (1965).
1272. I.S. Morozov, B.G. Korshunov, A.T. Simonich, Zh. Neorg. Khim., **1**:1646 (1956).
1273. V.V. Safonov, A.V. Konov, B.G. Korshunov, Zh. Neorg. Khim., **14**:2880 (1969).
1274. V.M. Ionov, B.G. Korshunov, V.V. Kokorev, I.S. Morozov, Izv. Vyssh. Uchebn. Zaved., Tsvetn. Metall., **3**:102 (1960).
1275. S. Tanii, Denki Kagaku, **32**:167 (1964).
1276. V.N. Desyatnik, N.M. Emel'yanov, V.P. Raspopin, Zh. Fiz. Khim., **48**:1880 (1974).
1277. K. Komarek, P. Herasymenko, J. Electrochem. Soc., **105**(4):216 (1958).
1278. Li-Si-Chan, Acta Metall. Sinica, **8**(2):187 (1965).
1279. I.V. Vasil'kova, I.I. Kozhina, M.P. Susarev, Z.L. Artemeva, Vestn. Leningrad. Gos. Univ., Ser. Fiz.-Khim., No. 4, Issue 1, 121 (1969).
1280. M.V. Kamenetskii, Nauchn.-Tekh. Inf. Byull. LPI, No. 3, 3 (1957).
1281. M.V. Kamenetskii, Tsvetn. Met., No. 2, 39 (1958).
1282. B.F. Markov, R.V. Chernov, Ukr. Khim. Zh., **25**:279 (1959).
1283. P. Ehrlich, G. Kaupa, K. Blankenstein, Z. Anorg. Allg. Chem., **299**:213 (1959).
1284. M.V. Kamenetskii, A.A. Kostyukov, V.A. Korchakov, in: The Physical Chemistry of Molten Salts and Slags, Metallurgizdat, Moscow (1962), p. 60.
1285. A.K. Bostandzhiyan, A.G. Bergman, Zh. Neorg. Khim., **4**:2564 (1959).

1286. C.R. Tipton, Reactor Handbook (Materials), 2nd ed., New York (1960), p. 442.
1287. V.N. Desyatnik, B.N. Dubinin, S.P. Raspopin, Zh. Fiz. Khim., **47**:2726 (1973).
1288. I. Kuroda, T. Suzuki, J. Electrochem. Soc. Jpn., **26**(7):140 (1958).
1289. A. Bogacz, W. Trzebiatowski, Roczn. Chem., **38**(5):729 (1964).
1290. V.N. Desyatnik, N.N. Kurbatov, S.P. Raspopin, Yu.F. Chervinskii, Zh. Fiz. Khim., **46**:2159 (1972).
1291. H.-I. Seifert, P. Ehrlich, Z. Anorg. Allg. Chem., **302**:284 (1959).
1292. E.K. Lupenko, I.V. Vasil'kova, I.I. Kozhina, Vestn. Leningrad. Gos. Univ., Ser. Fiz.-Khim., No. 16, Issue 3, 101 (1969).
1293. C. Grena, Bull. Soc. Chim. France, No. 4, 655 (1960).
1294. S.A. Shchukarev, I.A. Perfilova, Zh. Neorg. Khim., **8**:2106 (1963).
1295. B.G. Korshunov, D.V. Drobot, Zh. Neorg. Khim., **9**:222 (1964).
1296. I.S. Morozov, Z.N. Shevtsova, Li Chi-Fa, Zh. Neorg. Khim., **9**:2606 (1964).
1297. I.V. Vasil'kova, N.D. Zaitseva, P.I. Shapkin, Zh. Neorg. Khim., **8**:2360 (1963).
1298. B.G. Korshunov, V.I. Gol'din, L.A. Averkieva, Izv. Vyssh. Uchebn. Zaved., Tsvetn. Metall., No. 1, 101 (1962).
1299. I.V. Vasil'kova, N.D. Zaitseva, P.S. Shapkin, Zh. Neorg. Khim., **8**:2360 (1963).
1300. A.I. Efimov, I.V. Vasil'kova, E.K. Smirnova, N.D. Zaitseva, T.S. Shemyakina, I.A. Perfilova, in: The Chemistry of Rare Elements, Izd. Leningrad. Gos. Univ., Leningrad (1964), p. 38.
1301. D.V. Drobot, B.G. Korshunov, G.P. Borodulenko, Zh. Neorg. Khim., **13**:1635 (1968).
1302. I.V. Aleksanyants, N.M. Demido, A.S. Malyugin, A.D. Pogorelii, Zh. Prikl. Khim., **42**:1897 (1969).
1303. M.A. Kuvakin, Uch. Zap. Mordovsk. Univ., Ser. Khim. Nauk, No. 81, 20 (1971).
1304. B. Swaroop, S.N. Flengas, Can. J. Chem., **44**(2):199 (1966).
1305. I.S. Morozov, B.G. Korshunov, Zh. Neorg. Khim., **1**:145 (1956).
1306. L.J. Howell, R.C. Sommer, H.H. Kellog, J. Met., **9**, No. 1, Sec. 1, 193 (1957).
1307. R.L. Lister, S.N. Flengas, Can. J. Chem., **43**(11):2947 (1965).
1308. G.A. Abramov, M.A. Malinovskii, The Electrometallurgy of Nonferrous Metals, ONTI, Leningrad (1957), p. 79.

1309. M.A. Kuvakin, P.S. Kusakin, Zh. Neorg. Khim., **4**:2577 (1959); Problems of the Monitoring and Complex Use of Raw Materials in Metallurgy, Izd. Uralsk. Fil. Akad. Nauk SSSR, Sverdlovsk (1960), p. 149.
1310. I.N. Belyaev, A.K. Doroshenko, in: The Physicochemical Analysis of Salt Systems, Izd. Rostovsk. Gos. Univ., Rostov (1962), p. 21.
1311. K. Grjotheim, T. Halvorsen, Acta Chem. Scand., **21**(8):2300 (1967).
1312. C.E. Johnson, E. Hathaway, J. Electrochem. Soc., **118**(4):631 (1971).
1313. E.M. Levin, C.R. Robbins, H.F. McMurdie, Z. Phys. Chem., **102**(5-6):175 (1976).
1314. I.I. Il'yasov, A.G. Bergman, Zh. Obshch. Khim., **26**:1288 (1956).
1315. C.E. Johnson, E. Hathaway, J. Electrochem. Soc., **118**(4):631 (1971).
1316. E.M. Levin, C.R. Robbins, H.F. McMurdie, Z. Phys. Chem., **102**(5-6):175 (1976).
1317. R.E. Thoma, Rare Earth Research. III. Proceedings of IV Conf. on Rare Earth Research, New York (1964), p. 567.
1318. R.E. Thoma, H. Insley, G.M. Hebert, Inorg. Chem., **5**(7):1222 (1966).
1319. O. Schmitz-Dumont, G. Bergerhoff, Z. Anorg. Allg. Chem., **203**(1):314 (1956).
1320. R.E. Thoma, Rare Earth Research. III. Proceedings of IV Conf. on Rare Earth Research, New York (1964), p. 567.
1321. R.E. Thoma, H. Insley, G.M. Hebert, Inorg. Chem., **5**(7):1222 (1966).
1322. G.A. Bukhalova, N.A. Litovskaya, V.A. Lyutsedarskii, Izv. Akad. Nauk SSSR, Neorg. Mater., **5**(3):510 (1969).
1323. C.Y. Barton, I.D. Redman, R.A. Strehlow, J. Inorg. Nucl. Chem., **20**(1-2):45 (1961).
1324. V.A. Gladushchenko, A.G. Bergman, Zh. Neorg. Khim., **3**:1650 (1958).
1325. N.N. Volkov, Izv. Fiz.-Khim. Nauchn.-Issled. Inst. Irkutsk. Gos. Univ., **5**(2):158 (1961).
1326. I.N. Belyaev, A.K. Doroshenko, in: The Physicochemical Analysis of Salt Systems, Izd. Rostovsk. Gos. Univ., Rostov (1962), p. 21.
1327. Jan Lützow Holm, Acta Chem. Scand., **19**(3):638 (1965).
1328. É.P. Babaeva, in: Phase Diagrams of Some Salt Systems, Izd. Rostovsk. Gos. Univ., Rostov (1964), p. 42.
1329. N.Ya. Fedorov, S.I. Sklyarenko, E.S. Petrov, Izv. Sib. Otd. Akad. Nauk SSSR, Ser. Khim. Nauk, Issue 1, 104 (1964).
1330. É.P. Babaeva, G.A. Bukhalova, Zh. Neorg. Khim., **10**:1455 (1965).
1331. É.P. Babaeva, G.A. Bukhalova, Zh. Neorg. Khim., **11**:1959 (1966).
1332. R.E. Thoma, R.H. Karraker, Inorg. Chem., **5**(11):1933 (1966).

1333. R.E. Thoma, Rare Earth Research. III. Proceedings of IV Conf. on Rare Earth Research, New York (1964), p. 567.
1334. J.D. Donaldson, J.D. O'Donogue, R. Oteng, J. Chem. Soc., No. 6, 3876 (1965).
1335. V.T. Berezhnaya, G.A. Bukhalova, Zh. Neorg. Khim., **12**:2179 (1967).
1336. R.E. Thoma, Rare Earth Research. III. Proceedings of IV Conf. on Rare Earth Research, New York (1964), p. 567.
1337. R.E. Thoma, H. Insley, G.M. Hebert, Inorg. Chem., **5**(7):1222 (1966).
1338. V.S. Emel'yanov, A.I. Evstyukhin, At. Energ., **4**:80 (1956).
1339. R.E. Thoma, H. Insley, B.S. Landau, H.A. Friedman, W.R. Grimes, J. Phys. Chem., **63**(8):1266 (1959).
1340. R.E. Thoma, H. Insley, G.M. Hebert, H.A. Friedman, C.F. Weaver, Phase Equilibria in the System NaF-ThF_4-UF_4, ORNL-3304 (1963).
1341. R.E. Thoma, H. Insley, G.M. Hebert, H.A. Friedman, C.F. Weaver, J. Am. Ceram. Soc., **46**(1):37 (1963).
1342. R.V. Chernov, I.M. Ermolenko, Zh. Neorg. Khim., **18**:2238 (1973).
1343. R.E. Thoma, Rare Earth Research. III. Proceedings of IV Conf. on Rare Earth Research, New York (1964), p. 567.
1344. R.E. Thoma, H. Insley, G.M. Hebert, Inorg. Chem., **5**(7):1222 (1966).
1345. R.W.M. D'Eye, F.S. Martin, Repts. At. Energ. Res. Est. C/M292, 4 (1956).
1346. C.R. Tipton, Reactor Handbook (Materials), 2nd ed., New York (1960), p. 428.
1347. C.J. Barton, H.A. Friedman, W.R. Grimes, H. Insley, R.E. Moore, R.E. Thoma, J. Am. Ceram. Soc., **41**(2):63 (1958).
1348. R.E. Thoma, H. Insley, G.M. Hebert, H.A. Friedman, C.F. Weaver, J. Am. Ceram. Soc., **46**(1):37 (1963).
1349. J.F. Eichelberger, C.R. Hudgens, L.V. Jones, G. Pish, T.B. Rhinehammer, P.A. Tucker, L.J. Wittenberg, J. Am. Ceram. Soc., **46**(6):279 (1963).
1350. Yu.M. Korenev, V.B. Shol'ts, L.N. Sidorov, A.V. Novoselova, Zh. Fiz. Khim., **45**:281 (1971).
1351. R.E. Thoma, Rare Earth Research. III. Proceedings of IV Conf. on Rare Earth Research, New York (1964), p. 567.
1352. G.A. Bukhalova, É.P. Babaeva, Zh. Neorg. Khim., **11**:644 (1966).
1353. R.E. Thoma, H. Insley, G.M. Hebert, Inorg. Chem., **5**(7):1222 (1966).
1354. R.E. Thoma, Rare Earth Research. III. Proceedings of IV Conf. on Rare Earth Research, New York (1964), p. 567.
1355. R.E. Thoma, H. Insley, G.M. Hebert, Inorg. Chem., **5**(7):1222 (1966).
1356. G.A. Bukhalova, É.P. Babaeva, Zh. Neorg. Khim., **11**:624 (1966).

1357. O. Von Schmitz-Du Mont, H. Bornefeld, Z. Anorg. Allg. Chem., **287**:120 (1956).
1358. C.J. Barton, W.R. Grimes, H. Insley, R.E. Moore, R.E. Thoma, J. Phys. Chem., **62**:6665 (1958).
1359. C.E. Johnson, E. Hathaway, J. Electrochem. Soc., **118**(4):631 (1971).
1360. E.M. Levin, C.R. Robbins, H.F. McMurdie, Z. Phys. Chem., **102**(5-6):175 (1976).
1361. J. Kutscher, A. Schneider, Z. Anorg. Allg. Chem., **386**(1):38 (1971).
1362. I.I. Il'yasov, A.K. Bostandzhiyan, Zh. Obshch. Khim., **26**:2393 (1956).
1363. I.I. Il'yasov, A.K. Bostandzhiyan, Zh. Neorg. Khim., **2**:167 (1957).
1364. J. Kutscher, A. Schneider, Z. Anorg. Allg. Chem., **386**(1):38 (1971).
1365. V.G. Samuseva, V.E. Plyushchev, Zh. Neorg. Khim., **9**:2436 (1964).
1366. I.I. Il'yasov, Zh. Neorg. Khim., **11**:211 (1966).
1367. É.M. Moroz, S.V. Ketchik, S.S. Batsanov, Zh. Neorg. Khim., **17**:1775 (1972).
1368. J. Kutscher, A. Schneider, Z. Anorg. Allg. Chem., **386**(1):38 (1971).
1369. N.V. Karpenko, T.N. Sevast'yanova, Vestn. Leningrad. Gos. Univ., Ser. Fiz.-Khim., Issue 2, No. 10, 92 (1970).
1370. V.V. Safonov, O.V. Lemeshko, B.G. Korshunov, Zh. Neorg. Khim., **16**:2283 (1971).
1371. I.I. Il'yasov, G.G. Shchemeleva, A.G. Bergman, Zh. Neorg. Khim., **6**:699 (1961).
1372. V.V. Safonov, B.G. Korshunov, S.N. Stel'bovskaya, Zh. Neorg. Khim., **11**:2143 (1966).
1373. V.V. Safonov, B.G. Korshunov, T.N. Zimina, Zh. Neorg. Khim., **11**:906 (1966).
1374. V.V. Safonov, B.G. Korshunov, Z.N. Shevtsova, Zh. Neorg. Khim., **7**:1979 (1962).
1375. N.D. Chikanov, Zh. Neorg. Khim., **14**:1430 (1969).
1376. D.V. Drobot, B.G. Korshunov, V.A. Aleksandrova, Zh. Neorg. Khim., **13**:3354 (1968).
1377. N.I. Glistenko, N.D. Chikanov, Zh. Neorg. Khim., **13**:294 (1968).
1378. N.D. Chikanov, Zh. Neorg. Khim., **13**:2884 (1968).
1379. I.S. Morozov, Li Chi-fa, Zh. Neorg. Khim., **8**:2733 (1963).
1380. N.D. Chikanov, Zh. Neorg. Khim., **13**:2616 (1968).
1381. I.S. Morozov, Zh. Neorg. Khim., **1**:2792 (1956).
1382. N.D. Chikanov, Zh. Neorg. Khim., **13**:2884 (1968).
1383. V.V. Safonov, B.G. Korshunov, Izv. Vyssh. Uchebn. Zaved., Tsvetn. Metall., No. 6, 81 (1968).
1384. B.G. Korshunov, V.I. Ionov, T.A. Baklashova, V.V. Kokorev, Izv. Vyssh. Uchebn. Zaved., Tsvetn. Metall., **6**:114 (1960).

1385. I.S. Morozov, D.Ya. Toptygin, Zh. Neorg. Khim., 2:1915 (1957).
1386. L.A. Nisel'son, G.L. Perekhrest, Zh. Neorg. Khim., 3:2150 (1958).
1387. N.D. Chikanov, L.A. Kazberova, Zh. Neorg. Khim., 12:2509 (1967).
1388. N.I. Glistenko, N.D. Chikanov, Zh. Neorg. Khim., 13:293 (1968).
1389. B.G. Korshunov, V.N. Bezuevskaya, Zh. Neorg. Khim., 11:1482 (1966).
1390. N.D. Chikanov, Zh. Neorg. Khim., 14:1430 (1969).
1391. I.S. Morozov, B.G. Korshunov, Zh. Neorg. Khim., 1:145 (1956).
1392. L.A. Nisel'son, K.V. Tret'yakov, V.I. Tyurin, Zh. Neorg. Khim., 18:3063 (1973).
1393. Yu.A. Buslaev et al., Dokl. Akad. Nauk SSSR, 196:374 (1971).
1394. Yu.A. Buslaev, V.A. Shcherbakova, Zh. Strukt. Khim., 7:345 (1966).
1395. A.V. Nikolaev, A.A. Opalovskii, V.E. Fedorov, Thermal Analysis, Academic Press (1969), Nos. 3-4, p. 753.
1396. G.G. Fedorov, A.A. Tsvetkov, B.N. Sudarikov, Zh. Neorg. Khim., 18:2003 (1973).
1397. V.A. Legasov, B.B. Chaivanov, Zh. Fiz. Khim., 45:593 (1971).
1398. V.V. Safonov, A.G. Ivanchenko, B.G. Korshunov, Zh. Neorg. Khim., 19:852 (1974).
1399. Fam Ngok Tien, I.S. Morozov, Zh. Neorg. Khim., 14:1361 (1969).
1400. N.D. Chikanov, Zh. Neorg. Khim., 17:2558 (1972).
1401. Fam Ngok Tien, I.S. Morozov, Zh. Neorg. Khim., 16:2279 (1971).
1402. Fam Ngok Tien, I.S. Morozov, Zh. Neorg. Khim., 14:1361 (1969).
1403. E.G. Ippolitov, L.S. Garashina, B.N. Zhigarnovskii, Izv. Sib. Otd. Akad. Nauk SSSR, Ser. Khim. Nauk, No. 4, Issue 2, 49 (1968).
1404. A.N. Efimov, Z.P. Kudryashova, Zh. Neorg. Khim., 15:255 (1970).
1405. I.N. Belyaev, D.S. Lesnykh, I.G. Éikhenbaum, Zh. Neorg. Khim., 15:846 (1970).
1406. N.D. Chikanov, Zh. Neorg. Khim., 14:1430 (1969).
1407. I.N. Belyaev, D.S. Lesnykh, I.G. Éikhenbaum, Zh. Neorg. Khim., 15:846 (1970).
1408. J.D. Donaldson, R. Oteng, J. Chem. Soc. A, 18:2696 (1969).
1409. S.D. Dionis'ev, I.I. Il'yasov, A.G. Bergman, Zh. Neorg. Khim., 5:1135 (1960).
1410. V.I. Monastyrskaya, I.K. Tovmas'yan, in: Research in Thermography and Corrosion, Rostovsk. Knizhn. Izd., Rostov (1970), p. 156.

1411. I.I. Il'yasov, G.G. Shemeleva, A.G. Bergman, Zh. Neorg. Khim., **2**:2168 (1957).
1412. T. Modestova, T. Sumarokova, Zh. Neorg. Khim., **3**:1655 (1958).
1413. J. Delgery, Compt. Rend., **222**:886 (1946).
1414. Kalingerta, J. Am. Chem. Soc., **11**:3709 (1949).
1415. V.I. Monastyrskaya, in: Research in Thermography and Corrosion, Rostovsk. Knizhn. Izd., Rostov (1970), p. 169.
1416. T. Modestova, T. Sumarokova, Zh. Neorg. Khim., **3**:1655 (1958).
1417. I.I. Il'yasov, S.D. Dionis'ev, A.G. Bergman, Zh. Neorg. Khim., **9**:422 (1964).
1418. A.V. Storonkin, I.V. Vasil'kova, A.I. Efimov, V.A. Baruzdina, Vestn. Leningrad. Gos. Univ., Ser. Fiz.-Khim., No. 16, Issue 3, 137 (1975).
1419. M.P. Vorobei, A.E. Chernakov, O.V. Skiba, V.N. Desyatnik, Zh. Neorg. Khim., **16**:3388 (1971).
1420. Yu.G. Litvinov, I.I. Il'yasov, Zh. Neorg. Khim., **19**:812 (1974).
1421. N.D. Chikanov, Zh. Neorg. Khim., **18**:277 (1973).
1422. N.D. Chikanov, Zh. Neorg. Khim., **17**:2558 (1972).
1423. G. Brauer, M. Knausenberger, Z. Anorg. Allg. Chem., **320**:54 (1963).
1424. T.I. Drobasheva, A.G. Bergman, Zh. Neorg. Khim., **14**:1348 (1969).
1425. Yu.P. Afinogenov, A.E. Shulyakovskii, Yu.A. Litvinov, Zh. Neorg. Khim., **15**:1436 (1970).
1426. V.N. Desyatnik, Yu.T. Mel'nikov, S.P. Raspopin, I.F. Nichkov, Izv. Vyssh. Uchebn. Zaved., Tsvetn. Metall., No. 3, 99 (1969).
1427. I.I. Il'yasov, A.G. Bergman, Zh. Neorg. Khim., **2**:2771 (1957).
1428. C.R. Tipton, Reactor Handbook (Materials), 2nd ed., New York (1960), p. 442.
1429. D.J. Kelsey, K. Lewis, J. Phys. Chem., **68**(6):1584 (1964).
1430. I.I. Il'yasov, D.G. Barsegov, I.G. Berikashvili, L.P. Danilenko, Zh. Neorg. Khim., **14**:1484 (1969).
1431. V.I. Monastyrskaya, I.K. Tovmas'yan, in: Research in Thermography and Corrosion, Rostovsk. Knizhn. Izd., Rostov (1970), p. 156.
1432. S.I. Dionis'ev, A.G. Bergman, Zh. Neorg. Khim., **15**:2579 (1970).
1433. Yu.G. Litvinov, I.I. Il'yasov, Zh. Neorg. Khim., **19**:812 (1974).
1434. R. Rafaeloff, A. Roy, J. Djerassi, Israel J. Chem., **4**(1a):5 (1966).
1435. A. Roy, R. Rafaeloff, J.L. Borowitz, Israel At. Energ. Commiss. (Repts.), No. 1178, 121 (1967).
1436. V.N. Desyatnik, Yu.T. Mel'nikov, I.F. Nichkov, S.P. Raspopin, V.V. Makosov, At. Energ., **28**(3):247 (1970).

1437. V.N. Desyatnik, S.P. Raspopin, I.F. Nichkov, Yu.T. Mel'nikov, P.T. Porodnov, Izv. Vyssh. Uchebn. Zaved., Tsvetn. Metall., No. 5, 102 (1966).
1438. L.M. Sterlin, V.V. Artamonov, At. Energ., **22**:473 (1967).
1439. A. Roy, R. Rafaeloff, J.L. Borowitz, Israel At. Energ. Commiss. (Repts.), No. 1128, 121 (1967).
1440. V.N. Desyatnik, Yu.T. Mel'nikov, I.F. Nichkov, et al., The Chemistry of Rare Elements, Metallurgiya, Moscow (1971), p. 28.
1441. V.N. Desyatnik, Yu.T. Mel'nikov, I.F. Nichkov, S.P. Raspopin, Zh. Neorg. Khim., **17**:2789 (1972).
1442. N.G. Korzhukov, Vestn. Mosk. Gos. Univ., Ser. II, Khim., No. 3, 71 (1968).
1443. I.V. Aleksanyants, N.M. Demido, A.S. Malyugin, A.D. Pogorelyi, Izv. Vyssh. Uchebn. Zaved., Tsvetn. Metall., **6**:37 (1968).
1444. I.V. Aleksanyants, N.M. Demido, A.S. Malyugin, A.D. Pogorelyi, Zh. Neorg. Khim., **12**:1897 (1969).
1445. S.P. Baranov, V.D. Prisyazhnyi, A.S. Grigor'eva, Zh. Neorg. Khim., **18**:810 (1973).
1446. T. Sumarokova, T. Modestova, Zh. Neorg. Khim., **1**:2027 (1956).
1447. I.I. Il'yasov, A.G. Bergman, Zh. Obshch. Khim., **26**:981 (1956).
1448. I.I. Il'yasov, D.G. Barsegova, I.G. Berikashvili, L.P. Danilenko, Zh. Neorg. Khim., **14**:1484 (1969).
1449. V.I. Monastyrskaya, I.K. Tovmas'yan, in: Research in Thermography and Corrosion, Rostovsk. Knizhn. Izd., Rostov (1970), p. 169.
1450. I.V. Vasil'kova, A.I. Efimov, I.I. Kozhina, V.A. Baruzdina, Vestn. Leningrad. Gos. Univ., Ser. Fiz.-Khim., No. 10, Issue 2, 83 (1975).
1451. O. Schmitz-Dumont, G. Bergerhoff, Z. Anorg. Allg. Chem., **283**(1):277 (1956).
1452. S. Masao, U. Taizo, Denki Kagaku, **33**:136 (1965).
1453. J. Ravez, M. Vassiliadis, C. R. Acad. Sci., **C270**(2):219 (1970).
1454. J. Ravez, D. Dumora, C. R. Acad. Sci., **C269**(3):235 (1969).
1455. S. Masao, J. Ceram. Assoc. Jpn., **70**(804):343 (1962).
1456. I.N. Belyaev, E.A. Shurginov, A.S. Kudryashov, Zh. Neorg. Khim., **17**:2812 (1972).
1457. D.P. Belotskii, I.N. Antipov, V.F. Nadtochii, S.M. Dodik, Izv. Akad. Nauk SSSR, Neorg. Mater., **5**:1852 (1969).
1458. V.V. Safonov, I.V. Vasilishcheva, B.G. Korshunov, Zh. Neorg. Khim., **16**:2310 (1971).
1459. I.I. Il'yasov, A.G. Bergman, Zh. Neorg. Khim., **2**:2771 (1957).
1460. I.I. Il'yasov, D.G. Barsegov, I.G. Berikashvili, L.P. Danilenko, Zh. Neorg. Khim., **14**:1484 (1969).

1461. Fam Ngok Tien, I.S. Morozov, Zh. Neorg. Khim., **16**:2279 (1971).
1462. G.A. Bukhalova, N.A. Litovskaya, V.A. Lyutsedarskii, Izv. Akad. Nauk SSSR, Neorg. Mater., 5(3):510 (1969).
1463. R. Benz, R.M. Deuglass, J. Phys. Chem., **65**(8):1461 (1961).
1464. K.W.R. Johnson, M. Kahn, J.A. Leary, J. Phys. Chem., **65**(12):2226 (1961).
1465. M.P. Vorobei, A.E. Chernakov, O.V. Skiba, V.N. Desyatnik, Zh. Neorg. Khim., **16**:3388 (1971).
1466. V.N. Desyatnik, I.F. Nichkov, S.P. Porodnov, O.V. Skiba, Izv. Vyssh. Uchebn. Zaved., Tsvetn. Metall., No. 1, 101 (1970).
1467. N.P. Vorobei, V.N. Desyatnik, O.V. Skiba, N.M. Emel'yanov, Zh. Neorg. Khim., **18**:3060 (1973).
1468. M.P. Vorobei, O.V. Skiba, A.S. Bevz, V.N. Desyatnik, The Physical Chemistry and Electrochemistry of Salt Systems, Metallurgiya, Moscow (1973), p. 3.
1469. M.P. Vorobei, O.V. Skiba, A.S. Bevz, V.N. Desyatnik, Zh. Neorg. Khim., **18**:2506 (1973).
1470. L.E. Trevorrow, M.J. Steindler, D.V. Steidl, J.T. Savage, Inorg. Chem., **6**(5):1060 (1967).
1471. V.E. Plyushchev, S.B. Stepina, G.V. Zimina, V.G. Zhilyakov, Izv. Vyssh. Uchebn. Zaved., Tsvetn. Metall., No. 4, 112 (1964).
1472. P.I. Fedorov, E.S. Petrov, Izv. Sib. Otd. Akad. Nauk SSSR, Ser. Khim. Nauk, No. 14, Issue 6, 55 (1969).
1473. I.N. Belyaev, E.A. Shurginov, Zh. Neorg. Khim., **15**:883 (1970).
1474. E.A. Shurginov, in: Proceedings of IX Scientific Conference of Aspirants, Exact and Natural Science Series, Izd. Rostovsk. Gos. Univ., Rostov (1969), p. 37.
1475. R. Riccardi, C. Sinistri, V. Campari, A. Magistris, Z. Naturforsch., **25a**(5):781 (1970).
1476. V.A. Grin'ko, V.V. Safonov, V.I. Ksenzenko, B.G. Korshunov, Zh. Neorg. Khim., **17**:1755 (1972).
1477. S.A. Shchukarev, I.V. Vasil'kova, D.V. Korol'kov, Zh. Neorg. Khim., **8**:1933 (1963).
1478. D.V. Drobot, B.G. Korshunov, Nad B. Sharkadi, Zh. Neorg. Khim., **14**:3158 (1969).
1479. V.E. Plyushchev, S.B. Stepina, G.V. Zimina, V.G. Zhilyakov, Izv. Vyssh. Uchebn. Zaved., Tsvetn. Metall., No. 4, 112 (1964).
1480. N.Ya. Fedorov, E.S. Petrov, in: Rare Alkaline Elements, Nauka, Novosibirsk (1967), p. 185.
1481. N.Ya. Fedorov, E.S. Petrov, Izv. Sib. Otd. Akad. Nauk SSSR, Ser. Khim. Nauk, **2**(1):57 (1967).
1482. D.V. Drobot, B.G. Korshunov, Z.M. Shevtsova, in: The Physical Chemistry of Molten Salts, Metallurgiya, Moscow (1965), p. 48.
1483. I.K. Tovmas'yan, T.M. Tsushba, A.G. Bergman, Ukr. Khim. Zh., **36**:1007 (1970).

1484. G.G. Diogenov, T.E. Bykova, Zh. Neorg. Khim., **15**:1680 (1970).
1485. Nguen Ba-Chanh, I. Bouillaud, Bull. Soc. Chim. France, No. 7, 2513 (1971).
1486. I.I. Il'yasov, V.V. Volchanskaya, Ukr. Khim. Zh., **40**(7):732 (1974).
1487. T.E. Bykova, in: Biochemical and Technological Processes in the Food Industry, Buryatskoe Knizhn. Izd., Ulan-Udé (1974), p. 27.
1488. N.N. Volkov, M.N. Zakhvalinskii, in: Medical Raw Material Resources of the Irkutsk Region, Irkutsk. Obl. Gos. Izd., Irkutsk (1961), p. 216.
1489. M. Ahtee, A. Koski, Soumalais. Tiedeakat. Toimituks., Ser. AVI, No. 297, 1 (1968).
1490. I.I. Il'yasov, V.V. Volchanskaya, Ukr. Khim. Zh., **40**(7):732 (1974).
1491. B.F. Markov, R.V. Chernov, Ukr. Khim. Zh., **24**:143 (1958).
1492. Li Chi-Fa, I.S. Morozov, Zh. Neorg. Khim., **8**:708 (1963).
1493. G.A. Bukhalova, V.M. Burlakova, Zh. Neorg. Khim., **12**:1330 (1967).
1494. G.A. Bukhalova, V.M. Burlakova, Proceedings of Interuniversity Scientific-Theoretical Conf. of Aspirants, Izd. Rostovsk. Gos. Ped. Inst., Rostov (1967), p. 435.
1495. G.A. Bukhalova, V.M. Burlakova, Izv. Vyssh. Uchebn. Zaved., Khim. Khim. Tekhnol., **11**(3):259 (1968).
1496. A.N. Kirgintsev, V.A. Isaenko, V.I. Kosyakov, Izv. Akad. Nauk SSSR, Ser. Khim., No. 2, 439 (1970).
1497. V.V. Safonov, B.G. Korshunov, Z.N. Shevtsova, S.I. Bakum, Zh. Neorg. Khim., **9**:1687 (1964).
1498. V.V. Safonov, B.G. Korshunov, Z.N. Shevtsova, L.G. Shadrova, Zh. Neorg. Khim., **9**:1406 (1964).
1499. V.V. Safonov, A.V. Konov, B.G. Korshunov, Zh. Neorg. Khim., **14**:2880 (1969).
1500. P. Ehrlich, R. Schmitt, Z. Anorg. Allg. Chem., **308**:91 (1961).
1501. B.F. Markov, R.V. Chernov, Ukr. Khim. Zh., **25**:279 (1959).
1502. D.V. Korol'kov, G.N. Kudryashova, Zh. Neorg. Khim., **13**:1626 (1968).
1503. É.M. Moroz, S.V. Ketchik, S.S. Batsanov, Zh. Neorg. Khim., **17**:1775 (1972).
1504. C.R. Tipton, Reactor Handbook (Materials), 2nd ed., New York (1960), p. 441.
1505. V.N. Desyatnik, B.N. Dubinin, S.P. Raspopin, Zh. Fiz. Khim., **47**:2726 (1973).
1506. A. Bogacz, W. Trzebiatowski, Rocz. Chem., **38**(5):729 (1964).
1507. V.N. Desyatnik, N.N. Kurbatov, S.P. Raspopin, Yu.F. Chervinskii, Zh. Fiz. Khim., **46**:2159 (1972).
1508. V.M. Vdovenko, V.A. Volkov, I.I. Kozhina, I.G. Suglobova, Radiokhimiya, **14**:484 (1972).

1509. S.A. Shchukarev, I.L. Perfilova, Zh. Neorg. Khim., **8**:2106 (1963).
1510. A.I. Efimov, I.V. Vasil'kova, E.K. Smirnova, N.D. Zaitseva, T.S. Shemyakina, L.I. Perfilova, in: The Chemistry of Rare Elements, Izd. Leningrad. Gos. Univ., Leningrad (1964), p. 38.
1511. I.V. Vasil'kova, N.D. Zaitseva, V.A. Petrova, Zh. Neorg. Khim., **8**:2369 (1963).
1512. Fam Ngok Tien, I.S. Morozov, Zh. Neorg. Khim., **14**:1364 (1969).
1513. B.F. Markov, I.D. Panchenko, G.G. Kostenko, Ukr. Khim. Zh., **22**:287 (1956).
1514. I.N. Belyaev, A.K. Doroshenko, in: The Physicochemical Analysis of Salt Systems, Izd. Rostovsk. Gos. Univ., Rostov (1962), p. 21.
1515. V. Sholokhova, in: Proceedings of Interuniversity Scientific-Theoretical Conf. of Aspirants, Izd. Rostovsk. Gos. Ped. Inst., Rostov (1972), p. 434.
1516. I.K. Tovmas'yan, V.I. Sholokhova, G.A. Shegurova, Zh. Neorg. Khim., **17**:881 (1972); Ukr. Khim. Zh., **39**:576 (1973).
1517. I.I. Il'yasov, Yu.G. Litvinov, Ukr. Khim. Zh., **40**(5):476 (1974).
1518. É.P. Babaeva, G.A. Bukhalova, in: Phase Diagrams of Some Salt Systems, Izd. Rostovsk. Gos. Univ., Rostov (1964), p. 42; Zh. Neorg. Khim., **11**:1959 (1966).
1519. L.N. Sidorov, A.S. Alikhanyan, L.V. Novikova, V.B. Shol'ts, Zh. Fiz. Khim., **47**:2934 (1973).
1520. V.T. Berezhnaya, G.A. Bukhalova, Zh. Neorg. Khim., **12**:2179 (1967).
1521. R.V. Chernov, I.M. Ermolenko, Zh. Neorg. Khim., **19**:3096 (1974).
1522. R.E. Thoma, H. Insley, B.S. Landau, H.A. Friedman, W.K. Grimes, J. Am. Ceram. Soc., **41**(12):538 (1958).
1523. J.-C. Cousseins, J.-C. Cretenet, C. R. Acad. Sci., **C265**(25):1464 (1967).
1524. J.-C. Cretenet, C. R. Acad. Sci., **C267**(3):240 (1968).
1525. O. Schmitz-Du Mont, H. Bornefeld, Z. Anorg. Allg. Chem., **287**:120 (1956).
1526. C.R. Tipton, Reactor Handbook (Materials), 2nd ed., New York (1960), p. 431.
1527. V.E. Plyushchev, S.B. Stepina, G.V. Zimina, O.P. Molchanova, L.V. Savel'eva, D.A. Yashkov, Izv. Vyssh. Uchebn. Zaved., Tsvetn. Metall., No. 1, 65 (1970).
1528. V.G. Zyryanov, E.S. Petrov, Izv. Sib. Otd. Akad. Nauk SSSR, Ser. Khim. Nauk, No. 4, Issue 2, 109 (1974).
1529. T.N. Sevast'yanova, N.V. Karpenko, Zh. Neorg. Khim., **14**:3130 (1969).
1530. I.N. Belyaev, E.A. Shurginov, A.S. Kudryashov, Zh. Neorg. Khim., **17**:2812 (1972).

1531. V.V. Safonov, O.V. Lemeshko, B.G. Korshunov, Zh. Neorg. Khim., **16**:2283 (1971).
1532. M.A. Ol'skaya, T.I. Darvoid, P.I. Fedorov, L.A. Radushkevich, Study of the Process of Salt Formation and Single Crystal Growth in Halides of Thallium, Metallurgiya, Moscow (1970), p. 18.
1533. M.A. Ol'skaya, T.I. Darvoid, L.I. Gosteva, Study of the Process of Salt Formation and Single Crystal Growth in Halides of Thallium, Metallurgiya, Moscow (1970), p. 18.
1534. D.V. Drobot, V.A. Aleksandrova, B.G. Korshunov, Zh. Neorg. Khim., **15**:1144 (1970).
1535. D.V. Drobot, B.G. Korshunov, V.A. Aleksandrova, Zh. Neorg. Khim., **14**:859 (1969).
1536. D.V. Drobot, B.G. Korshunov, V.A. Aleksandrova, Zh. Neorg. Khim., **13**:3354 (1968).
1537. V.A. Grin'ko, V.V. Safonov, V.I. Ksenzenko, Zh. Neorg. Khim., **19**:853 (1974).
1538. N.D. Chikanov, Zh. Neorg. Khim., **17**:277, 2558 (1972).
1539. V.V. Safonov, B.G. Korshunov, Zh. Neorg. Khim., **15**:2300 (1970).
1540. N.I. Glistenko, N.D. Chikanov, Zh. Neorg. Khim., **13**:294 (1968).
1541. V.V. Safonov, B.G. Korshunov, Zh. Neorg. Khim., **13**:2804 (1968).
1542. L.N. Éingorn, Ukr. Khim. Zh., **16**:404 (1950).
1543. N.D. Chikanov, Zh. Neorg. Khim., **18**:277 (1973).
1544. N.D. Chikanov, Zh. Neorg. Khim., **18**:277 (1973).
1545. B.G. Korshunov, V.N. Bezuevskaya, Zh. Neorg. Khim., **12**:1304 (1967).
1546. N.D. Chikanov, Zh. Neorg. Khim., **17**:2558 (1972).
1547. N.D. Chikanov, Zh. Neorg. Khim., **17**:277 (1972).
1548. V.V. Safonov, A.V. Konov, B.G. Korshunov, Zh. Neorg. Khim., **14**:3147 (1969).
1548a. B.A. Voitovich, N.F. Lozovskaya, Ukr. Khim. Zh., **31**:1137 (1965).
1549. B.G. Korshunov, V.N. Bezuevskaya, Zh. Neorg. Khim., **12**:3280 (1967).
1550. V.K. Ezhov, Zh. Neorg. Khim., **17**:2043 (1972).
1551. O.D. Maslov, V.A. Legasov, V.N. Prusakov, B.B. Chaivanov, Zh. Fiz. Khim., **41**:1832 (1967).
1552. V.A. Legasov, V.N. Prusakov, B.B. Chaivanov, Zh. Fiz. Khim., **44**:2629 (1970).
1553. V.V. Safonov, I.V. Vasilishcheva, V.M. Khazan, B.G. Korshunov, P.I. Fedorov, Izv. Vyssh. Uchebn. Zaved., Tsvetn. Metall., No. 4, 82 (1971).
1554. N.D. Chikanov, Zh. Neorg. Khim., **13**:2884 (1968).
1555. V.V. Safonov, B.G. Korshunov, Izv. Vyssh. Uchebn. Zaved., Tsvetn. Metall., No. 6, 81 (1968).
1556. N.D. Chikanov, Izv. Vyssh. Uchebn. Zaved., Khim. Khim. Tekhnol., **13**:1047 (1970).

1557. Fam Ngok Tien, I.S. Morozov, Zh. Neorg. Khim., **16**:2279 (1971).
1558. I.Z. Babievskaya, I.S. Morozov, Zh. Neorg. Khim., **17**:571 (1972).
1559. E.G. Ippolitov, L.S. Garashina, B.N. Zhigarnovskii, Izv. Sib. Otd. Akad. Nauk SSSR, Ser. Khim. Nauk, No. 4, Issue 2, 49 (1968).
1560. F. Thevet, Ch. Dargon, Bull. Soc. Chim. France, No. 11, 2448 (1974).
1561. I.S. Morozov, Li Chi-Fa, Zh. Neorg. Khim., **8**:2733 (1963).
1562. C.R. Tipton, Reactor Handbook (Materials), 2nd ed., New York (1960), p. 442.
1563. Yu.P. Afinogenov, A.D. Lemesh, Zh. Neorg. Khim., **12**:1705 (1967).
1564. I.S. Morozov, Z.N. Shevtsova, Li Chi-Fa, Zh. Neorg. Khim., **9**:2606 (1964).
1565. L.A. Nisel'son, B.N. Ivanov-Émin, L.E. Larionova, Zh. Neorg. Khim., **6**:186 (1961).
1566. I.N. Belyaev, A.K. Doroshenko, in: The Physicochemical Analysis of Salt Systems, Izd. Rostovsk. Gos. Univ., Rostov (1962), p. 69.
1567. B. Frit, M.M. Chbany, J. Inorg. Nucl. Chem., **31**(9):2685 (1969).
1568. N.D. Chikanov, Zh. Neorg. Khim., **13**:2616 (1968).
1569. L.V. Gorbunov, V.N. Desyatnik, S.P. Raspopin, K.I. Trifonov, Zh. Neorg. Khim., **19**:3093 (1974).
1570. V.N. Desyatnik, N.M. Emel'yanov, Yu.T. Mel'nikov, S.P. Raspopin, A.M. Tsybizov, Zh. Fiz. Khim., **46**:1886 (1972).
1571. J. Ravez, M. Vassiliadis, P. Hagenmuller, C. R. Acad. Sci., **C268**(21):1876 (1969).
1572. R.W.M. D'Eye, F.S. Martin, Repts. At. Energ. Res. Est., No. C/M 292, 4 (1956);
1573. J.-C. Cretenet, C. R. Acad. Sci., **C268**:945 (1969).
1574. B.P. Sobolev, E.G. Ippolitov, B.M. Zhigarnovskii, L.S. Garashina, Izv. Akad. Nauk SSSR, Neorg. Mater., **1**(3):362 (1965).
1575. V.V. Safonov, B.G. Korshunov, Izv. Vyssh. Uchebn. Zaved., Tsvetn. Metall., **6**:81 (1968).
1576. B.G. Korshunov, V.I. Ionov, T.A. Baklashova, V.V. Kokorev, Izv. Vyssh. Uchebn. Zaved., Tsvetn. Metall., **6**:114 (1960).
1577. I.S. Morozov, D.Ya. Toptygin, Zh. Neorg. Khim., **2**:1915 (1957).
1578. L.A. Nisel'son, G.L. Perekhrest, Zh. Neorg. Khim., **3**:2150 (1958).
1579. N.D. Chikanov, L.A. Kazberova, Zh. Neorg. Khim., **12**:2509 (1967).
1580. A.G. Krokhina, I.S. Morozov, Zh. Neorg. Khim., **14**:1402 (1969).
1581. N.I. Glistenko, N.D. Chikanov, Zh. Neorg. Khim., **13**:293 (1968).

1582. B.G. Korshunov, V.N. Bezuevskaya, Zh. Neorg. Khim., **11**:1482 (1966).
1583. N.D. Chikanov, Zh. Neorg. Khim., **14**:1430 (1969).
1584. L.A. Nisel'son, K.V. Tret'yakova, V.I. Tyurin, Zh. Neorg. Khim., **18**:3063 (1973).
1585. V.K. Ezhov, Zh. Neorg. Khim., **17**:2043 (1972).
1586. V.A. Legasov, B.B. Chaivanov, Zh. Fiz. Khim., **45**:593 (1971).
1587. V.V. Safonov, V.A. Grin'ko, M.B. Varfolomeev, E.S. Malysheva, V.I. Ksenzenko, Zh. Neorg. Khim., **18**:2827 (1973).
1588. V.V. Safonov, N.F. Kuklina, B.G. Korshunov, Zh. Neorg. Khim., **20**:554 (1975).
1589. V.V. Safonov, A.V. Konov, B.G. Korshunov, Zh. Neorg. Khim., **14**:3147 (1969).
1590. Yu.P. Afinogenov, Z.N. Khaustova, Zh. Neorg. Khim., **14**:2594 (1969).
1591. V.V. Safonov, A.V. Konov, B.G. Korshunov, Zh. Neorg. Khim., **15**:1150 (1970).
1592. V.V. Safonov, B.G. Korshunov, Zh. Neorg. Khim., **13**:2804 (1968).
1593. V.V. Safonov, O.V. Lemeshko, Zh. Neorg. Khim., **18**:1857 (1973).
1594. V.V. Safonov, I.V. Vasilishcheva, B.G. Korshunov, Zh. Neorg. Khim., **16**:2310 (1971).
1595. V.N. Desyatnik, I.F. Nichkov, P.T. Porodnov, S.P. Raspopin, O.V. Skiba, Izv. Vyssh. Uchebn. Zaved., Tsvetn. Metall., No. 1, 101 (1970).
1596. C.F. Weaver, R.E. Thoma, H. Insley, H.A. Friedman, J. Am. Ceram. Soc., **43**(4):213 (1960).
1597. G.P. Luchinskii, Zh. Fiz. Khim., **46**:2959 (1972).
1598. R.V. Chernov, V.M. Moshnenko, Zh. Neorg. Khim., **18**:1964 (1973).
1599. I.S. Morozov, D.Ya. Toptygin, Zh. Neorg. Khim., **1**:2601 (1956).
1600. B.G. Korshunov, V.N. Bezuevskaya, Zh. Neorg. Khim., **12**:3280 (1967).
1601. T.A. Polivanova, Nauchn. Tr., Voronezh. Otd. Vses. Khim. Ova., Issue 1, 173 (1957).
1602. I.I. Il'yasov, L.V. Rozhkovskaya, A.G. Bergman, Zh. Neorg. Khim., **2**:1883 (1957).
1603. M.A. Ol'skaya, O.N. Postnikova, P.I. Fedorov, et al., Research in the Processes of Salt Formation and Growth of Single Crystals of Thallium Halides, Vol. 29, Metallurgiya, Moscow (1970), p. 3.
1604. M.A. Ol'skaya, T.I. Darvoid, E.T. Khokhrina, O.N. Postnikova, Research in the Processes of Salt Formation and Growth of Single Crystals of Thallium Halides, Vol. 29, Metalluriya, Moscow (1970), p. 9.
1605. I.I. Il'yasov, S.D. Dionis'ev, A.G. Bergman, Zh. Neorg. Khim., **7**:626 (1962).

1606. M.A. Ol'skaya, M.A. Gurevich, O.N. Postnikova, T.I. Darvoid, Research in the Processes of Salt Formation and Growth of Single Crystals of Thallium Halides, Vol. 29, Metallurgiya, Moscow (1970), p. 13.
1607. G. Thiele, W. Rink, Z. Anorg. Allg. Chem., **414**(1):47 (1975).
1608. V.N. Fadeev, P.I. Fedorov, Zh. Neorg. Khim., **9**:2028 (1964).
1609. A.P. Palkin, I.P. Palyura, Zh. Neorg. Khim., **4**:2717 (1959).
1610. A.P. Palkin, I.P. Palyura, Zh. Neorg. Khim., **5**:160 (1960).
1611. V.I. Monastyrskaya, in: The Physical Chemistry and Electrochemistry of Molten Salts and Slags, Naukova Dumka, Kiev (1969), p. 324.
1612. S.P. Baranov, V.D. Prisyazhnyi, A.S. Grigor'eva, Zh. Neorg. Khim., **18**:810 (1973).
1613. I.I. Il'yasov, A.G. Bergman, Zh. Neorg. Khim., **2**:2771 (1957); **4**:913 (1959).
1614. J.-C. Cretenet, C. R. Acad. Sci., **C267**(3):240 (1968).
1615. V.N. Desyatnik, Yu.A. Izmodenov, Yu.T. Mel'nikov, I.F. Nichkov, S.P. Raspopin, At. Energ., **26**:549 (1969).
1616. V.N. Desyatnik, S.P. Raspopin, I.I. Trifonov, Izv. Vyssh. Uchebn. Zaved., Tsvetn. Metall., No. 1, 100 (1972).
1617. L.A. Khripin, Yu.V. Gagarinskii, G.M. Zadneprovskii, L.A. Luk'yanova, At. Energ., **19**:437 (1965).
1618. R.W.M. D'Eye, F.S. Martin, Repts. At. Energ. Res. Est., No. C/M292, 4 (1956).
1619. L.A. Khripin, S.A. Poduzova, G.M. Zadneprovskii, Zh. Neorg. Khim., **13**:2796 (1968).
1620. R.E. Thoma, H. Insley, G.D. Brunton, J. Inorg. Nucl. Chem., 36(5):1095 (1974).
1621. C.J. Barton, W.R. Grimes, H. Insley, K.E. Moore, R.E. Thoma, J. Phys. Chem., 62(6):665 (1958).
1622. V.K. Ezhov, Zh. Neorg. Khim., **17**:2043 (1972).
1623. V.A. Legasov, A.S. Märinin, Zh. Fiz. Khim., **46**(9):2420 (1972).
1624. N.D. Chikanov, Izv. Vyssh. Uchebn. Zaved., Khim. Khim. Tekhnol., **13**:1047 (1970).
1625. V.A. Legasov, A.S. Marinin, Zh. Fiz. Khim., 46(10):2649 (1972).
1626. V.A. Legasov, A.S. Marinin, Zh. Neorg. Khim., **18**:565 (1973).
1627. L.A. Nisel'son, B.N. Ivanov-Émin, L.E. Larionova, Zh. Neorg. Khim., **6**:186 (1961).
1628. L.A. Nisel'son, Zh. Neorg. Khim., **7**:693 (1962).

MIX
Papier aus verantwortungsvollen Quellen
Paper from responsible sources
FSC® C105338

If you have any concerns about our products,
you can contact us on
ProductSafety@springernature.com

In case Publisher is established outside the EU,
the EU authorized representative is:
Springer Nature Customer Service Center GmbH
Europaplatz 3, 69115 Heidelberg, Germany

Printed by Libri Plureos GmbH
in Hamburg, Germany